Proceedings of the 10th International Symposium on Insect-Plant Relationships

SERIES ENTOMOLOGICA

VOLUME 56

*The titles published in this series are listed at the end of this volume.*

# Proceedings of the 10th International Symposium on Insect-Plant Relationships

*Edited by*

Stephen J. Simpson, A. Jennifer Mordue (Luntz) and Jim Hardie

Reprinted from *Entomologia Experimentalis et Applicata* 91(1) 1999 with additional indices

SPRINGER-SCIENCE+BUSINESS MEDIA, B.V.

**Library of Congress Cataloging-in-Publication Data**

International Symposium on Insect-Plant Relationships (10th : 1998 :
Oxford, England)
Proceedings of the 10th International Symposium on Insect-Plant
Relations / edited by Stephen J. Simpson, A. Jennifer Mordue
(Luntz), and Jim Hardie.
p. cm. -- (Series entomologica ; v. 56)
ISBN 978-90-481-5247-6 ISBN 978-94-017-1890-5 (eBook)
DOI 10.1007/978-94-017-1890-5
1. Insect-plant relationships Congresses. I. Simpson, Stephen J.
II. Mordue, A. Jennifer. III. Hardie, Jim. IV. Title. V. Series.
QL496.I427 1998
577.8'5--dc21 99-27067

ISBN 978-90-481-5247-6

*Printed on acid-free paper*

# Contents

# Preface

Over the past 40 years the SIP meetings have played a central role in the development of the field of insect-plant relationships, providing both a show-case for current research as well as a forum for the airing and development of important new ideas. The 10th symposium followed that tradition. The week 4–10 July 1998 saw an exhilarating exchange of data and ideas across a wide range of related disciplines, from cutting-edge neuroscience in the opening session through to phylogenetic analysis of broad-scale patterns of insect-plant co-evolution at the close. Since the June 1860 meeting of the British Association for the Advancement of Science (now known as the 'Wilberforce-Huxley debate'), the University Museum of Natural History at Oxford has been host to numerous formative scientific conferences. Thanks to the efforts of the 163 participants from 25 countries, SIP 10 lived up to that reputation. It was a pleasure for us to organise the meeting.

The symposium comprised 12 keynote addresses, 60 oral presentations and 71 posters. Only a small proportion of the papers presented at the symposium are published in full here, however. We have taken one step further a procedure first adopted in 1978 by Reg Chapman and Liz Bernays for SIP-4, and next used by our predecessors Erich Städler, Martine Rahier and Robert Bauer for SIP-9: namely, a selection of full-length papers has been published as a special issue of *Entomologia*. All papers have been through the full rigour of the normal peer-review procedure of the journal. Many valuable contributions from the meeting are not present, nor in many cases were they submitted, but they will no doubt soon be published elsewhere. What does appear in the volume is representative of the structure of the symposium, and gives an indication of the scope of the meeting. Our feeling was that a book of fully refereed papers would provide more of a service to the field than the unrefereed compendium that is more typical of the proceedings volume from a scientific conference. The publisher, Kluwer Academic, and the Editors of *Entomologia* concurred fully with that view.

We would like to express our thanks and gratitude to all those who attended the symposium, helped with the organisation and running of the meeting, submitted papers and refereed manuscripts. In particular we would like to thank Catherine Hughes and Susannah Barry from Oxford International. Additionally, we owe many thanks to staff at Kluwer, notably Natasha Bonnevalle, and to Louis Schoonhoven and Steph Menken from the Editorial Board of *Entomologia* for their help and advice in the preparation of the volume to a tight schedule.

Finally, we would like to dedicate this volume to Dick Southwood, who missed the meeting due to ill health.

STEVE SIMPSON
JENNY MORDUE (LUNTZ)
JIM HARDIE

# Acknowledgements

The organisers wish to acknowledge the support of the following institutions and organisations for their contributions and sponsorship:

- ADC BioScientific
- Birkhäuser Verlag AG
- Blackwells Publishing
- Board of Visitors, Oxford University Museum of Natural History
- Cambridge University Press
- Carfax Publishing
- Hoechst Schering AgrEvo GmbH
- Novartis Crop Protection AG
- Oxford University Press
- Stanley Thornes Publishers
- Tracksys Ltd.
- Zeneca Agrochemicals

Entomologia Experimentalis et Applicata **91:** 1–6, 1999.



*Opening address*

# Insects and plants: two worlds come together

L. M. Schoonhoven
*Department of Entomology, Agricultural University, P.O. Box 8031, 6700 EH Wageningen, The Netherlands*

It is a privilege and great pleasure for me to welcome you to this 10th Symposium on Insect-Plant Relationships, especially as I see so many friends and dear colleagues. In particular, however, I would like to welcome those of you who are attending a SIP meeting for the first time, because scientific progress depends primarily on a continuous input of new blood and fresh ideas. As more than 50% of the scheduled communications will be presented by first-timers in this symposium series, an ample supply of fresh views seems to be ensured.

Newcomers may be interested to learn that this series of meetings was initiated, now about 40 years ago, by Professor Jan de Wilde in Wageningen, who felt that there is more between insects and plants than the latter just being eaten from time to time by the former. His imagination could hardly have fathomed the intricacies of the relationships between insects and plants, as have gradually been revealed during the past nine symposia, and which, I am sure after looking at the programme, will amaze us again during the next few days. These meetings are now being held at a regular interval of three years. Their attendance totalled between 40 and 80 participants at the first meetings, but has since grown and stabilized (with some fluctuations) at around 150. At the present meeting we are together with 163 colleagues, representing 25 different countries.

It is the second time that we convene in the United Kingdom. Twenty years ago SIP4 was held in Slough, well-organized by Reg Chapman and Liz Bernays. I am very glad that, although they have deserted Europe, they are again with us today. This country is a very appropriate location for such a jubilee-meeting, as it has for many years done more than its share in the development of this area of research. Among its many important contributors it suffices to mention only two names, the late Professor John Kennedy, and that of today's unfortunate absentee Professor Dick Southwood, to realize that the field of insect-plant relationships has profited tremendously from British inputs.

International meetings form an interesting element of scientific communication. Even at the present time, when everybody has a telephone on his desk or in his pocket, an e-mail box in his computer and a fax-machine around the corner, the urge to attend conferences does not decrease. Is it, in addition to obtaining specific information from colleagues working on the same problems, to refresh and broaden our view on the field as a whole? Is it to realize once again the very special position which green plants with their associated herbivorous insects occupy in the living world? Indeed, insects and plants do form the two most speciose taxa of organisms which constitute the thin but vast living film enveloping our planet Earth. This living mantle acts as a truly unique geological force, which has shaped and will continue to shape the Earth's crust, as Vladimir Vernadsky (1997) tells us so eloquently and convincingly in his recently translated book 'The Biosphere'. Not only in number of species do tracheophyte plants and insects seem to dominate the envelope of life extending over most of the continents, but also in terms of quantity of matter (biomass) green plants are unsurpassed, whereas arthropods form the principal part of animal living matter. This biosphere, the thin film between our planet's originally barren surface and its cosmic milieu, has transformed our planet over the aeons to a probably unique spaceship in the universe. The plants and insects we talk about these days emphatically contribute to this uniqueness.

In my view the field of insect-plant interactions is overwhelmingly interesting because of the enormous diversity of both groups, as well as their apparently intimate relationships. Its complexity is already evident from the sheer numbers of species. The plant *Tanacetum vulgare*, just as an example, harbours 143 different

## Insect fauna of *Tanacetum vulgare* L. (tansy)

| | |
|---|---|
| 143 phytophagous insect species: | 13% monophages<br>45% oligophages (i.e. restricted to Asteraceae)<br>40% polyphages |
| | 62% chewers (incl. 'grating' species)<br>38% sucking insects |
| | 65% ectophytic species<br>35% endophytic species<br>30% miners<br>5% gall makers |

*Figure 1.* Feeding patterns of phytophagous insect fauna present on *Tanacetum vulgare* (tansy) in Central Europe. Data from Schmitz, 1998.

phytophagous insect species (plus an accompanying set of natural enemies) (Schmitz, 1998) (Figure 1). There exist 300 000 or more plant species, each with its own insect fauna, that is partly unique, partly shared with other plants. The evolutionary success of both sizeable taxa, insects and plants, is based on *mutual dependency*, as was clearly expressed by Southwood (1973): 'The insect/plant relationship may basically be in terms of shelter, food or transport. (...) Plants provide food and/or shelter for insects, whilst the mobility of the latter has led to the evolution of plants' dependence on them for the transport of pollen and propagules.' This mutual dependency tends to be forgotten as a result of our overspecialized and compartimentalized daily scientific practice. Insect-plant scientists focus on insects as plant consumers and also on the mechanisms plants possess to limit feeding damage, whereas floral biologists interested in the marvellous subtleties of pollination, have only a vague knowledge of the principles underlying herbivory. Curiously, floral biology and insect-plant relationships seem to be two completely separate research areas, and their investigators rarely meet, though I am sure that a joint meeting of these two groups would lead to fruitful interactions and unexpected new vistas.

This compartmentalization of present-day scientific research naturally has also affected our field of interest, that is the domain of plants with their insect herbivores. Thus we recognize topics, such as behaviour, sensory physiology, tritrophic relationships, ecology, genetics and evolution, and, as a relatively new subject, the role of endosymbionts and endophytes in insect-plant relationships, a classification which has also helped the organizers of the present meeting to offer us a logical and absorbable programme. Of course, as in other scientific disciplines, specialization makes it possible to find pearls of knowledge which never would have been discovered with wide-angle approaches. These different subdisciplines, as crystallized during the past decades, resulted in the elucidation of several principles basic to our field of research.

Hardly more than 50 years ago the exact sites of an insect's smell and taste organs were still unknown, but now we record routinely nerve impulses from individual olfactory and gustatory cells in many species. The neural codes transmitted to the brains have partly been unravelled and allow predictions on the insect's behaviour as governed by its sensory messages. Hardly more than 30 years ago insect feeding preferences were still thought to be wholly innate and stereotypic. At present we know that they are not fully fixed, but possess some plasticity and are modified in response to previous experience or nutritional needs. Decision-making may be affected, as it turned out also in insects, by 'selective attention' (Bernays, 1996).

Knowledge of plant chemistry has advanced spectacularly, thus presenting important cues in our search for factors determining host-plant recognition in insects. The high degree of food specialization of most herbivorous insects, whereby even so-called generalists show distinct and refined food preferences, have largely been attributed to chemical differences between plants, or rather plant parts. (The effects of physical factors, such as toughness, hairiness etc, are underrated). This notion is supported by the wealth of qualitative and quantitative information on plant chemistry, which, due to rapid advances in chemical analysis techniques, becomes available at an ever increasing rate. Whereas Harborne in 1977 listed approximately 12 000 plant compounds which had been identified, this figure was recently raised to an astounding 100 000 (Hartmann, 1996). Most of these identifications concern compounds extracted from whole plants or plant parts. However, as the plant surface forms a critical barrier between plants and insects, another breakthrough was made when, among others, Erich Städler and his colleagues (e.g., Städler & Roessingh, 1991) succeeded in identifying chemicals present at the plant surface. It may be concluded that fairly detailed knowledge of a plant's chemical composition has now come within reach. Sophisticated techniques presently available for identification of volatiles which plants emit to the surrounding air, reflect another important step forward in the search of plant compounds affecting insect behaviour. These few facts should suffice to illustrate the progress made within hardly more than a quarter of a century. The volume of new information and new insights resulting from it, is still growing, as shown in Figure 2.

I referred earlier to the amazing subtleties and intricacies which appear to underlie insect-plant relations and stir the imagination. I cannot refrain from illustrating this by one or two examples, which to me exemplify the beauty of functional mechanisms. Both cases relate to a built-in physiological plasticity to adapt to previous experiences in herbivores and plants respectively.

In several herbivorous insects food preferences can be modified by previous experience, indicating some form of memory. Interestingly, this memory is partly located in peripheral elements of the nervous system, because the sensitivity of gustatory (and olfactory) receptors may be altered dramatically, depending on the type of food to which the insect has become accustomed (Schoonhoven, 1969; Blaney et al., 1986). In locusts nutritional needs, as reflected in blood sugar and amino acid levels, affect the responsiveness of their taste cells (Simpson & Raubenheimer, 1996), another form of short-term receptor changes. An insect's sensory environment may change not only receptor sensitivity, but also influence the number of newly formed sensilla. Increased chemical diversity in the insect's environment causes an increase in the number of chemoreceptory sensilla, thus indicating the presence of both physiological and morphological adaptability (Rogers & Simpson, 1997).

On the plant side, physiological and phenotypic plasticity is not less spectacular. Ever since Green & Ryan (1972) showed that plants may undergo chemical changes following herbivory, thereby increasing their resistance to insect attack, this previously unrecognized protection system has captured the imagination of plant physiologists, entomologists and ecologists alike. Studies to determine the mechanism and function of this system have shown an explosive growth, as appears from the recently published fascinating book by Karban & Baldwin (1997). Again, not only physiological changes may be induced, but prior herbivory may also induce structural changes. In *Urtica dioica*, stinging nettle, the density of stinging hairs was higher for newly emerged leaves following insect damage than in the absence of herbivores (Pullin & Gilbert, 1989) and in cotton plants the number of terpenoid containing pigment glands increased when the plants were fed upon by caterpillars (McAuslane et al., 1997). Both species thus can modify their morphology in response to insect attack. Obviously the underlying physiological mechanisms offer many challenges to plant physiologists.

Although important and fascinating insights in the ecology of insect-plant relationships have been gained in the past decades, and a lively debate on evolutionary mechanisms is going on (see, e.g., Thompson, 1994; Schoonhoven et al., 1998), it seems that ecological and evolutionary insights into insect-plant relationships are developing at a lower rate than our understanding of mechanisms at the level of the individual insect or plant. That is fully understandable in view of the fact that ecological relationships are usually studied in the field, whereas physiological and behavioural studies are often done in the lab. But after all, the field is the place 'where the action is', where nature shows its full complexity, and it is the field situation that we strive to understand. Moreover, it seems that the study of ecological and evolutionary aspects was started notably later than, for instance, behavioural analysis, as may be concluded from the attention these fields have received at previous SIP-meetings (Figure 3). Whereas at the first meetings a mere 6% of all presentations reported on evolutionary and ecological studies, this figure rose to 30% for the three most recent meetings.

Another interesting trend is the recent spectacular increase of reports on tritrophic interactions. These studies have shown that (under laboratory conditions) some herbivore-infested plants emit volatiles which attract the herbivore's natural enemies. Some cases have been reported of plants which produce specific signals

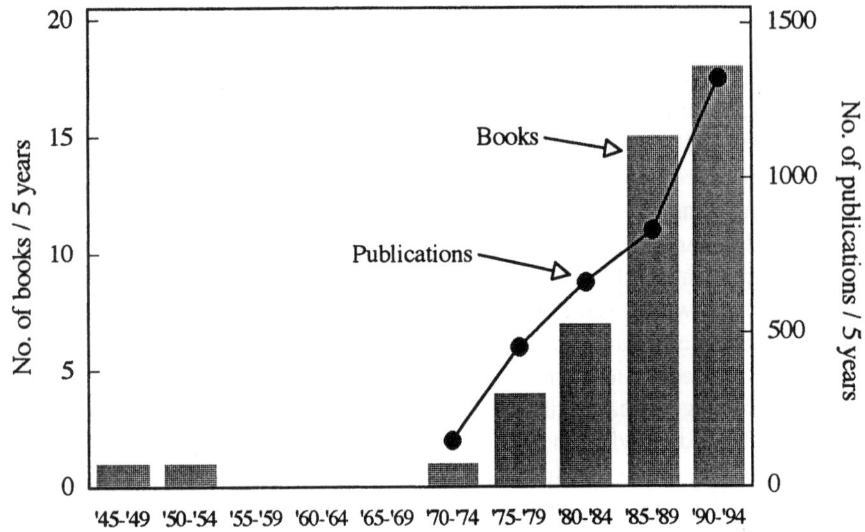


*Figure 2.* Numbers of books and publications on insect-plant relationships published in the period 1945–1994.

depending on herbivore species, thereby specifically alerting natural enemies of that particular herbivore (De Moraes et al., 1998). Another insect-plant combination was found to produce only parasitoid-alerting signals when infested by young herbivores, but not when it was damaged by older stages of the same species, presumably because it is then 'too late' to exert some functional effect (Takabayashi et al., 1995). Findings of this type suggest that plants are equipped with intricate response mechanisms, finely tuned to attack by different herbivores at different developmental stages. It remains to be shown, however, how common such mobilization signals are among different plant species, and whether under natural conditions these systems act as protecting mechanisms at all.

Not less important are the developments at the other end of the scale, i.e. at the molecular biology level. Plant physiologists employ at an increasing rate modern molecular tools to introduce into the plant's genome factors which increase their resistance to insects. We are witnessing the first results being marketed!

When thinking of some striking aspects of insect-plant biology, I may, as a personal choice, briefly mention two concepts, which typically belong to the domain of biology, and have proved to be extremely fruitful also in unravelling certain insect-plant relationships. The first is natural variation, a basic characteristic of life. It reaches truly sensational dimensions in plant chemistry, where a seemingly endless variety in space and time exists. On the insect's side, host-plant preference even under strictly standardized conditions, shows ample variability due to genotypic and experiential factors (Schoonhoven et al., 1998). Interactions between insects and plants are obviously obfuscated by the combined effects of natural variation in both partners. Although this variability may frustrate our results, it is the essence of most if not all biological phenomena, and deserves our full attention. I wholeheartedly agree with Bernie Roitberg's wording: 'vive la variance!'

As a second biological principle which I consider quite rewarding I may mention approaches based on cost-benefit models. This method has been shown to be very successful in understanding various aspects of insect-plant relationships, for instance in comparisons of production costs of plant defence chemicals with risks of losses to herbivory. This method, likewise, has produced interesting results in risk assessment studies on variation of nectar production as related to pollinator activity (Klinkhamer & de Jong, 1993). Undoubtedly many more insect-plant relationships will gain transparency when subjected to cost-benefit analysis.

Are the few aspects of insect-plant biology I have briefly touched upon only interesting and of importance to advance our understanding of Nature, or can we honestly try to convince our fund providers (i.e. ultimately the taxpayers), that our research will sooner or later yield indispensable knowledge for securing the production of our daily food? Or provide vital information for saving the greatest treasure on Earth: natural diversity? Lipke & Fraenkel's (1956) phrasing: the study of insect-plant relationships constitutes 'the very heart of agricultural entomology' beautifully

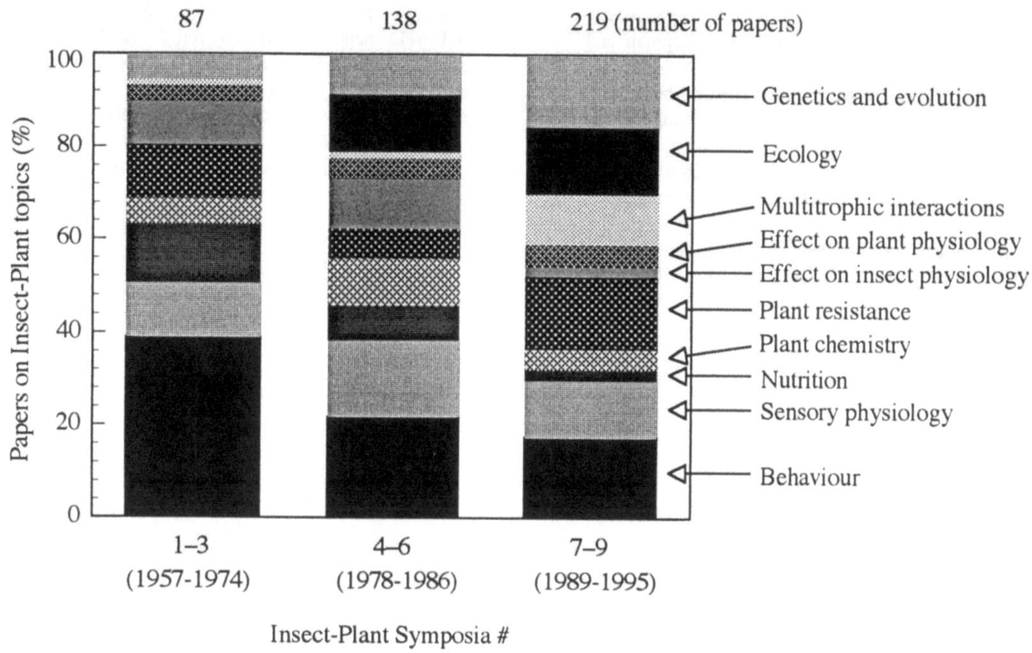


*Figure 3.* Relative frequencies of topics discussed at SIP 1–9 (1957–1995).

catches the pivotal role of our research for the well-being of mankind. There is little doubt that agricultural cntomology without this typc of input would appcar to be in a cul-de-sac. The fact that a truly alarming worldwide trend of a decrease of basic entomological research at the organismal level has set in, will also bear negative effects on insect-plant studies and thus eventually on agricultural entomology.

We have set ourselves to take part in elucidating the intricate tapestry of two interwoven biological empires, insects and plants. It is not an easy task, but a satisfying expedition in Vernadsky's 'envelope of life' which blankets planet Earth. In view of the titles of the communications scheduled for this week, there is every reason to believe that this meeting will fulfill its goal: to obtain a better understanding of the principles ruling insect-plant relationships.

With these words I declare this symposium open.

## References


Bernays, E. A., 1996. Selective attention and host-plant specialization. Entomologia Experimentalis et Applicata 80: 125–131.

Blaney, W. M., L. M. Schoonhoven & M. S. J. Simmonds, 1986. Sensitivity variations in insect chemoreceptors: a review. Experientia 42: 13–19.

De Moraes, C. M., W. J. Lewis, P. W. Paré, H. T. Alborn & J. H. Tumlinson, 1998. Herbivore-infested plants selectively attract parasitoids. Nature 393: 570–573.

Green, T. R. & C. A. Ryan, 1972. Wound-induced proteinase inhibitor in plant leaves: A possible defense mechanism against insects. Science 175: 776–777.

Harborne, J. B., 1977. Introduction to Ecological Biochemistry. Academic Press, London.

Hartmann, Th., 1996. Diversity and variability of plant secondary metabolism: a mechanistic view. Entomologia Experimentalis et Applicata 80: 177–188.

Karban, R. & I. T. Baldwin, 1997. Induced Responses to Herbivory. University of Chicago Press, Chicago.

Klinkhamer, P. G. L. & T. J. de Jong, 1993. Attractiveness to pollinators, a plant's dilemma. Oikos 66: 180–184.

Lipke, H. & G. S. Fraenkel, 1956. Insect nutrition. Annual Review of Entomology 1: 17–44.

McAuslane, H. J., H. T. Alborn, & J. P. Toth, 1997. Systemic induction of terpenoid aldehydes in cotton pigment glands by feeding of larval *Spodoptera exigua*. Journal of Chemical Ecology 23: 2861–2879.

Pullin, A. S. & J. E. Gilbert, 1989. The stinging nettle, *Urtica dioica*, increases trichome density after herbivore and mechanical damage. Oikos 54: 275–280.

Rogers, S. M. & S. J. Simpson, 1997. Experience-dependent changes in the number of chemosensory sensilla on the mouthparts and antennae of *Locusta migratoria*. The Journal of Experimental Biology 200: 2313–2321.

Schmitz, G., 1998. The phytophagous insect fauna of *Tanacetum vulgare* L. (Asteraceae) in Central Europe. Beiträge zur Entomologie 48: 219–235.

Schoonhoven, L. M., 1969. Sensitivity changes in some insect chemoreceptors and their effect on food selection behaviour. Proceedings of the Koninklijke Nederlandse Akademie van Wetenschappen C72: 491–498.

Schoonhoven, L. M., T. Jermy & J. J. A. van Loon, 1998. Insect-Plant Biology. From Physiology to Evolution. Chapman and Hall, London.

Simpson, S. J. & D. Raubenheimer, 1996. Feeding behaviour, sensory physiology and nutrient feedback: a unifying model. Entomologia Experimentalis et Applicata 80: 55–64.

Southwood, T. R. E., 1973. The insect/plant relationship – an evolutionary perspective. Symposia of the Royal Entomological Society of London 6: 3–30.

Städler, E. & P. Roessingh, 1991. Perception of surface chemicals by feeding and ovipositing insects. Symposia Biologica Hungarica 39: 71–86.

Takabayashi, J., S. Takahashi, M. Dicke & M. A. Posthumus, 1995. Developmental stage of herbivore *Pseudaletia separata* affects production of herbivore-induced synomone by corn plants. Journal of Chemical Ecology 21: 273–287.

Thompson, J. N., 1994. The Coevolutionary Process. University of Chicago Press, Chicago.

Vernadsky, V. I., 1997. The Biosphere. Copernicus, New York.

*Entomologia Experimentalis et Applicata* **91:** 7–18, 1999.



# Dynamic representation of odours by oscillating neural assemblies

Gilles Laurent, Katrina MacLeod, Mark Stopfer & Michael Wehr
*California Institute of Technology, Division of Biology, 139-74, Pasadena, CA 91125, USA*

Accepted: December 3, 1998

*Key words:* oscillations, coding, olfaction, locust, honey bee


## Abstract

Stimulus evoked oscillatory synchronization of neural assemblies has been most clearly documented in the olfactory and visual systems. Recent results with the olfactory system of locusts show that information about odour identity is contained in spatial and temporal aspects of an oscillatory population response. This suggests that brain oscillations may reflect a common reference for messages encoded in time. Although stimulus-evoked oscillatory phenomena are reliable, their roles in perception, memory and pattern recognition remain to be demonstrated. Using honey bees, we demonstrated that odour encoding involves, as in locusts, the oscillatory synchronization of assemblies of neurons, and that this synchronization is, here also, selectively abolished by the GABA receptor antagonist picrotoxin. In collaboration with Dr Brian Smith's laboratory, we showed, using a behavioural learning paradigm, that picrotoxin-induced desynchronization impairs the discrimination of molecularly similar odourants, but not that of dissimilar odours. It appears, therefore, that oscillatory synchronization of neuronal assemblies is relevant, and essential for fine odour discrimination. Finally, experiments with locust mushroom body neurons, two synapses downstream from the antennal lobe, indicate that their responses to odours become less specific when antennal lobe neurons are desynchronized by picrotoxin injection. These results suggest that oscillatory synchronization and the kind of temporal encoding it affords provide an additional dimension by which the brain can segment spatially overlapping stimulus representations.


## Introduction

A major goal of modern neuroscience is to understand the various facets of neural coding by brain circuits. Since the pioneering work of Adrian, it has become unambiguously clear that neuronal firing rates usually represent certain attributes of a stimulus. Other results (including Adrian's own; 1942, 1950), however, indicated that neuronal responses to static sensory stimuli are often temporally structured, suggesting that the time structure of spike trains might also play a role in information coding. Early work in the mammalian olfactory bulb, for example, showed prominent (30–80 Hz) oscillations in the EEG, riding on slower (5–10 Hz) waveforms coupled to sniffing (Adrian, 1942; Freeman, 1975, 1978, 1992; Gray & Skinner, 1988). More recent work in cat and primate visual cortex indicates that pairs of neurons often synchronize and oscillate under very specific stimulation conditions (Eckhorn et al., 1988; Gray & Singer, 1989; Engel et al,. 1991; Jagadeesh et al., 1992; Singer & Gray, 1995). Similarly, work in primate striate cortex suggests that stimulus-evoked neuronal responses contain complex temporal features that can be described with a small number of principal parameters (Optican & Richmond, 1987; Geisler et al., 1991; McClurkin et al., 1991a,b; Heller et al., 1995; Gawne et al., 1996). In primate frontal cortex, Abeles and co-workers described exquisitely precise sequences of neural activation distributed over several neurons (Abeles & Goldstein, 1977; Abeles & Gerstein, 1988; Abeles, 1991). In some cases, the occurrence of these statistically improbable events is well correlated with the monkey's behavioural task (Vaadia et al., 1995). In auditory cortex, it was recently shown that the degree of pairwise correlation between neurons is sometimes steadily increased during presentation of a pure tone, even though the firing rate of those neurons is no different than that outside of the stimulation period (deCharms & Merzenich, 1996). Although neurons, especially in mammalian cortex, appear to be noisy – because their responses to a constant stimulus are variable – recent *in vitro* physiological results show that the mechanisms underlying spike generation can be very precise (Mainen & Sejnowski, 1995), providing parts of a potential mechanistic substrate for high-resolution temporal codes (Victor & Purpura, 1996).

If information about a stimulus is indeed contained in temporal features of neural activity, is it useful to the animal? In other words, can it be decoded by downstream circuits, and can it be shown that the animal or its brain 'care' about these features? These questions need to be answered before we can convincingly argue that temporal codes are relevant for perception. We have recently developed and studied a preparation in which olfactory stimuli appear to be represented by ensembles of neurons that display both stimulus-evoked oscillatory synchronization and slower, but precise and stimulus-specific temporal response patterns (Laurent & Davidowitz, 1994; Laurent et al., 1996; Wehr & Laurent, 1996; Laurent, 1996b). We have identified one essential synaptic mechanism responsible for the synchronization of the principal neurons during odour response (MacLeod & Laurent, 1996) and can use this knowledge to disrupt specifically and *in vivo* the temporal features of these neurons' responses while recording from downstream circuits or testing the animal's perceptual state. Using a combination of *in vivo* electrophysiology and behavioural experiments, we are attempting to determine whether neural synchronization and temporal features of neural activity indeed play a significant role in olfactory perception.

## Olfactory maps

In 1991, Linda Buck and Richard Axel cloned, from rats, the first putative odourant receptor (OR) genes, thus opening the way to a remarkably detailed description of the projection patterns between nasal epithelium and olfactory bulb (Vassar et al., 1993, 1994; Ressler et al., 1993, 1994; Sullivan et al., 1995; Axel, 1995; Mombaerts et al., 1996; Buck, 1996). ORs belong to a large superfamily of G protein-coupled receptors and contain 7 hydrophobic domains likely to act as transmembrane -helices. Homologous families have been identified in humans, mice, salamander and teleost fish (Ngai et al., 1993; Buck, 1996). Non-homologous families of putative ORs have since been identified in the mammalian vomeronasal system (Dulac & Axel, 1995) and in the worm *C. elegans* (Sengupta et al., 1996). Insect OR genes remain, to this day, frustratingly elusive.

The molecular characterization of these genes has not yet been accompanied by a definitive functional proof that they indeed encode ORs. *In situ* hybridization and transgenic mouse technology, however, have allowed a detailed description of OR expression patterns and projections to the olfactory bulb. These studies established the following features. Individual OR neurons (ORNs) appear to express at most a handful of ORs (possibly a single one). ORNs expressing the same OR can be found randomly distributed over large 'zones' or horizontal stripes of the nasal epithelium. These zonal expression patterns are bilaterally symmetrical, and virtually identical across conspecifics. ORNs that express ORs of the same subfamily are situated in the same nasal expression zone (Ressler et al., 1993, 1994). All ORNs that express the same OR gene converge to the same one (or at most two) glomerulus ($-i$) in the ipsilateral olfactory bulb (Ressler et al., 1994; Vassar et al., 1994; Mombaert et al., 1996). Because we do not yet know the functional tuning of any of the ORs, the functional 'logic' of these projections still remains unknown. It is possible, for example, that the OR projection maps represent 'odotopic' or epitope maps (Mori, 1987; Mori & Shepherd, 1994; Mori & Yoshihara, 1995; Hildebrand 1995), i.e., maps of the structural features of the odourants. In this respect, recent physiological work by Yokoi et al. (1995), indicating that certain mitral cells in the rabbit olfactory bulb show a preferred 'tuning' to aliphatic alcohols of a certain carbon chain length and lateral inhibition for lengths shorter and longer than the preferred one, suggests that neighbouring glomeruli might even represent closely related structural features of odourant molecules. The patterns of projection from epithelium to bulb are bilaterally symmetrical and nearly identical across conspecifics.

These striking results have led some investigators to propose that these OR projection maps represent a 'code' for odours (Lewin, 1994; Shepherd, 1994; Buck, 1996). Although possible in principle, we remain skeptical that maps are necessarily an intrinsic part of codes in the nervous system (Laurent, 1997). It is indeed difficult to prove that the decoding of neural signals by downstream neurons makes explicit use of the *spatial* arrangement of their inputs (as a

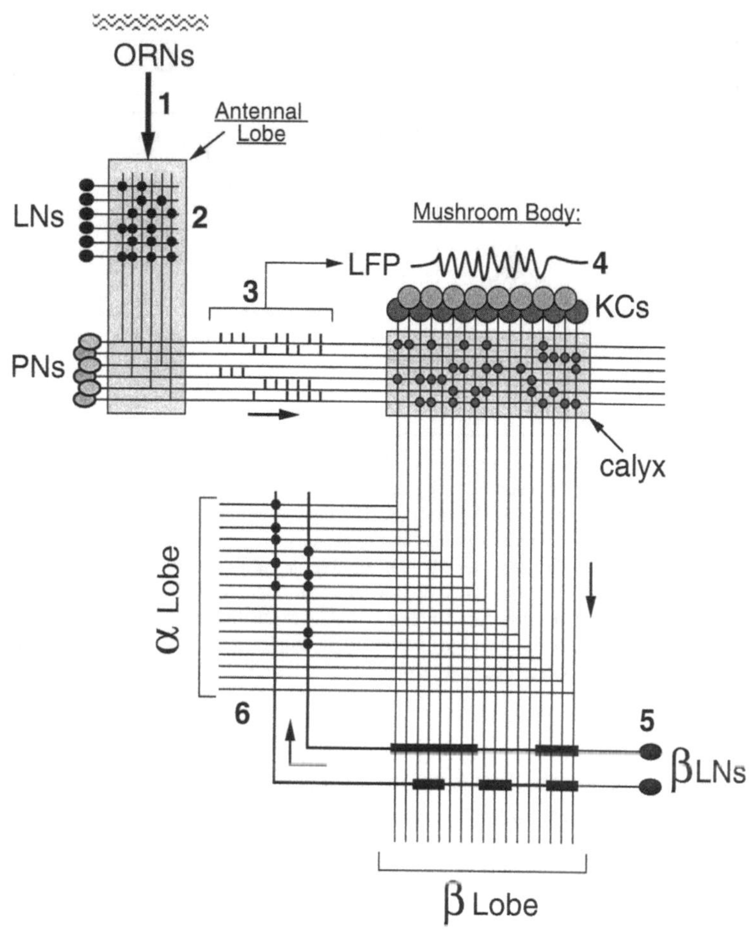


*Figure 1.* Circuit diagram of the olfactory pathway in the locust brain. Odors are transduced by arrays of olfactory receptor neurons (ORNs) in the antenna (1). ORNs activate ensembles of local and projection neurons (LNs and PNs) in the antennal lobe. LNs inhibit PNs (2) via a fast picrotoxin-sensitive GABAergic synapse. This inhibitory synapse is (at least in part) responsible for the oscillatory synchronization of the PNs that respond to an odor. PNs responding to a given odor usually display specific and reliable slow temporal activity pattern (3), superimposed on these oscillatory responses. These patterns are neuron- and odor-specific. The co-activation of PNs during an odor response causes synchronized and rhythmic EPSPs in Kenyon cells (KCs), the intrinsic neurons of the mushroom body. These synchronized EPSPs (caused by direct connections between PN axon collaterals and KC spiny dendrites in the calyx) can be detected extracellularly as an odor-evoked 20 Hz oscillatory local field potential (LFP) in the calyx (4). Arrays of KCs are so activated and send synchronized action potentials down their axons. These axons bifurcate into the so-called alpha and beta lobes. In the beta lobe, KC axons contact extrinsic neurons (betal lobe neurons, 5), which send axonal terminals to the alpha lobe (6). It is not known whether a reciprocal arrangement of projections exist between alpha and beta lobes. Beta and alpha lobe neurons can thus be used as a 'read-out' of this olfactory coding circuit.

ribosome does when it linearly decodes a strand of mRNA for example). As more information about the odour 'tuning' of the ORs becomes available, it will be fascinating to correlate the tuning of receptors to that of their post synaptic neurons in the bulb, and to understand, at least, the logic of these OR maps.

## Olfactory processing

The basic problem that is solved by the olfactory system is to detect (generally) volatile chemicals over a very wide dynamic range (often with thresholds <1 fM), to form 'unique' representations of complex, combinatorial multi-molecular 'objects' (odours), to store these representations, and to allow recognition upon presentation of 'noisy' stimuli (pattern completion). In essence, deciphering olfactory codes would amount to understanding one of the most remarkable pattern recognition devices. Our investigations focus on the neural representation of odours in the first and second relay neuropils of the olfactory pathway. Considerable work on the cellular and systems aspects of

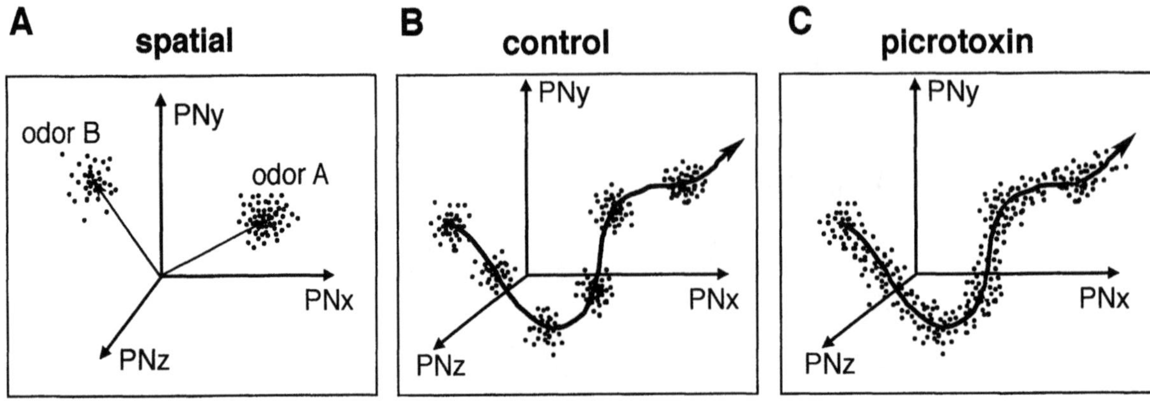


*Figure 2.* Possible odour encoding schemes. Each axis represents activity along 3 of the $n$ dimensions used for encoding, where $n$ can be thought of as the number of independent projection neurons (PNs) in one antennal lobe. Each diagram is therefore an idealized projection of the $n$-dimensional odour representation onto a 3-D (or 3-PN) space. A. The representation is static, and each odour is represented by a cloud of points in a region of PN space. The clouds of points represents (inter-trial) noise in the representation of each odour. B. The representation of each odour is dynamic and is updated at regular time intervals – in our case, at each cycle of the population oscillation. An odour is thus represented by a (discretized) trajectory in this space (see Wehr & Laurent, 1996; Laurent et al., 1996; Laurent, 1997). C. In the presence of the GABA receptor antagonist picrotoxin (PCT), the PNs desynchronize (leading to the disappearance of the 20 Hz LFP oscillations) but retain their slow and odour-specific response patterns (see MacLeod & Laurent, 1996; Stopfer et al., 1997). The representation thus remains dynamic, but is not precisely updated by neurons whose spikes occur at regular 50 ms intervals. Under these conditions, fine odour discrimination is impaired (Stopfer et al., 1997).

these circuits in both vertebrates and invertebrates has established the following facts. Odour representation is distributed amongst many neurons. Imaging experiments in the OB of lower vertebrates, for example, indicate that large numbers of glomeruli are activated by most odours, and that the foci of activation depend little on odour concentration (Kauer, 1974; Thommesen, 1978; Satou, 1992; Cinelli & Kauer, 1992; Cinelli et al., 1995; see also Joerges et al., 1997 in honeybees, and Friedrich & Korsching, 1997 in zebrafish). The activation patterns in the OB are the same whether the odour is puffed over the entire nasal epithelium or over restricted zones of it (Cinelli et al., 1995). The 'tuning' of mitral cells of the generalist olfactory system is generally broad, hard to define in molecular terms, although recent results suggest the existence of a form of molecular tuning to C chain length, at least for aliphatic alcohols (Mori, 1987; Yokoi et al., 1995). Mitral cells often display complex multiphasic temporal response patterns (Kauer, 1974, 1981; Kauer & Moulton, 1974; Meredith & Moulton, 1978; Chaput & Holley, 1980; Meredith 1981, 1986, 1992) that may depend on odour concentration. Similar findings have been made with insect pheromone-sensitive neurons (Burrows et al., 1982; Christensen & Hildebrand, 1987; Christensen et al., 1989, 1993, 1995). Odour presentation often leads to oscillatory field potential responses. This has now been established in the OB of vertebrates (Adrian, 1942; Gray & Skinner, 1988; Freeman, 1992; Gray, 1994), in the procerebral lobe of the terrestrial mollusc *Limax maximus* (Gelperin & Tank, 1990; Delaney et al., 1994; Kleinfeld et al., 1994) and in the antennal lobe of locusts (Laurent & Davidowitz, 1994).

Among these features, one of the most intriguing is the almost universal finding of stimulus-induced oscillatory synchronization (Gray, 1994). Such a phenomenon, also observed in the visual (Eckhorn et al., 1988; Singer & Gray, 1995) and sensori-motor (Murthy & Fetz, 1992) system of mammals, was first proposed by Milner (1974) to be a means to 'bind' the neural elements of cell assemblies (as defined theoretically by Hebb in the 1940s). Further theoretical work by von der Malsburg & Schneider (1986) suggested that oscillations might offer a substrate for figure-ground segmentation in the visual system, whereby neurons encoding figure and ground could be segregated by their respective phase relative to a common oscillatory 'clock'. Though much experimental work followed this idea (Eckhorn et al., 1988; Singer & Gray, 1995), little direct support for this hypothesis has been published thus far. Gelperin's work with *Limax* procerebral lobe (Gelperin & Tank, 1990; Delaney et al., 1994; Kleinfeld et al., 1994) shows that odours cause a collapse of a phase gradient that is observed at rest over the width of the procerebral lobe, possibly related to an odour-evoked change in the attentive state of the animal. A precise role for oscillatory synchronization of neural assemblies (if any), however, still remains to be firmly established.

## Structure of the insect olfactory system and relations to the vertebrate system

Though fascinating in its own right, work with arthropod and molluscan olfactory systems is made all the more relevant for mammalian (including human) neuroscience by the striking similarity between the vertebrate and invertebrate olfactory systems. Indeed, both architectural design and neuronal physiology appear similar in most respects, leading Hildebrand & Shepherd (1997) to state that 'it seems that a common set of neural mechanisms have evolved across phyla for detecting and discriminating among olfactory stimuli'. Among these similarities are the following: (1) Absorption and transport of odourants by lymph or mucus (nose or antenna), and common presence of olfactory binding proteins, whose effect is to enhance ORN response. (2) Transduction via activation of G-protein-coupled second-messenger pathways: cAMP in vertebrates, $IP_3$ and DAG in insects, and possibly (though not convincingly proven yet) both in certain vertebrate and invertebrate species. (3) Generalist and specialist odourant pathways. Moths, for example, possess a macroglomerular complex in their antennal lobe that receives exclusive input from pheromone-tuned ORNs on the antenna (Homberg et al., 1988; Kanzaki et al., 1989; Lee & Strausfeld, 1990; Almaas et al., 1991; Hansson et al., 1991; Christensen et al., 1993). Similarly, pheromonal olfactory signals in vertebrates project from a vomeronasal organ to a specialized 'accessory' olfactory bulb. (4) Convergence of ORNs to a first relay formed of a glomerular neuropil: olfactory bulb (OB) in vertebrates; antennal lobe (AL) in insects (Masson & Mustaparta, 1990) and crustaceans (Mellon et al., 1992). (5) Local dendrodendritic circuits between inhibitory local interneurons (granule and periglomerular cells in vertebrates; local neurons, LNs, in insects) and projection neurons (mitral/tufted cells in vertebrates; PNs in insects) (Jahr & Nicoll, 1982, Christensen et al., 1993; Ache et al., 1993; MacLeod & Laurent, 1996). (6) Odour-evoked field potential oscillations in OB and AL (see below). (7) Divergent projection patterns from OB/AL to second relay, associated with olfactory memory functions: piriform cortex in vertebrates, mushroom body in insects (Homberg et al., 1988; Haberly & Bower, 1989; Davis, 1993; Hammer & Menzel, 1995). (8) Odour-evoked temporal response patterns in principal cells of the OB/AL (Kauer, 1974; Meredith, 1992; Laurent & Davidowitz, 1994).

It thus seems that most of the integrative principles gathered from any of these systems have an analogue in most of (if not all) the others. For example, our recent demonstration of the role of LN-mediated inhibition in generating oscillatory synchronization of PNs in the insect antennal lobe (MacLeod & Laurent, 1996) directly supports mechanistic models (as yet not directly confirmed experimentally) of circuit dynamics in the mammalian olfactory bulb (Rall et al., 1966; Freeman, 1992; Gray, 1994). We believe, therefore, that despite mechanistic differences (such as, for example, the precise nature of the neurotransmitter systems used), the integrative principles derived from work with insects will prove to be directly applicable to an understanding of vertebrate olfaction.

## Spatio-temporal odour responses of locust antennal lobe neurons

*Spatial features.* When an airborne odour is puffed onto one antenna of a locust, a subgroup of projection (PNs) and local (LNs) neurons in its ipsilateral antennal lobe is activated (Figure 1, 2A). The size of the PN subgroup is estimated to be about 10–15% of the total complement (i.e., 80–100 of 830 PNs), and appears not to depend greatly on whether the odour is mono- or multi-molecular (Laurent & Davidowitz, 1994). Individual PNs can respond to, and hence participate in representing, different odours, even if these odours do not share molecular compounds. These 'spatial' representations are thus overlapping and combinatorial (Laurent, 1996ab). If the antennal lobe were a noise-free system, if PNs could be described simply as *responding* or *not responding* to an odour, and if downstream decoding networks could differentiate two PN ensembles that differed only by one PN, the number of possible combinations (and therefore odour representations) would be astronomical: $830!/(730! \times 100!)$, i.e., $17 \times 10^{130}$. The three conditions listed above, however, are unlikely to apply to brain circuits.

*Temporal features.* Odour responses of locust antennal lobe PNs contain the following overlapping temporal features.

*Oscillations.* The responses of PNs to odours contain 20–30 Hz membrane potential oscillations composed of fast alternating EPSPs and IPSPs (Laurent & Davidowitz, 1994). The spikes produced by a PN are thus often periodic (see below). The IPSPs are caused by GABA-mediated inputs (Christensen et al., 1993; Leitch & Laurent, 1996) from connected LNs which oscillate at the same frequency, with a 90° phase lead over the PNs (MacLeod & Laurent, 1996). Blocking the GABAergic synapse between LNs and PNs with picrotoxin selectively desynchronizes all antennal lobe neurons. Individual PNs receive inputs from many LNs, for hyperpolarizing a LN during an odour response does not significantly alter the oscillatory response of a postsynaptic PN (Laurent & Davidowitz, 1994; MacLeod & Laurent, 1996).

*Synchronization.* The PNs that respond to the same odour fire in synchrony. Their synchronized spikes provide periodic input to the postsynaptic mushroom body Kenyon cells, producing 20–30 Hz local field potential (LFP) oscillations in the mushroom body calyx. Every odour, but not air alone, evokes such bursts of LFP oscillations, whose frequency is independent of the odour. These oscillations can also be seen as subthreshold 20–30 Hz oscillations of membrane potential in the Kenyon cells (Laurent & Naraghi ,1994; Laurent & Davidowitz, 1994).

*Phase.* The phase of PN spikes relative to the LFP oscillations (or to other responding PNs) varies around a constant mean that is stimulus-independent. Spike phase, therefore, contains no information about odour identity (Laurent & Davidowitz, 1994; Laurent, 1996b; Wehr & Laurent, 1996).

*Phase-locking.* Not all the successive spikes produced by a PN in response to an odour phase-lock to the LFP. The ones that do, however, occur in relatively precise temporal windows during the odour response. The length and timing of these windows differ across neurons for an odour, and across odours for one neuron. Consequently, any two neurons that respond to the same odour may produce phase-locked spikes during a few cycles only of the ensemble response, or even never fire synchronized spikes together (Laurent et al., 1996).

*Slow temporal patterns.* Superimposed on these fast odour-evoked oscillatory responses are slow temporal patterns. Indeed, individual PNs usually respond to an odour with a specific spike profile, composed of several successive periods of activity and silence. Different PNs respond differently to the same odour, and individual PNs respond differently to different odours (Laurent et al., 1996).

*Precision.* The ordering of PN spikes in these oscillatory temporal patterns is very precise. Indeed, the response of a PN to two odours A and B may differ only by the rank order of these spikes: They can, for example, occur with non-zero probabilities in cycles 1, 2, 3 of the LFP oscillation for odour A and cycles 2, 3, 4 for odour B. This indicates that a rate code that considers spike number alone misses information contained in the timing of the spikes within the oscillatory response (Wehr and Laurent, 1996).

*Inter-PN coupling.* The firing probability of a PN during a given cycle of its response is coupled to its firing probability in a different cycle of the same trial. Similarly, the firing probability of a PN during a given cycle is linked to the firing probability of other PNs recorded simultaneously in this or a different cycle of the same trial (Wehr & Laurent, 1996). In other words, the firing behaviours of different PNs are not independent during an odour response, and the encoding of odours must therefore be studied in the context of a complex circuit dynamics in which PNs probably influence each other.

## A spatio-temporal coding hypothesis

A simple mean rate decoding scheme which assigns significance only to mean firing rates and discards spike timing, yields a fraction only of the information about odour identity that is contained in the spike trains (Wehr & Laurent, 1996). In other words, we demonstrated that odour-specific information is contained in precise *spatial and temporal* aspects of PN firing. We therefore propose that space *and* time (i.e., *which* neurons and *when* they are activated) are essential dimensions of the code for odours. The *spatial* aspects are contained in the *identities* of the activated PNs (10–15% of total number of available PNs for most odours). The *temporal* aspects involve two interlocked phenomena: (1) The *transient and periodic synchronization* of active PNs and (2) the '*evolution*' of the odour-coding assemblies along an odour-specific 'trajectory' during an odour response (Figure 2B). In this hypothesis, the oscillation can be seen as a 'clock' at whose rate the spatial representation is updated during a single odour response

(Laurent et al., 1996; Wehr & Laurent, 1996; Laurent, 1996b). What is now needed is evidence that both spatial and temporal aspects of the representation are indeed used by the brain for odour learning, recognition or discrimination.

While the idea that sensory stimuli can be represented combinatorially by distributed ensembles of coarsely tuned neurons is not new (see, for example, Georgopoulos, 1995 and Salinas & Abbott, 1994 for representation of hand movement in primate motor cortex or Wilson & McNaughton, 1993 for representation of behavioural space in rat hippocampus), the hypothesis that spike timing and/or oscillations might play a role in stimulus encoding has received rather little hard experimental support (Abeles, 1981; Bialek et al., 1991; O'Keefe & Recce, 1993; Gray, 1994; Heller et al., 1995; Abbott et al., 1996). We think, therefore, that a convincing demonstration (or falsification) of this proposed scheme will require, among other things, that the *temporal* features of the stimulus representation be manipulated, and that the effects of specific manipulations be tested on downstream circuits and behaviour.

## Mechanistic features and their practical consequences

The synchronization and phase-locking of antennal lobe PNs during odour responses can be blocked selectively by local injection of the GABA receptor chloride channel blocker picrotoxin into the antennal lobe (MacLeod & Laurent, 1996). This manipulation blocks the fast IPSPs caused by LNs onto PNs *but not* the slow and odour-specific temporal patterns seen in PNs during odour responses, even when these appear to be caused by long-lasting synaptic inhibition (MacLeod & Laurent, 1996). This suggests that the slow temporal patterns are shaped by mechanisms which are independent of the activation of this GABA receptor type. Other mechanisms include: other types of GABA receptors (note that GABAB receptors are not known in insects and that phaclophen, a GABAB receptor antagonist, had no effect on the temporal patterns; MacLeod & Laurent, in preparation); other neurotransmitters – such as histamine (McClintock & Ache, 1989), though cimetidine and pirylamine had effects only at very high concentrations; ionic pumps (though long-lasting hyperpolarizations do not always follow spike bursts; Laurent et al., 1996); nitric oxide (Müller & Buchner, 1993; Gelperin, 1994); temporal patterning of the antennal afferent input; and finally, complex dynamical behaviour of the highly interconnected antennal lobe circuits.

Our finding that fast local inhibition is essential to produce oscillatory synchronization of antennal lobe networks agrees with recent *in vitro* results showing that inhibitory feedback shapes oscillatory synchronization in mammalian hippocampus (Whittington et al., 1995; Traub et al., 1996a,b; Wang & Buzsaki, 1996), thalamus (van Vreeswijk et al., 1995; Bal & McCormick, 1996) and cerebral cortex (Lytton & Sejnowski, 1991; Connors & Amitai, 1997). Our results in locusts, however, were obtained *in vivo* and in non anaesthetized animals, and showed that the response tuning of PNs was not altered by picrotoxin-mediated desynchronization. It is to our knowledge the only preparation thus far in which oscillations in a specific circuit can be abolished *in vivo*, conditions essential for testing the role of neural synchronization for stimulus learning and discrimination.

## Behavioural role of oscillatory synchronization for odour discrimination

To test the possible importance of oscillatory synchronization (and thus, of the temporal codes they appear to carry) for odour discrimination, we used honey bees (Stopfer et al., 1997), which can be trained to extend their proboscis in response to specific odours after a few associative forward pairings of these odours with a sucrose reward [Proboscis Extension (PE) conditioning] (Kuwabara, 1957; Bitterman et al., 1983; Smith & Menzel, 1989; Menzel, 1990; Hammer & Menzel, 1995).

A first set of experiments was designed to establish whether odour presentation to the antenna of a bee produces the same neural phenomena in its antennal lobes and mushroom bodies as those observed in locusts. These experiments indeed demonstrated that many of the phenomena described above for locusts also apply to bees. Namely, odours evoke synchronized (30 Hz) oscillations of antennal lobe neurons and phase-locked oscillations of field potential in the ipsilateral mushroom body calyx; these synchronized oscillations are blocked by picrotoxin application, while the slow temporal response patterns of PNs are not altered by picrotoxin (Stopfer et al., 1997). Picrotoxin can therefore be used in the honey bee as a selective agent to abolish oscillatory synchronization of PNs during odour conditioning (Figure 2C).

In a second set of experiments, performed in collaboration with Drs Brian Smith and Seetha Bhagavan (Ohio State University), we tested the importance of oscillatory synchronization for odour learning and discrimination, using picrotoxin injection into, or selective application onto, the antennal lobes. We used a proboscis extension (PE) conditioning assay to test whether picrotoxin could disrupt olfactory discrimination (Bitterman et al., 1983; Smith & Menzel, 1989). When forager honey bees experience forward pairing of an odour (conditioned stimulus or CS) with sucrose reinforcement, their PE response to that specific odour increases dramatically for 48 h or more. This increase is selectively due to associative learning mechanisms (Menzel, 1990). In addition, the conditioned response generalizes to some extent to odours that are structurally similar to the conditioned odourant (Smith & Menzel, 1989). For example, once conditioned to an aliphatic alcohol (e.g., 1-hexanol), bees show a somewhat heightened response to other, structurally similar alcohols (e.g., 1-octanol). This generalization response is never as strong as the response to the CS itself, but it is higher than the generalization response to structurally dissimilar odourants (e.g., terpenes). We reasoned that, if oscillatory synchronization plays a role in odour learning or discrimination, picrotoxin (PCT) application to the antennal lobe should diminish the ability of bees to discriminate odours and therefore, should increase odour generalization. Animals were divided in two groups: a control (saline-treated) group and a test (PCT-treated) group and the drugs were applied blind (Stopfer et al., 1997). We used a recovery interval of 10 min between drug treatment and conditioning. Both PCT- and saline-treated groups learned the CS-sucrose pairing equally well, showing a maximal response by conditioning trial 5. Sixty minutes after conditioning, the two groups were given extinction trials with the CS, a similar odour (S) and a dissimilar odour (D). The number of animals in each group that responded with a proboscis extension to CS, S or D was measured. Saline-treated animals responded significantly more often to the CS than they did to S, indicating discrimination of the two related odours. PCT-treated animals, by contrast, failed to discriminate the CS from S. Like the saline-treated controls, however, PCT-treated bees could discriminate the dissimilar odour, geraniol, from the two aliphatic alcohols (CS and S). We concluded that PCT had no significant nonspecific effects on learning but rather, that PCT affected certain odour discrimination tasks specifically (Stopfer et al., 1997).

## Conclusions

These results suggest that neural synchronization plays a role in fine sensory discrimination tasks, i.e., ones that require the separation of stimuli whose (spatial) neural representations probably overlap greatly (see results of imaging experiments in honeybee antennal lobes by Joerges et al., 1997, for example). Synchronization – and hence the associated temporal features of the codes for odours –, however, appears unimportant for the discrimination of dissimilar stimuli (i.e., ones less likely to have spatially overlapping neural representations). Because the effect of PCT was limited to the conditioning period, and because discrimination of dissimilar odours was always possible 60 min after conditioning, it is concluded that PCT-induced desynchronization did not impair odour learning *per se*. Rather, PCT (and thus desynchronization) appeared to impair the separation of the neural representations of two related odour stimuli, of which one was stored in memory in a form that lacked its natural oscillating temporal features. Synchronization of PN assemblies may therefore be important to help reduce the overlap between the neural representations of related stimuli, possibly by using the temporal aspects of their representations (Laurent et al., 1996; Wehr & Laurent, 1996) as separable features. These behavioural results provide direct support for the hypothesis presented above, proposing that spike timing and synchronization are important parameters for odour sensation or perception. How these features are read-out, however, remains unknown. It is thus now important to further document the nature of these temporal representations and to determine the mechanisms which underlie their putative decoding by downstream networks. This will require the identification of neuronal populations in or downstream of the mushroom bodies, and the description of their sensitivity to the temporal structure of their inputs from the antennal lobes. The realization that temporal features of spike trains might play a significant role in information coding in the brain is becoming more widespread than it once was. This insect preparation (and hopefully soon others) will, we hope, allow direct experimental tests of the causal relationship that may exist between dynamical codes and perception. If stimulus representations in the brain rely on spatio-temporal codes, it is likely that the mechanisms for pattern learning and recognition should be adapted to such dynamic encoding and decoding tasks too. Finding what these mechanisms are may not be so utopian and should be fascinating.

## Acknowledgements

This work has been supported by grants from the National Science Foundation (USA) and the Sloan Center for Theoretical Neuroscience at Caltech to GL. Many thanks also to our friends Brian H. Smith and Seetha Bhagavan for their crucial collaboration on the honeybee experiments.

## References

Abbott, L. F., E. T. Rolls & M. J. Tovee, 1996. Representational capacity of face coding in monkeys. Cerebral Cortex 6: 498–505.

Abeles, M., 1981. Corticonics. Neural Circuits of the Cerebral Cortex. Cambridge University Press, Cambridge, UK.

Abeles, M. & G. Gerstein, 1988. Detecting spatiotemporal firing patterns among simultaneously recorded single neurons. Journal of Neurophysiology 60: 909–924.

Abeles, M., & M. H. Goldstein, 1977. Multispike train analysis. Proceedings of the Institute of Electrical and Electronics Engineers 65: 762–773.

Ache, B. W., H. Hatt, H. Breer, I. Boekhoff & F. Zufall, 1993. Biochemical and physiological evidence for dual transduction pathways in lobster olfactory receptor neurons. Chemical Senses 18: 523.

Adrian, E. D., 1950. The electrical activity of the mammalian olfactory bulb. Electroencephalography and Clinical Neurophysiology 2: 377–88.

Adrian, E. D., 1942. Olfactory reactions in the brain of the hedgehog. Journal of Physiology, London 100: 459–473.

Almaas, T. J., T. A. Christensen & H. Mustaparta, 1991. Encoding of different features of an olfactory stimulus by the sex pheromone receptors in *Heliothis zea*. Journal of comparative Physiology A 169: 249–258.

Axel, R., 1995. The molecular logic of smell. Scientific American 1273: 54–159.

Bal, T., & D. A. McCormick, 1996. What stops synchronized thalamocortical oscillations? Neuron 17: 297–308.

Bhagavan, S. & B. H. Smith, 1997. Olfactory conditioning in the honey bee, *Apis mellifera*: Effects of odor intensity. Physiology and Behavior 61: 107–117.

Bialek, W., F. Rieke, R. R. de Ruyter van Steveninck & D. Warland, 1991. Reading a neural code. Science 252: 1854–1857.

Bitterman, M. E., R. Menzel, A. Fietz & S. J. Schäfer, 1983. Classical-conditioning of probiscus extension in honeybee (*Apis-mellifera*). Journal of Comparative Psychology 97: 107–119.

Buck, L., 1996. Information coding in the vertebrate olfactory system. Annual Review of Neursocience 19: 517–544.

Buck, L., & R. Axel, 1991. A novel multigene family may encode odorant receptors: a molecular basis for odorant recognition. Cell 65: 175–187.

Burrows, M., J. Boeckh & J. Esslen, 1982. Physiology and morphology properties of interneurons in the deutocerebrum of male cockroaches which respond to female pheromone. Journal of comparative Physiology A 145: 447–457.

Chaput, M., & A. Holley, 1980. Single unit responses of olfactory bulb neurons to odour presentation in awake rabbits. Journal of Physiology (Paris) 76: 551–558.

Christensen, T. A., I. D. Harrow, C. Cuzzocrea, P. W. Randolph & J. G. Hildebrand, 1995. Distinct projections of two populations of olfactory receptor axons in the antennal lobe of the sphinx moth *Manduca sexta*. Chemical Senses 20: 313–323.

Christensen, T. A., & J. G. Hildebrand, 1987. Male-specific, sex pheromone-selective projection neurons in the antennal lobes of the moth *Manduca sexta*. Journal of comparative Physiology A 160: 553–569.

Christensen, T. A., J. G. Hildebrand, J. H. Tumlinson, R. E. Doolittle, 1989. Sex pheromone blend of *Manduca sexta*: responses of central olfactory interneurons to antennal stimulation in male moths. Archives of Insect Biochemistry and Physiology 10: 281–291.

Christensen, T. A., B. R. Waldrop, I. D. Harrow & J. G. Hildebrand, 1993. Local interneurons and information processing in the olfactory glomeruli of the moth *Manduca sexta*. Journal of Comparative Physiology A 173: 385–399.

Cinelli, A. R., K.A. Hamilton & J. S. Kauer, 1995. Salamander olfactory bulb neuronal activity observed by video-rate voltage-sensitive dye imaging. 3. Spatial and temporal properties of responses evoked by odorant stimulation. Journal of Neurophysiology 73: 2053–2071.

Cinelli, A. R. & J. S. Kauer, 1992. Voltage-sensitive dyes and functional activity in the olfactory pathway. Annual Review of Neuroscience 15: 321–351.

Connors, B. W. & Y. Amitai, 1997. Making waves in the neocortex. Neuron 18: 347–349.

Davis, R. L., 1993. Mushroom bodies and *Drosophila* learning. Neuron 11: 1–14.

DeCharms, R. C. & M. M. Merzenich, 1996. Primary cortical representation of sounds by the coordination of action potential timing. Nature 381: 610–613.

Delaney, K. R., A. Gelperin, M. S. Fee, J. A. Flores, R. Gervais, D. W. Tank & D. Kleinfeld, 1994. Waves and stimulus modulated dynamics in an oscillating olfactory network. Proceedings of the National Academy of Sciences USA 91: 669–674.

Dulac, C. & R. Axel, 1995. A novel family of genes encoding putative pheromone receptors in mammals. Cell 83: 195–206.

Eckhorn, R., W. Bauer, W. Jordan, M. Brosch, W. Kruse, M. Munk & H. J. Reitboeck, 1988. Coherent oscillations: A mechanism of feature linking in the visual cortex? Biological Cybernetics 60: 121–130.

Engel, A. K., P. König, A. K. Kreiter & W. Singer, 1991. Interhemispheric synchronization of oscillatory neuronal responses in cat visual cortex. Science 252: 1177–1179.

Freeman, W. J., 1975. Mass Action in the Nervous System. Academic Press, New York.

Freeman, W. J., 1978. Spatial properties of an EEG event in the olfactory bulb and cortex. Electroencephalography and Clinical Neurophysiology 44: 586–605.

Freeman, W. J., 1992. Nonlinear dynamics in olfactory information processing. In: J. L. Davis & H. Eichenbaum (eds) Olfaction, a Model System for Computational Neuroscience. MIT Press, Cambridge, Mass., pp. 225–249.

Friedrich, R. W. & S. I. Korsching, 1997. Combinatorial and chemotopic odorant coding in the zebrafish olfactory bulb visualized by optical imaging. Neuron 18: 737–752.

Gawne, T. J., T. W. Kjaer, J. A. Hertz & B. A. Richmond, 1996. Adjacent visual cortical complex cells share about 20% of their stimulus-related information. Cerebral Cortex 6: 482–489.

Geisler, W. S., D. G. Albrecht, R. J. Salvi & S. S. Saunders, 1991. Discrimination performance of single neurons: rate and temporal-pattern information. Journal of Neurophysiology 66: 334–362.

Gelperin, A., 1994. Nitric oxide mediates network oscillations of olfactory interneurons in a terrestrial mollusc. Nature 369: 61–63.

Gelperin, A., D. W. Tank, 1990. Odour-modulated collective network oscillations of olfactory interneurons in a terrestrial mollusc. Nature 345: 437–440.

Georgopoulos, A. P., 1995. Current issues in directional motor control. Trends in Neuroscience 18: 506–510.

Granger, R., J. Ambros-Ingerson, H. Henry & G. Lynch, 1988. Partitioning of sensory data by a cortical network. In: D. Anderson (ed.), Neural Information Processing Systems. American Institute of Physics, New York, pp. 317–337.

Gray, C. M., 1994. Synchronous oscillations in neuronal systems: mechanisms and functions. Journal of computational Neuroscience 1: 11–38.

Gray, C. M. & W. Singer, 1989. Stimulus-specific neuronal oscillations in orientation columns of cat visual cortex. Proceedings of the National Academy of Sciences USA 86: 1698–1702.

Gray, C. M. & J. E. Skinner, 1988. Centrifugal regulation of neuronal activity in the olfactory bulb of the waking rabbit as revealed by reversible cryogenic blockade. Experimental Brain Research 69: 378–386.

Haberly, L. B. & J. M. Bower, 1989. Olfactory cortex: Model circuit for study of associative memory. Trends in Neuroscience 12: 258–64.

Hammer, M. & R. Menzel, 1995. Learning and memory in the honey-bee. Journal of Neuroscience 15: 1617–1630.

Hansson, B. S., T. A. Christensen & J. G. Hildebrand, 1991. Functionally distinct subdivisions of the macro glomerular complex in the antennal lobe of the male sphinx moth *Manduca sexta*. Journal of comparative Neurology 312: 264–278.

Heller, J., J. A. Hertz, T. W. Kjaer & B. J. Richmond, 1995. Information flow and temporal coding in primate pattern vision. Journal of computational Neuroscience 2: 175–193.

Hildebrand, J. G., 1995. Analysis of chemical signals by nervous systems. Proceedings of the National Academy of Sciences USA 92: 67–74.

Hildebrand, J. G. & G. M. Shepherd, 1997. Mechanisms of olfactory discrimination: Converging evidence for common principles across phyla. Annual Review of Neuroscience 20: 595–631.

Homberg, U., R. Montague & J. G. Hildebrand, 1988. Anatomy of antenno-cerebral pathways in the brain of the sphinx moth *Manduca sexta*. Cell and Tissue Research 254: 255–281.

Jagadeesh, B., C. M. Gray & D. Ferster, 1992. Visually evoked oscillations of membrane potential in cells of cat visual cortex. Science 257: 552–554.

Jahr, C. E. & R. A. Nicoll, 1982. An intracellular analysis of dendrodendritic inhibition in the turtle in vitro olfactory bulb. Journal of Physiology (London) 326: 213–234.

Joerges, J., A. Kuttner, C. G. Galizia & R. Menzel, 1997. Representation of odours and odour mixtures visualized in the honeybee brain. Nature 387: 285–288.

Kanzaki, R., E. A. Arbas, N. J. Strausfeld & J. G. Hildebrand, 1989. Physiology and morphology of projection neurons in the antennal lobe of the male moth *Manduca sexta*. Journal of comparative Physiology A 165: 427–453.

Kauer, J. S., 1991. Contributions of topography and parallel processing to odour coding in the vertebrate olfactory pathway. Trends in Neuroscience 14: 79–85.

Kauer, J. S., 1974. Response patterns of amphibian olfactory bulb neurones to odour stimulation. Journal of Physiology (London) 243: 695–715.

Kauer, J. S. & D. G. Moulton, 1974. Responses of olfactory bulb neurones to odour stimulation of small nasal areas in the salamander. Journal of Physiology (London) 243: 717–737.

Kleinfeld, D., K. R. Delaney, M. S. Fee, J. A. Flores, D. W. Tank & A. Gelperin, 1994. Dynamics of propagating waves in the olfactory network of a terrestrial mollusc: An electrical and optical study. Journal of Neurophysiology 72: 1402–1419.

Kuwabara, M., 1957. Bildung des bedingten Reflexes von Pavlovs Typus bei der Honigbiene, *Apis mellifica*. Journal of the Faculty of Science of Hokkaido University. (Series VI, Zoology) 13: 458–464.

Laurent, G., 1996a. Odour images and tunes. Neuron 16: 473–476.

Laurent, G., 1996b. Dynamical representation of odours by oscillating and evolving neural assemblies. Trends in Neuroscience 19: 489–496.

Laurent, G., 1997. Olfactory processing: Maps, time and codes. Current Opinions in Neurobiology 7: 547–553.

Laurent, G. & H. Davidowitz, 1994. Encoding of olfactory information with oscillating neural assemblies. Science 265: 1872–1875.

Laurent, G. & M. Naraghi, 1994. Odorant-induced oscillations in the mushroom bodies of the locust. Journal of Neuroscience 14: 2993–3004.

Laurent, G., M. Wehr & H. Davidowitz, 1996. Temporal representation of odors in an olfactory network. Journal of Neuroscience 16: 3837–3847.

Lee, J. K. & N. J. Strausfeld, 1990. Structure, distribution and number of surface sensilla and their receptor cells on the olfactory appendage of the male moth *Manduca sexta*. Journal of Neurocytology 19: 519–538.

Leitch, B. & G. Laurent, 1996. GABAergic synapses in the antennal lobe and mushroom body of the locust olfactory system. Journal of comparative Neurology 372: 487–514.

Leon, M., 1987. Plasticity of olfactory output circuits related to early olfactory learning. Trends in Neuroscience 10: 434–438.

Lewin, B., 1994. On neuronal specificity and the molecular basis of perception. Cell 79: 935–943.

Libet, B. & R. W. Gerard, 1939. Control of the potential rhythm of the isolated frog brain. Journal of Neurophysiology 2: 153–169.

Lytton, W. W. & T. J. Sejnowski, 1991. Simulations of cortical pyramidal neurons synchronized by inhibitory interneurons. Journal of Neurophysiology 66: 1059–1079.

McClintock, T. S. & B. W. Ache, 1989. Histamine directly gates a chloride channel in lobster olfactory receptor neurons. Proceedings of the National Academy of Sciences USA 86: 8137–8141.

MacKay-Sim, A. & S. Kesteven, 1994. Topographic patterns of responsiveness to odorants in the rat olfactory epithelium. Journal of Neurophysiology 71: 150–160.

MacLeod, K. & G. Laurent, 1996. Distinct mechanisms for synchronization and temporal patterning of odour-encoding neural assemblies. Science 274: 976–979.

Macrae, A. W., 1971. On calculating unbiased information measures. Psychological Bulletin 75: 270–277.

Mainen, Z. & T. J. Sejnowski, 1995. Reliability of spike timing in neocortical neurons. Science 268: 1502–1506.

Masson, C. & H. Mustaparta, 1990. Chemical information processing in the olfactory system of insects. Physiological Reviews 70: 199–245.

McClurkin, J. W., T. J. Gawne, L. M. Optican & B. J. Richmond, 1991. Lateral geniculate neurons in behaving primates. II. Encoding information in the temporal shape of the response. Journal of Neurophysiology 66: 794–808.

McClurkin, J. W., T. J. Gawne, B. J. Richmond, L. M. Optican & D. L. Robinson, 1991. Lateral geniculate neurons in behaving primates. I. Responses to two-dimensional stimuli. Journal of Neurophysiology 66: 777–793.

Mellon, D., V. Alones & M. D. Lawrence, 1992. Anatomy and fine structure of neurons in the deutocerebral projection pathway of the crayfish olfactory system. Journal of comparative Neurology 321: 93–111.

Menzel, R., 1990. Learning, memory and 'cognition' in honey bees. In: R. P. Kesner & D. S. Olton (eds) Neurobiology of Comparative Cognition. Erlbaum, New Jersey, pp. 237–292.

Meredith, M., 1981. The analysis of response similarity in single neurons of the goldfish olfactory bulb using amino acids as odor stimuli. Chemical Senses 6: 277–293.

Meredith, M., 1986. Patterned response to odor in mammalian olfactory bulb: The influence of intensity. Journal of Neurophysiology 56: 572–597.

Meredith, M., 1992. Neural circuit computation: Complex patterns in the olfactory bulb. Brain Research Bulletin 29: 111–117.

Meredith, M. & D. G. Moulton, 1978. Patterned response to odour in single neurons of goldfish olfactory bulb: influence of odor quality and other stimulus parameters. Journal of General Physiology 71: 615–643.

Milner, P., 1974. A model for visual shape recognition. Psychological Review 81: 521–535.

Mombaert, P., F. Wang, C. Dulac, S. K. Chao, A. Nemes, M. Mendelshon, J. Edmondson & R. Axel, 1996. Visualizing an olfactory sensory map. Cell 87: 675–686.

Mori, K., 1987. Membrane and synaptic properties of identified neurons in the olfactory bulb. Progress in Neurobiology 29: 275–320.

Mori, K. & G. M. Shepherd, 1994. Emerging principles of molecular signal processing by mitral/tufted cells in the olfactory bulb. Seminars in Cell Biology 5: 65–74.

Mori, K. & Y. Yoshihara, 1995. Molecular recognition and olfactory processing in the mammalian olfactory system. Progress in Neurobiology 45: 585–619.

Müller, U. & E. Buchner, 1993. Histochemical localization of NADPH-Diaphorase in the adult *Drosophila* brain: Is nitric oxide a neuronal messenger also in insects? Naturwissenschaften 80: 524–526.

Murthy, V. N. & E. E. Fetz, 1992. Coherent 25–35 Hz oscillations in the sensorimotor cortex of the awake behaving monkey. Proceedings of the National Academy of Sciences USA 89: 5670–5674.

Ngai, J., A. Chess, M. M. Dowling, N. Necles, E. R. Macagno & R. Axel, 1993. Coding of olfactory information: topography of odorant receptor expression in the catfish olfactory epithelium. Cell 72: 667–680.

O'Keefe, J. & M. L. Recce, 1993. Phase relationship between hippocampal place units and the EEG theta-rhythm. Hippocampus 3: 317–330.

Optican, L. M. & B. J. Richmond, 1987. Temporal encoding of two-dimensional patterns by single units in primate inferior temporal cortex. III. Information theoretic analysis. Journal of Neurophysiology 57: 162–178.

Rall, W., G. M. Shepherd, T. S. Reese & M. W. Grightman, 1966. Dendro-dendritic synaptic pathway for inhibition in the olfactory bulb. Experimental Neurology 14: 44–56.

Ressler, K. J., S. L. Sullivan & L. B. Buck, 1994. Information coding in the olfactory system: evidence for a stereotyped and highly organized epitope map in the olfactory bulb. Cell 79: 1245–1255.

Ressler, K. J., S. L. Sullivan & L. B. Buck, 1993. A zonal organization of odorant receptor gene expression in the olfactory epithelium. Cell 73: 597–609.

Salinas, E. & L. F. Abbott, 1994. Vector reconstruction from firing rates. Journal of computational Neuroscience 1: 89–107.

Satou, M., 1992. Synaptic organization, local neuronal circuitry, and functional segregation of the teleost olfactory bulb. Progress in Neurobiology 34: 115–142.

Shepherd, G. M., 1994. Discrimination of molecular signals by the olfactory receptor neuron. Neuron 13: 771–790.

Sengupta, P., J. H. Chou & C. J. Bargman, 1996. *odr-10* encodes a seven transmembrane domain olfactory receptor required for responses to the odorant diacetyl. Cell 84: 899–909.

Singer, W. & C. M. Gray, 1995. Visual feature integration and the temporal correlation hypothesis. Annual Review of Neuroscience 18: 555–586.

Smith, B. H. & S. Cobey, 1994. The olfactory memory of the honeybee *Apis mellifera*. II. Blocking between odourants in binary mixtures. Journal of Experimental Biology 195: 91–108.

Smith, B. H. & W. M. Getz, 1994. Nonpheromonal olfactory processing in insects. Annual Review of Entomology 39: 351–375.

Smith, B. H. & R. Menzel, 1989. An analysis of variability in the feeding motor program of the honey bee – the role of learning in releasing a modal action network. Ethology 82: 68–81.

Stengl, M., H. Hatt & H. Breer, 1992. Peripheral processes in insect olfaction. Annual Review of Physiology 54: 665–681.

Stopfer, M., S. Bhagavan, B. H. Smith & G. Laurent, 1997. Impaired odour discrimination on desynchronization of odour encoding neural assemblies. Nature 390: 70–74.

Sullivan, S. L., S. Bohm, K. J. Ressler, L. F. Horowitz & L. B. Buck, 1995. Target-independent pattern specification in the olfactory epithelium. Neuron 15: 779–789.

Thommesen, G., 1978. The spatial distribution of odour induced potentials in the olfactory bulb of char and trout (Salmonidae). Acta Physiologica Scandinavica 102: 414–426.

Traub, R. D., M. A. Whittington, I. M. Stanford & J. G. R. Jefferys, 1996. A mechanism for generation of long-range synchronous fast oscillations in the cortex. Nature 383: 621–624.

Traub, R. D., M. A. Whittington, S. B. Colling, G. Buzsaki & J. G. R. Jefferys, 1996. Analysis of gamma rhythms in the rat hippocampus in vitro and in vivo. Journal of Physiology (London) 493: 471–484.

Vaadia, E., I. Haalman, M. Abeles, H. Bergman, Y. Prut, H. Slovin & A. Aertsen, 1995. Dynamics of neuronal interactions in monkey cortex in relation to behavioural events. Nature 373: 515–518.

Vassar, R., S. K. Chou, R. Sitcheran, J. M. Nuñez, L. B. Vosshall & R. Axel, 1994. Topographic organization of sensory projections to the olfactory bulb. Cell 79: 981–991.

Vassar, R., J. Ngai & R. Axel, 1993. Spatial segregation of odorant receptor expression in the mammalian olfactory epithelium. Cell 74: 309–318.

Victor, J. D. & K. P. Purpura, 1996. Nature and precision of temporal coding in visual cortex: A metric-space analysis. Journal of Neurophysiology 76: 1310–1326.

Von der Malsburg, C. & W. Schneider, 1986. A neural cocktail-party processor. Biological Cybernetics 54: 29–40.

Vreeswijk, C. van, L. F. Abbott & G. B. Ermentrout, 1995. When inhibition, not excitation synchronizes neural firing. Journal of computational Neuroscience 1: 313–322.

Wang, X.-J. & G. Buzsáki, 1996. Gamma oscillation by synaptic inhibition in a hippocampal interneuronal network model. Journal of Neuroscience 16: 6402–6413.

Wehr, M. & G. Laurent, 1996. Odour encoding by temporal sequences of firing in oscillating neural assemblies. Nature 384: 162–166.

Whittington, M. A., R. D. Traub & J. G. R. Jeffreys, 1995. Synchronized oscillations in interneuron networks driven by metabotropic glutamate receptor activation. Nature 373: 612–615.

Wilson, M. A. & B. L. McNaughton, 1993. Dynamics of the hippocampal ensemble code for space. Science 261: 1055–1058.

Yokoi, M., K. Mori & S. Nakanishi, 1995. Refinement of odor molecule tuning by dendrodendritic synaptic inhibition in the olfactory bulb. Proceedings of the National Academy of Sciences USA 92: 3371–3375.

*Entomologia Experimentalis et Applicata* **91:** 19–28, 1999.



# Chemo-discriminatory neurones in the sub-oesophageal ganglion of *Locusta migratoria*

Stephen M. Rogers* & Stephen J. Simpson
*Department of Zoology, South Parks Rd, Oxford OX1 3PS, UK; *Current address: School of Biological Sciences, Bassett Crescent East, Southampton, SO16 7PX, UK*

Accepted: October 28, 1998

*Key words:* Locust, sub-oesophageal ganglion, chemoreception, gustation, CNS, maxillary palps

## Abstract

The sub-oesophageal ganglion of *Locusta migratoria* was searched for neurones responsive to stimulation of the maxillary palps using intracellular recording techniques. Two plant stimuli were used: wheat, a host plant and cabbage, an unacceptable non-host plant. The stimuli were presented to the palp as both intact leaf tissue and as droplets of aqueous solutions of plant extracts. All the sampled neurones that responded to stimulation of the palp also responded to simple mechanical stimulation. However, 25% of the neurones exhibited consistent differences in response to the two plants when presented as both leaf tissue and droplets, strongly suggesting that they also received a chemosensory input. These differential responses most commonly took the form of differences in the duration of cell activity and/or variation in the latency of the onset of response. The receptive fields of differentially responding neurones were confined to the maxillary palp, or at most to the maxillary and labial palps.

## Introduction

Despite the undoubted importance of contact-chemoreception in the lives of many, if not most, insects, very little indeed is known about the integration of gustatory signals in the Central Nervous System (CNS) and this stands in contrast to the progress made in the understanding of olfactory processing (Masson & Mustaparta, 1990; Hildebrand & Shepherd, 1997). The present paper describes the properties of some neurones in the Sub-Oesophageal Ganglion (SOG) of locusts that are responsive to stimulation of receptors on the maxillary palps; structures that are important in food selection and feeding behaviour. These neurones exhibited differential responses to wheat, an acceptable host plant, and cabbage, an unacceptable non-host plant to *Locusta migratoria*.

The ends of each palp (the 'palp-dome') bears an array of approximately 400 sensilla in adult locusts (Chapman, 1982), by far the majority of which are uniporous contact chemoreceptors, each containing 5–9 chemosensory neurones and a single mechanosensory afferent (Haskell & Schoonhoven, 1969; Blaney *et al.*, 1971). The primary destination of most of these afferents, several thousand in total, is the SOG. During normal exploratory behaviour the ends of the palps are vibrated rapidly against the substrate ('palpation'), sampling the mechanical and chemical environment at a frequency of 10–15 Hz (Blaney & Chapman, 1970).

## Materials and methods

The neuropil of the SOG was searched for neurones responsive to stimulation of the sensilla on the end of the right maxillary palp using standard intracellular recording techniques (see Simpson, 1992). The two plant stimuli were each presented to the palp in two distinct ways. In the first instance, intact plant tissue was used. A piece of wheat blade, or a similarly sized and shaped piece of cabbage leaf was wrapped tightly around the barrel of a short piece of capillary glass (1.2 mm diameter and 3 mm long) and secured in a rocking device as depicted in Figure 1. When activated by a 100 ms square wave pulse from a signal generator, this device moved the plant stimulus 3–4 mm forward

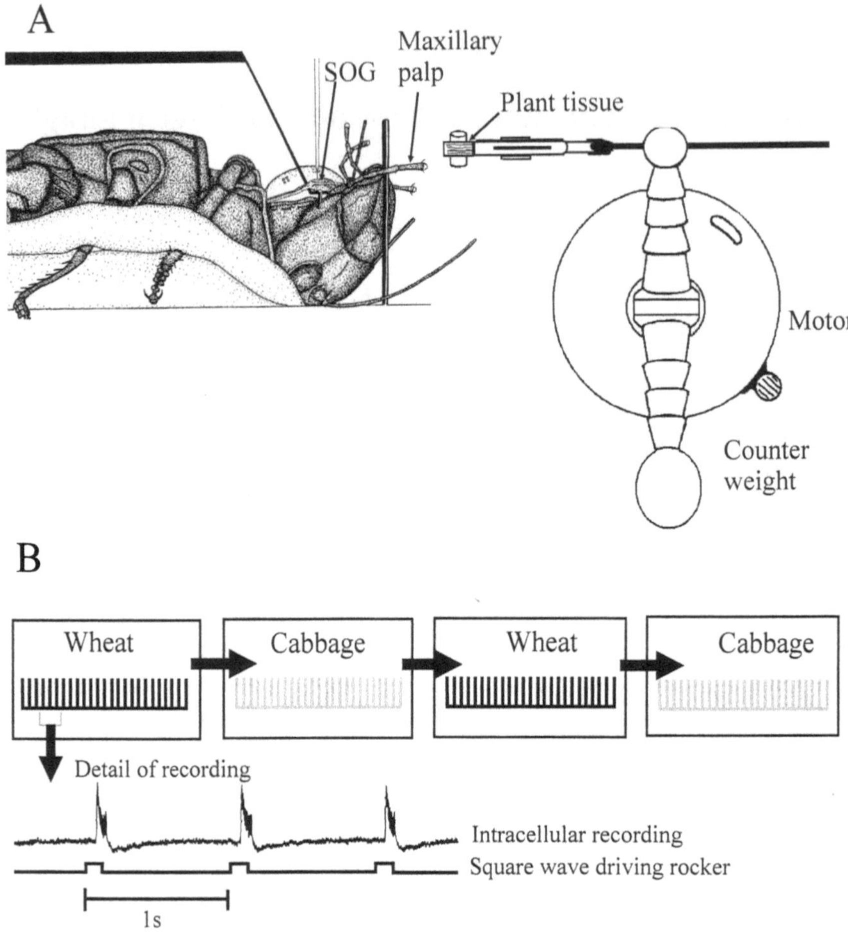


*Figure 1.* Stimulus and recording regime. A rocking device made from a small DC motor was used to stimulate the right maxillary palp of the locust. (A) The pin bearing the leaf-tissue stimulus was brought forward and downwards until it just touched the palp dome and was held there until the counterweight caused the device to recoil to resting position (shown) when released by the end of the the100ms square-wave pulse driving the device (B) Schematic of the stimulating regime used. The palp was repeatedly stimulated at 1s intervals by one of the plant stimuli, before the being exchanged for the other. Two alternate blocks of response data for each plant type were usually gathered. An expanded section of data, showing three separate stimulating square waves and an intracellular recording of the response of a neurone is shown below.

such that it just contacted the sensilla on the palp-dome. It was held there until released by the end of the square wave, when the counterweight on the device caused it to recoil rapidly to its resting position. The stimulation was repeated at 1-s intervals, building up a series of response data to the plant stimulus, consisting, in most instances, of 14-16 separate contacts. Usually, two alternate bouts of stimulation by each plant species were performed. This allowed a comparison of the response to the two different plants whilst allowing some control for factors such as the quality of the recording, the way in which the stimulus contacted the palp and neural processes such as adaptation. As the palp was not free to move, each contact with the plant tissue was made to the same subset of sensilla on the palp-dome. Allowing for the curvature of the palp-dome, this probably amounted to no more than 40 sensilla, or 10% of the total number. For the purposes of comparison and analysis, the individual cell responses were filtered to remove the spikes, a major source of variation between successive stimulations, leaving the underlying post-synaptic potentials (psps), which were, in most instances, a more consistent feature of the response to a particular plant stimulus. The individual responses in a particular bout were aligned to the start of the square wave driving the rocker and averaged. These averages were over-layed to produce a data summary for each of the sampled neurones.

The second way in which the wheat and cabbage stimuli were presented was as aqueous solutions of methanolic plant extracts (100 g of fresh plant material was freeze dried and steeped for 24 h in 200 ml methanol; the methanol was evaporated using a rotary evaporator and the residue re-dissolved in 100 ml

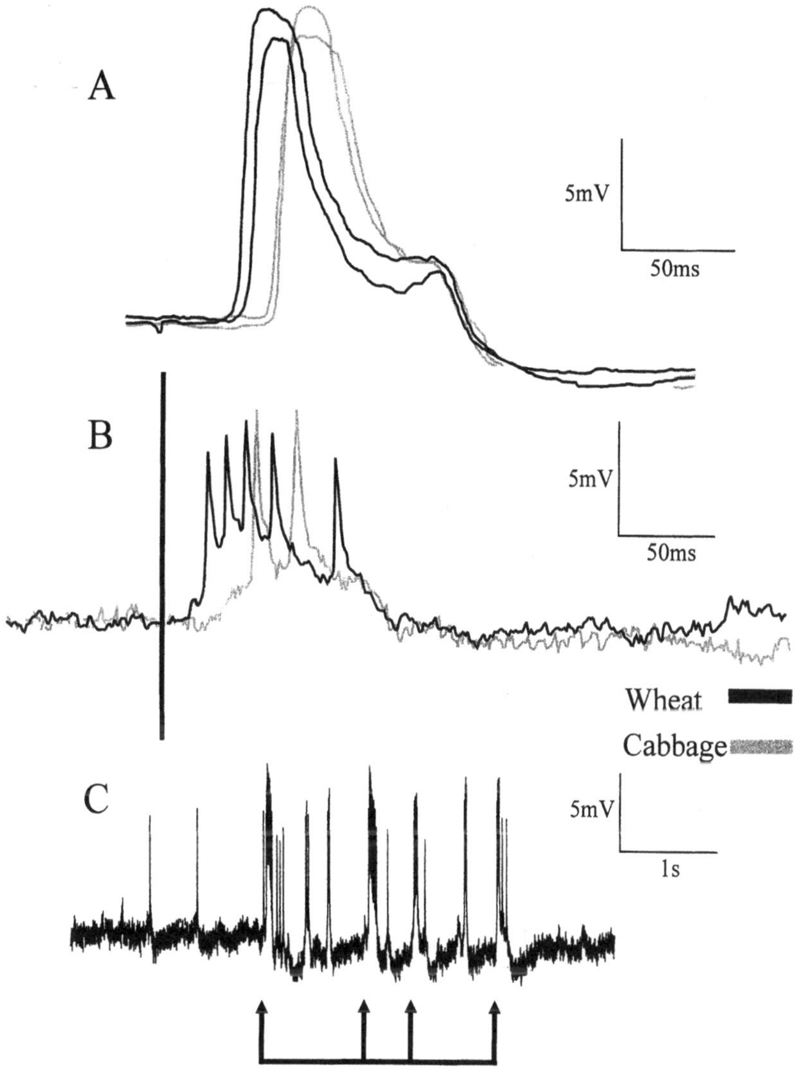


*Figure 2.* Data from XN 2. (A) Filtered and averaged responses for each of the four blocks of gathered response data, two alternate wheat (black) and cabbage (grey) series. Each of the superimposed traces is an average of sixteen separate touches of the stimulus to the palp. (B) Superimposed responses of the neurone to representative wheat and cabbage droplets applied to the palp dome, aligned to the stimulus artefact (the vertical line). (C) The mechanosensory response of the neurone produced by touching the palp with a piece of tissue paper.

water). Droplets of each solution were applied alternately using a syringe and left in contact with the palp dome for at least 1 s, after which the palp was wiped clean. Finally the purely mechano-responsive properties of the sampled neurones were tested by stimulating the palp and other body regions, namely the contralateral maxillary palp, the labial palps, the antennae, labrum and tarsi, with a piece of tissue paper or a pin. Attempts (not always successful) were made to stain impaled neurones with Lucifer-Yellow whilst gathering physiological data by passing a constant hyperpolarizing current of approximately 1nA through the recording electrodes, which were backfilled with the dye. The reconstructed neurones are shown from the ventral aspect of the SOG, with the nerves numbered as in Tyrer & Gregory (1982). Neurones were named according to the scheme devised by Robertson & Pearson (1983) and following from the descriptions in Simpson (1992) and Rogers (1998). Where no anatomical data were available, neurones are named XN followed by a numeral, purely as a matter of convenience.

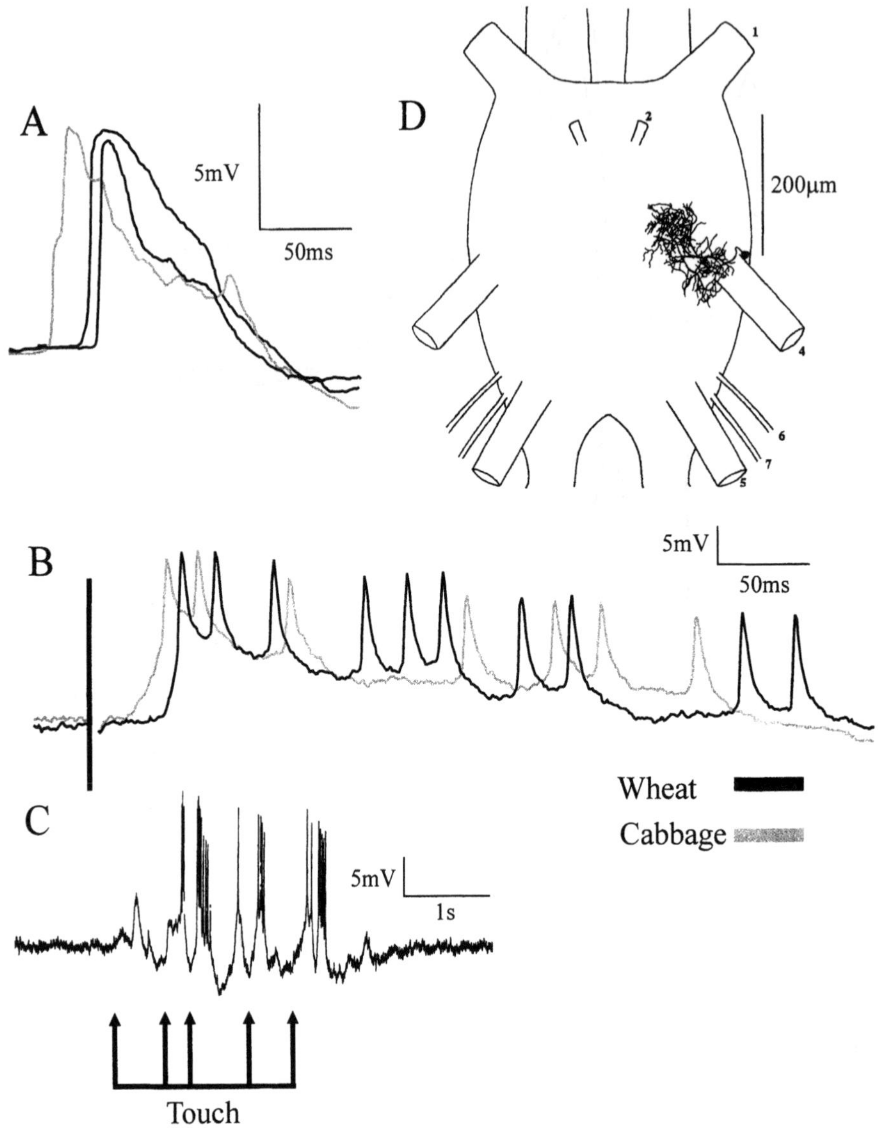


*Figure 3.* Data from SOG IN 112. (A) Superimposed averaged responses to sixteen separate touches of leaf tissue for two bouts of wheat (black) and a single bout of cabbage stimulation (grey). (B) Two superimposed representative droplet responses, aligned to the stimulus artefact. (C) The mechanosensory response of the neurone to touching the palp-dome. (D) Ventral view of the SOG with the circum-oesophageal connectives orientated to the top of the page, showing the anatomy of the neurone.

**Results**

All the neurones that responded to stimulation of the maxillary palp, and were recorded from for long enough to be fully characterised ($n = 32$), were responsive to purely mechanosensory stimulation when the palp was touched by a neutral stimulus such as tissue paper or pin. Three-quarters of the sampled neurones displayed no difference in response to the two plant stimuli when presented as either intact leaf tissue or droplets; in half of this number the neurone responded weakly, if at all, to either applied droplet type. All these neurones are tacitly assumed to have been purely mechano-responsive in function. The remaining neurones all displayed some qualitatively consistent difference in response to the two different plants when presented as both intact leaf tissue and as droplets, strongly suggesting that these neurones received different chemosensory inputs activated from each plant type.

Neurone XN3 (Figure 2) was representative of approximately half of the neurones that exhibited dif-

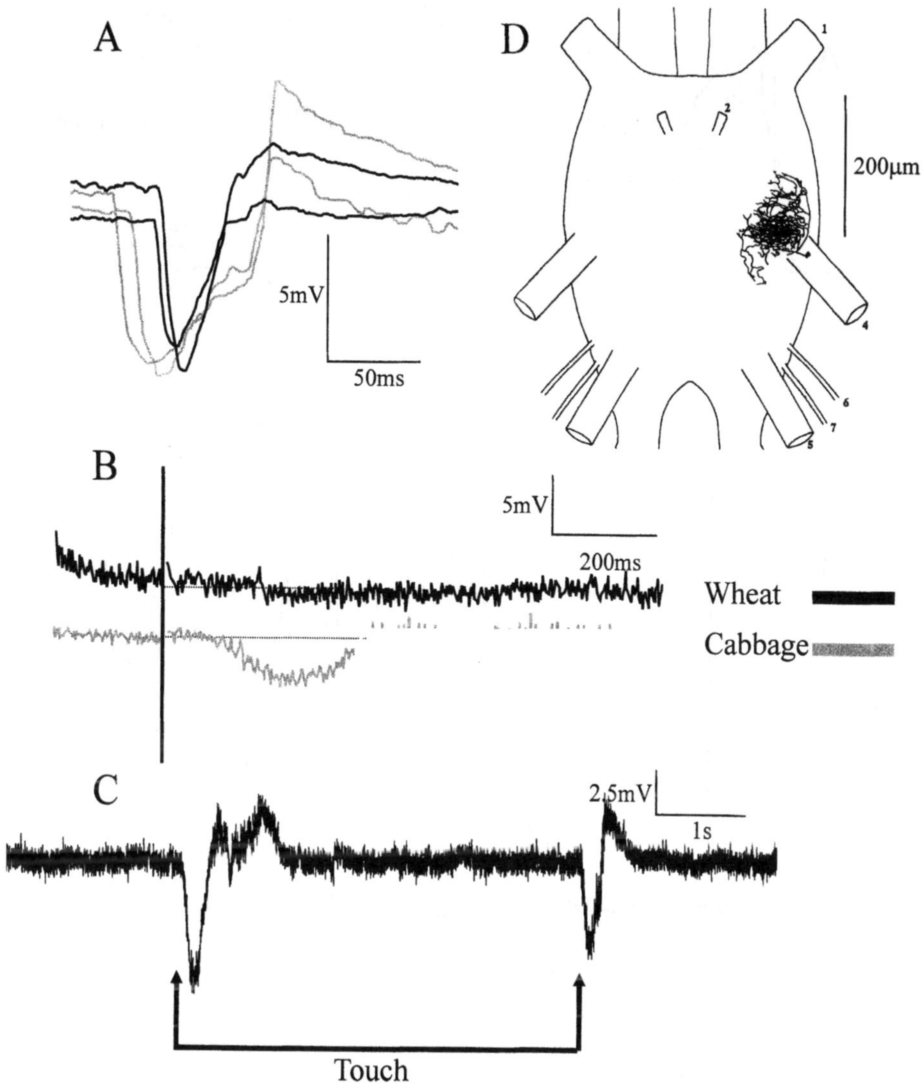


*Figure 4.* SOG IN 114. (A) Filtered and averaged responses for each of the four blocks of gathered response data, two wheat (black) and two cabbage (grey). Each trace is an average of 14–16 separate contacts with the plant tissue. (B) Superimposed responses of the neurone to representative wheat and cabbage droplets applied to the palp dome. (C) The mechanosensory response of the neurone as produced by touching the palp with a piece of tissue paper. (D) The anatomy of the neurone.

ferential responses to wheat and cabbage. Each of the four superimposed traces in Figure 2A is an average of 16 separate contacts of the plant tissue with the palp-dome. All four averages are shown aligned on the start of the square wave driving the rocker that presented the stimulus to the palp. When stimulated by wheat the response started on average 9ms earlier than when stimulated by cabbage, and this was closely repeated for both blocks of data. All four averages shared the same coincident end point, brought about by the withdrawal of the stimulus at the end of the square wave. Figure 2B depicts the superimposed responses of the neurone to representative wheat and cabbage extract droplets aligned to the stimulus artefact. A similar pattern of response was seen here as well. When stimulated by cabbage droplets, the neural response was much slower developing compared to the rapid onset of the psp when stimulated by wheat droplets, leading to a considerable difference (8 ms) in the onset of spiking, similar to that observed for the plant tissue stimulation. The strong response of this neurone to purely mechanical stimulation of the maxillary palp dome is shown in Figure 2C.

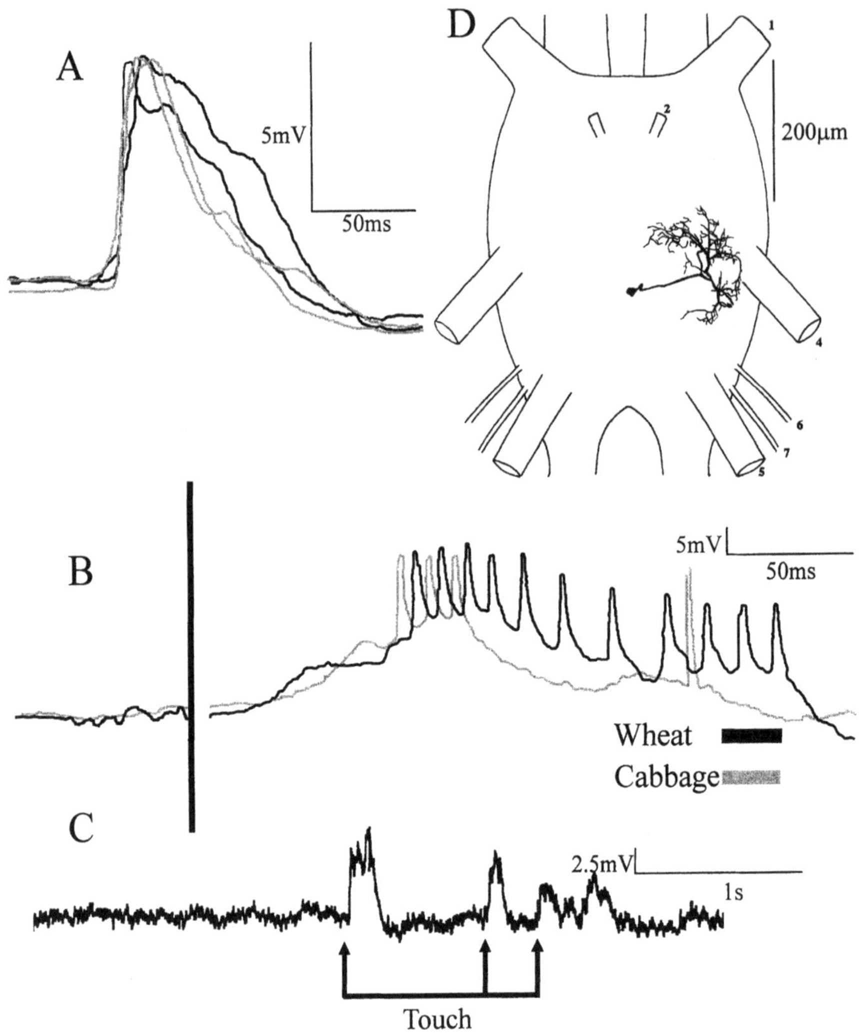


*Figure 5.* Data from SOG IN 113. (A) Superimposed averaged responses to sixteen separate touches of leaf tissue for two alternate bouts of cabbage (grey) and wheat stimulation (black). (B) Two representative droplet responses. (C) The mechanosensory response of the neurone to touching the palp-dome with a piece of tissue paper. Note the difference in the vertical scale between this figure as compared to those responses to chemical stimuli shown in A. and B. (D) The anatomy of the neurone.

Figure 3 depicts data from neurone SOG IN 112, which had response properties similar to those of XN3, but in contrast, the response started earlier and persisted for longer when this cell was stimulated by cabbage compared to wheat. Again, qualitatively similar differences were seen when stimulated by both plant tissue (mean difference 11 ms) and droplets (mean difference 10 ms). As described above, Figure 3a depicts the averaged responses to 16 separate touches for two bouts of cabbage and a single bout of wheat stimulation, Figure 3B shows two representative droplet responses, and Figure 3C the mechanosensory response of the neurone to touching the palp-dome. Some anatomical data were gathered for this neurone (Figure 3D). The anatomy reveals the neurone to be a small local interneurone confined entirely to the ipsilateral maxillary neuromere.

SOG IN 114 (Figure 4) belonged to a different class of neurone to the two cells described above. These data are from an apparently non-spiking local interneurone (Figure 4D) that received inhibitory inputs when the palp dome was stimulated. Nevertheless, differences in the onset of response when stimulated by the two plants can also be seen in this neurone. The inhibitory post-synaptic potential (ipsp) evoked by stimulation with cabbage started earlier (mean 11 ms) and persisted for longer (mean 18ms) than the more transient response to wheat (Figure 4A).

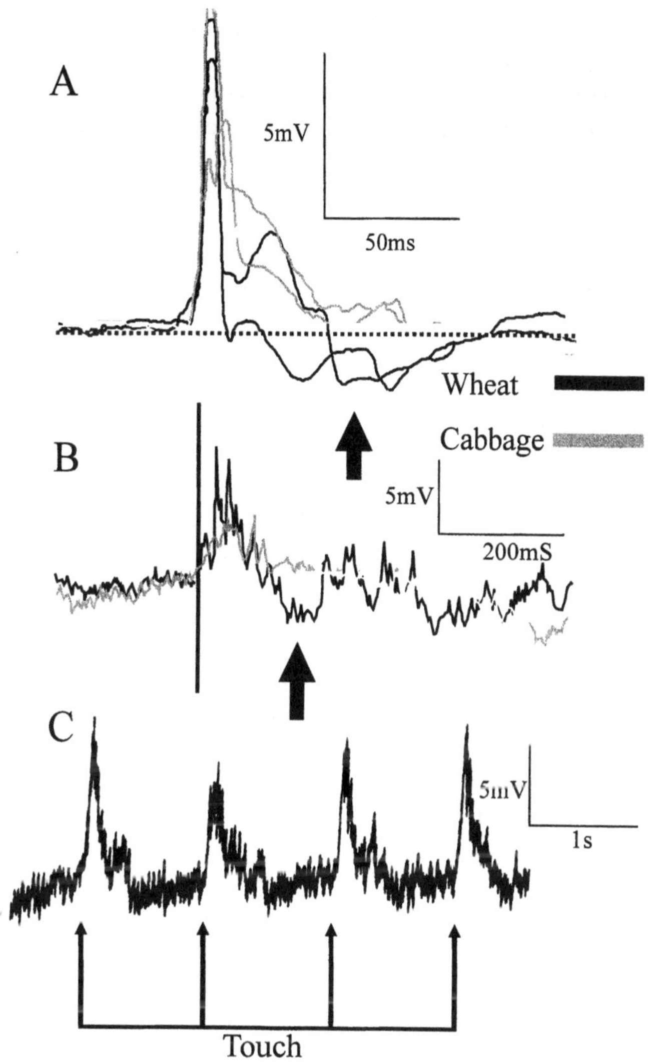


*Figure 6.* Data from XN 7. (A) Filtered and averaged responses for each of the four blocks of gathered response data, two wheat (black) and two cabbage (grey). Each trace is an average of 13–16 separate contacts with the plant tissue. When stimulated by wheat, there is a delayed inhibitory component to the response (arrow) that is missing when stimulated by cabbage. (B) Superimposed responses of the neurone to a wheat and cabbage droplet applied to the palp dome. Again, a delayed inhibitory component (arrow) is apparent when stimulated by the wheat droplet, which is absent from the cabbage droplet response. (C) The mechanosensory response of the neurone as produced by touching the palp-dome with a pin.

The droplet responses in this neurone do not so closely match the responses to touches of the plant tissue. Cabbage droplets, when applied to the palp, induced a slow developing and long lasting hyperpolarization, but stimulation by wheat droplets had no observable effect on the neurone. When stimulated mechanically, rapidly developing ipsps were produced (Figure 4C).

Whilst variable latencies in the onset of response where the most commonly observed differences in chemo-discriminatory neurones, other forms of differential response were also sometimes encountered. For example, SOG IN 113, illustrated in Figure 5, was very variable in the onset of its response, but this was not linked to plant type. The averaged responses in Figure 5A are shown aligned to the start of response, not the square wave driving the rocker as previously. However, there were differences in the dynamics of the response to the two different plants that were repeated in both modes of stimulus presentation. When stimulated by wheat, applied as either leaf (Figure 5A) or droplet (Figure 5B), a greater degree of excitation was maintained for longer, compared to the responses to cabbage, which repolarized more rapidly from maximal excitation. In the unfiltered responses to leaf tissue, as in the droplet responses, the sustained depolarization evoked by wheat bore spikes for its entire duration, whilst only a very brief burst of spiking was produced when stimulated by cabbage. This was the only one of the sampled neurones in which the response to purely mechanical stimulation was demonstrably weaker than when touched by plant tissue (Figure 5C), each touch of a piece of tissue paper producing only sub-threshold epsps. As for the other chemo-discriminatory neurones for which anatomical data were gathered, this cell was a small local interneurone with its arborization entirely confined to the ipsilateral maxillary neuromere (Figure 5D).

Another variety of discrimination was shown by XN7 (Figure 6). In this neurone there was an initial excitatory component that was common to both plant types (Figure 6A, B), but when stimulated by wheat there was a subsequent, delayed inhibitory input (arrow) that was absent when stimulated by cabbage or by a purely mechanical stimulus (Figure 6C). The droplet responses were qualitatively similar, but the duration and relative timing of the excitatory and inhibitory (wheat) components were greatly increased.

Although all the sampled neurones responded to mechanical stimulation, there were differences in the extent of the receptive field shown by chemo-discriminatory and non-discriminatory neurones. All those neurones that exhibited differential responses to wheat and cabbage had receptive fields confined to the ipsilateral maxillary palp, or at most to the ipsilateral maxillary and labial palps (Table 1). The anatomy of the successfully stained chemo-discriminatory neurones was consistent with these restricted sensory fields. These neurones had their major arborizations entirely confined to the ipsilateral maxillary neuromere. Neurones in which no difference in response to the two different plant stimuli was apparent, all had receptive fields which extended to other regions on the head and body. These receptive fields commonly extended to the ipsilateral antenna and labrum and/or to the ipsilateral tarsi, or to all four palps, or more rarely to all the tested areas, both ipsilateral and contralateral, on the head and body (Table 1). Many, though not all, the successfully staining non-discriminating neurones had extensive arborizations, in the SOG and frequently also had ascending and/or descending projections (Rogers, 1998; Simpson, 1992).

## Discussion

Interpreting and acting appropriately to the stream of chemosensory information that reaches the CNS is a formidably complex task. Therefore it is perhaps unsurprising that, even in a very brief survey such as this, a variety of neurone types were encountered, which represented the different chemical stimuli provided by a host and non-host plant in a number of different ways. These encompassed variability in the latency of the onset of response (XN 3, SOG IN 112, and SOG IN 114), differences in the duration of spike bursts (SOG IN 113), and the presence of a delayed inhibitory input when stimulated by the host plant, that was absent when stimulated by the non-host plant (XN 7). Amongst this diversity, these neurones shared all shared two common features, namely local receptive fields, and the convergence of chemosensory and mechanosensory inputs in the same neurones. Neurones responding differentially to the two plant stimuli had very local receptive fields compared to non-differentially responding neurones (Simpson, 1992; Rogers, 1998). On an insect, such as a locust, there is a hierarchy of different chemosensory fields on the body and head that sample a potential food, either sequentially or in conjunction with others (Bernays & Simpson, 1982). Clearly, within the CNS there must be some co-ordination of inputs from these different sensory fields to produce an integrated behavioural outcome; however, from these data it appears that at least the early processing of chemosensory information is carried out at a very local level and in isolation.

The second shared characteristic shown by these neurones, the conjoining of mechanosensory and chemosensory inputs, meant that an unadulterated chemosensory input was never seen in any of the neurones, as contact with a leaf or droplet would invariably have elicited a mechanosensory response as well. Thus chemosensory inputs were only seen through differences they made to the response of a neurone, not by whether the neurone responded or not. This raises the possibility that some of those neurones in which no difference in response was apparent, were receiving chemosensory inputs, but that these were not differentially activated by the plants used.

Perceiving the chemical environment through a sensory system that incorporates both chemical and mechanical information could potentially lead to a confounding of different strengths of input from the two modalities, and make the task of chemosensory as-

*Table 1.* Receptive fields of sampled neurones in the SOG. The right maxillary palp-dome was the default stimulus site and this number gives the total number of neurones tested in each category. Only data from neurones held for long enough to fully characterise the receptive field are included. The numbers in each field indicate the number of neurones with receptive fields extending to these regions

| | Chemo-discriminatory neurones | | Non-differentially responsive neurones | |
|---|---|---|---|---|
| | right | left | right | left |
| Maxillary palps | 5 | | 16 | 11 |
| Labial palps | 2 | | 13 | 10 |
| Antennae | | | 9 | 6 |
| Labrum | | | 7 | 6 |
| Tarsi[1] | | | 9 | 7 |

[1]Tarsal responses invariably extended to all three ipsilateral tarsi.

sessment more ambiguous. It is possible, however, that variability in mechanical input is minimised through controlling the placement of the palp during palpation; perhaps through using feedback from mechanosensory interneurones to standardise the strength of contact with the substrate. In other respects, the joining of the two modalities may be advantageous. It is hypothesised that insects make assessments of potential foods by integrating inputs from many chemosensory neurones, which may be housed in spatially separate sensilla using cross-fibre coding (Chapman, 1995). Gating chemosensory inputs through mechanoresponsive neurones may be a way of ensuring that a weak input from a particular sensillum is because of a lack of suitable stimulating chemicals and not because it is not actually contacting the substrate. The ways in which the differences between the plant stimuli were represented by the discriminatory neurones could in themselves serve as ways in which different kinds of information, both taste and touch could be unambiguously conveyed by the same neurones. For example, the psps produced in response to contact with leaf tissue were commonly greater than those evoked by applied droplets, but latency differences (the most commonly encountered form of chemosensory differentiation) were similar for both modes of stimulus presentation.

Locusts habitually palpate at a high frequency (10–15 Hz), and each palp will only be in contact with the substrate for a short period (33–50 ms, if the palp is in contact for 50% of each cycle, although Blaney & Chapman (1970) suggest the period of contact may be as little as 10 ms). Given these very brief contacts with the substrate, representing different chemical qualities as variations in the latency of the start of response is potentially a very effective way of rapidly encoding chemical qualities. During rapid palpation the end of activity in a neurone is more likely to be governed by the lifting of the palp, rather than the intrinsic duration of sensory inputs into the cell. Furthermore, a window of 10ms is a very brief interval in which to use spike frequency to code chemical quality unless the firing rate of the chemosensory afferents is exceptionally high. Latency coding, however, only allows an assessment to be made immediately following contact. This may be one reason why palpation, and other behaviours such as tarsal drumming, are commonly associated with chemical sampling behaviour, as it is the repeated contact which allows a stream of chemosensory information to reach the CNS. Blaney & Duckett (1975) demonstrated that palpation could serve as a means of minimising adaptation of the chemosensory afferents. Brief, repeated series of contacts provided a higher overall sensory input over longer periods than a single sustained placement of the palp. Neither of these possible explanations of the significance of palpation is mutually exclusive.

It is unclear how the chemosensory mediated differences in response dynamics are generated. A particular problem arises in trying to explain the mechanism of producing differences in the latency of the onset of response, when both plant stimuli contacting the palp shared a common mechanosensory component. The data in this study gives no indication of how the response dynamics seen may arise. The differences in latency may be a consequence of the nature of convergent inputs from gustatory sensilla onto these neurones, or perhaps be derived via other interneuronal inputs.

The present work provides the first description of neurones in the CNS of locusts that display clear differences in response to host and non-host plant chemical stimuli as detected by gustatory sensilla on the mouthparts. The challenge now is to gain an understanding of how these neural differences may be translated into the behavioural acceptance or rejection of different plants.

**Acknowledgements**

We would like to thank Stuart Warwick for his programming skills and his help in the presentation and analysis of the data.

## References

Bernays, E. A. & S. J. Simpson, 1982. Control of food intake. Advances in Insect Physiology 16: 59–102.

Blaney, W. M. & R. F. Chapman, 1970. The functions of the maxillary palps of Acrididae (Orthoptera). Entomologia Expermentalis et Applicata 13: 363–376.

Blaney, W. M., R. F. Chapman & A. G. Cook, 1971. The structure of the terminal sensilla on the maxillary palps of *Locusta migratoria* (L.), and changes associated with moulting. Zeitschrift für Zellforschung 121: 48–68.

Blaney, W. M. & A. M. Duckett, 1975. The significance of palpation of the maxillary palps of *Locusta migratoria* (L): an electrophysiological and behavioural study. Journal of Experimental Biology 63: 701–712.

Chapman, R. F., 1982. Chemoreception: The significance of receptor numbers. Advances in Insect Physiology 16: 247–333.

Chapman, R. F., 1995. Chemosensory regulation of feeding. In: R. F. Chapman & G. de Boer (eds), Regulatory Mechanisms in Insect Feeding. Chapman & Hall, New York, pp. 101–136.

Haskell, P. T. & L. M. Schoonhoven, 1969. The function of certain mouthpart receptors in relation to feeding in *Schistocerca gregaria* and *Locusta migratoria migratorioides*. Entomologia Experimentalis et Applicata 12: 423–440.

Hildebrand, J. G. & G. M Shepherd, 1997. Mechanisms of olfactory discrimination: converging evidence for common principles across Phyla. Annual Review of Neuroscience 20: 595–631.

Masson, C. & H. Mustaparta, 1990. Chemical information processing in the olfactory system of insects. Physiological Reviews 70: 199–245.

Robertson, R. M. & K. G. Pearson, 1983. Interneurones in the flight system of the locust: distribution, connections and resetting properties. Journal of Comparative Neurology 215: 33–50.

Rogers, S. M., 1998. Chemosensory development and integration in *Locusta migratoria*. D.Phil thesis, University of Oxford, UK.

Simpson, S. J., 1992. Mechanoresponsive neurones in the suboesophageal ganglion of the locust. Physiological Entomology 17: 351–369.

Tyrer, N. M & G. E. Gregory, 1982. A guide to the neuroanatomy of locust suboesophageal and thoracic ganglia. Philosophical Transactions of the Royal Society London, series B 297: 91–123.

*Entomologia Experimentalis et Applicata* **91:** 29–35, 1999.



# Specialist deterrent chemoreceptors enable *Pieris* caterpillars to discriminate between chemically different deterrents

Joop J. A. van Loon & Louis M. Schoonhoven
*Laboratory of Entomology, Wageningen Agricultural University, P.O. Box 8031, 6700 EH Wageningen, The Netherlands*

Accepted: October 19, 1998

*Key words:* chemoreception, deterrents, *Pieris*, Brassicaceae, cardenolides, host-plant selection, evolution

## Abstract

*Pieris* butterflies (Lepidoptera: Pieridae) are specialist herbivores of cruciferous plants. They exploit glucosinolates, secondary plant metabolites chemotaxonomically characteristic for this plant family, as token stimuli. In addition to particular glucosinolates, some genera of the Cruciferae contain cardenolides, steroidal allelochemicals that act as potent feeding and oviposition deterrents to several *Pieris* species. We investigated the sensory mechanisms by which these compounds are perceived in larvae. *Pieris* caterpillars and many other lepidopterous species are endowed with so-called generalist deterrent receptors, that respond to a broad spectrum of secondary plant substances. In *Pieris* caterpillars we found a second type of deterrent chemoreceptor in maxillary styloconic taste sensilla. This neuron is very sensitive to cardenolides (threshold 0.1–0.3 $\mu$M). The generalist deterrent receptor also responds to these substances but its threshold lies at 50–100× higher concentrations. In behavioural preference experiments *Pieris brassicae* L. caterpillars preferred cardenolide-treated cabbage leaf discs when confronted with a choice between them and a deterrent substance that does not occur in the Brassicaceae. The cardenolides acted as potent deterrents when offered against untreated cabbage leaf discs. This demonstrates that the balance of activity elicited in the two types of deterrent chemoreceptors determines the behavioural decision.

## Introduction

Over 80% of the herbivorous insect species are specialized feeders (Schoonhoven et al., 1998). Their behavioural decision to accept a plant as food or oviposition substrate is predominantly based on sensory information. In caterpillars, plant chemicals released from the leaf interior upon biting have a strong effect on the decision whether to accept the plant for sustained feeding (Schoonhoven, 1987). Stimulation of feeding behaviour is exerted by two groups of plant substances that can be distinguished on the basis of their general *vs.* limited taxonomic occurrence. The first group is made up of primary metabolites that are present in all plant species, such as sugars and amino acids. The second group are the so-called 'token stimuli' (Fraenkel, 1959), secondary plant metabolites that are taxonomically characteristic for certain plant taxa. For a growing number of oligophagous species it has been demonstrated that host-plant recognition is governed by contact-chemoreception of token stimuli (Schoonhoven et al., 1998). Conversely, rejection of non-host plants for feeding or oviposition has in a considerable number of cases been ascribed to contact-chemoreception of secondary plant metabolites that are designated as deterrents. Many plant families produce deterrent compounds that are chemotaxonomically characteristic. Compounds present in a certain family that are 'token-stimuli' to herbivores specialized on this family can be deterrent to non-associated oligophagous or generalist herbivores. The plant family of Brassicaceae provides a well-studied example: glucosinolates strongly stimulate feeding and oviposition of the specialized *Pieris* butterflies (Chew & Renwick, 1995) while they act as strong deterrents to several generalist herbivorous insect species (Chew, 1988). Larvae and adults possess chemoreceptors that specifically respond to these compounds

(Schoonhoven, 1967; Ma & Schoonhoven, 1973; Du et al., 1995). However, some genera of the Brassicaceae, e.g., *Cheiranthus* and *Erysimum*, are not acceptable to some *Pieris* species, although they do possess glucosinolates. Renwick and coworkers (reviewed by Chew & Renwick, 1995) have amply demonstrated that rejection is due to the production of cardenolides by these crucifers. There are no other examples of oligophagous insects for which the compounds from the host-plant family causing rejection behaviour are identified. All cardenolides isolated from crucifers that have been tested on either larvae or adult females act as deterrents, albeit to different degrees (Chew & Renwick, 1995). In adult females of *P. rapae* L. and two subspecies of *P. napi* L. chemoreceptors sensitive to cardenolides have been localised in tarsal sensilla (Städler et al., 1995; Du et al., 1995). Chemosensory sensitivity in *P. brassicae* L. larvae to one of the cardenolides, helveticoside (*syn.* erysimin), was reported by van Loon (1990) but specificity range of the adult and larval taste neurons was not investigated. Specificity of cardenolide chemoreception is of interest because in *P. brassicae* caterpillars two maxillary taste neurons responding to deterrents have been found. One is in the medial styloconicum and is a so-called 'generalist' deterrent neuron that is responsive to a wide range of chemically diverse secondary plant substances such as alkaloids, phenolics and steroids, including cardenolides (Ma, 1972; van Loon, 1990, Schoonhoven et al., 1992). The second, present in the lateral maxillary sensillum styloconicum, is a much more specific neuron and is characterized in this paper. We also link the chemosensory responses to behaviour by analysing the role of the two deterrent neurons in the food selection behaviour of the larvae, which is of interest in view of the rejection by the larvae of cardenolide-containing crucifers.

## Materials and methods

*Insects.* Laboratory colonies of *P. brassicae* and *P. rapae* were reared on greenhouse-grown Brussels sprouts plants *Brassica oleracea* var. *gemmifera* cv. Titurel (Sluis & Groot, Enkhuizen, The Netherlands) under *ad libitum* conditions under a L15:D9 photoperiod, at 23–25 °C and 60–70% r.h. Non-feeding caterpillars approaching their final larval ecdysis were collected from the mass culture and individually put in glass petri dishes. Upon ecdysis they were given a piece of excised leaf. Caterpillars used for electrophysiological and behavioural assays had finished ecdysis 24–48 h before testing, had fresh weights of 225–325 mg (*P. brassicae*) and 100–150 mg (*P. rapae*), and had been starved for 2 h.

*Chemicals.* Azadirachtin was obtained from Dr E. D. Morgan, Chemistry Department, Keele University, UK. Erysimoside was purchased from ICN Biomedicals Cleveland Ohio, USA. All other steroidal compounds were obtained from Sigma Chemical Co. Purity of the chemicals was >98%. Due to their limited solubility in water, chemicals were presolubilized in a small volume of analytical grade methanol and stepwise diluted with 1% Tween-80 in distilled water (behavioural assays) or 0.005 M KCl in distilled water (electrophysiology). The highest concentration tested (3 mM) had a methanol concentration of 2%, which was also present in the control solution in both behavioural and electrophysiological experiments.

*Behavioural bioassay.* The dual-choice cabbage leaf disc assays have been described in detail by Luo et al. (1995). Ten $\mu$l of cardenolide, azadirachtin or control solutions of the concentrations indicated in the figures were applied to and distributed over the surface of a cabbage leaf disc, that had an average fresh weight of 100 mg. In the assays in which the effect of the cardenolides was tested against a control to which only the solvent was applied, a deterrency index $[(C - T) \times 100/(C + T)$, where $C$ and $T$ represent leaf surface areas consumed from control and treated leaf discs respectively] was used to quantify strength of deterrence. For assays aimed at comparing deterrency of two deterrent compounds (cardenolide *vs.* azadirachtin) the deterrency index was calculated as $(C - A) \times 100/(C + A)$, where $C$ represents leaf surface area consumed from cardenolide treated and $A$ from azadirachtin treated leaf discs respectively.

*Electrophysiology.* Tip-recording (Hodgson et al., 1955), modified as described by van Loon (1990) was used to record responses from the medial and lateral contact-chemosensory sensilla styloconica on the maxillary galea of the *Pieris* species (Ma, 1972). The input bias current of the AD 515K operational preamplifier was 0.1 pA. Electrophysiological responses were quantified using the number of spikes in the first 1500 ms after the start of stimulation. Neural activity was sampled at 0.1 ms intervals with an Intel 486 based personal computer equipped with a DAS16 A/D conversion board (Metrabyte Co.). An

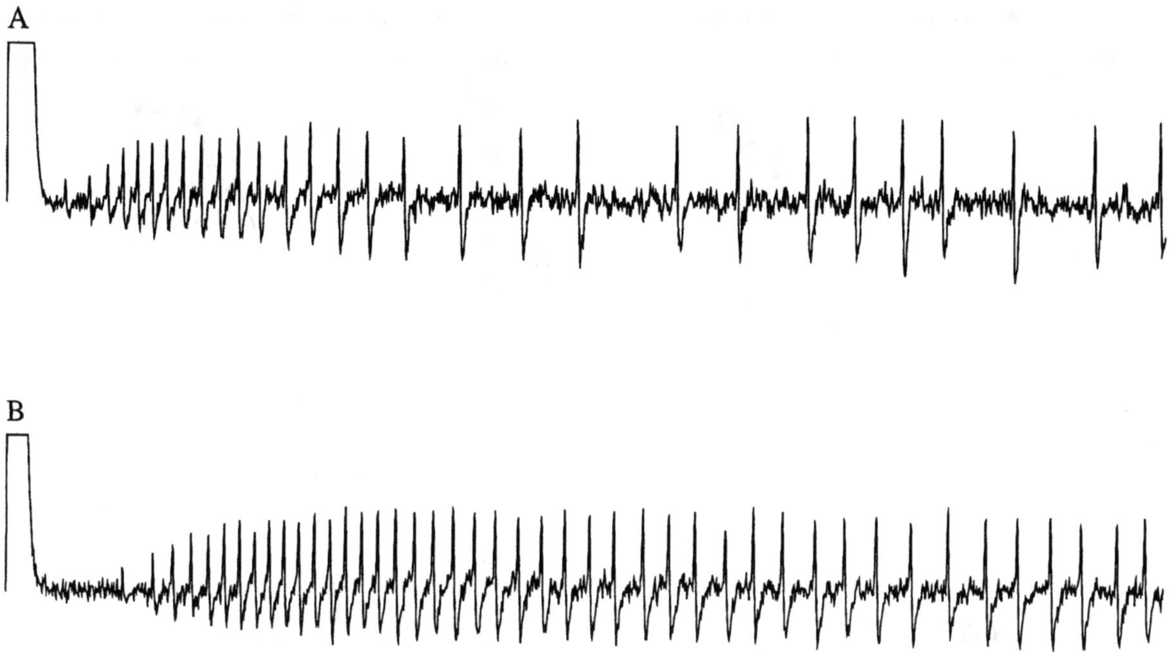


*Figure 1.* Electrophysiological recordings obtained by tip-recording from the lateral maxillary galeal styloconic sensillum of *P. brassicae* in response to A: 330 nM and B: 330 $\mu$M concentrations of helveticoside. Duration of recordings is 500 ms.

*Table 1.* Electrophysiological results of stimulating lateral (L) and medial (M) maxillary styloconic taste hairs of caterpillars of *Pieris brassicae* and *P. rapae* with steroids at 0.01 mM in 5 mM KCl

| Compound | *P. brassicae* | | *P. rapae* | |
|---|---|---|---|---|
| | L | M | L | M |
| Cholesterol | – | – | – | – |
| β-sitosterol | – | – | – | – |
| Ecdysteron | – | + | – | + |
| Ponasteron A | – | + | n.t. | n.t. |
| Ouabain | – | + | – | + |
| Digitoxin | + | + | + | + |
| Helveticoside | ++ | + | ++ | + |
| Strophantin-K | ++ | + | ++ | + |
| Erysimoside | ++ | + | ++ | + |

+: 30–50 a.p./1.5 s; ++: 51–100 a.p./1.5 s.
n.t.: not tested; –: inactive.

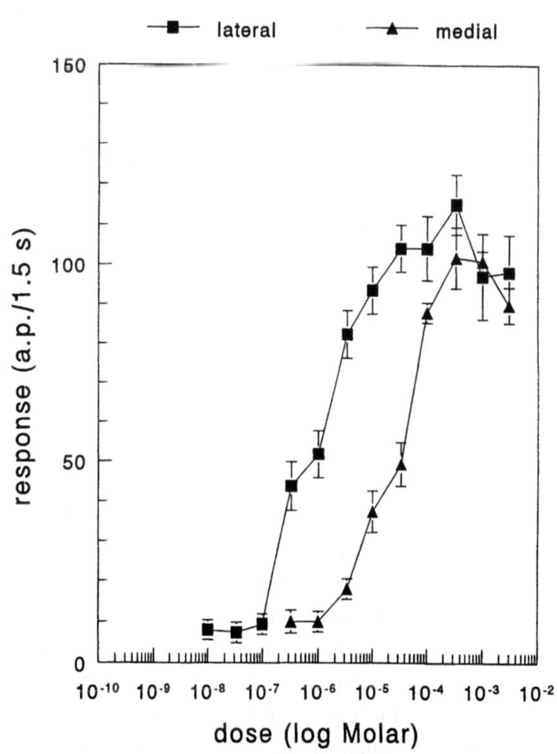


*Figure 2.* Dose-response relationships for the intensity of excitation (action potentials (a.p.) per 1500 ms; mean ± SEM) of single neurons present in either the lateral or medial sensillum styloconicum of *P. brassicae* in response to helveticoside solutions (n = 12).

interface was used ('GO-box') for signal conditioning. Digitized traces were analyzed by means of Sapid Tools version 3.5 software (Smith et al., 1990).

## Results

*Electrophysiology.* In both *P. brassicae* and *P. rapae* final instar caterpillars, of the nine steroids tested

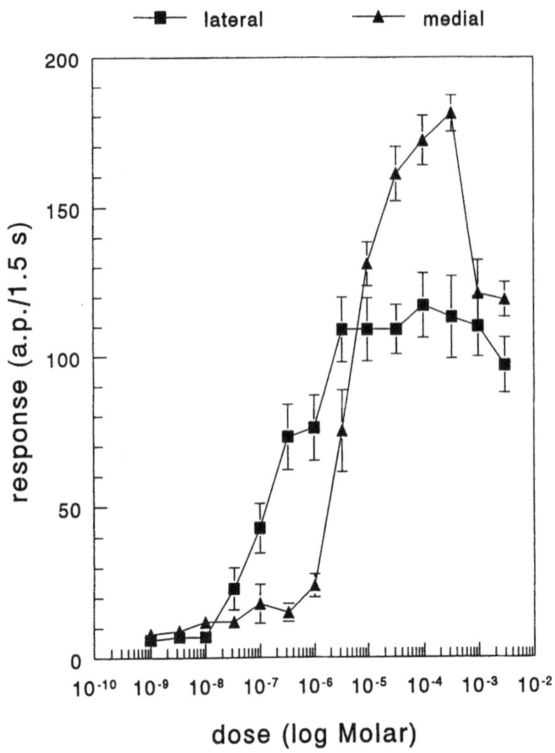


*Figure 3.* Dose-response relationships for the intensity of excitation (action potentials (a.p.) per 1500 ms; mean ± SEM) of single neurons present in either the lateral or medial sensillum styloconicum of *P. brassicae* in response to erysimoside solutions (n = 12).

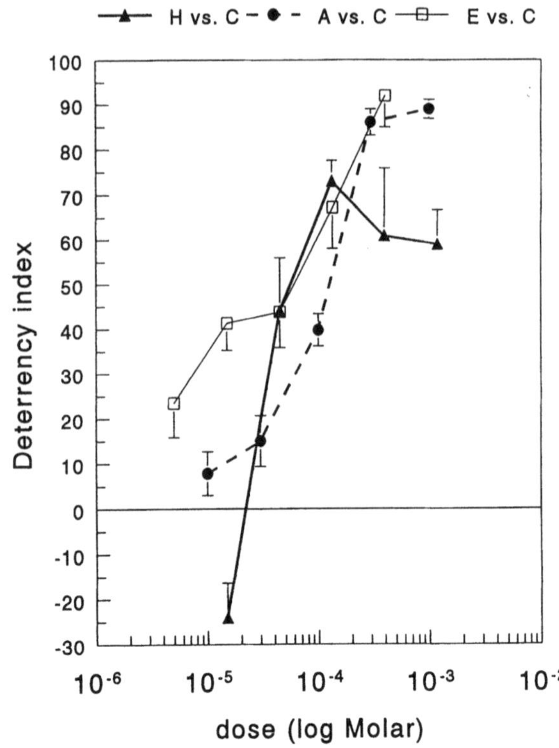


*Figure 4.* Relationships between dose of a cardenolide or azadirachtin (A) applied to cabbage leaf discs (10 $\mu$l/disc) and deterrency index (mean ± SEM; see text for calculation) in dual choice situations between cardenolide treated (helveticoside (H) and erysimoside (E)) and control leaf discs (H vs C and E vs C). Data for azadirachtin (A vs C) were included for direct comparison and were taken from Luo et al. (1995). Values of the deterrency index higher than 30 are statistically significant (two-sided Wilcoxon's matched pair signed rank test, $P < 0.02$; n = 15).

only the four cardenolides that have been isolated from Brassicaceae elicited responses from the lateral sensillum styloconicum (Table 1). The lateral sensillum styloconicum responded with a unicellular tonic response type to the cardenolides (Figure 1). The responses were often, but not always, preceded by relatively long latency times ranging between 50–150 ms. The amplitude of the first action potentials recorded was small and increased over the next 50–100 ms and subsequently remained stable. Erysimoside and helveticoside were selected for more detailed dose-response studies as they are dominant compounds in cardenolide-containing Brassicaceae. The threshold of response is between 100 and 300 nM and similar for helveticoside and erysimoside (Figures 2 and 3). The temporal characteristics of the response of the single neuron in the medial sensillum styloconicum when stimulated by the cardenolides were similar to that of the neuron contained in the lateral sensillum, but occurred at ca. 50–100 times higher stimulus concentrations. For erysimoside the maximal response reached higher levels than for helveticoside.

*Dual-choice assays.* Concentrations applied to the leaf disc surface exceeding 0.008 mM (erysimoside) or 0.02 mM (helveticoside) led to significant deterrency (Figure 4). The higher deterrent effectiveness of erysimoside corresponds with the higher action potential frequencies evoked by this compound. For comparative purposes the dose-response curve for azadirachtin is also shown in Figure 4. It is apparent that for *P. brassicae* erysimoside is at least as strong a deterrent as azadirachtin when the alternative leaf discs are devoid of exogeneously applied deterrents. In contrast, caterpillars significantly preferred leaf discs coated with the cardenolides over those treated with azadirachtin at equivalent concentrations (Figure 5), except at the lowest dose assayed, at which no preference was noted. It appears that azadirachtin is equally deterrent whether the cardenolides are coated on cabbage or not, indicating that the deterrency of

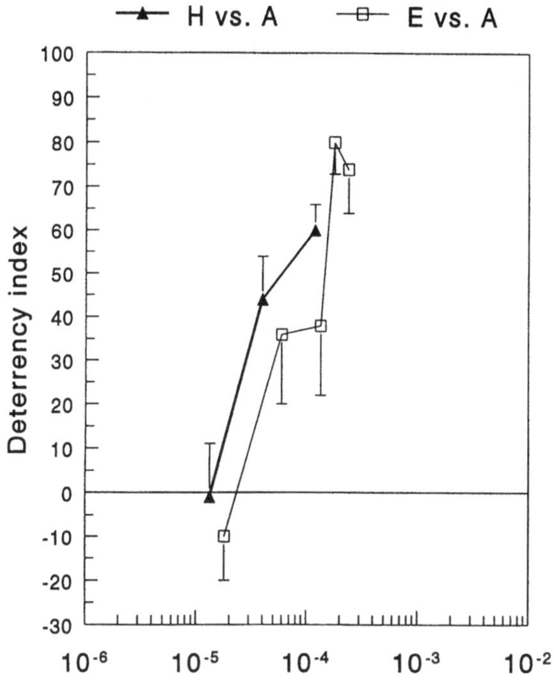


*Figure 5.* Relationships between dose of the cardenolides helveticoside (H) and erysimoside (E) and azadirachtin (A) applied to cabbage leaf discs (10 $\mu$l/disc) and deterrency index (mean $\pm$ SEM; see text for calculation) in dual choice situations between cardenolide-treated and azadirachtin-treated discs at equimolar concentrations (H vs A and E vs A). Values of the deterrency index higher than 30 are statistically significant (two-sided Wilcoxon's matched pair signed rank test, $P < 0.02$; $n = 15$).

cardenolides disappears when paired with azadirachtin as the alternative choice.

## Discussion

*Sensitivity.* The dose-response relationships of the taste neuron responding to cardenolides present in the lateral styloconic sensillum on the maxillary galea reveal a threshold that ranks among the lowest reported for insect taste cells (Städler et al., 1994, 1995). The threshold concentration required for exciting the deterrent neuron in the medial sensillum is one to two orders of magnitude higher. The medial deterrent neuron of *P. brassicae* is the classical example of a generalist taste cell. Since its discovery (Ma, 1969) it has been shown to respond to over 50 pure compounds that all act as feeding deterrents (Schoonhoven et al., 1992; Luo et al., 1995; Messchendorp et al., 1996). Moreover, for different chemical categories it was demonstrated that the activity level of this cell correlates in a straightforward way with the level of deterrency (Blom, 1978; Schoonhoven & Blom, 1988; Luo et al., 1995; Messchendorp et al., 1996). In this respect it acts as a 'labelled line' in coding of food preference (Dethier, 1982; Schoonhoven, 1987). Although the focus in this study is on *P. brassicae*, the electrophysiological tests reported in Table 1 demonstrate that the related *P. rapae* also is endowed with a cardenolide-sensitive neuron in the lateral sensillum styloconicum. Deterrency of cardenolides to *P. rapae* was documented by Sachdev-Gupta et al. (1993).

*Specificity.* The lateral deterrent neuron responds only to cardenolides and is completely insensitive to all other secondary plant metabolites that excite the medial generalist deterrent neuron. The designation 'specialist deterrent cell' is appropriate for the lateral deterrent neuron and as such it is the first example of its sort in insects. At 1000 times higher concentrations it also responds to several phenolic acids and anthocyanins (van Loon, 1990), which at these concentrations also excite the medial deterrent neuron.

*Sensory physiology and behaviour.* The cardenolide-specific taste neuron allows the specialist *Pieris* larvae to discriminate between deterrents present in cruciferous plants and compounds of other chemical categories which occur in other plant families. This capacity was revealed in dual-choice tests between cardenolides and azadirachtin which belong to two different categories of deterrent compounds. This indicates that the balance of activity elicited in the two types of deterrent chemoreceptors determines the behavioural decision. This observation suggests that at some point in the evolutionary history of the *Pieris*-crucifer relationship adaptive value was and possibly still is associated with this discriminatory ability, that remarkably resides in a complement of only two taste cells.

*Evolution of deterrent chemoreceptors.* Evolution of taste receptor types in herbivorous insects is a largely unexplored area (Schoonhoven et al., 1992; Jermy, 1993). Several new parsimonious scenarios can be put forward based on the discovery of a specialized deterrent neuron in *Pieris* caterpillars. In one putative scenario, generalist deterrent neurons, considered to be an ancestral type of gustatory receptor neuron (Dethier, 1980) with an inhibitory effect on feeding, are thought to be endowed with a limited number of receptor sites, each sensitive to a particular chemical category of secondary plant metabolites (van Loon, 1996). Specialized deterrent neurons can be envisaged to have evolved from generalist neurons through a gradual loss of some receptor sites. Token stimulus receptors, specialist taste neurons stimulating feeding and ovipostion, may likewise have evolved from generalist deterrent receptors but in this case a change of the synaptic connection between receptor neuron and first-order interneurons in the suboesophageal ganglion of the central nervous system from inhibitory to excitatory has to be assumed. Alternatively, specialist deterrent neurons might have evolved from token stimulus receptors, in which the synaptic connection changed in the opposite direction. This would imply that cardenolides in evolutionary history might have acted as stimulants to *Pieris* species in addition to glucosinolates. In *Plutella xylostella* L., a crucifer specialist also exploiting glucosinolates as token stimuli, cardenolides are neutral or possibly stimulatory for oviposition behaviour (Renwick et al., 1990).

*Adult deterrent chemoreceptors.* Adult females of *P. brassicae* and *P. rapae* are deterred from ovipositing by cardenolides and both species possess cardenolide-sensitive taste neurons in medially located tarsal B-sensilla (Rothschild et al., 1988; Du et al., 1995; Städler et al., 1995). In contrast, the lateral distal B-sensilla are insensitive to cardenolides but respond to azadirachtin (J. J. A. van Loon, unpubl.). Perhaps adult females have a similar specialist/generalist deterrent cell system. At present we are investigating the adult female tarsal deterrent chemoreceptors in more detail.

### Acknowledgements

We would like to thank Wim Frentz for his diligent contribution to the behavioural experiments.

### References

Blom, F., 1978. Sensory activity and food intake: a study of input-output relationships in two phytophagous insects. Netherlands Journal of Zoology 28: 277–340.

Chew, F. S., 1988. Searching for defensive chemistry in the Cruciferae, or, do glucosinolates always control interactions of Cruciferae with their potential herbivores and symbionts? No! In: K. C. Spencer (ed), Chemical Mediation of Coevolution. Academic Press, San Diego, pp. 81–112.

Chew, F. S. & J. A. A. Renwick, 1995. Chemical ecology of host-plant choice in *Pieris* butterflies. In: R. T. Cardé & W. J. Bell (eds), Chemical Ecology of Insects, 2nd ed., Chapman & Hall, New York, pp. 214–238.

Dethier, V. G., 1980. Evolution of receptor sensitivity to secondary plant substances with special reference to deterrents. The American Naturalist 115: 45–66.

Dethier, V. G., 1982. Mechanisms of host plant recognition. Entomologia Experimentalis et Applicata 31: 49–56.

Du, Y.-J., J. J. A. van Loon & J. A. A. Renwick, 1995. Contact chemoreception of oviposition stimulating glucosinolates and an oviposition deterrent cardenolide in two subspecies of *Pieris napi*. Physiological Entomology 20: 164–174.

Fraenkel, G. S., 1959. The raison d'être of secondary plant substances. Science 129: 1466–1470.

Hodgson, E. S., J. Y. Lettvin & K. D. Roeder, 1955. Physiology of a primary receptor unit. Science 122: 417–418.

Jermy, T. , 1993. Evolution of insect-plant relationships – a devil's advocate approach. Entomologia Experimentalis et Applicata 66: 3–12.

Loon, J. J. A. van, 1990. Chemoreception of phenolic acids and flavonoids in larvae of two species of *Pieris*. Journal of Comparative Physiology A 166: 889–899.

Loon, J. J. A. van, 1996. Chemosensory basis of oviposition and feeding behaviour in herbivorous insects. Entomologia Experimentalis et Applicata 80: 7–13.

Luo, L.-E., J. J. A. van Loon & L. M. Schoonhoven, 1995. Behavioural and sensory responses to some neem compounds by *Pieris brassicae* larvae. Physiological Entomology 20: 134–140.

Ma, W.-C., 1969. Some properties of gustation in the larva of *Pieris brassicae*. Entomologia Experimentalis et Applicata 12: 584–590.

Ma, W.-C., 1972. Dynamics of feeding responses in *Pieris brassicae* L. as a function of chemosensory input. Mededelingen Landbouwhogeschool Wageningen 72–11.

Ma, W.-C. & L. M. Schoonhoven, 1973. Tarsal chemosensory hairs of the large white butterfly *Pieris brassicae* and their possible role in oviposition behaviour. Entomologia Experimentalis et Applicata 16: 343–57.

Messchendorp, L., J. J. A. van Loon & G. J. Z. Gols, 1996. Behavioural and sensory responses to drimane antifeedants in *Pieris brassicae* larvae. Entomologia Experimentalis et Applicata 79: 195–202.

Renwick, J. A. A., C. D. Radke & K. Sachdev-Gupta, 1990. Tolerance of cardenolides in *Erysimum cheiranthoides* by the diamondback moth, *Plutella xylostella*. Symposia Biologica Hungarica 39: 527–528.

Rothschild, M., H. Alborn, G. Stenhagen & L. M. Schoonhoven, 1988. A strophanthidin glycoside in Siberian wallflower: A contact deterrent for the large white butterfly. Phytochemistry 27: 101–108.

Sachdev-Gupta, K., C. D. Radke, J. A. A. Renwick & M. B. Dimock, 1993. Cardenolides from *Erysimum cheiranthoides*: feeding deterrents to *Pieris rapae* larvae. Journal of Chemical Ecology 19: 1355–1369.

Schoonhoven, L. M., 1967. Chemoreception of mustard oil glucosides in larvae of *Pieris brassicae*. Proceedings of the Koninklijke Nederlandse Akademie van Wetenschappen Serie C 70: 556–568.

Schoonhoven, L. M., 1987. What makes a caterpillar eat? The sensory code underlying feeding behaviour. In: R. F. Chapman, E. A. Bernays & J. G. Stoffolano (eds), Advances in Chemoreception and Behaviour, Springer-Verlag, New York, pp. 69–97.

Schoonhoven, L. M. & F. Blom, 1988. Chemoreception and feeding behaviour in a caterpillar: towards a model of brain functioning in insects. Entomologia Experimentalis et Applicata 49: 123–129.

Schoonhoven, L. M., W. M. Blaney & M. S. J. Simmonds, 1992. Sensory coding of feeding deterrents in phytophagous insects. In: E. A. Bernays (ed), Insect-Plant Interactions Vol. 4, CRC Press, Boca Raton, pp. 59–79.

Schoonhoven, L. M., T. Jermy & J. J. A. van Loon, 1998. Insect-Plant Biology: from Physiology to Evolution. Chapman & Hall, London. 409 + X pp.

Smith, J. J. B., B. K. Mitchell, B. M. Rolseth, A. T. Whitehead & P. J. Albert, 1990. SAPID Tools: microcomputer programs for analysis of multi-nerve recordings. Chemical Senses 15: 253–270.

Städler, E., B. Ernst, J. Hurter & E. Boller, 1994. Tarsal contact chemoreceptor for host marking pheromone of the cherry fruit fly, *Rhagoletis cerasi*: responses to natural and synthetic compounds. Physiological Entomology 19: 139–151.

Städler, E., J. A. A. Renwick, C. D. Radke & K. Sachdev-Gupta, 1995. Tarsal contact chemoreceptor response to glucosinolates and cardenolides mediating oviposition in *Pieris rapae*. Physiological Entomology 20: 175–187.

Entomologia Experimentalis et Applicata **91:** 37–50, 1999.



# Olfactory responses and sensilla morphology of the blackcurrant leaf midge *Dasineura tetensi*

D. J. Crook & A. J. Mordue (Luntz)
*Department of Zoology, University of Aberdeen, Tillydrone Avenue, Aberdeen, AB24 2TZ, UK*

Accepted: November 27, 1998

*Key words: Dasineura tetensi*, olfaction, plant volatiles, antennal sensilla


## Abstract

Olfactory responses of *Dasineura tetensi* (Rubs) (Diptera; Cecidomyiidae) to leaf volatiles of blackcurrant (*Ribes nigrum*) were tested in a 4-way olfactometer. Newly emerged virgin females showed no response to the leaf volatiles emitted from a blackcurrant shoot. Newly emerged males (which are known to respond to a pheromone released by the female) also showed no response to the leaf volatiles. Two hours after mating females responded positively, indicating that leaf volatiles may play a role in host plant finding. Scanning (SEM) and transmission (TEM) electron microscopy of the antennae of *D. tetensi* showed that males and females share five sensillum types; sensilla chaetica (mechanoreceptors), sensilla trichodea, sensilla basiconica, uniporous peg sensilla and circumfila (chemoreceptors). The sensilla chaetica and sensilla trichodea resemble those found on other insects. Sensilla basiconica were found on all antennal subsegments except the tip. These are multiporous receptors with five unbranched dendrites filling the lumen. Small peg sensilla located on the tips of both male and female antennae may function as contact-chemoreceptors. The circumfila, which are a unique type of sensilla found only on cecidomyiid antennae, form loops around each of the antennal subsegments, being attached to the surface by a series of stalks. TEM revealed that each stalk consisted of one sensillum containing a single highly branched dendrite. The distal regions of the walls of each sensillum are fused together to form the circumfila. Circumfila have multiporous walls and a lumen filled with multiple branches of dendrites. Their structure suggests that they are important olfactory receptors in both the male and female.


## Introduction

The blackcurrant leaf midge (*Dasineura tetensi*) has become a pest of blackcurrant plantations in eastern Europe and the UK, particularly in the warmer southern areas (Alford, 1984). Larval feeding causes severe distortion of young leaves which can then become contaminated by secondary fungal infections. Up to 90% of shoots can be damaged by *D. tetensi* in an infestation (Sass, 1972; Simova-Tosic & Dobrivojevic, 1981; Vazyulya, 1982; Metlitskii & Zeinalov, 1993). Damage to newly planted cuttings may be particularly serious and render them useless. In addition, the leaf curling and twisting in a severe attack mask reversion symptoms and may prevent the recognition of rogue cultivars when inspecting nursery stocks (Barnes, 1948). Since it was first reported as a pest in the UK by Bagnall & Harrison (1917), research has been based mainly on studying its life cycle, host-plant resistance and chemical control. Pheromones and plant volatiles have, however, proved to play an important role in the ecology of several cecidomyiids with sex pheromones involved in the mating behaviour (Harris & Foster, 1999). Males of several species including *D. tetensi*, have been shown to respond to 'calling' females (Garthwaite et al., 1986) and the main component of the Hessian fly. *Mayetiola destructor* (Say), sex pheromone has now been identified as 25-(*E*)-10-tridecen-2-yl acetate (Harris & Foster, 1999). As cecidomyiid adults must move from pupation sites in the soil to oviposition sites on plants, mated females presumably have well-developed host-finding behaviour. Furthermore, as cecidomyiid midges tend to have a specialised or narrow host range, host-finding behaviour of cecidomyiid females is likely to include flight responses to host-plant chemicals. For example, females of the brassica pod midge, *Dasineura brassicae* (Winn) which specialises on plants in the family Brassicaceae (Pettersson, 1976; Murchie et al., 1997) and the apple leafcurling midge, *Dasineura mali* (Keiffer) which specialises on apple foliage (Galanihe & Harris, 1997) have been shown to respond to host-plant chemicals. The Hessian fly, *M. destructor* (Harris & Rose, 1989) and *D. tetensi* (Crook, 1998) are both known to antennate before ovipositing suggesting that plant chemicals may be important after alighting. To date, however, there is no evidence to show that *D. tetensi* responds to host-plant chemicals.

The antennae of Cecidomyiidae have been of particular interest to entomologists since they were first studied at the end of the last century (Janet, 1896; Rübsaamen, 1896; Felt, 1897; Keiffer, 1900), mainly due to the unusual arrangement of various sensory hairs and the possession of the looped sense organs known as circumfila. Despite several cecidomyiid species being important crop pests, to date there have been few detailed studies of the antennal sensilla. Slifer & Sekhon (1971) used scanning and transmission electron microscopy to study the antennal sensilla of sorghum midge, *Contarinia sorghicola*, an important pest of sorghum and other grasses in many parts of the world. Solinas & Nuzzaci (1987) studied in detail the antennal sensilla of another cecidomyiid midge, *Mycodiplosis erysiphes*. This study examines the responses of *D. tetensi* to host-plant odours and describes the morphology of both male and female antennal sensilla.

## Materials and methods

*Insects.* Newly-emerged male and female adult midges, from cocoons collected from blackcurrant plantation soil (Ewell Farm, Kent) were used for all purposes. When mated female insects were required, newly emerged male and female midges were placed together in a plastic vial (5 × 2 cm) and monitored until mating took place. Emergence occurred between 7.00–8.00 BST and mating usually took place between 10.00–12.00 BST when the midges were most active.

*Behavioural bioassays.* A 4-way olfactometer (Pettersson, 1970; Vet et al., 1983) was used to test the behavioural response of adult *D. tetensi* to host-plant odours. The olfactometer comprised of two glass sections measuring 115 mm × 115 mm × 18 mm (width × length × height). Four odour fields were created in the experimental chamber by drawing air through the outlet situated in the middle of the arena, using a vacuum pump (WISA 110 Model). Airflow (400 ml $min^{-1}$) was regulated and measured with a flow meter (GAP 500 ml $min^{-1}$ capacity). Each arm of the chamber was connected to a set of three 250 ml glass conical flasks. The flask closest to the chamber contained a shoot (at the six leaf stage) of the midge susceptible blackcurrant cultivar 'Baldwin' or was left empty if it was being used as a control. The second flask contained 100 ml of distilled water through which the incoming air was bubbled to create a uniform humidity. This was connected to the third flask which contained 60 g of charcoal to filter the incoming outside air. All tube connections were made of silicon rubber (Norlab, 8 mm bore, 2 mm i.d). Visual cues from the plant material were obscured by placing black card screens between the olfactometer and the glass flasks. To limit other external visual cues all experiments took place in an observation box measuring 39 cm × 39 cm × 17 cm at 18–20 °C. A 35 cm × 25 cm hole was cut from the ceiling of the observation box directly above the olfactometer, above which was positioned a 9 W halogen warm white colour temperature lamp (Philips PL Electronic-C). An illumination level of 20.8 W $m^{-2}$ was measured directly above the experimental arena. In each experiment, single adult midges were introduced into the olfactometer. Each trial consisted of a choice between filtered, humidified air, passed over a shoot of 'Baldwin', or three filtered, humidified air controls. In each experiment an insect was introduced into the centre of the olfactometer. Air was then drawn through the apparatus for one minute after which the test plant odour was introduced into the arena via one of the inlets. The time spent in each of the four arms in the olfactometer during a 20 min observation period was recorded using a stopwatch. If a midge remained inactive in the central zone for more than 10 min the run was discontinued and a new one started after cleaning the apparatus. After each run the apparatus was washed in detergent solution (BP detergent, 1% by volume) and then rinsed with distilled water. The olfactometer and flasks were rotated through 90° after each run to prevent any directional bias. Olfactometer bioassays were analysed by paired t-tests comparing the mean time in seconds spent in the treated arm with the total mean time spent in the three control arms (Campbell et al., 1993).

*Scanning electron microscopy.* Isolated midge heads were mounted on specimen stubs with either carbon glue or double-sided tape and coated with platinum in an Emscope SC 500A sputter coater. Specimens were coated three times, being tilted and rotated between each run to maximise coating and reduce subsequent charging. Specimens were coated without using freeze drying or critical point drying techniques as these methods damaged the antennal sensilla. Ten midges of each sex were examined with a Cambridge Instruments S600 microscope at an accelerating voltage of 15 kV.

*Transmission electron microscopy.* Three different fixation regimes were employed. First, whole midge heads were fixed for 1 h in Karnovsky's (1965) fixative: 5% gluteraldehyde and 2.5% paraformaldehyde, buffered in 0.1 M sodium cacodylate. To ease subsequent handling of the antennae, the heads were embedded in blocks of agar (1.5% solution of low-gelling-temperature agarose; Sigma, St. Louis, MO) before being fixed for a further 2 h. The heads were then washed overnight in buffer, post-fixed in aqueous 1% osmium tetroxide ($OsO_4$) for 2 h, dehydrated in a series of alcohols, and embedded as above in L.R. white resin, hard grade (Agar Sci. Ltd., Stansted, Essex, UK). Secondly, whole midge heads were fixed for 1 h in a solution of 10% acrolein and 2.5% glutaraldehyde, buffered in 0.1 M sodium cacodylate. The heads were then washed, post-fixed and embedded as above. In the third regime, whole midge heads were fixed for 18 h in 5% glutaraldehyde. The heads were then washed in Pipes buffer, and placed in 0.1 M sodium cacodylate buffer before being post-fixed for 18 h in 0.2% $OsO_4$ in cacodylate buffer (1:1 ratio). The preparations were then washed in distilled water and embedded in agar blocks as above, before being embedded in araldite resin for 72 h at 70 °C. Silver sections were cut with a diamond knife (Micro Star, Taab Labs, Reading, Berks, UK), picked up on formvar or pyroxylin coated grids, stained with uranyl acetate and lead citrate, and examined with a Philips 301 microscope.

## Results

*Olfactometer bioassays.* The responses of *D. tetensi* to host-plant odours in the olfactometer showed that mated females spent a significantly greater amount of time in the arm containing blackcurrant foliage (Table 1). Unmated females and adult males did not respond to the host-plant odour.

*Antennal sensilla morphology.* The female antennae were shorter than the male, with a mean antennal length of $542.5 \pm 3.0$ $\mu$m ($n = 10$) compared to $930 \pm 4.5$ $\mu$m ($n = 10$). Unless stated otherwise, the following description of the antennal morphology applies to both sexes studied.

The antennal flagellum, consisting of thirteen subsegments, is attached to the head via the cup shaped pedicel and scape (Figure 1A). The pedicel of the female measures some 40 $\mu$m in diameter, the male being slightly larger at 50 $\mu$m. The first 12 subsegments are globular shaped and uniform in length, measuring some 35 $\mu$m in diameter and 80 $\mu$m in length. The antennal subsegments of the male are separated by a stalk measuring some 30 $\mu$m in length, thus making the male antenna longer than the female where the subsegments are fused directly (Figures 1A, 1B, 2A, 2B). The thirteenth subsegment is approximately 20 $\mu$m shorter than the others and has a slight constriction below the tip. This feature has been described as 'indistinctly trindose' (Hardy, 1960; Slifer & Sekhon, 1971) for other cecidomyiid species. Microtrichia (minute, hair-like, non-innervated cuticular structures) are present on all antennal subsegments but absent on the stalks between the male subsegments.

At least five sensillum types occur on both the male and female antennae: sensilla chaetica, sharp-tipped sensilla trichodea, sensilla basiconica, circumfila and terminal sensory pegs. The flagella of male and female *D. tetensi* each bear $657 \pm 6.3$ and $584.4 \pm 24.9$ sensilla, respectively ($n = 5$). The distributions and mean lengths of these sensilla are represented in Tables 2 and 3.

*Sensilla chaetica.* Only one type of aporous hair occurs on *D. tetensi* antennae, the sensilla chaetica (Figures 2A, 2B). These mechanoreceptors are present on both the male and female antennae, occurring in whorls at the distal region of each subsegment. On subsegments 1–12 the sensilla are found within the loops of circumfila (Figures 2C, 2D). Numbers of sensilla per loop range from 5–15 and 12–17 for the female and male antennae respectively. The distal area of subsegment 13 has the sensilla arranged randomly around its circumference. Significantly more sensilla chaetica were found on the male antennae than on the female ($P < 0.05$; Table 2). These blunt tipped hairs, between 1.5–2 $\mu$m in width, possess an irregular pat-

*Table 1.* Responses of adult *D. tetensi* to plant odour from the susceptible *Ribes nigrum* cultivar, Baldwin, in an olfactometer

| Adult tested | Mean time spent in centre of olfactometer (s) ±s.e. | Mean time spent in odour laden air (s) ±s.e. | Mean time spent in clean air in (s) ±s.e. | Significance |
|---|---|---|---|---|
| Unmated male ($n = 10$) | 280 (±86.0) | 162 (±86.7) | 246 (±56.9) | N.S. |
| Unmated female ($n = 10$) | 69 (±23.4) | 167 (±55.3) | 285 (±62) | N.S. |
| Mated female ($n = 7$) | 154 (±86.2) | 669 (±96.4) | 123 (±29.8) | $P < 0.01$ |

N.S. – not significantly different ($P > 0.05$, *t*-test).

*Table 2.* Distribution and numbers (range) of sensilla on the 13 antennal flagellum subsegments of male (M) and female (F) *D. tetensi*

| Subsegment | Sensilla chaetica | | Sensilla trichodea | | Sensilla basiconica | | Terminal sensory pegs | | Circumfila stalks | |
|---|---|---|---|---|---|---|---|---|---|---|
| | M | F | M | F | M | F | M | F | M | F |
| 1 | 12–15* | 5–7 | 13–14 | 12–18 | 1–2 | 0–2 | 0 | 0 | 12–15 | 18–20* |
| 2 | 16–24* | 7–12 | 12–16 | 9–16 | 1–3 | 1–3 | 0 | 0 | 15–20 | 18–20* |
| 3 | 16–24* | 8–13 | 13–15* | 10–14 | 3–5 | 1–5 | 0 | 0 | 17–19 | 21–24* |
| 4 | 20–25* | 9–13 | 10–14 | 8–15 | 4–5 | 3–5 | 0 | 0 | 17–21 | 21–24 |
| 5 | 19–24* | 11–14 | 12–14* | 8–13 | 4–5* | 2–5 | 0 | 0 | 17–20 | 19–23 |
| 6 | 20–24* | 9–14 | 11–13* | 8–11 | 2–4 | 2–4 | 0 | 0 | 14–18 | 21–24* |
| 7 | 20–24* | 9–13 | 9–13* | 9–11 | 2–4 | 2–4 | 0 | 0 | 13–20 | 21–23* |
| 8 | 21–26* | 9–14 | 9–12* | 7–10 | 1–4 | 2–3 | 0 | 0 | 14–18 | 21–23* |
| 9 | 21–25* | 8–16 | 10–13* | 6–10 | 2–3 | 2–3 | 0 | 0 | 15–20 | 20–23* |
| 10 | 20–26* | 8–15 | 10–11* | 6–8 | 2–3* | 1–3 | 0 | 0 | 15–19 | 20–22* |
| 11 | 18–25* | 8–15 | 8–11* | 6–9 | 1–3 | 1–2 | 0 | 0 | 14–19 | 20–21* |
| 12 | 20–27* | 7–15 | 7–9* | 4–6 | 0–3 | 0–1 | 0 | 0 | 14–19 | 19–20* |
| 13 | 14–27* | 7–15 | 3–8* | 3 | 0–3 | 0–3 | 1–3 | 1–3 | 0–4 | 16–18* |
| Mean total ±s.e. | 276 ±4.8 | 137 ±4.9 | 145 ±2.7 | 120 ±2.5 | 36 ±1.3 | 32 ±1.9 | 1.8 ±0.5 | 1.6 ±0.4 | 199 ±9.4 | 299 ±28.5 |

For sensilla chaetica, sensilla trichodea, sensilla basiconica, and terminal sensory pegs $n = 8$. For circumfila stalks $n = 5$. Subsegment 1 is proximal. An individual circumfilum is defined as one stalk attachment to the antenna.
*Indicates a significantly greater number of sensilla in one sex ($P < 0.05$, *t*-test).

*Table 3.* Mean lengths ($\mu$m) ±s.e. ($n$) of different sensillum types on the antennae of *D. tetensi*

| | Sensilla chaetica | Sensilla trichodea | Sensilla basiconica | Terminal sensory pegs |
|---|---|---|---|---|
| Male | 126.2 ± 2.7* (65) | 26.5 ± 0.6 (69) | 4.8 ± 0.1 (31) | 2.2 ± 0.1 (12) |
| Female | 71.4 ± 0.9 (65) | 32.3 ± 0.7* (70) | 4.9 ± 0.03 (45) | 2.1 ± 0.1 (11) |

Sensilla were measured along the whole length of the flagellum.
*Indicates significantly longer sensilla in that sex ($P < 0.05$, *t*-test).

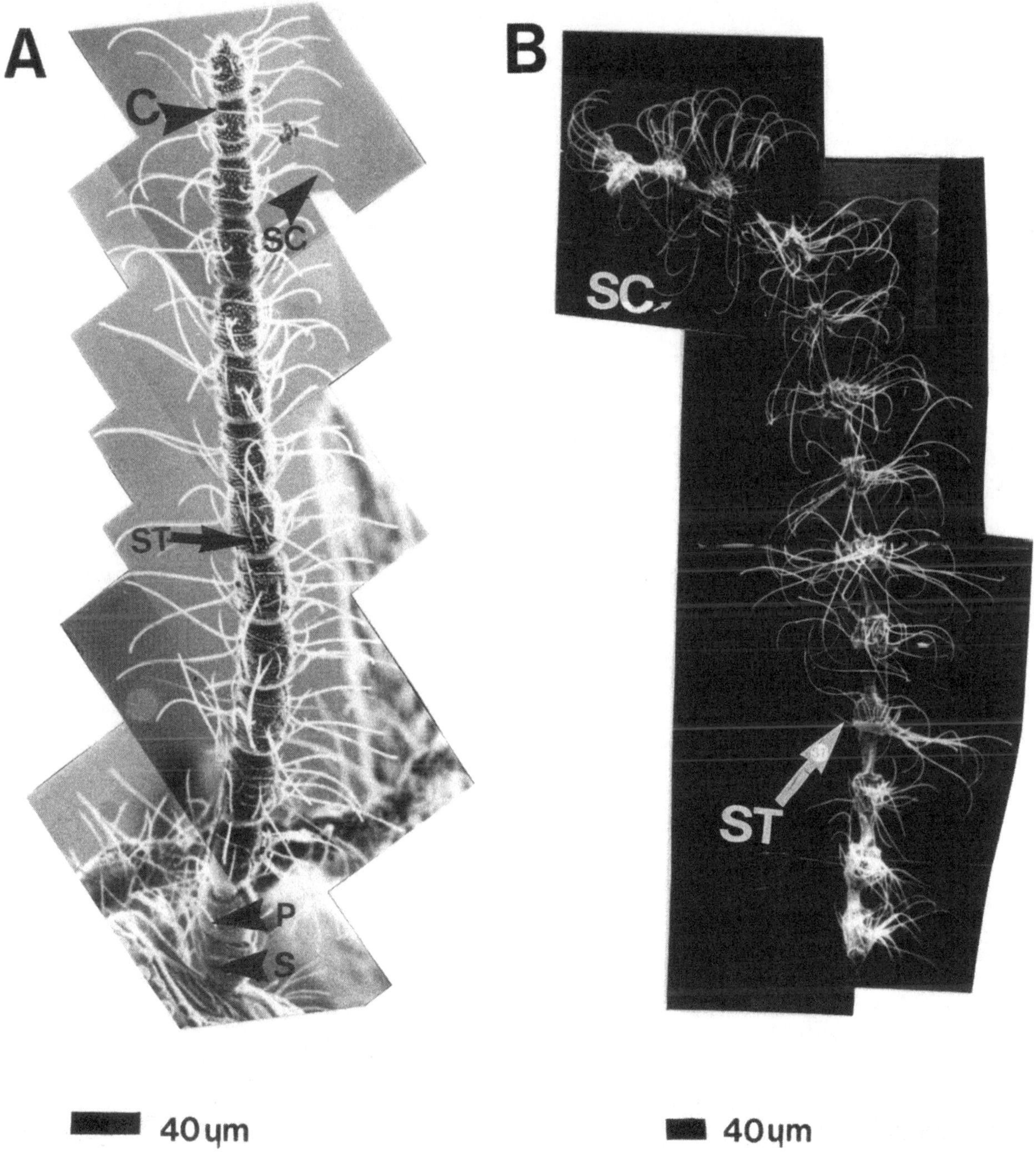


*Figure 1.* Scanning electron micrographs of *D. tetensi.* (A) Montage of whole female antenna. (B) Montage of whole male antenna (without pedicel and scape). P, pedicel; S, scape; ST, sensilla trichodea; SC, sensilla chaetica; C, circumfila.

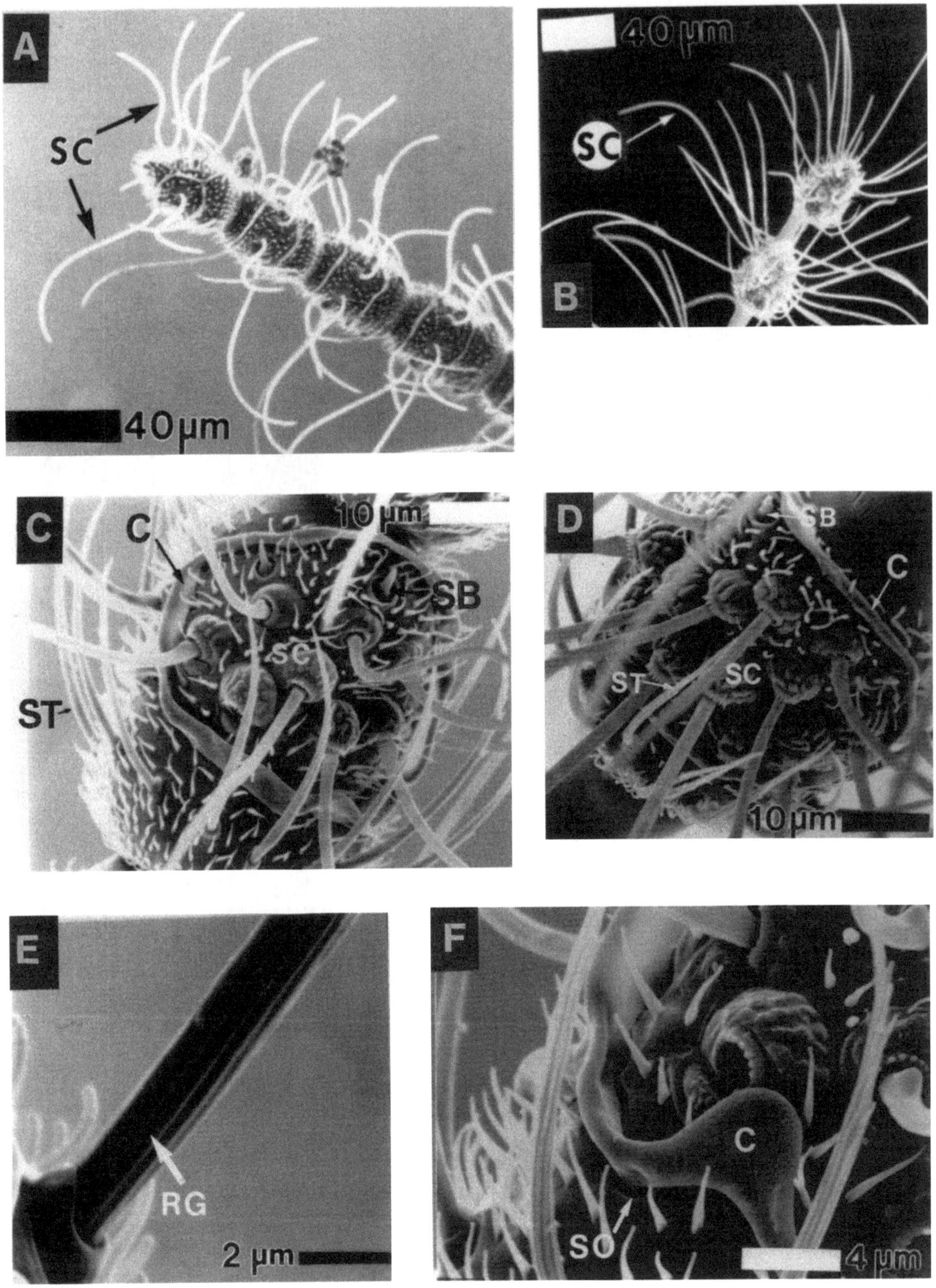
A
SC
40 µm
40 µm
SC
B
C
C
10 µm
SB
SC
ST
D
SB
C
ST
SC
10 µm
E
RG
2 µm
F
C
SO
4 µm

*Figure 2.* Scanning electron micrographs of various sensilla on the antennae of *D. tetensi.* (A) Sensilla chaetica on subsegments 10–13 of the female antenna. (B) Sensilla chaetica on subsegments 12–13 of the male antenna. (C) Third subsegment of a female antenna showing the typical arrangement of the various sensilla types. (D) Third subsegment of a male antenna showing a similar arrangement of sensilla to that of the female. (E) Socket and proximal end of a trichoid sensillum from the third subsegment of a male antenna. Note the deep ridges running longitudinally along the hairshaft. (F) Circumfila situated on the third antennal subsegment of a female antenna. Note the numerous sockets attaching the circumfila to the antennal surface. C, circumfila; RG, ridge; SB, sensilla basiconica; SC, sensilla chaetica; SO, socket; ST, sensilla trichodea.

tern of longitudinal ridges running along their length. Small lateral branches measuring approximately 2 $\mu$m in length are also present on the proximal side of these hairs. Each hair possesses a bulbous body at the base where it meets the socket opening (Figures 2C, 2D). The socket of the tactile hair is circular in outline measuring 6 $\mu$m in diameter, its opening having a ridged structure on the distal rim.

In section the sensilla chaetica cellular components consist of a single sensory neurone whose dendrite ends with a tubular body at the base of the hair shaft, which in turn interacts with the inner socket (Figures 3A, 3B). A mass of microtubules emerge from the proximal side of the tubular body, through the proximal dendritic segment towards a ciliary structure where it can be clearly seen to expand in diameter. The neurone is associated with three accessory (or sheath) cells. The innermost thecogen cell envelops the proximal dendritic segment which is also partially enveloped by the intermediate accessory cell, the trichogen cell, and the outermost tormogen cell both of which border the sensillar sinus with numerous microvilli. The ultrastructure of the male and female sensilla chaetica is similar, with the exception that the male sensilla are significantly longer ($P < 0.05$), being nearly twice the length of those of the female (Table 3).

*Sharp-tipped sensilla trichodea.* Sharp-tipped sensilla trichodea are chemoreceptors which occur on all subsegments of both male and female antennae. There were significantly more sensilla trichodea on the male antennae when compared to the female ($P < 0.05$; Table 2). Each subsegment bears a single whorl of up to 18 sensilla at the proximal end, being situated evenly around the circumference beneath the sensilla chaetica and loops of circumfila (Figures 2C, 2D).

The sharp-tipped sensilla trichodea consist of a hair shaft which is slightly longer in the female than the male becoming thinner towards the tip (Table 3). A series of longitudinal ridges run parallel along the length of the hair shaft from the smooth horseshoe shaped socket (Figure 2E).

In section (Figures 3C, 3D) two sensory neurones can clearly be seen to be surrounded by a trichogen and a tormogen cell. Their dendritic distal segments are encased in a thick sheath until they enter the shaft lumen. The two dendrites then run unbranched to the hair tip. Within the ciliary body there appears to be a '9 + 0' arrangement of microtubule doublets, but it is difficult to tell whether this arrangement is continuous within the dendrites as they run up the hair shaft to its tip. In transverse section the 2 dendrites within the lumen of the hair can be seen clearly along with pores within the thin cell wall (Figure 3C).

*Circumfila.* Each subsegment of the male and female antennae are encircled by looped structures. The loops of the circumfila are attached to the antennal surface by a series of stalks situated within inverted sockets (Figure 2F). Each stalk splits into two approximately 2–3 $\mu$m above the antennal surface where it joins its two neighbouring stalks to form a continuous looped structure. The female circumfila are more complex than the male. Each female subsegment has 2–3 loops of circumfila running around its circumference in a figure of eight pattern. Each male circumfilum consists of one loop with a single band running around the proximal circumference of each subsegment. The female had significantly more stalks per subsegment than the male antennae ($P < 0.05$; Table 2).

The surface of the circumfila appears to be smooth under the scanning electron microscope. In section, however, the surface is made up of many small ridges, in between which are numerous pores (Figures 4A, 4B). The surface is covered in a thin layer of epicuticle. No pores are located in the stalk portion of the circumfila (Figure 4A).

The cellular components are made up of a single sensory neurone whose distal dendritic segment is encased in a sheath until it reaches the base of a circumfila socket (Figure 4A) when the dendrite forms several branches, which run parallel up the length of the stalk (Figure 4C). Multiple branching takes place approximately 1 $\mu$m along the length of the stalk filling the lumen of the circumfila with numerous

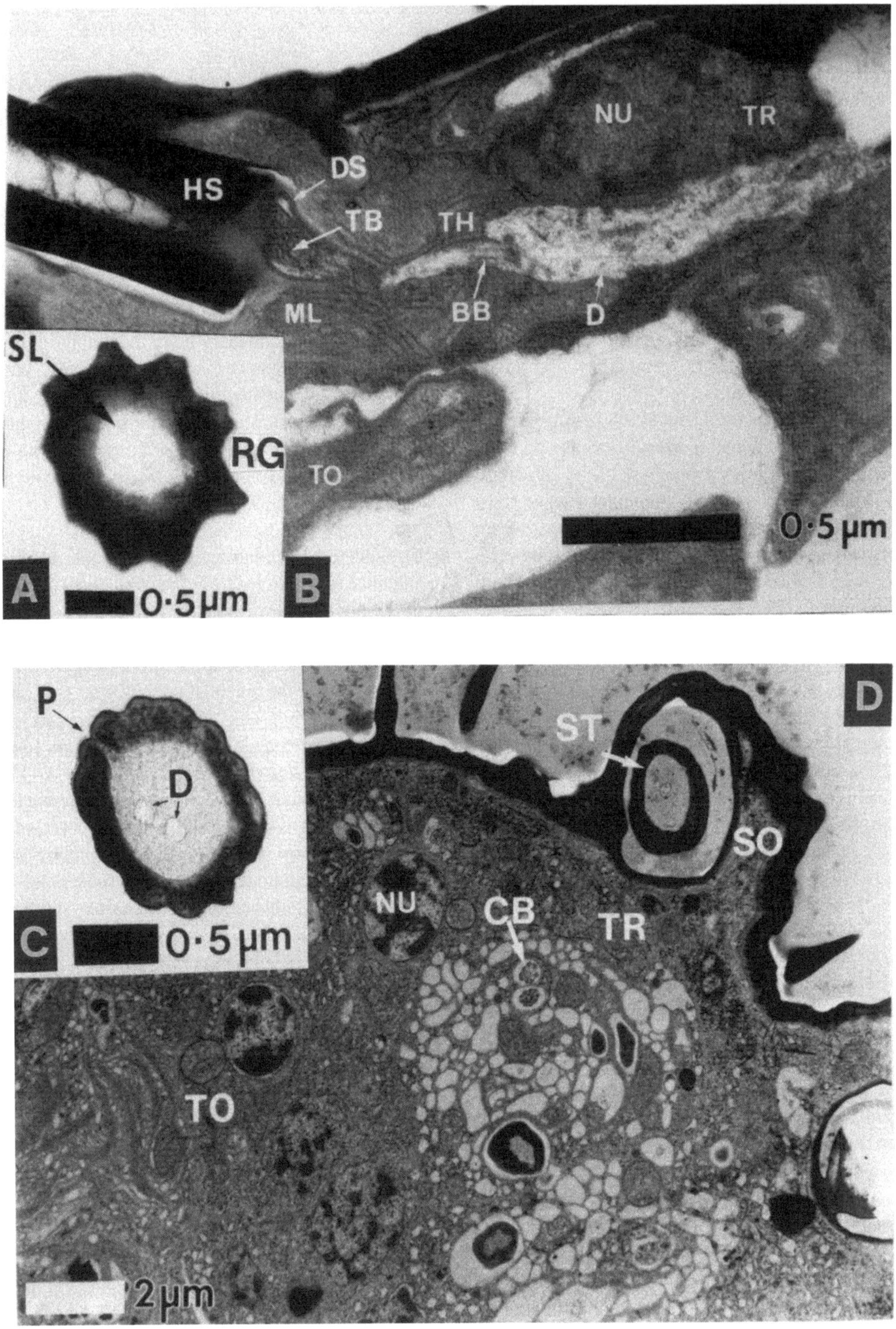
NU
TR
DS
HS
TB
TH
ML
BB
D
SL
RG
TO
0·5 μm
A
0·5μm
B
P
D
ST
SO
NU
CB
TR
C
0·5μm
TO
2μm

*Figure 3.* Transmission electron micrographs of sensilla chaetica (A,B) and sensilla trichodea (C,D) in *D. tetensi* female. (A) fixed in Karnovsky's fixative, (B, C and D) fixed in 5% gluteraldehyde. (A) Cross section showing grooves in the thick wall. Note the lack of dendritic material in the lumen. (B) Longitudinal section of the hair shaft base and tubular body. Note the basal body of the single sensory neurone. (C) Cross section of sensilla trichodea with two dendritic branches. (D) Transverse section showing the ciliary bodies of the two sensory cells (see text for accessory cells) which run unbranched up into the hairshaft of the sensilla. BB, basal body; CB, ciliary body; D, dendrite; DS, dendritic sheath; HS, hairshaft; ML, microlamellae; NU, nucleus; P, pore; RG, ridge; SL, sub-lumen; SO, socket; ST, sensilla trichodea; TB, tubular body; TH, thecogen cell; TO, tormogen cell; TR, trichogen cell.

dendritic branches, each containing a single microtubule. A thecogen cell encases the proximal dendritic segment of the sensory neurone. A trichogen and tormogen cell, identifiable by the presence of numerous microvilli, encase the distal portion of the dendritic segment up to the proximal area of the circumfila stalk.

*Sensilla basiconica.* The sensilla basiconica are peg shaped structures between 3–4 $\mu$m in length. They have between 9–10 grooves running parallel to the hair shaft axis which can be clearly seen under the scanning electron microscope (Figure 5A). There was no significant difference in the number of sensilla basiconica found on male and female antennae (Table 2). Up to five may be found on any subsegment although the usual number was three. Between 1–3 sensilla were usually situated within the loop of a circumfila at the distal end of each subsegment. Occasionally between 1–2 sensilla were also located on the proximal end of each subsegment, situated below the site where two circumfila loops were joined. Although these sensilla are very similar to the surrounding microtrichia they can be distinguished by the fact that they possess a socket.

In longitudinal section these sensilla possess five non-branching sensory neurones which are encased in a thick cuticular sheath (Figure 5B). These ascend the full length of the hair shaft lumen, to an opening in the tip. The five sensory neurones are accompanied by three sheath cells; a thecogen, a trichogen and a tormogen cell before they enter the hair shaft. Lighter areas occur in the cuticular wall which appear to be 'channels', although they are not clear in section (Figure 5B). These channels lead from the lumen containing the dendrites to the open base between the grooves in the outer wall. These could possibly be pore channels (often referred to as spoke channels) which are found in the sensilla basiconica of other insects (Altner & Prillinger, 1980; Cribb & Jones, 1995).

*Terminal sensory pegs.* The tips of both the male and female antennae bear 2–3 stubby, smooth walled pegs set into unevenly edged sockets. The pegs are between 1–2 $\mu$m in length. The sensilla appear to possess a single pore in the tip, which is very difficult to see under the scanning electron microscope. The ultrastructure of these sensilla has not been investigated further.

## Discussion

Adult female *D. tetensi* responded to host-plant odour in a similar manner to that of *D. brassicae* (Diptera: Cecidomyiidae) (Pettersson, 1976). In both gall midge species only mated females were attracted suggesting that olfaction plays a crucial role in finding a suitable oviposition site. The lack of attraction of male *D. tetensi* to leaf volatiles is probably due to the fact that the life cycle of *D. tetensi* is already closely associated with the host plant (adult insects usually emerge from soil directly below the host plant). Males emerge underneath the shelter of the plant canopy, in close proximity to virgin females of the same generation. Although it is unclear where mating occurs, it is important to note that mate location by males is facilitated by a sex pheromone released by the females. Males of the closely related apple leafcurling midge *Dasineura mali* also respond to sex pheromones produced by virgin females but do not respond to host-plant odours (Harris et al., 1996). Unmated females were also not attracted to the host-plant odour stimulus suggesting that mating is the trigger for a switch in behaviour and is a pre-requisite to searching for an oviposition site. In the Hessian fly, *M. destructor* this change from virgin to mated behaviour is believed to be induced by two factors (Bergh et al., 1990). Firstly, there is the mechanical act of mating which has a short-term effect on the receptivity of the female. Secondly, there is an unknown male-produced substance that is transferred to the female during mating, which triggers the permanent transition to mated behaviour.

The present observations confirm that *D. tetensi* has a similar array of sense organs to two other Cecidomyiid species, *Contarinia sorghicola* (Slifer & Sekhon, 1971) and *Mycodiplosis erysiphes* (Solinas & Nuzzaci, 1987). The general function or functions of the sensilla are inferred from their ultrastruc-

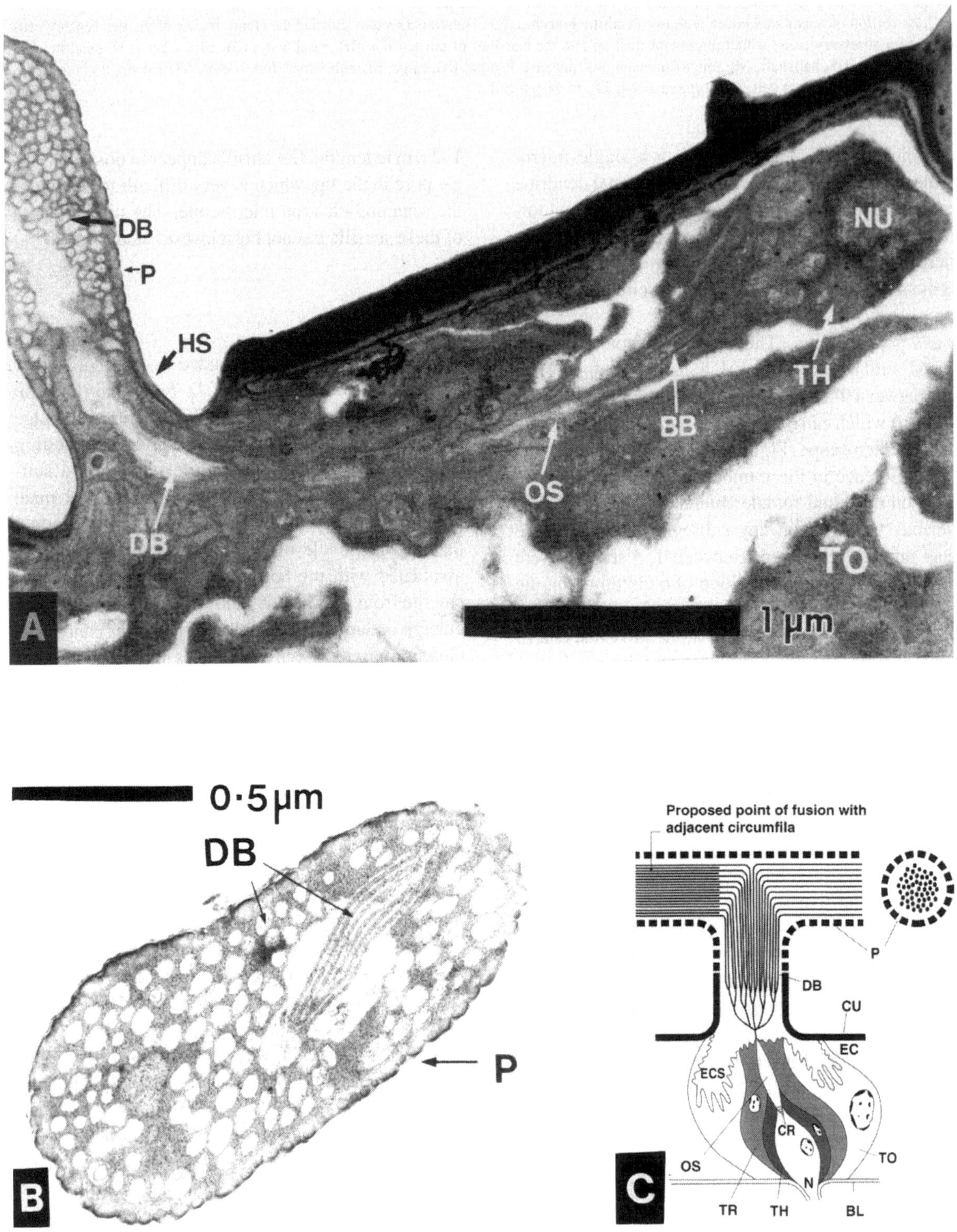
DB
P
HS
NU
TH
BB
OS
DB
TO
A
1 µm
0·5µm
DB
P
B
Proposed point of fusion with
adjacent circumfila
P
DB
CU
EC
ECS
CR
OS
TO
N
TR
TH
BL
C

*Figure 4.* Transmission electron micrographs of the circumfilum on the female antenna (fixed in Karnovsky's fixative). (A) A longitudinal section of circumfilum, showing the outer dendritic segment of a single sensory neurone running to the base of the sensillum socket before branching into numerous dendrites. (B) Cross section of circumfilum showing numerous dendritic branches within the lumen and multiple pores in the outer wall. (C) Diagrammatic representation of the ultrastructural components of circumfilum found on the antennae of male and female *D. tetensi*. The circumfilum is a multiporous sensillum possessing a single sensory neurone which branches into numerous dendrites in the hairshaft, which in turn leads to or may interact with branching dendrites of the adjacent circumfilum stalk. BB, basal bodies; BL, basal lamina; CU, cuticle; CR, ciliary region; DB, dendritic branch; EC, epidermal cell; ECS, extracellular space; HS, hair shaft; N, nerve cell; NU, nucleus; OS, outer dendritic segment; P, wall pore; TH, thecogen cell; TO, tormogen cell; TR, trichogen cell.

ture, although verification will require behavioural and electrophysiological studies.

In both the male and female antenna the sensilla chaetica are most prominent. They have the typical appearance of mechanoreceptors and this is supported by their fine structure, a single sensory neurone whose dendrite ends in a tubular body at the base of a thick-walled hair (McIver, 1975; Keil & Steinbrecht, 1984). The morphology of these sensilla closely resembles sensilla chaetica found on other cecidomyiid species (Slifer & Sekhon, 1971; Solinas & Nuzzaci, 1987). The bulbous body at the hair shaft base, along with the ridged socket opening may act to limit the hair's range of movement. These sensilla may protect the more delicate underlying structures found on the antennae or may act to determine whether the antennae were in contact with a substrate, such as a leaf. The sensilla chaetica of the male are more numerous and up to twice the length of those on the antennae of the female. This suggests that the male may also utilise these sensilla to orientate to sound as is common in male mosquitoes which are tuned to the flight sound of the female (Slifer & Sekhon, 1971; McIver & Hudson, 1972).

Sensilla trichodea have been identified as the main olfactory sensilla for several insect species (Schneider, 1964; Schneider & Steinbrecht, 1968; Zacharuk, 1980). Sensilla trichodea of *D. tetensi* make up 22% and 20% of the total number of sensilla found on male and female antennae respectively. These sensilla closely resemble the sharp-tipped sensory hairs on both *C. sorghicola* (Slifer & Sekhon, 1971) and *M. erysiphes* (Solinas & Nuzzaci, 1987). Like *D. tetensi* the sensilla trichodea of *M. erysiphes* also contain two dendrites, which do not branch either before or after ascending the hair shaft (Solinas & Nuzzaci, 1987). Slifer & Sekhon (1971), however, found that the sensilla trichodea of *C. sorghicola* were innervated by five neurones that upon entering the lumen, branched to form between 21–36 dendrites. Solinas & Nuzzaci (1987) recorded that there were slightly more sensilla trichodea on the female antennae than on the male, so they postulated that these olfactory sensilla played an important role in finding oviposition sites on plants and/or fungi. However, on ten of the thirteen antennal subsegments on *D. tetensi* there were significantly more of these sensilla on males than on females, suggesting that they may be more important in the detection of a female released sex pheromone.

The morphology of the circumfila of *D. tetensi* reveals cellular components which are characteristic of olfactory receptors. The circumfila of *D. tetensi* share many features with those of other cecidomyiid species (Slifer & Sekhon, 1971; Solinas & Nuzzaci, 1987); thin multiporous walls and a lumen filled with multiple branches of dendrites, arising from sensory neurones below each socket. In contrast to the single sensory neurone found in *D. tetensi*, *M. erysiphes* has two sensory neurones which appear to branch before entering the socket of each circumfilum stalk (Solinas & Nuzzaci, 1987). The surface of *D. tetensi* male and female circumfila closely resembles that of *M. erysiphes* (Solinas & Nuzzaci, 1987). Both, when examined under the scanning electron microscope, appear smooth, their ridged surface only becoming apparent when viewed in section. Male circumfila of *C. sorghicola* (Slifer & Sekhon, 1971), however, have a very pronounced ridged surface which can be seen under the scanning electron microscope, and contrasts with the smooth surface of the female circumfila.

In the two previous studies of cecidomyiid circumfila (Slifer & Sekhon, 1971; Solinas & Nuzzaci, 1987) the threadlike sensilla were seen to encircle the flagellomeres quite differently in males and females. For example, in *C. sorghicola* (Slifer & Sekhon, 1971) the male circumfila consist of a series of elongated loops, measuring up to 36 $\mu$m in height compared to the female circumfila which consist of two circular loops surrounding each subsegment, which are at a height of 2–3 $\mu$m above the antennal surface. Due to the fact that the circumfila have features commonly found in olfactory receptors and that the circumfila of the male have a much more elaborate structure than the female, Solinas & Nuzzaci (1987) suggested that

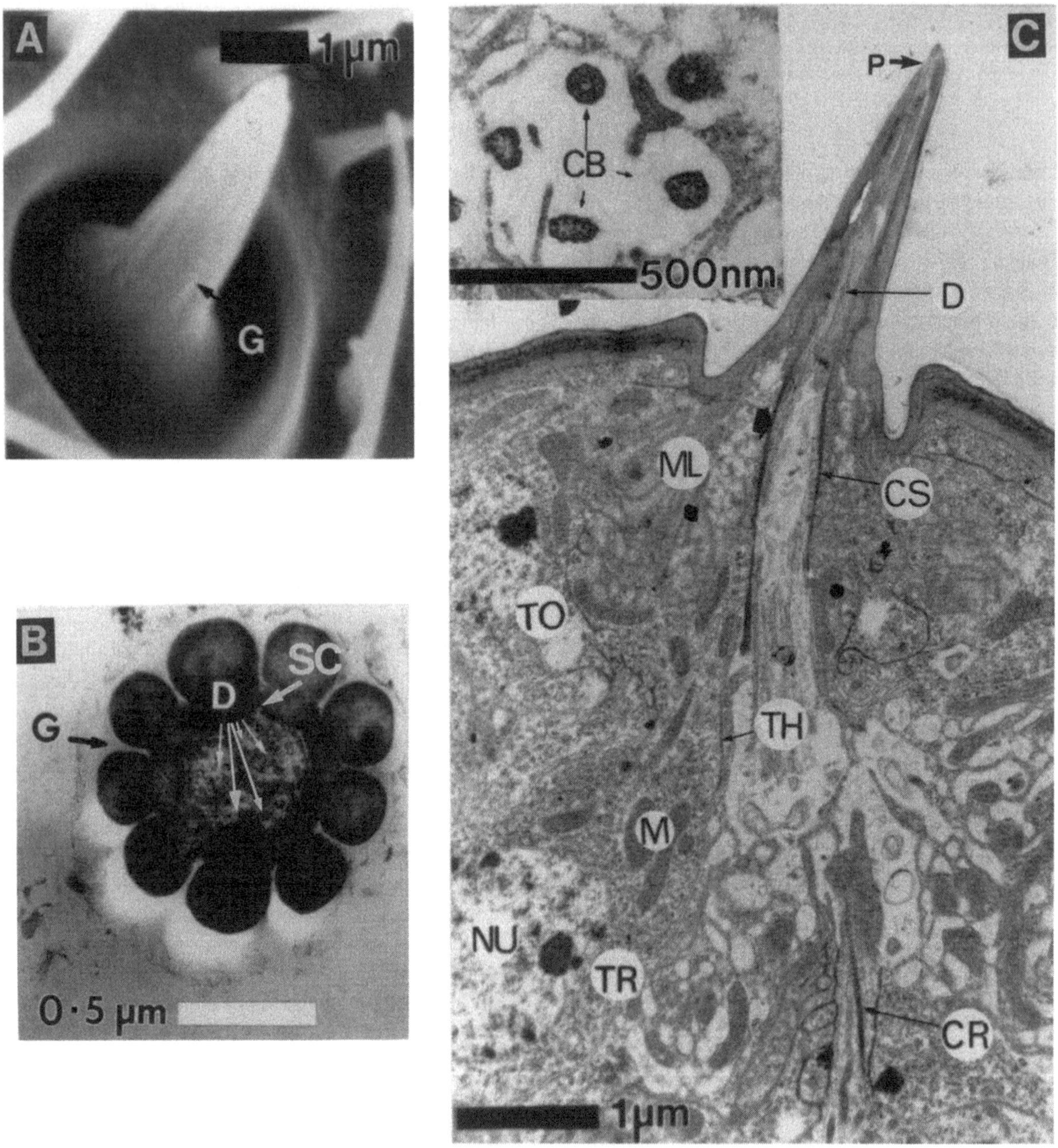


*Figure 5.* Sensilla basiconica on the antenna of female *D. tetensi.* (A) Scanning electron micrographs showing the peg-like structure and longitudinally running grooves. (B) Cross section through a sensilla basiconica fixed in 10% acrolein/2.5% glutaraldehyde, showing five dendrites. (C) Longitudinal section showing the five sensory neurones innervating a basiconic sensillum, fixed in 5% glutaraldehyde, some of which have been cut through the ciliary region. Note the cuticular sheath surrounding the distal dendritic segments. Insert: Cross section of five dendrites innervating a basiconic sensillum, one cut through the basal body and four cut through the ciliary body. CB, ciliary body; CR, ciliary region; CS, cuticular sheath; D, dendrite; G, groove; M, mitochondria; ML, microlamellae; NU, nucleus; P, pore; SC, spoke channel; TH, thecogen cell; TO, tormogen cell; TR, trichogen cell.

the circumfila of *M. erysiphes* may be important as sex pheromone receptors. The structure of *D. tetensi* circumfila conflicts with this assumption in that the female sensilla appear to be more elaborate than the male. The circumfila of *D. tetensi* however, no doubt play an important olfactory role.

The sensilla basiconica of *D. tetensi* closely resemble the 'grooved pegs' of *M. erysiphes* (Solinas & Nuzzaci, 1987) in their morphology. Both have a grooved hairshaft, contain unbranched sensory neurones whose proximal dendritic segments are encased in a cuticular sheath. Solinas & Nuzzaci (1987), however, did not observe any spoke channels within the peg walls and referred to the wall as being aporous. For the majority of grooved pegs that have been studied the conventional model requires pore channels (or spoke channels) for stimulant entry (Altner & Prillinger, 1980; Cribb & Jones, 1995). Insects which possess 'grooved pegs' with a single pore at the tip are extremely rare (Blaney & Chapman, 1969; Toh, 1977). Pore channels have a low electron density similar to that of the cuticle of the peg, very narrow dimensions, and an often convoluted structure within the cuticle. This makes them very difficult to see and they can therefore be easily overlooked (Slifer, 1970; Cribb & Jones, 1995). The 'grooved pegs' of *Aedes aegypti* for example were observed to possess electron dense strands running from the central lumen of the peg to the base of the grooves. These were discounted as the stimulant entrances because they were rarely observed (4 sections in every 150) (McIver, 1974). McIver (1974) also stated that a single pore occurring at the tip of the peg was the point of entry for stimulant molecules. In a reappraisal of the pore channel system in the 'grooved pegs' of *Aedes aegypti* Cribb & Jones, (1995) found that *A. aegypti* lacked a terminal pore but possessed 1–2 pore channels in every second or third section.

The sensilla basiconica on *D. tetensi* certainly appear to possess pore channels in transverse section although they are not clearly visible when viewed in longitudinal section, where a pore tip is clearly visible (Crook, 1998). The sensilla basiconica on *D. tetensi* therefore appear to be multiporous sensilla which could act as olfactory or humidity receptors (Zacharuk, 1980; Altner & Loftus, 1985).

The terminal sensory pegs appear to have a similar external structure to the sensilla found on the distal subsegments of *C. sorghicola* (Slifer & Sekhon, 1971), male *Culicoides impunctatus* and *Culicoides nubeculosus* (Blackwell et al., 1992), and the apical labial sensilla of the Aleyrodidae (Walker & Gordh, 1989). The function of these sensilla on *C. sorghicola* remains unknown but the sensilla found on both *Culicoides* and the Aleyrodidae are believed to be contact chemoreceptors, the latter probably being important in host-plant selection.

In conclusion, mated females respond to specific leaf volatiles emitted from the leaves and buds of blackcurrant plants and detailed descriptions of the olfactory sensilla provide the basis for future electrophysiological studies of the sensilla involved in chemical ecology.

## Acknowledgements

This study was supported by a NERC studentship to DC. The technical advice and assistance of Kevin McKenzie (University of Aberdeen), Jerry Cross, Colin Campbell and Ms Penny Greeves (Horticultural Research International, East Malling, Kent), Nick Birch and Rex Brennan (Scottish Crop Research Institute, Invergowrie, Dundee) are gratefully acknowledged.

## References

Alford, D. V., 1984. A colour atlas of fruit pests: their recognition, biology and control. Crosby Lockwood, London.

Altner, A. & R. Loftus, 1985. Utrastructure and function of insect thermo- and hygroreceptors. Annual Review of Entomology 30: 273–295.

Altner, H. & C. Prillinger, 1980. Utrastructure of invertebrate chemo-, thermo- and hygro-receptors and its functional significance. International Review of Cytology 67: 69–139.

Bagnall, R. S. & J. W. H. Harrison, 1917. A preliminary catalogue of British Cecidomyiidae (Diptera) with special reference to the Gall Midges of the North of England. Transactions of the Entomological Society of London 65: 346–426.

Barnes, H. F., 1948. Gall midges of Economic Importance. Vol. III. Gall Midges of Fruit. Crosby Lockwood, London, pp. 63–69.

Bergh, J. C., M. O. Harris & S. Rose, 1990. Temporal patterns of emergence and reproductive behaviour of the Hessian fly (Diptera: Cecidomyiidae). Annals of the Entomological Society of America 83: 998–1004.

Blackwell, A., A. J. Mordue (Luntz) & W. Mordue, 1992. Morphology of the antennae of two species of biting midge: *Culicoides impunctatus* (Goetghebuer) and *Culicoides nubeculosus* (Meigen) (Diptera, Ceratopogonidae). Journal of Morphology 213: 85–103.

Blaney, W. M. & R. F. Chapman, 1969. The fine structure of the terminal sensilla on the maxillary palps of *Schistocerca gregaria* (Forskål) (Orthoptera, Acrididae). Zeitschrift für Zellforschung und Mikroskopische Anatomie 99: 74–97.

Campbell, C. A. M., J. Pettersson, J. A. Pickett, L. J. Wadhams & C. M. Woodcock, 1993. Spring migration of damson-hop aphid, *Phorodon humuli* (Homoptera, Aphididae) and summer host plant-derived semiochemicals released on feeding. Journal of Chemical Ecology 19: 1569–1576.

Cribb, B. W. & M. K. Jones, 1995. Reappraisal of the pore channel system in the grooved pegs of *Aedes aegypti*. Tissue & Cell 1: 47–53.

Crook, D. J., 1998. Interactions of the blackcurrant leaf curling midge, *Dasineura tetensi* with resistant and susceptible host plant cultivars. Ph.D thesis. University of Aberdeen.

Felt, E. P., 1897. Antennal structure of certain diplosids. Psyche 8: 3–5.

Galanihe, L. D. & M. O. Harris, 1997. Plant volatiles mediate host-finding behaviour of the apple leafcurling midge. Journal of Chemical Ecology 23: 2639–2655.

Garthwaite, D. G., C. Wall & L. Wardlow, 1986. Further evidence for a female sex pheromone in the blackcurrant leaf midge *Dasineura tetensi*. British Crop Protection Conference. Pests and Diseases. Vol. 1, pp. 355–357.

Hardy, D. E., 1960. Insects of Hawaii. University of Hawaii Press, Honolulu, Hawaii. Vol. 10, pp. 1–368.

Harris, M. O. & S. P. Foster, 1999. Gall midges. In: J. Hardie & A. Minks (eds), Pheromones of Non-lepidopteran Insects Associated with Agricultural Plants. CABI, Wallingford. in press.

Harris, M. O., S. P. Foster, K. Age & S. Dana, 1996. Sex pheromone communication in the apple leaf curling midge (*Dasineura mali*). Proceedings of the New Zealand Plant Protection Conference 49: 52–58.

Harris, M. O. & S. Rose, 1989. Temporal changes in the egglaying behaviour of the Hessian fly. Entomologia Experimentalis et Applicata 53: 17–29.

Janet, C. H., 1896. Sur les Filets arqués des Antennes des Xylodiplosis (Dipt.) Bulletin Societé Entomologique de France (séance du 8 avril 1896): 183–185.

Karnovsky, M. J., 1965. A formaldehyde-glutaraldehyde fixative of high osmolarity for use in electron microscopy. Journal of Cell Biology 27: 137A.

Keiffer, J. J., 1900. Monographie des Cecidomyiidae d'Europe et d'Algerie. Annales de la Societé Entomologique de France 69: 182–471.

Keil, T. A. & R. A. Steinbrecht, 1984. Mechanoreceptive and olfactory sensilla in insects. In: R. C. King & H. Akai (eds) Insect Utrastructure. Vol. 2, Plenum Press, New York & London, pp. 477–512.

McIver, S. B., 1974. Fine structure of antennal grooved pegs of the mosquito, *Aedes aegypti*. Cell Tissue Research 153: 327–337.

McIver, S. B., 1975. Structure of cuticular mechanoreceptors of arthropods. Annual Review of Entomology 20: 381-397.

McIver, S. B. & A. Hudson, 1972. Sensilla on the antennae and palps of selected Wyeomyia mosquitoes. Journal of Medical Entomology 9: 337–345.

Metlitskii, O. Z. & A. S. Zeinalov, 1993. Non chemical control of gall-midges on currant. Sadovodstvo-i-Vinogradarstvo 3: 16–17.

Murchie, A. K., L. E. Smart & I. H. Williams, 1997. Responses of *Dasineura brassicae* and its parasitoids *Platygaster subuliformis* and *Omphale clypealis* to field traps baited with organic isothiocyanates. Journal of Chemical Ecology 23: 917–926.

Pettersson, J., 1970. An aphid sex attractant. I. Biological Studies. Entomologia Scandinavica 1: 63–73.

Pettersson, J., 1976. Ethology of *Dasineura brassicae* Winn. (Dipt., Cecidomyiidae) I. Laboratory studies of olfactory reactions to the host plant. Symposia Biologica Hungarica 16: 203–208.

Rübsaamen, E. H., 1896. Zurückweisung der Angriffe in J. J. Kieffer Abhandlung: Die Unterscheidungsmerkmale der Gallmücken. Entomologische Nachrichten 22: 119–127.

Sass, B., 1972. The mass appearance of a new blackcurrant pest, *Dasineura tetensi* in Hungary. Kerteszeti Kutato Intezet Kozlemenyei 6: 40–45.

Schneider, D., 1964. Insect Antennae. Annual Review of Entomology 9: 103–122.

Schneider, D. & R. A. Steinbrecht, 1968. Check list of insect olfactory sensilla. In Invertebrate receptors. Symposia of the Zoological Society of London 23: 279–297.

Simova-Tosic, D. & K. Dobrivojevic, 1981. Contribution to the knowledge of the blackcurrant leaf midge *Dasineura tetensi*, Rübsaamen (Cecidomyiidae, Diptera). Zaštita Bilja 32: 233–240.

Slifer, E., 1970. The structure of arthropod chemoreceptors. Annual Review of Entomology 15: 121–142.

Slifer, E. H. & S. S. Sekhon, 1971. Circumfila and other sense organs on the antenna of the sorghum midge (Diptera, Cecidomyiidae). Journal Morphology 133: 281–302.

Solinas, M. & G. Nuzzaci, 1987. Antennal sensilla of *Mycodiplosis erysiphes* Ruebs. (Cecidomyiidae, Diptera). Bollettino Isituto di Entomologia. Universita degli Studi di Bologna 41: 173–194.

Toh, Y., 1977. Fine structure of antennal sense organs of the male cockroach *Periplaneta americana*. Journal of Utrastructure Research 60: 373–394.

Vazyulya, A. G., 1982. Pests and diseases of blackcurrant. Zaschchita Rastenii. NIZI Sadovodstva Nechernozemnoi Polesy 7: 54–55.

Vet, L. E. M., J. C. van Lenteren, M. Heymans & E. Meelis, 1983. An airflow olfactometer for measuring olfactory responses of hymenopterous parasitoids and other small insects. Physiological Entomology 8: 97–106.

Walker, G. P. & G. Gordh, 1989. The occurrence of apical labial sensilla in the Aleyrodidae and evidence for a contact chemosensory function. Entomologia Experimentalis et Applicata 51: 215–224.

Zacharuk, R. Y., 1980. Ultrastructure and function of insect chemosensilla. Annual Review of Entomology 25: 27–47.

*Entomologia Experimentalis et Applicata* **91:** 51–58, 1999.



# Experience-based food consumption by larvae of *Pieris rapae*: addiction to glucosinolates?

J. A. A. Renwick & Kimberly Lopez
*Boyce Thompson Institute, Ithaca, NY 14853, USA*

Accepted: October 18, 1998

*Key words: Pieris rapae*, feeding stimulants, glucosinolates, crucifers, dietary experience, dependence, addiction, deterrents


## Abstract

When newly hatched larvae of *P. rapae* were transferred to cowpea foliage, they readily accepted this non-host as food, whereas later instars that had fed on cabbage rejected cowpea. However, when cowpea leaf discs were treated with aqueous extracts of cabbage foliage, they were accepted by cabbage-reared larvae. Experiments were conducted to determine whether larvae reared on one host plant would be stimulated to feed by extracts of other hosts. Larvae reared on *Brassica juncea, Cleome spinosa, Tropaeolum majus, Sinapis alba, Alliaria petiolata, Barbarea vulgaris* and cabbage (*Brassica oleracea*) were offered extracts of each of the other host plants on cowpea discs in choice assays. Larvae were generally stimulated to feed by extracts of all the alternate hosts, but quantitative differences in consumption occurred. In most cases, levels of discrimination between treatment and control cowpea discs showed no significant preference for extracts of the previously experienced plant. Since the test plants (and their extracts) contain glucosinolates of widely different structures, a general addiction to glucosinolates was suggested. A single glucosinolate, sinigrin, was sufficient to elicit feeding by cabbage-reared larvae. The time required for individual neonates to become addicted to glucosinolates as they fed on cabbage, as measured by refusal of cowpea, varied from 6 to 30 hours. Bioassays of cowpea extracts failed to show any deterrent activity and, therefore, supported the conclusion that addiction to glucosinolates is responsible for the fixation of *P. rapae* larvae on their host plants.


## Introduction

Plant chemistry is well known to be a major guiding force in the acceptance or rejection of plants by phytophagous insects. Host selection often involves recognition of stimulatory compounds that indicate suitability, or of deterrent compounds that serve as a warning of unsuitability of a plant for feeding and development (Bernays & Chapman, 1994). Stimulants and deterrents at the leaf surface play a primary role in the selection of a site for oviposition by the adult female (Städler, 1994; Renwick & Chew, 1994). But the hatching larvae may also be capable of making a choice, depending on their mobility and the need to search for suitable food. The basis for larval choice may often be the same as for adults, but in some cases, the larvae and adults of a species have different sensitivities to individual chemicals in the plant (Renwick, 1996). However, in either case, a balance of positive and negative signals perceived by the insect determines whether the plant is accepted or rejected (Dethier, 1982; Miller & Strickler, 1984; Renwick & Huang, 1994).

Feeding by polyphagous insects is generally stimulated by nutrients, such as sugars, amino acids and other primary metabolites that are present in all plants. However, oligophagous insects usually require additional, more specific chemicals for stimulation of feeding, and several such compounds, or classes of compounds, have been identified (Bernays & Chapman, 1994). But despite the fact that specialists depend on the presence of specific compounds for recognition of their hosts, the larvae of some species will feed on non-hosts at the time of hatching, and may even com-

*Table 1.* Test plants and their predominant glucosinolate constituents

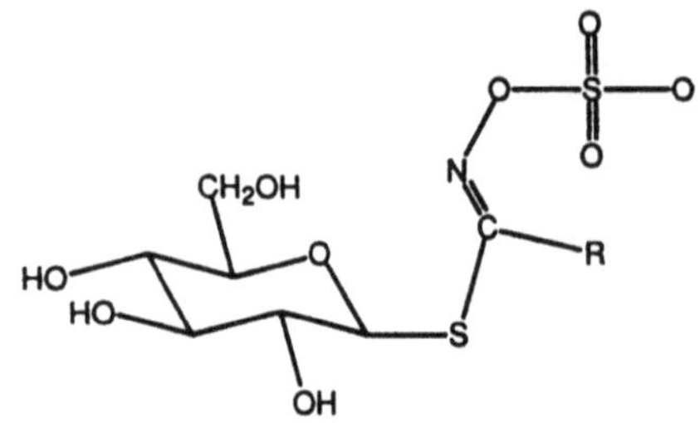


| Plant species | Predominant Glucosinolate | R-Group |
|---|---|---|
| *Brassica juncea* (brown mustard) | sinigrin | $—CH_2—CH{=}CH_2$ |
| *Cleome spinosa* (spider plant) | glucocapparin | $—CH_3$ |
| *Tropaeolum majus* (nasturtium) | glucotropaeolin | 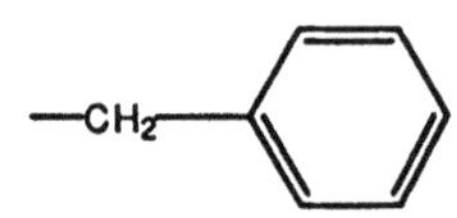 |
| *Sinapis alba* (white mustard) | glucosinalbin | 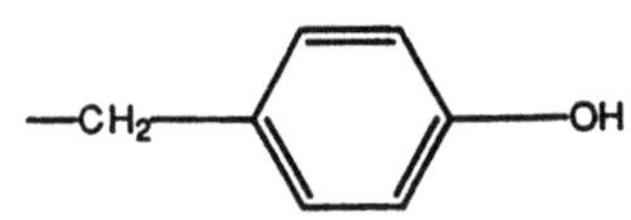 |
| *Allaria petiolata* (garlic mustard) | sinigrin | $—CH_2—CH{=}CH_2$ |
| *Barbarea vulgaris* (wintercress) | glucobarbarin | 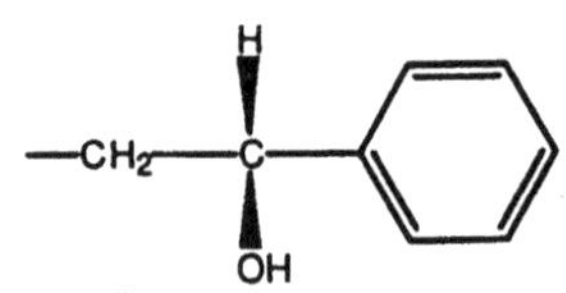 |
| *Brassica oleracea* (cabbage) | glucobrassicin | 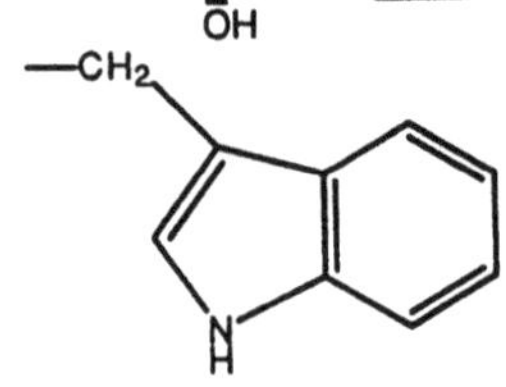 |

plete their development on these plants (Yamamoto, 1974). Larvae of *Manduca sexta*, which is a specialist on members of the Solanaceae, will feed and develop on foliage of cowpea, *Vigna sinensis* (Fabaceae), although this is not a host plant in nature (de Boer, 1992). As a result, cowpea has served as a convenient test plant for studying behavioral and physiological aspects of host acceptance, particularly for *M. sexta* (de Boer 1992; M. L. del Campo & J. A. A. Renwick, unpubl.).

The behavioral response of a specialist insect to plant chemicals may be dependent to a large degree on the previous dietary experience of that insect. This may involve habituation, sensitization, imprinting, learning or induction. Induced preference for a particular plant has been shown for many insects (Jermy, 1987; Szentesi & Jermy, 1990), and may be explained by changes at the peripheral sensory level and/or in the central nervous system (Schoonhoven, 1977). In a few extreme cases, an insect will starve to death rather than accept a new host (Ma,1972; Hanson,1976; Scriber,1982; Renwick & Huang, 1995).

Much of the work on induced preferences has focused on larvae of *Pieris* (Pieridae) butterflies, which

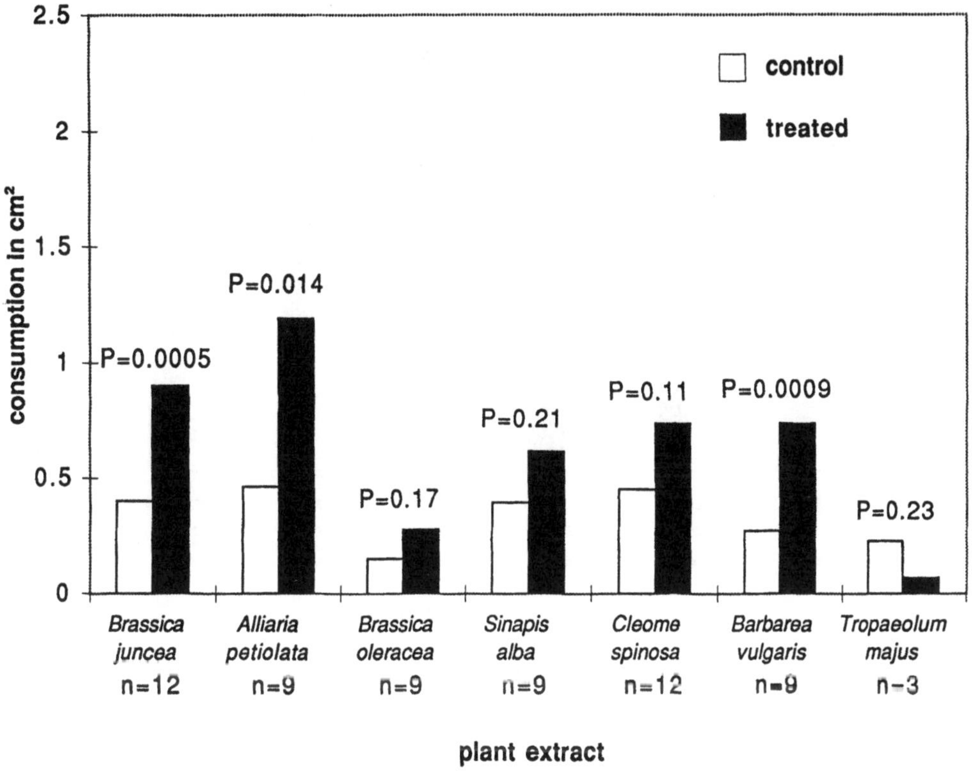


*Figure 1* Feeding response of 4th instar *Pieris rapae* larvae reared on *Brassica juncea* to aqueous extracts of other host plants.

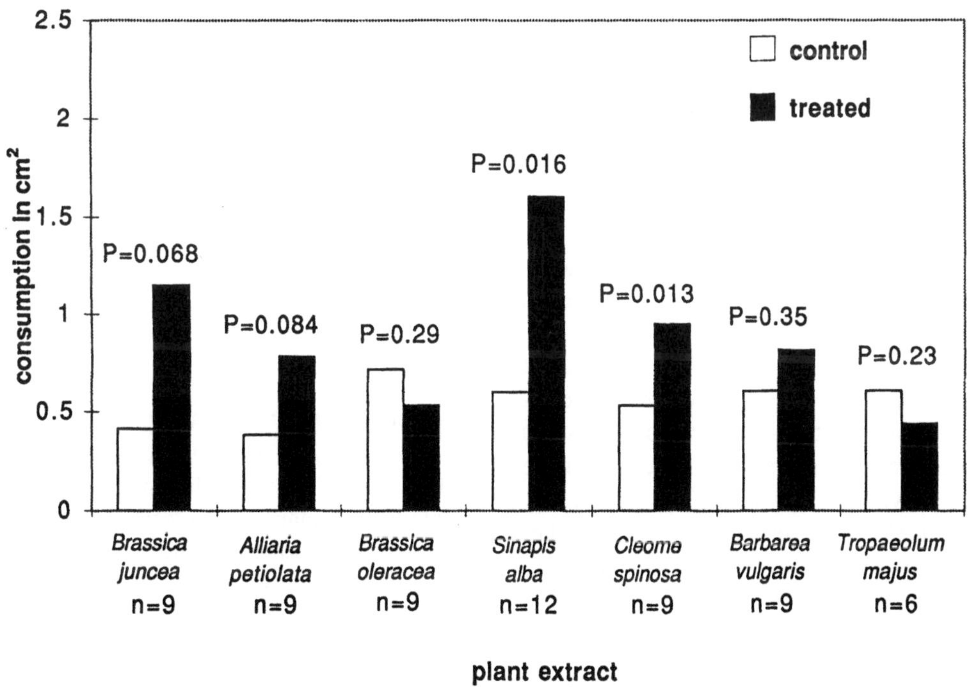


*Figure 2* Feeding response of 4th instar *Pieris rapae* larvae reared on *Sinapis alba* to aqueous extracts of other host plants.

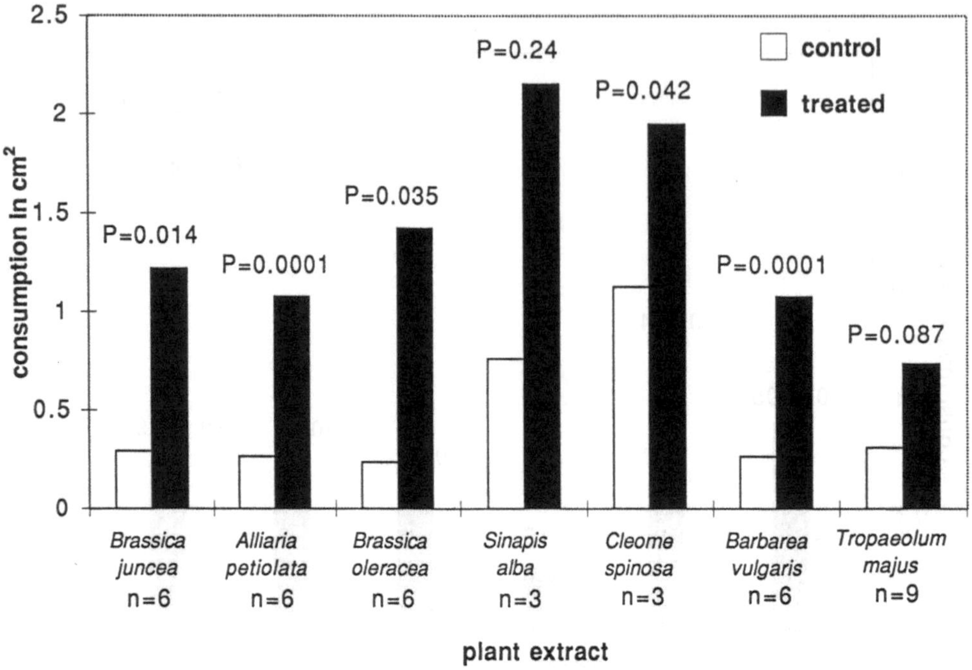


*Figure 3* Feeding response of 4th instar *Pieris rapae* larvae reared on cabbage (*Brassica oleracea*) to aqueous extracts of other host plants.

are specialists on the Brassicaceae (= Cruciferae) (Johansson, 1951; Ma, 1972). However, the host range of these insects includes members of other plant families, including the Capparidaceae, Resedaceae and Tropaeolaceae, which are linked to the crucifers by their production of glucosinolates (Verschaffelt, 1910). The involvement of glucosinolates in stimulating oviposition by *P. brassicae* and *P. rapae* has been clearly shown (van Loon et al., 1992; Renwick et al., 1992), and the activity of these glycosides as feeding stimulants for the larvae has been demonstrated (Verschaffelt, 1910; David & Gardiner, 1965). However, the relative activities of different glucosinolates or the involvement of glucosinolates in induced preferences have not been reported.

Preliminary work in our laboratory has shown that newly hatched larvae of *Pieris rapae* will feed on cowpea foliage, but later instar larvae (2nd to 5th instars) that have been reared on cabbage refuse to feed on the cowpea. Since we have shown that previous dietary experience can affect the sensitivity of *P. rapae* larvae to deterrents (Renwick & Huang, 1995; Huang & Renwick, 1995a), the possible presence of a deterrent in cowpea was suggested. However, when an aqueous extract of cabbage is applied to discs of cowpea foliage, the cabbage-reared larvae are stimulated to feed (J. A. A. Renwick & K. D. Lopez, unpubl.). Since this extract is known to contain glucosinolates (Renwick et al., 1992), we asked whether induced preference in response to specific glucosinolates can be responsible for acceptance of the cowpea. Since samples of individual glucosinolates are not generally available, we selected plants that produce a predominance of one glucosinolate (Table 1) to test whether larvae become 'fixed' on a particular compound, when reared on these plants. In addition, we designed experiments

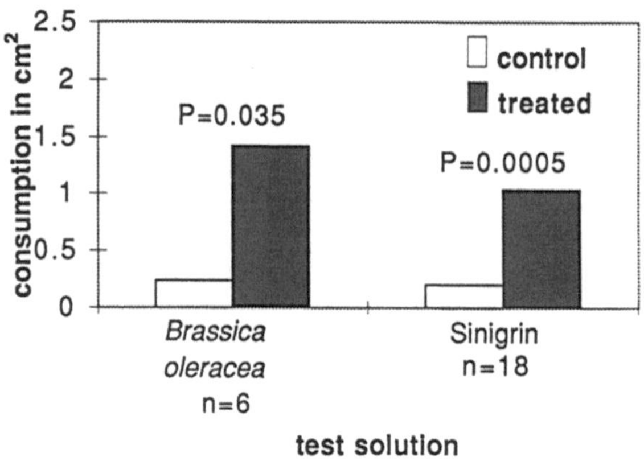


*Figure 4* Feeding response of 4th instar cabbage-reared *Pieris rapae* larvae to cabbage aqueous extract or a solution of sinigrin alone.

to test for the possible role of deterrents in cowpea to explain the refusal of foliage by cabbage-reared larvae.

## Materials and methods

*Insects.* Insects for bioassays were from a colony of *Pieris rapae* L. maintained in a greenhouse at ca. 25 °C with a photoperiod of L16:D8 and in a laboratory under the same conditions. Larvae were reared on *Brassica oleracea* L. var. Golden Acre. The colony was supplemented annually with field-collected butterflies from Ithaca, NY.

*Plants.* All plants for rearing insects and chemical extraction were grown in an air-conditioned greenhouse at ca. 25 °C, with a photoperiod of L16:D8 provided by 400-W multivapor high intensity discharge lamps. The plants were grown in artificial soil (Cornell Mix A with Osmocote® ) and fertilized weekly with 300 ppm Peters Excel Cal-Mag 15-5-15. The following eight plants were used for comparative studies: *Alliaria petiolata* B. Ehrh. (Cruciferae), seedlings collected in Ithaca, N.Y.; *Barbarea vulgaris* R. Br. (Cruciferae), seeds collected in Brooktondale, N.Y.; *Brassica oleracea* L. var. Golden Acre (Cruciferae), *Brassica juncea* (L.) Czern. et Coss (Cruciferae); *Cleome spinosa* Jacq. (Capparidaceae); *Sinapis alba* L. (Cruciferae); *Tropaeolum majus* L. (Tropaeolaceae). The plants were four to six weeks old when they were used for rearing insects or for extractions. Cowpea plants (*Vigna sinensis* var. Pink Eye Purple Hull) for bioassays were grown under the same conditions.

*Chemical.* Sinigrin monohydrate from horseradish (CAS # 3952-98-5) was obtained from Sigma Chemical Company , St. Louis, MO.

*Extraction of plant material.* Fresh foliage from each plant was extracted in boiling ethanol for 10 min, cooled, homogenized, and the extract was filtered and evaporated to dryness *in vacuo*. The dried extract was washed with hexane to remove lipids, and the residue was dissolved in water. The aqueous extract was partitioned 3 times with 1-butanol, and the water fraction was evaporated to dryness and stored in the refrigerator. Cowpea foliage was extracted in the same way, but the butanol fraction was saved for deterrent assays.

*Bioassays.* Stimulant assays were conducted using 4th instar *P. rapae* that were reared on one of the eight test plants. Each experience group of insects was tested for feeding response to extracts of each of the eight test plants. Assays were conducted in 250 ml ice cream cups with a layer of paraffin wax on the bottom, covered with moist filter paper. Cowpea leaf discs, 3.1 cm diam. were cut from fully expanded leaves with a cork borer and treated with 0.1 gle (gram leaf equivalents) of plant extract or 0.04 mg sinigrin in 70% methanol solution. Control discs were treated with solvent alone. Two treated and two control discs were pinned in alternating fashion to the bottom of each assay cup. Four larvae were placed at the center of each cup, which was covered with a plastic lid, and the cups were placed in a sealed box at high humidity, and kept in an incubator at 26° for 24 h. The areas of the leaf discs were then measured with a LI-COR model 3100 area meter to determine the amount of feeding. Blank discs without larvae were subjected to the same conditions to determine the amount of disc shrinkage. Deterrent assays were conducted in the same way, using cabbage leaf discs, to test for possible deterrent activity of cowpea extracts. However, only one larva per cup was used in these bioassays, and whether the test insect fed on a treated or control disc was recorded.

*Data analysis.* Assays were replicated from 6 to 12 times for tests of each group of insects on each extract. Consumption was measured in $cm^2$ by subtracting the area of each disc remaining from the average blank disc area after shrinkage. Statistical analysis was performed using Minitab, and a one-sample *t*-test was used to compare consumption of treated and control discs in each assay. A two sample *t*-test was performed to compare levels of discrimination between treated and control dics for the different extracts.

*Development of dependence.* Experiments were conducted to measure the time required for first instar *P. rapae* to become dependent on constituents of cabbage. Neonate larvae obtained from eggs collected on Parafilm (Webb & Shelton, 1988) were placed on a cabbage leaf and transferred after different time periods to cowpea leaf discs in ice cream cups. Acceptance (feeding) or rejection (no visible feeding) of the cowpea was determined after 24 h.

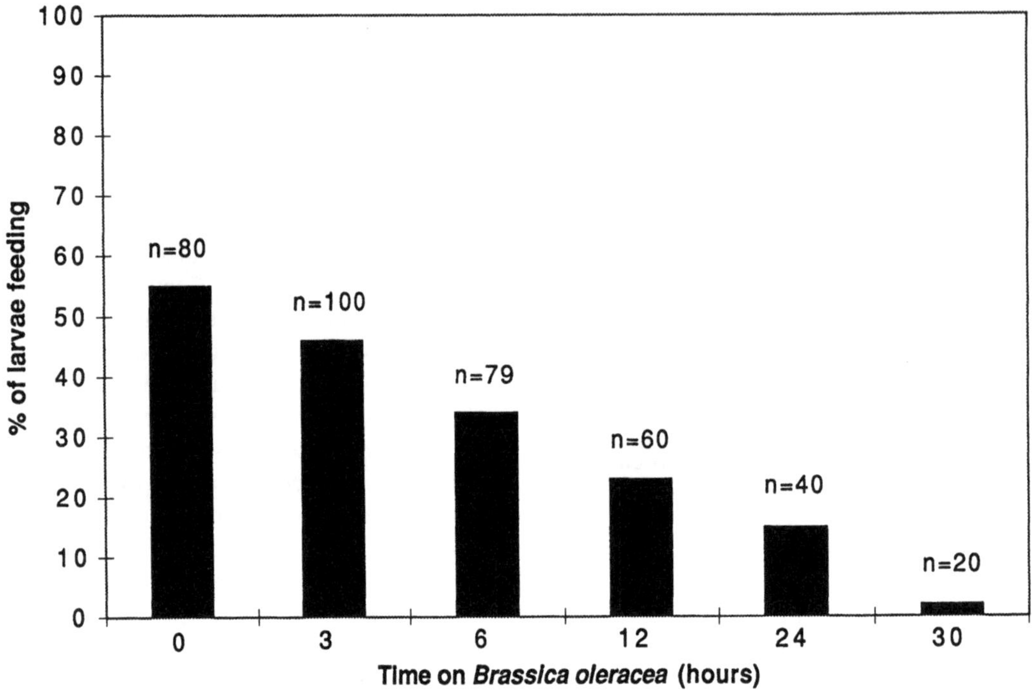


*Figure 5* Acceptance of cowpea discs by neonate larvae of *Pieris rapae* that fed for different periods of time on cabbage.

## Results and discussion

When neonate larvae of *P. rapae* were placed on leaf discs of cowpea, almost 70% of the individuals readily fed on this plant tissue. Although continued feeding and development was slower than on cabbage foliage, most of the feeding larvae molted to 2nd instars within a period of 4 days. In contrast, 4th instar larvae that were transferred from cabbage to cowpea discs did not feed. However, when cabbage aqueous extract was applied to cowpea discs, 81% of the cabbage-reared 4th instars accepted the treated discs as food. These results are consistent with the observations of Verschaffelt (1910), who showed that *Pieris* larvae are stimulated to feed by glucosinolate-containing sap or extracts of crucifer hosts. But specific host stimulants do not appear to be necessary for feeding by inexperienced neonates. This would suggest that the response of larvae to stimulants is dependent on dietary experience. Previous results in our laboratory have shown that neonates of *P. rapae* are relatively insensitive to deterrents (Renwick & Huang, 1995; Huang & Renwick, 1995a), so dietary experience may affect responses to both stimulants and deterrents.

The responses of larvae reared on different species to extracts of other host species indicated that constituents of any host plant will stimulate feeding by larvae from other hosts (Table 2). However, considerable variation in consumption and in discrimination between treated and control discs was evident. Larvae reared on *A. petiolata* and *S. alba* appeared to consume large quantities in response to their own extract and showed high levels of discrimination between treated and control discs. However, in general larvae reared on other hosts showed no distinct preference for extracts of their dietary plants. Larval experience on *T. majus* resulted in higher levels of consumption in response to other host extracts, but discrimination between treated and control discs was low. Larvae reared on *T. majus* have previously been shown to have reduced sensitivity to deterrents (Renwick & Huang, 1995). Increased acceptance of cowpea discs may, therefore, be a result of suppressed development of sensitivity to deterrents or suppressed development of dependence on stimulants. The response of larvae from other plants to *T. majus* extracts was generally very weak. This may be due to the presence of deterrents, including chlorogenic acid, in that plant (Huang & Renwick, 1995b).

Larvae reared on *B. juncea* significantly discriminated between treated and control cowpea discs when offered extracts of *B. juncea*, *A. petiolata* or *B. vulgaris* (Figure 1), but total consumption was relatively low. Similarly, larvae reared on *S. alba* significantly

*Table 2.* Percent area of treated and control discs of cowpea consumed by 4th instar larvaeof *Pieris rapae* that were reared on different host plants. Treated discs (3.1 cm diam) received 0.1 gle aqueous extract of the test plant, whereas control discs received solvent only

| Larval experience | *Brassica juncea* | | *Alliaria petiolata* | | *Brassica oleracea* | | *Sinapis alba* | | *Cleome spinosa* | | *Barbarea vulgaris* | | *Tropaeolum majus* | | Total avg. % |
|---|---|---|---|---|---|---|---|---|---|---|---|---|---|---|---|
| | C | T | C | T | C | T | C | T | C | T | C | T | C | T | |
| *B. juncea* | 6.3 | 14.1 | 7.7 | 19.5 | 2.5 | 4.5 | 6.4 | 10.1 | 7.0 | 11.9 | 4.3 | 12.0 | 3.7 | 1.1 | 7.9 |
| *A. petiolata* | 10.4 | 18.6 | 0.0 | 17.3 | 3.2 | 4.8 | 4.6 | 6.2 | 3.3 | 6.9 | 6.1 | 21.5 | 3.8 | 4.3 | 7.9 |
| *B. oleracea* | 4.9 | 21.2 | 4.3 | 17.3 | 4.0 | 22.8 | 12.2 | 35.8 | 18.1 | 32.4 | 6.9 | 54.9 | 4.6 | 10.9 | 17.9 |
| *S. alba* | 6.6 | 18.4 | 5.8 | 12.9 | 12.0 | 9.7 | 9.6 | 30.2 | 8.3 | 14.6 | 18.5 | 18.6 | 9.7 | 6.9 | 13.0 |
| *C. spinosa* | 0.4 | 12.0 | 1.4 | 10.8 | 3.4 | 4.1 | 3.0 | 27.6 | 9.8 | 23.6 | 2.0 | 24.9 | 1.8 | 11.8 | 9.8 |
| *B. vulgaris* | 32.2 | 43.5 | 10.5 | 8.3 | 5.2 | 1.7 | 2.5 | 5.5 | 2.5 | 6.0 | 0.3 | 6.1 | 10.5 | 6.1 | 10.1 |
| *T. majus* | 20.3 | 42.4 | 15.2 | 12.8 | 8.7 | 48.0 | 21.4 | 35.2 | 30.5 | 32.5 | 36.0 | 35.7 | 24.5 | 25.2 | 27.7 |
| Total avg. % | 11.3 | 24.3 | 6.4 | 14.1 | 5.6 | 13.7 | 8.5 | 21.5 | 11.4 | 18.3 | 10.6 | 24.8 | 8.4 | 9.5 | |

discriminated between extract and control for *S. alba* and *C. spinosa*, but consumption was greater than for the *B. juncea*-reared larvae (Figure 2). However, cabbage-reared larvae consumed more than either of these groups, and showed good discrimination in response to extracts of *B. juncea, A. petiolata, B. oleracea, C. spinosa* and *B. vulgaris* (Figure 3). Since these plants produce a range of aliphatic and aromatic glucosinolates (Table 1), no general pattern of preference could be seen. But the fact that cabbage-reared larvae discriminate between control discs and those treated with many other host extracts might suggest that exposure to the multiple glucosinolates present in cabbage (Renwick et al., 1992) renders them more capable of recognizing a range of stimulants. The degree of response to extracts may be affected by quantity as well as quality of glucosinolates present. No attempt was made in these experiments to adjust samples for concentration.

The responses of cabbage-reared larvae to sinigrin was similar to their responses to the total aqueous extract of cabbage (Figure 4). This means that exposure to a range of glucosinolates, but primarily glucobrassicin in cabbage, results in the ability of the insects to recognize and respond to a single glucosinolate. Since many of the test plants were chosen for their production of predominantly one glucosinolate, it appears likely that exposure to a single glucosinolate results in a recognition response of the larvae to any other glucosinolate. Both aromatic and aliphatic glucosinolates seem to have the same effect.

The time required to develop this apparent dependence on glucosinolates was measured by allowing neonate larvae to feed on cabbage for different lengths of time before transferring them to cowpea discs. Initially, 56% of the naive larvae accepted cowpea, but after feeding for 6 h on cabbage, only 35% of these would still accept the cowpea foliage (Figure 5). A steady reduction in numbers accepting cowpea with time feeding was evident, and after 30 h, only 2% made any attempt to feed on cowpea. These results indicate considerable individual variation in time needed for the physiological changes that lead to rejection behavior. This decline in acceptance of cowpea might be compared with the increased levels of rejection of *Tropaeolum majus* by *Pieris brassicae* (Ma, 1972) and *P. rapae* (Renwick & Huang, 1995) after feeding on cabbage.

Deterrent assays of cowpea extracts were conducted with 4th instar larvae that were reared on cabbage. When the total aqueous extract was tested at 0.1 gle per disc, no deterrent effect was detected. In choice bioassays with 18 individual larvae, 13 of these fed on the treated discs while 11 fed on the control dics within a 24 h observation period.

Since the results of these deterrent assays showed no deterrent effect, and larvae readily accepted cowpea discs (with their complete chemical complement) that were treated with host extracts, we can conclude that discriminatory behavior of larvae is not due to deterrents in the cowpea. Instead, it seems that *P. rapae* larvae become fixed on glucosinolates in their host plants. After a period of 6 to 30 h of feeding on a host, they become dependent on these constituents of crucifers and a few other plant families that serve as hosts. The development of dependence on host constituents has already been described for *Manduca sexta*, which is strongly stimulated to feed by chemicals in solanaceous plants (M. L. del Campo & J. A. A. Renwick, unpubl.). However, a single compound seems to be responsible for this effect (M. L. del Campo & J. A. A. Renwick, unpubl.). The fixation of *P. rapae* on glucosinolates appears to occur in response to a wide range of different glucosinolates. The basic glucosinolate molecule, regardless of the aromatic or aliphatic substitution, must possess the necessary characteristics to trigger the physiological changes involved in development of dependence. This phenomenon may be considered as a type of addiction to food constituents. Electrophysiological and biochemical experiments will be necessary to determine the physiological mechanism and degree of involvement of the insect's sensory system.

## Acknowledgements

This research was supported in part by NSF Grant no. DEB-9419797.

## References

Bernays, E. A. & R. F. Chapman, 1994. Host-Plant Selection by Phytophagous Insects. Chapman & Hall, New York, 312 pp.

Bernays, E. A. & R. F. Chapman, 1987. The evolution of deterrent responses in plant-feeding insects. In: R. F. Chapman, E. A. Bernays & J. G. Stoffolano, Jr. (eds.), Perspectives in Chemoreception and Behavior. Springer–Verlag, New York, pp. 159–173.

David, W. A. L. & B. O. C. Gardiner, 1965. The effect of sinigrin on the feeding of *Pieris brassicae* L. larvae transferred from various diets. Entomologia Experimentalis et Applicata 9: 95–98.

de Boer, G., 1992. Diet-induced food preference by *Manduca sexta* larvae: acceptable non-host plants elicit a stronger induction than host plants. Entomologia Experimentalis et Applicata 63: 3–12.

Dethier, V. G., 1982. Mechanisms of host-plant recognition. Entomologia Experimentalis et Applicata 31: 49–56.

Hanson, F. E., 1976. Comparative studies on induction of food choice preferences in lepidopterous larvae. Symposia Biologica Hungarica 16: 71.

Huang, Xinpei & J. A. A. Renwick, 1995a. Cross habituation to feeding deterrents and acceptance of a marginal host plant by *Pieris rapae* larvae. Entomologia Experimentalis et Applicata 76: 295–302.

Huang, Xin Pei & J.A.A. Renwick, 1995b. Chemical and experiential basis for rejection of *Tropaeolum majus* by *Pieris rapae* larvae. Journal of Chemical Ecology 21: 1601–1617.

Jermy, T., 1987. The role of experience in the host selection of phytophagous insects. In: R. F. Chapman, E. A. Bernays & J. G. Stoffolano, Jr. (eds.), Perspectives in Chemoreception and Behavior. Springer–Verlag, New York, pp. 143–157.

Johansson, A. S., 1951. The food plant preference of the larvae of *Pieris brassicae* L. Norsk Entomologisk Tidsskrift. B, 8: 187–195.

Loon, J. J. A. van, A. Blaakmeer, F. C. Griepink, T. A. van Beek, L. M. Schoonhoven & A. de Groot, 1992. Leaf surface compound from *Brassica oleracea* (Cruciferae) induces oviposition by *Pieris brassicae* (Lepidoptera: Pieridae). Chemoecology 3: 39–44.

Ma, W. C., 1972. Dynamics of feeding responses in *Pieris brassicae* Linn. as a function of chemosensory input: a behavioural, ultrastructural and electrophysiological study. Mededelingen Landbouwhogeschool Wageningen, 72–11: 1–162.

Miller, J. R. & K. L. Strickler, 1984. Finding and accepting host plants. In: W. Bell. & R. Cardé (eds), Chemical Ecology of Insects. Chapman & Hall, London, pp. 127–155.

Renwick, J. A. A., 1996. Diversity and dynamics of crucifer defenses against adults and larvae of cabbage butterflies. In: J. T. Romeo, J. A. Saunders & P. Barbosa (eds), Recent Advances in Phytochemistry 30: Plenum Press, New York, pp. 57–79.

Renwick, J. A. A. & F. S. Chew, 1994. Oviposition behavior in Lepidoptera. Annual Review of Entomology 39: 377–400.

Renwick, J. A. A. and Xinpei Huang, 1994. Interacting chemical stimuli mediating oviposition by Lepidoptera. In: T. N. Ananthakrishnan (ed.), Functional Dynamics of Phytophagous Insects. Oxford & IBH Publishing Co., New Delhi, pp. 79–94.

Renwick, J. A. A. and Xin Pei Huang, 1995. Rejection of host plant by larvae of cabbage butterfly: diet-dependent sensitivity to an antifeedant. Journal of Chemical Ecology 21: 465–475.

Renwick, J. A. A., C. D. Radke, K. Sachdev-Gupta & E. Städler, 1992. Leaf surface chemicals stimulating oviposition by *Pieris rapae* (Lepidoptera: Pieridae) on cabbage. Chemoecology 3: 33–38.

Schoonhoven, L. M., 1977. On the individuality of insect feeding behaviour. Proceedings of the Koninklijke Nederlandse Akademie van Wetenschappen C80: 341–350.

Scriber, J. M., 1982. The behaviour and nutritional physiology of southern armyworm larvae as a function of plant species consumed in earlier instars. Entomologia Experimentalis et Applicata 31: 359–369.

Städler, E., 1994. Behavioral responses of insects to plant secondary compounds. In: G. A. Rosenthal & M. R. Berenbaum (eds), Herbivores: their Interactions with Secondary Plant Metabolites. 2nd edn, vol. II, pp. 45–88.

Szentesi, A. & T. Jermy, 1990. The role of experience in host plant choice by phytophagous insects. In: E. A. Bernays (ed.), Insect-Plant Interactions II, CRC Press, Boca Raton, pp. 39–74.

Verschaffelt, E., 1910. The cause determining the selection of food in some herbivorous insects. Proceedings of the Koninklijke Nederlandse Akademie van Wetenschappen 13: 536–542.

Webb, S. E. & A. M. Shelton, 1988. Laboratory rearing of imported cabbageworm. Bulletin of New York's Food and Life Sciences 122: 1–6.

Yamamoto, R. T., 1974. Induction of host plant specificity in the tobacco hornworm, *Manduca sexta*. Journal of Insect Physiology 20: 641–650.

*Entomologia Experimentalis et Applicata* **91:** 59–65, 1999.




# Role of visual cues and interaction with host odour during the host-finding behaviour of the cabbage moth

Julio C. Rojas[1,2] & Tristram D. Wyatt[1]
[1]*Department of Zoology, University of Oxford, South Parks Road, Oxford OX1 3PS, UK;* [2]*El Colegio de la Frontera Sur, apdo postal 36, Tapachula, Chiapas, CP 30470, Mexico*

Accepted: October 15, 1998

*Key words:* Lepidoptera, Noctuidae, *Mamestra brassicae*, host-finding behaviour, visual cues, host-choice, interaction, odour, volatiles

## Abstract

The approach and landing responses of female *Mamestra brassicae* (L.) (Lepidoptera: Noctuidae) to visual cues from artificial plant leaves of different shapes and presence/absence of cabbage plant odour were investigated in a laboratory wind tunnel. The leaves were painted with cadmium yellow colour and observed under dim red light. Females showed oriented flight towards plant odours but landed significantly more often when the odour was presented with an artificial leaf. In three-choice tests, the shape of the leaf targets (circle, square or triangle) did not influence the female response. However, the size of the target did influence the insect response: the females preferred landing on square targets with sides of 5 or 10 cm rather than on the largest target, with sides of 15 cm. The orientation of the target influenced the insects' response: females landed significantly more often on the target positioned vertically than horizontally.

## Introduction

The role of visual cues in the host-finding process has been established in different insect species, including visual location by pollinators, herbivores, parasitoids, and predators (Prokopy & Owens, 1983). However, the majority of studies of the role of vision during host-finding behaviour have been done with diurnal insects, and little is known about the role of visual cues in nocturnal species.

The importance of the interaction between different stimuli during the host-finding process has been reported in several insect species, particularly in Homoptera and Diptera (Harris & Miller, 1988; Harris & Rose, 1990; Todd et al., 1990; Aluja & Prokopy, 1993). In moths, the interaction between visual cues and host odour during the host-finding has been studied only in three species so far (Yamamoto et al., 1969; Brantjes, 1976; Ockenfels & Schmidt, 1992). For instance, in the nocturnal *Hadena bicruris* Hufn. (Noctuidae), which lays its eggs in the white flowers of *S. latifolia*, search behaviour and orientation to the flowers are induced by floral volatiles (Brantjes, 1976). After floral visits, a moth may approach non-fragrant white objects, but she lands only in the presence of olfactory stimuli. In contrast, the attraction of the diurnal moth *Zygaena trifolii* (Esper) (Zygaenidae) to artificial flower heads is dependent on both visual and scent stimuli. In the absence of either, settling behaviour and feeding were not observed (Ockenfels & Schmidt, 1992).

In this paper we investigate the influence of visual cues and their interaction with plant odour during the host-finding behaviour of the cabbage moth under laboratory conditions.

## Materials and methods

*Moth rearing.* The laboratory colony was started with adults collected from a light trap at Tackley, Oxfordshire, during June 1995, and wild insects were added to the colony every summer. All insects were reared in controlled conditions under L16:D8 pho-

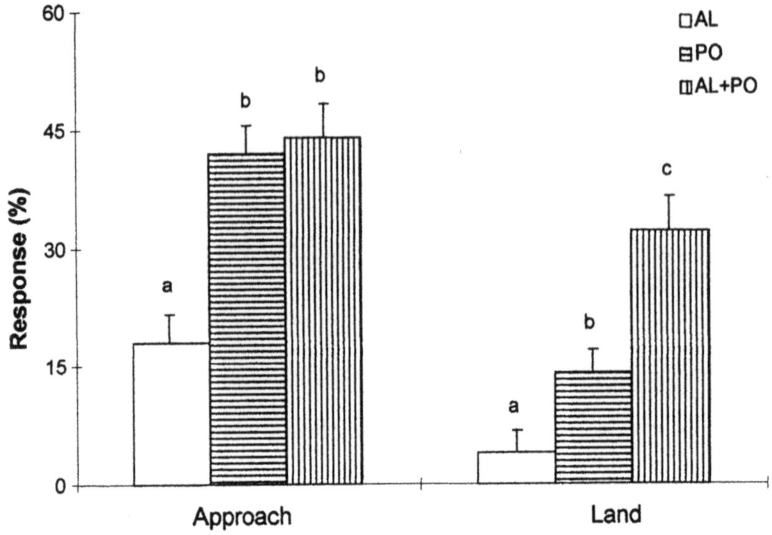


*Figure 1.* Approach and landing responses (mean ± SEM) of female *M. Brassicae* to artificial leaf (AL), plant odour (PO), and artificial leaf + plant odour (AL + PO) in non-choice tests under red light. Data are average of 10 replications; in each, five insects were tested individually. In this and later figures, bars marked by the same letter are not significantly different ($P < 0.05$; Tukey tests). Ten replicates each of 5 insects.

toperiod cycle at 25 ± 1 °C, 30–50% r.h.. Larvae were reared on artificial diet (Poitout et al., 1972). Mated adult females 4–6 days of age were used for the experiments.

*Plant material.* Cabbage (Xmas Drumhead) plants 3–5 months old were used for making extracts. The plants were grown in a greenhouse with no artificial lighting and were fertilised every 15 days with a NPK (7:7:7) fertiliser.

*Plant extracts.* The extracts were made in a Panasonic NM-6803B microwave oven operating at 2450 MHz and a maximum power of 800 W (Ganzler et al., 1986). Cabbage leaves (25 g) were cut and placed in 50 ml of hexane in a glass flask and irradiated for 10 s (low level). Then, the flask was allowed to cool for 10 min to ambient temperature (15–20 °C) by placing a dish of water under it. This procedure was repeated three times. The extracts were dried with $MgSO_4$, filtered and concentrated to 0.5 g leaf equivalents/0.1 ml hexane.

*Wind tunnel.* The observations were carried out in a wind tunnel, 160 cm long and 75 cm high and wide, as described by Wyatt et al. (1997). A detachable cabinet containing a Fischbach speed controller fan was used to push air through the tunnel with a velocity of 35 cm $s^{-1}$. An activated-charcoal filter cleaned the air entering the tunnel. Illumination, provided by two 85-W red fluorescent tubes (Thorn EMI, UK) mounted 5 cm above the wind tunnel, gave a light intensity of 4.54 $\mu$E $m^{-2}$ at the point of release platform 20 cm from the downwind screen and about 25 cm from the tunnel floor. Black and white dots immediately beneath the tunnel floor provided optomotor feedback for the moths.

*Response to artificial leaves.* The influence of the visual cues on the female response to host plants was evaluated using artificial cabbage leaves made of paper and painted with Liquitex acrylic paint (cadmium yellow medium, Munsell hue: 1.8Y, value 8.2). The artificial leaf was fixed to a stick (3 mm diameter × 30 cm high). The stick was inserted into a rubber stopper (2.5 cm diameter × 3 cm high) and placed in the centre of the upwind end of the tunnel. The female's response to this artificial leaf was evaluated in non-choice tests. Each observation was begun by releasing an insect from a release cylinder on a 45 cm high platform at the downwind end of the tunnel. Each insect was observed for 3 min. In this and following experiments, the females were scored for approaching (when the insect approached to within 10 cm of the artificial leaf) and landing on the target. The observation period was 3 min and each insect was flown individually. The experiment was replicated 10 times, with 5 insects in each replicate.

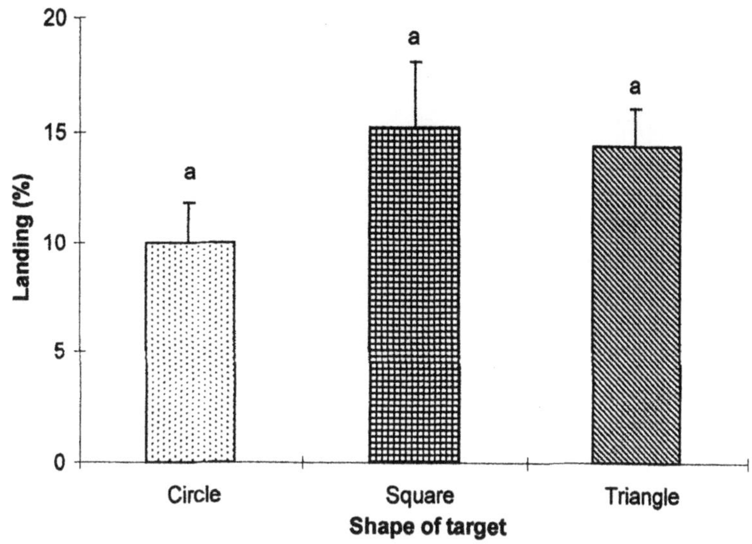


*Figure 2.* Landing (mean ± SEM) of female *M. brassicae* on targets of different shapes but same area compared in three-choice tests under red light. Data are the average of 23 replications; in each, five insects were tested individually.

*Response to plant odour.* The importance of the host odour for female orientation and landing on the host plant was investigated using cabbage extract (0.5 g leaf equivalents in 0.1 ml hexane) released from a cotton wool wick (1.5 cm diameter × 2 cm long). The cotton wool wick was fixed to a stick inserted into a rubber stopper (as above). The wick was placed in the centre of the upwind end of the tunnel and the insect response was evaluated in non-choice tests after the solvent had evaporated. The experiment was replicated 10 times, with 5 insects in each replicate.

*Response to combination of artificial leaves and plant odour.* The interaction between visual cues and host odour was studied using the artificial leaves described above and cabbage extract. In this case the plant extract was released from a cotton wool wick (as above), which was placed at the upwind top side of the artificial leaf, so that only 0.5 cm of the wick was exposed. The female's response to the target was evaluated in non-choice tests with insects observed individually. The experiment was replicated 10 times, with 5 insects in each replicate.

*Response to artificial leaves of different shape.* The possible influence of the shape of the artificial leaves on the female's response was studied using a circle, triangle and square of equal area (45 $cm^2$) as targets (made as described above) in a three-way test. A slot of 2 $cm^2$ was cut through the targets for placing the cotton wool wick, which was positioned at the same height (30 cm) on the back of the targets. This was made to avoid modifying the profile of the targets and to keep the odour stimulus at the same height in different experiments. To avoid positional bias, the target positions were changed after 5 observations. Twenty-three replications were made with ten insects in each replicate.

*Response to artificial leaves of different area.* The influence of the target size on the female response to plant odour was studied using square models, with sides 5, 10 and 15 cm made of paper painted with cadmium yellow colour in a three way test. As above, a slot of 2 $cm^2$ was made in the target for the wick, positioned at a height of 30 cm. The target positions were changed after each replicate to avoid positional bias. Seventeen replications were made with five insects in each block.

*Response to artificial leaves in different orientations.* The influence of target orientation on the insects' response to plant odour was investigated using square models (10 × 10 cm) made of paper as described above placed in a horizontal (presenting no profile to the insect) or vertical position. The plant odour was released from a cotton wool wick placed at 30 cm height. In the case of the horizontal target, a slot of 2 $cm^2$ was made through the paper for placing the wick. A similar hole was made in the vertical target and the wick inserted through it. In both cases only 0.5 cm of the wick was exposed to wind. Eleven replicates were made, 5 in-

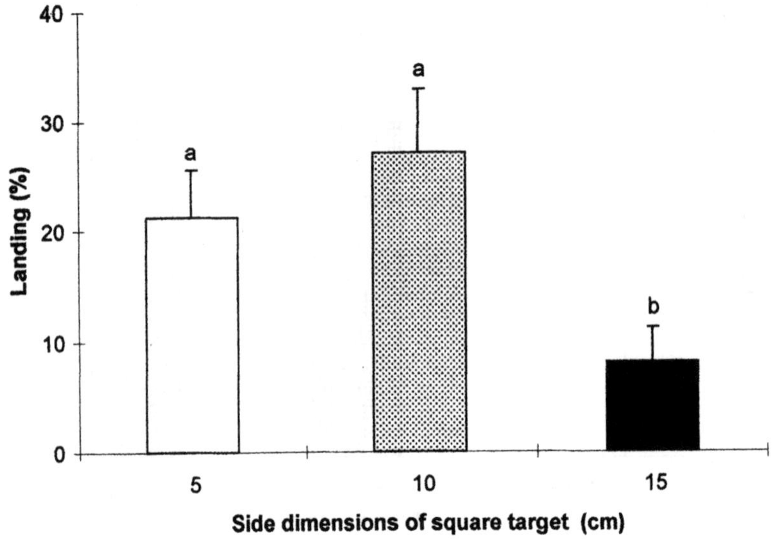


*Figure 3.* Landing (mean ± SEM) of female *M. brassicae* on square targets of different area compared in three-choice tests under red light. Data are the average of 17 replications; in each, five insects were tested individually.

sects for each replicate. As above the target positions were changed after each replicate.

*Statistical analysis.* All data were tested for homogeneity of variances (Bartlett's test, Minitab release 10.50 xtra) and normality (Anderson-Darling normality test, Minitab release 10.50 xtra). If the evidence indicated that any of these assumptions could not be supported for an analysis of variance (ANOVA) or a $t$-test, the data were transformed (Sokal & Rohlf, 1995). Once the data met these assumptions, they were subjected to $t$-tests assuming homogeneity of variances when two treatments were compared or one-way ANOVA when more than two treatments were compared. When ANOVA showed significant treatment effects, Tukey tests were used for separating means. An $\alpha$ of $P < 0.05$ was used.

## Results

*Response to artificial leaves, plant odour, and artificial leaves + plant odour.* Once the females were released, they began to fly in a random fashion, and no zigzagging flight was observed when an artificial leaf was used. Almost 20 percent of the females approached the target, but only a few females landed on it (Figure 1). The female response to plant odour alone and the artificial leaf + plant odour was quite different because the females flew in an oriented fashion and approached the targets more frequently ($F = 12.39$; df = 2, 27; $P < 0.0001$). However, females landed significantly more often on the artificial leaf + plant odour than in response to plant odour alone ($F = 16.77$; df = 2, 27; $P < 0.0001$) (Figure 1).

*Response to artificial leaves of different shape.* The results showed that the shape of the targets did not influence the female response (Figure 2) since there were no significant differences in numbers landing on circular, square or triangular targets when they were compared in three-choice tests ($F = 1.157$; df = 2, 66; $P = 0.320$).

*Response to artificial leaves of different area.* The size of the target did influence the insect response to the source: the females preferred landing on targets with sides of 5 or 10 cm rather on the largest target, with sides of 15 cm ($F = 4.405$; df = 2, 24; $P = 0.017$) (Figure 3).

*Response to artificial leaves in different orientations.* The results showed that position of the target did influence the insect response (Figure 4). The females landed significantly more often on the target positioned vertically than horizontally ($t = 5.796$; df = 20; $P < 0.001$).

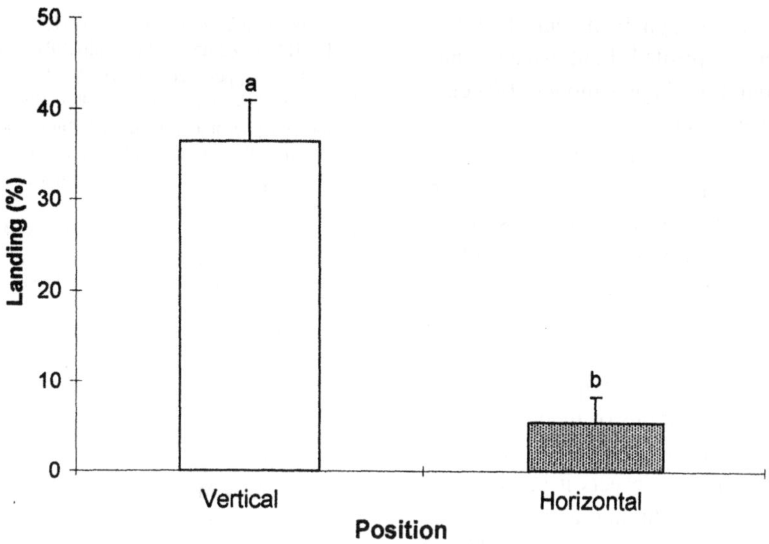


*Figure 4.* Landing (mean ± SEM) of female *M. brassicae* on targets placed vertically and horizontally compared in two-choice tests under red light. Data are the average of 11 replications; in each, five insects were tested individually.

## Discussion

Visual detection of host plants by phytophagous insects may involve the use of colour stimuli (hue, value, and saturation) or other physical cues such as size and shape (Prokopy & Owens, 1983). The present study shows that despite its nocturnal flight behaviour some visual cues may play a role in the host-finding behaviour of the cabbage moth but odour cues remain very important and there is an interaction between these cues (Figure 1). It is just possible that the greater landing in response to plant odour when offered on the artificial leaf is because the leaf offers something to land on but in the natural situation one would not occur without the other. Yamamoto et al. (1969) found that *Manduca sexta* females approached artificial plants but they did not land on them unless these artificial plants were sprayed with host-plant extracts. Similarly, after floral visits, *H. bicruris* females may approach non-fragrant white objects, but they only land in the presence of olfactory stimuli (Brantjes, 1976).

With respect to the influence of the model shape on landing behaviour, it was found that females landed equally on artificial leaves of different shapes but the same surface area (Figure 2). Städler (1974) found that oviposition of *Choristoneura fumiferana* (Clem.) was influenced by the shape of substrate. Female *C. fumiferana* clearly preferred needle-shaped models over the broad-leaf shaped ones. Foster et al. (1997), studying the ovipositional responses of the generalist moth, *Epiphyas postvittana* (Walker) found there were no significant differences between the mean number of eggs laid on model plants with rectangular, circular, or triangular leaves of the same surface area. However, no direct comparison between these results and those found in the present study can be carried out since Städler (1974) and Foster et al. (1997) evaluated oviposition and not landing. In experiments comparing the response of male *Cadra cautella* (Walker) to figures varying in shape, Levinson & Hoppe (1983) found that males flew preferentially toward vertical rectangles and exhibited no response to open conical or prismatic figures. Similar results to those found in the present study have been reported in other phytophagous insects. For example, Prokopy et al. (1983) reported that leaf pattern played only a minor role as a cue in the orientation and alighting of *D. radicum* females on host plants. Harris et al. (1993) found that female Hessian flies, *Mayetiola destructor* (Say), landed on targets with different shapes (one circular and other square) with equal frequency.

The size of the artificial leaves did influence the landing of *M. brassicae* since the females landed more often on square leaf models with sides of 5 and 10 cm than 15 cm (Figure 3). No other studies have evaluated the influence of this factor during host-finding behaviour of moths. However, studies of other phytophagous insects have shown that this factor seems to have significant effects on landing and oviposition. Female *M. destructor* landed more often on target size of 8 × 8 cm than smaller targets (2 × 2, and 4 × 4 cm)

(Harris et al., 1993). Roessingh & Städler (1990) reported that *D. radicum* oviposited about twice as many eggs in the sand around the bigger model (180 $cm^2$) than in the smaller one (45 $cm^2$).

*M. brassicae* females landed more often on vertical models than on horizontal ones (Figure 4). In other moth species it has been shown that the position of the leaf influences oviposition (Van Leerdam et al., 1984). These authors found that *E. loftini* laid more eggs on vertical rather than horizontal leaves. In another moth, *E. postvittana* there were no significant differences in total number of eggs oviposited between models with different orientations of model leaf (45° down, horizontal, or 45° up) were found (Foster et al., 1997). In other phytophagous insects it has been reported that females preferred to land and oviposit on vertical than on horizontal models (Harris & Miller, 1984; Roessingh & Städler, 1990; Harris et al., 1993). For example, Hessian fly females landed 1.7 times more frequently on targets with vertical rather than horizontal bars (Harris et al., 1993).

Several studies have shown that vertical cylinders or targets influence the flight or landing of male moths, of some species, on sex-pheromone sources (Foster & Ayers, 1995, and references therein). However, how the object affects the behaviour is difficult to identify, because of the confounding problem of the visual properties of the object and its effect on the odour plume (Foster & Ayers, 1995; Wyatt et al., 1993, 1997).

## Acknowledgements

We thank Dr Martin C. Birch for encouragement during the realisation of this study. Dr Lester J. Wadhams of IACR-Rothamsted advised on microwave extraction. JCR was supported by CONACYT (Mexico) and ORS (UK) during the course of this study. Additional financial support was provided by Lady Margaret Hall (Oxford).

## References

Aluja, M. & Prokopy, R. J., 1993. Host odor and visual stimulus interaction during intratree host finding behavior of *Rhagoletis pomonella* flies. Journal of Chemical Ecology 19: 2671–2696.

Brantjes, N. B. M., 1976. Riddles around the pollination of *Melandrium album* (Mill.) Grarke (Caryophyllaceae) during the oviposition by *Hadena bicruris* Hufn. (Noctuidae, Lepidoptera). I + II Proceedings of the Koninklijke Nederlandse Akademie van Wetenschappen Series C. 79: 1–12, 127–141.

Foster, S. P. & Ayers, R. H., 1995. Effect of a vertical object on the sex pheromone-mediated flight and landing behaviours of the male lightbrown apple moth, *Epiphyas postvittana* (Walker) (Lepidoptera: Tortricidae). Journal of Insect Behavior 8: 465–483.

Foster, S. P., Howard, A. J. & Harris, M. O., 1997. The influence of tactile and other non-chemical factors on the ovipositional responses of the generalist herbivore *Epiphyas postvittana*. Entomologia Experimentalis et Applicata 83: 147–159.

Ganzler, K., Salgo, A. & Valko, K., 1986. Microwave extraction. A novel sample preparation method for chromatography. Journal of Chromatography 371: 299–306.

Harris, M. O. & Miller, J. R., 1984. Foliar form influences oviposition behavior of the onion fly. Physiological Entomology 9: 145–155.

Harris, M. O. & Miller, J. R., 1988. Host-acceptance behaviour in an herbivorous fly, *Delia antiqua*. Journal of Insect Physiology 34: 179–190.

Harris, M. O. & Rose, S., 1990. Chemical, color, and tactile cues influencing oviposition behavior of the Hessian fly (Diptera: Cecidomyiidae). Environmental Entomology 19: 303–308.

Harris, M. O., Rose, S. & Malsch, P., 1993. The role of vision in the host plant-finding behaviour of the Hessian fly. Physiological Entomology 18: 31–42.

Levinson, H. Z. & Hoppe, T., 1983. Preferential flight of *Plodia interpunctella* and *Cadra cautella* toward figures of definitive shape and position with notes on the interaction between optical and pheromone stimuli. Journal of Applied Entomology 96: 491–500.

Ockenfels, P. & Schmidt, F., 1992. Olfactory and visual cues in host-finding in the Burnet moth, *Zygaena trifolii*, In: S. B. J. Menken, J. H. Visser & P. Harrewijn (eds), Proceedings of the 8th International Symposium Insect-Plant Relationships. Kluwer Academic Publications, Dordrecht, pp. 125–126.

Poitout, S., Bues, R. & Le Rumeur, C., 1972. Elevage sur milieu artificiel simple de deux noctuelles parasites du coton *Earias insulana* et *Spodoptera littoralis*. Entomologia Experimentalis et Applicata 15: 341–350.

Prokopy, R. J., Collier, R. H. & Finch, S., 1983. Visual detection of host plant by cabbage root flies. Entomologia Experimentalis et Applicata 34: 85–89.

Prokopy, R. J. & Owens, E. D., 1983. Visual detection of plants by herbivorous insects. Annual Review of Entomology 28: 337–364.

Roessingh, P. & Städler, E., 1990. Foliar form, colour, and surface characteristics influence oviposition behaviour in the cabbage root fly *Delia radicum*. Entomologia Experimentalis et Applicata 57: 93–100.

Sokal, R. R. & Rohlf, F. J., 1995. Biometry: the principles and practice of statistics in biological research. 3d Edition. W. H. Freeman & Company, New York.

Städler, E., 1974. Host plant stimuli affecting oviposition behavior of the Eastern spruce budworm. Entomologia Experimentalis et Applicata 17: 176–188.

Todd, J. L., Phelan, P. L. & Nault, L. R., 1990. Interaction between visual and olfactory stimuli during host-finding by leafhopper, *Dalbulus maidis* (Homoptera: Cicadellidae). Journal of Chemical Ecology 16: 2121–2133.

Van Leerdam, M. B., Johnson, K. J. R. & Smith, J. W. 1984. Effects of substrate physical characteristic and orientation on oviposition by *Eoreuma loftini* (Lepidoptera: Pyralidae). Environmental Entomology 13: 800–802.

Wyatt, T. D., Phillips, A. D. J. & Gregoire, J. C. 1983. Turbulence, trees and semiochemicals – wind-tunnel orientation of the predator, *Rhizophagous grandis*, to its barkbeetle prey, *Dendroctonus micas*. Physiological Entomology 18: 204–210.

Wyatt, T. D., Vastiau, K. & Birch, M. C., 1997. Orientation of male *Anobium punctatum* (Coleoptera: Anobiidae): use of visual and turbulence cues to its own pheromone. Physiological Entomology 22: 191–196.

Yamamoto, R. T., Jenkins, R. J. & McClusky, R. K., 1969. Factors determining the selection of plants for oviposition by the tobacco hornworm *Manduca sexta*. Entomologia Experimentalis et Applicata 12: 504–508.

*Entomologia Experimentalis et Applicata* **91:** 67–82, 1999.



# Integrating nutrition: a geometrical approach

D. Raubenheimer & S. J. Simpson
*Department of Zoology, University of Oxford, South Parks Road, Oxford, OX1 3PS, UK*

Accepted: October 18, 1998

*Key words:* nutritional theory, nutrient balancing, locusts, protein intake, carbohydrate intake, salt intake


## Abstract

We present and illustrate using data from insects an integrative approach to modelling animal nutrition. This framework enables the unification within simple geometrical models of several nutritionally relevant measures. These include: the optimal balance and amounts of nutrients required to be ingested and allocated to growth by an animal over a given time period (the intake and growth targets, respectively); the animal's current state in relation to these requirements; available foods and the consequences for the animal's state of ingesting them; the amounts of ingested nutrients that are retained and eliminated; and animal performance. Data are presented on intake targets in insects, illustrating how they change over various time-scales (physiological, developmental, and evolutionary). Most importantly, the geometrical approach enables a clear description to be made of the trade-offs reached by animals in regulating their nutritional balance. Animals given a nutritionally balanced food, or two or more imbalanced but complementary foods, can satisfy their nutrient requirements. However, those eating non-complementary imbalanced foods must reach a suitable compromise between over-ingesting some nutrients and under-ingesting others. Data are presented comparing the rules of compromise for protein and carbohydrate ingestion in grass-feeding (*Locusta migratoria*) and polyphagous (*Schistocerca gregaria*) locusts.


## Introduction

A recurring theme in the study of animal behaviour has been integration. This has developed largely on two fronts. First, there have been repeated exhortations that an aim of ethologists should be to work towards an understanding of behaviour that integrates across different *levels of analysis*. Following Tinbergen (1951, 1963) four levels, or some variants thereof, are usually recognised: the mechanisms, the development, the evolutionary history and the functions of behaviour. See Dawkins (1989) and Dewsbury (1992) for further discussion of the levels of analysis in behavioural studies.

A second field of study where integration has been a primary aim is in attempts to understand behaviour as the outcome of a multiplicity of interacting causes. This is, in other words, an attempt to understand how animals integrate causal factors to produce behaviour that is appropriate to a given circumstance. For instance, a foraging animal detecting some signs of potential predators must decide whether to flee or to feed; the resulting behaviour is contingent on the integration of the external (e.g., strength of predator and food cues) and internal (nutritional state, history of interactions with predators, etc.) factors. The major approaches to studying such *organismal integration* have been the use of control theory to model the interactions among physiology, behaviour and the animal's external environment (e.g., Calow, 1976; Toates, 1986), and the development of state space models (McFarland & Houston, 1981).

Among the categories of animal behaviour, nutrition poses particular challenges to workers wishing to achieve an integrative understanding of their subject. The above issues of the levels of analysis and the organismal integration of nutritional decisions with other behavioural systems are of course relevant. In addition, the process of nutritional regulation is itself a complex challenge of integration for the organism. Firstly, animals usually require several nutrients simultaneously, at individual levels that may change with recent nutritional history, developmental stage, levels of activity, etc. The various requirements are met by eating foods which might seldom contain the nutrients at the required levels – and which are, furthermore, frequently (e.g., noxious allelochemicals deleterious synthesized by plant tissue). Secondly, in its passage from the environment to the anus, there are various levels at which a food is processed to minimise the discrepancy between its chemical composition and the nutritional requirements of the animal (e.g., habitat selection, food selection, food switching, and the regulation of absorption, assimilation and excretion). There are several impressive examples of how animals integrate the various stages of nutritional processes, those closest to home concerning locusts (e.g., Simpson et al., 1995; Simpson & Raubenheimer, 1993a, b, 1996; Raubenheimer & Simpson, 1993, 1996, 1997, 1998).

Over the past few years we have developed and applied to the study of insects and vertebrates a powerful framework for integrative research into nutritional processes (Raubenheimer & Simpson, 1993, 1994, 1997; Simpson & Raubenheimer, 1993a, 1997; Simpson et al., 1995). The framework is centred on simple geometrical models that enable the consideration in a single scheme of all of Tinbergen's levels of analysis, as well as various categories of organismal integration, including the issues of multiple nutrients and the serial stages of nutritional processing. Here we provide an overview of our approach, developing the key concepts and presenting data illustrating their application to insect nutrition.

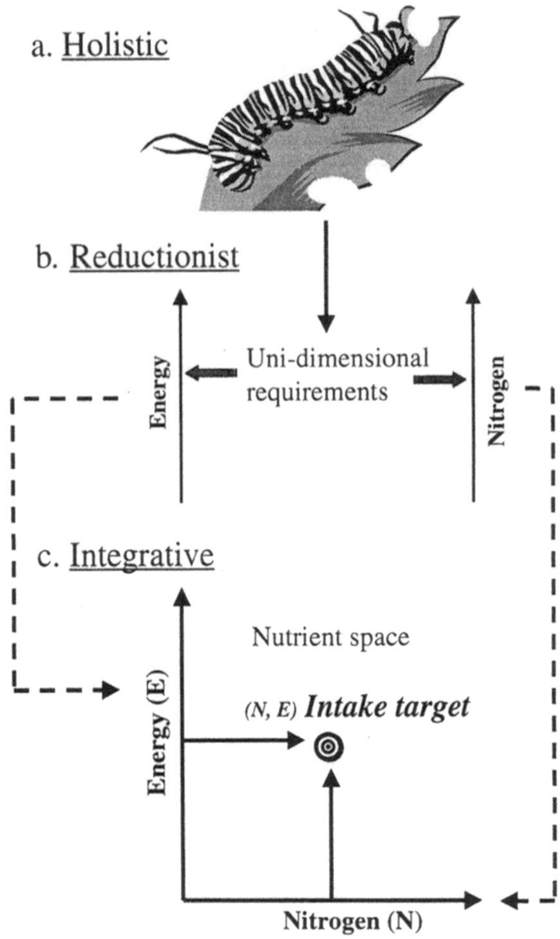


*Figure 1.* Three approaches to characterising nutritional requirements.

## Intake requirements

*The need for an integrative approach.* A central reference point to our explorations of nutritional processes is the ingestive requirements of the animal. Hitherto, there have been three major approaches to characterising these requirements. The food-centred approach considers food types the relevant currency of animal feeding behaviour (Figure 1a). While this approach is important in understanding ecological processes, on its own it has severe limitations in organism-centred studies: foods are merely the vehicles to animals for the nutrients they contain. Therefore, the primary target of natural selection is at the level of nutrients, and any patterns concerning food intake are most readily explicable at this level.

Nutrient-centred approaches have developed along two lines (Figure 1b). The one, most explicitly articulated in the optimal foraging paradigm (Stephens & Krebs, 1986), considers energy (i.e., energy-yielding nutrients) the primary currency. An elaboration of this approach has been to consider non-energetic nutrients, only in so far as they impose constraints on energy intake (Pulliam, 1975; Belovsky, 1990). The other, most prevalent in the literature on the nutritional ecology of insects, considers nitrogen as primary (McNeill & Southwood, 1978; Mattson, 1980; White, 1993). While both approaches have yielded impressive insights, they are limited in generality since a single nutrient will constitute a strong predictor of behaviour only under a restricted range of circumstances. For instance, animals undergoing somatic or reproductive growth might prioritise the intake of nitrogen, and those feeding predominately to fuel energy metabolism would weight more strongly energy intake. For those feeding on nutritionally balanced foods, energy and nitrogen can be considered a single resource and therefore either or both will predict behaviour equally (see below).

A major limitation of the single-nutrient approach is therefore that it deals only peripherally, or not at all, with the central nutritional question of how animals integrate the intake of various nutrients. There are, however, at least two *a-priori* reasons why nutritional integration should be considered a central issue in attempts to understand animal nutrition. Firstly, even though a single nutrient might well be the most limiting at a given time, in most instances animals simultaneously require other nutrients. Growing animals, for example, require energy to fuel metabolic requirements, while non-growing ones require a source of nitrogen for enzyme synthesis, tissue repair and so forth. Secondly, the single nutrient approach places undue emphasis on the deficient nutrient being the 'limiting' one. However, since animals (and particularly herbivores) seldom feed on foods that are perfectly balanced nutritionally, they frequently are forced to ingest excesses of other nutrients in order to limit the shortfall of the deficient one (Raubenheimer, 1992; Simpson & Raubenheimer, 1993a; Raubenheimer & Simpson, 1993). The ability to ingest surplus quantities of excessive nutrients might thus be an important limitation on animals – indeed, our research suggests that for some insects and vertebrates this might be a primary limiting factor in nutritional regulation (Raubenheimer & Simpson, 1997). An adequate explanation of nutritional behaviour should thus incorporate not only the deficient nutrient, but also the excessive ones.

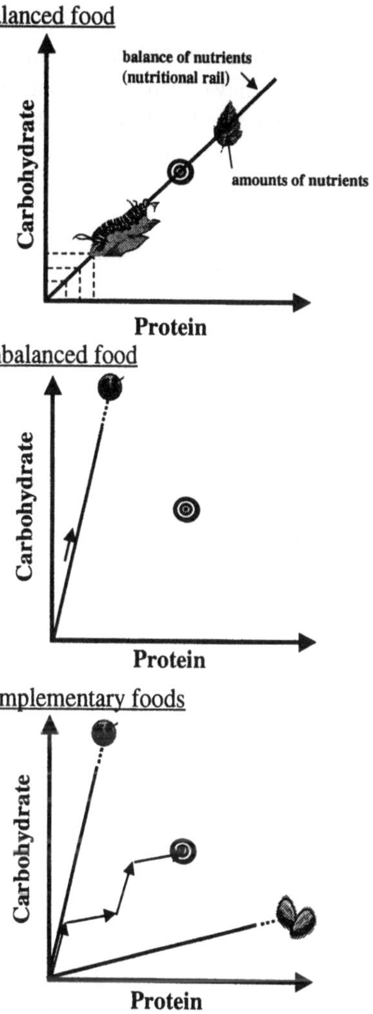


*Figure 2.* The amounts of nutrients in a food item can be represented as a point in nutrient space (e.g., the leaf in a.), while the balance of nutrients is described by a linear trajectory (nutritional rail) that project into a nutrient space at an angle describing the ratio of nutrients. If this ratio is the same as is required by an animal (i.e., it is a nutritionally balanced food), the animal's intake target can be reached by feeding on the food (a). It cannot, by contrast, be reached by feeding on an imbalanced food (b). It can, however, be reached by switching between two complementary imbalanced foods (c). In b. and c. (and subsequent figures) the foods are described only by their balance of nutrients (nutritional rails), since the amounts of nutrients in each are assumed to exceed the scale of the axes.

*The geometrical approach.* Our approach has therefore been to combine the axes representing various nutrients into one (or more – see below) two-dimensional *nutrient spaces* (Figure 1c). In this nutrient space appears a point representing the individual amounts of the various nutrients (and hence their balance) that should optimally be ingested by the animal over a given time, the *intake target*. Foods too can be depicted in the nutrient space, and so can the relationship between a given food item and the animal's current requirements. For example, the leaf in Figure 2a is defined in the scheme by the amount it contains of the two nutrients, both of which are present in quantities that exceed the respective target co-ordinates. Inherent in this description is a further, critical, parameter describing the leaf, the balance of the two nutrients it contains. This is depicted as a linear trajectory from the origin to the point describing the amounts of the nutrients in the leaf. Such trajectories representing foods have been called *nutritional rails*, to capture the fact that as the animal eats this food its nutritional state changes along a path coincident with this trajectory (Raubenheimer & Simpson, 1993). In this case, the nutritional rail passes through the intake target, and the animal's nutritional requirements can therefore be satisfied by eating the food. This leaf is, in other words, an adequate food in two respects: it contains sufficient quantities of both nutrients to satisfy the animal's requirements, and the nutrients are balanced with respect to these requirements.

The rail in Figure 2b, by contrast, describes a nutritionally imbalanced food, containing carbohydrates in excessive concentration relative to proteins. This rail does not pass through the intake target, and an animal with access only to this food is thus unable to satisfy its optimal nutrient requirements. The options open to an animal in this predicament are discussed in some detail below.

The animal could, however, reach the intake target if in addition to this food it had available a second nutritionally imbalanced food, provided the rail describing the second food lay on the opposite side of the intake target to the first (i.e., containing an excess of protein relative to carbohydrate). As shown in Figure 2c, this could be done by switching between the foods, and thus moving alternately along a trajectory parallel with the one then the other of the rails. Combinations of nutritionally imbalanced foods that allow an animal to 'zigzag' its way through nutrient space in this manner to the intake target are termed *complementary* foods.

*Experimentally locating the intake target.* Since our approach is primarily an empirical one, it is necessary to give due consideration to obtaining data on the position of intake targets [see Simpson & Raubenheimer (1995) for further discussion on such methodological issues]. An important step in estimating this position is to measure the free-running output of the regulatory mechanisms for the nutrients in consideration. The assumption here is that nutritional regulatory mechanisms have been configured by natural selection to ensure that when the environment permits they provision the animal with the optimal amounts and balance of nutrients. Experimentally, this approach requires that the animal has available complementary foods whose nutritional rails jointly span a sufficiently large area in nutrient space to encompass the intake target, as shown in Figure 2c. The point actually selected by the animal should, by this reasoning, provide a good estimate of the position of the target.

An important consideration in planning such studies, is the nature of the foods to be used. There exists, potentially, a trade-off between the ecological realism achieved using natural foods, and the analytical power available using synthetic foods. While much of our work to date has used chemically defined synthetic foods [see Simpson & Abisgold (1985) and Raubenheimer & Simpson (1993) for details of these foods], recent work in the field (Wright, 1998; G. A. Wright, S. J. Simpson & D. Raubenheimer, unpubl.) and using genetically modified *Arabidopsis* mutants (Wright, 1998; G. A. Wright, D. Raubenheimer & S. J. Simpson, unpubl.) illustrate the power of the geometrical approach when applied to more complex systems. For clarity, the illustrative data presented in this paper all concern synthetic foods, in which the number of relevant nutritional dimensions can readily be experimentally controlled. In the final part of this paper we discuss some important considerations in expanding the models to deal with a greater number of nutritional dimensions.

Whatever the number of dimensions, there are some caveats to heed in using the above approach to identifying the position of intake targets. Firstly, the foods used should not be so extreme in their imbalance that they are not recognised by the animal as food (*pathological foods*), but at the same time should be sufficiently divergent to provide a high probability that they encompass the target position. Fortunately, first estimates of the positions of the intake target rail are available for a wide range of insect species (Simpson & Raubenheimer, 1993a; see also below), and this should enable the selection of moderately divergent experimental foods in most cases. Furthermore, a good indicator that inappropriate rails have been used is obtained when an animal takes a substantial majority of its intake from one of the foods. Most likely, in this case the target lies outside of the accessible area of nutrient space, on the side of the favoured food (Chambers et al., 1995, 1997).

A second caveat is that in order to have some degree of confidence that a selected point in nutrient space does represent the intake target, it is necessary to demonstrate that this is the outcome of nutritional regulation rather than the result of some process independent of the nutrients in consideration (e.g., random feeding on the two foods). This is achieved by including in the experiment several food pairings, such that the animals in the different treatment groups would have to alter the proportions of the two foods ingested to achieve the same point in nutrient space. Such *target defence* was demonstrated for protein and carbohydrate regulation in *Locusta migratoria* by using treatment pairings in which one of the foods, both foods or neither of the foods were diluted with in-

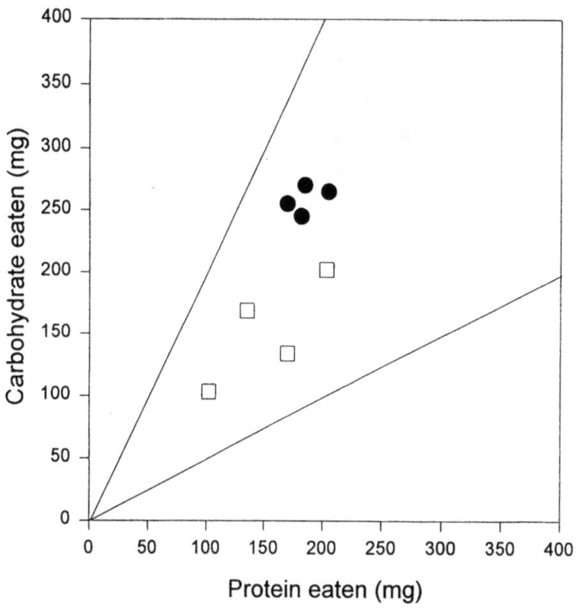


*Figure 3.* Selected intake points of *L. migratoria* over the first 6 days of the 5th stadium. Circles represent observed two-dimensional means of four treatments ($n = 10$) given food pairings (%protein:%carbohydrate) 28 : 14 + 14 : 28; 28 : 14 + 7 : 14; 14 : 7 + 14 : 28; 14 : 7 + 7 : 14. Squares represent hypothetical outcome in the absence of target defence.

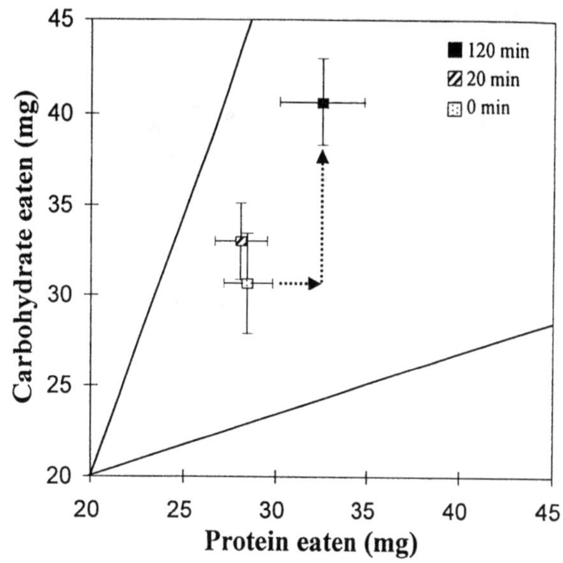


*Figure 4.* Nutrient selection by adult female *L. migratoria* given foods 14:28 and 28:14 (% protein:% digestible carbohydrate) for 24 h after flying for varying periods.

digestible cellulose bulk (Figure 3) Chambers et al., 1995. Where it is not feasible to use diluents, an alternative approach would be to use one more- and one less-extreme rail on either side of the suspected target position. Treatments would then consist of complementary pairings either of the two more-extreme rails, the two moderate rails or one of each (e.g., see Trumper & Simpson, 1993).

Thirdly, it is useful to ensure that the foods used in each treatment pairing are not too dissimilar in respects other than the nutrients of interest. Otherwise, there is a danger that the proportion of the two foods ingested (hence the point achieved in nutrient space) represents regulation not only for the nutrients in question, but also for other constituents. This is a further context in which experiments to test defence of the target, as described above, are useful. If animals regulate to the same point in, for example, a protein-carbohydrate nutrient space despite the fact that the relative levels of other nutrients have been altered, then this suggests very strongly that there are no hidden nutritional dimensions influencing the results.

Finally, an important source of subsidiary information on the position of intake targets comes from performance measures. Intake targets are functional optima, so a situation where animals show poor performance across a number of fitness measures at the selected point of intake (relative to groups experimentally confined to foods that restrict them to elsewhere in nutrient space) suggests that confounding factors are operative. For example, the food pairings might not encompass the target rail, pathological foods might be involved, and/or there might be important dimensions that are omitted from the model. On the other hand, performance at the intake target need not be optimal by all measures, since animals might be naturally selected to occupy positions in nutrient space that represent a favourable balance among conflicting nutrient requirements and ecological pressures. An example of this is presented below.

## Intake targets are dynamic

While intake targets provide a central reference point in our geometrical models of nutrition, it is important to emphasise that this is a dynamic, rather than static reference point. There are, in fact, several levels at which targets move, including across physiological, developmental, and evolutionary time-scales.

The optimal nutrient requirements at a given time depend on the physiological demands that are placed on the animal, and the regulation of nutrient intake should reflect this. Figure 4 shows, for example, the selected intake points by adult *Locusta migratoria*

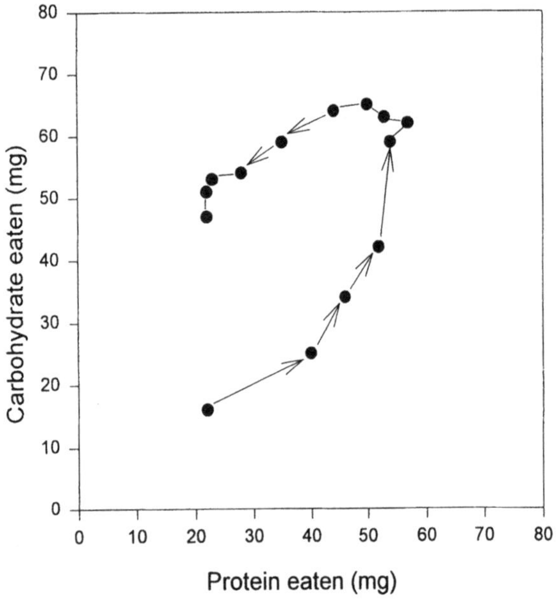


*Figure 5.* Changes in daily intake of protein and carbohydrate in adult *L. migratoria* allowed to select between foods from the time of moulting until the end of the somatic growth phase.

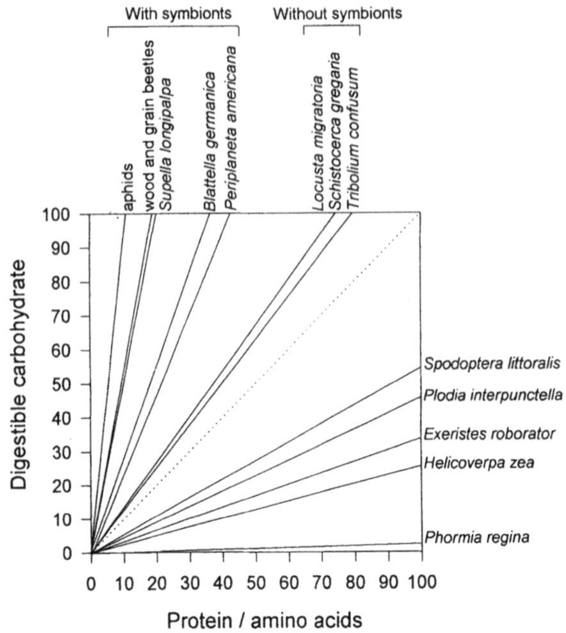


*Figure 6.* Comparison of the optimal dietary protein:carbohydrate ratio of various insect groups. Insects with the lowest protein:carbohydrate ratio are those with endosymbiotic bacteria that contribute to nitrogen metabolism.

flown for 20 min, 120 min and unflown controls (D. Raubenheimer & L. Firth, unpubl.). Although there was no appreciable difference in the selected point for control insects and those flown for 20 min, those flown for 120 min selected a substantially different point. Most interesting is that the movement of the target following 120 min flight was significant along the carbohydrate axis, but not the protein axis. Presumably, this reflects the fact the energy-providing role of carbohydrates in these insects (Jutsum & Goldworthy, 1976).

On a longer time-scale, intake targets move with development, reflecting the different nutrient requirements of various stages of growth, reproduction and senescence. Figure 5 for example shows the changes in daily intake of protein and carbohydrate in adult female *Locusta migratoria* allowed to switch between foods from the time of moulting until the end of the somatic growth phase (data from Chyb & Simpson, 1990). Total intake increased for the first four days following the moult, but the ratio of protein to carbohydrate remained fairly constant. On the fifth day, the amount of protein ingested began to decrease relative to carbohydrate. On subsequent days this relative decrease continued, and the total amounts of each nutrient eaten also declined, as is evident from the backwards loop of the target trajectory. This relative decrease in protein intake, and the absolute decrease in the levels ingested of both nutrients, coincided with the decline of tissue growth towards the end of the somatic growth phase.

Intake targets also change in evolutionary time, as animals are selected to utilise different sources of nutriment and evolve differing life history strategies. Figure 6 shows, for example, data from a comparative analysis of the optimal protein:carbohydrate ratio of 117 insect species (Simpson & Raubenheimer, 1993a). This study revealed that insects with the steepest target rail (lowest P:C ratio) were those with endosymbiotic bacteria that contribute to nitrogen metabolism.

Finally, there are cases where different species have a similar intake target, but they have evolved to reach this in very different ways. A hypothetical example is presented for a nutrient generalist and a specialist feeder in Figure 7. Below we consider a real example, in the comparison of *Schistocerca gregaria*, a polyphagous grasshopper, with *Locusta migratoria*, a grass-feeding specialist.

We stress that the example in Figure 7 describes a *nutrient generalist*, being an animal whose diet encompasses a wide area of nutrient space, as opposed to *food* generalists, which feed on a wide range of food types. This is a potentially important distinction, since food generalists sometimes might not also be nutrient

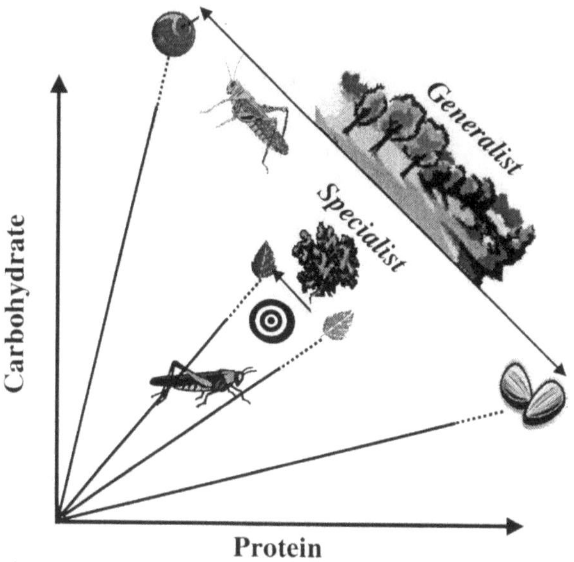


*Figure 7.* Hypothetical example where the intake target remains unchanged over evolutionary time, but the means of reaching it diverges between species. A nutrient generalist species might use a substantially different nutritional landscape to achieve the same nutrient intake as a specialist.

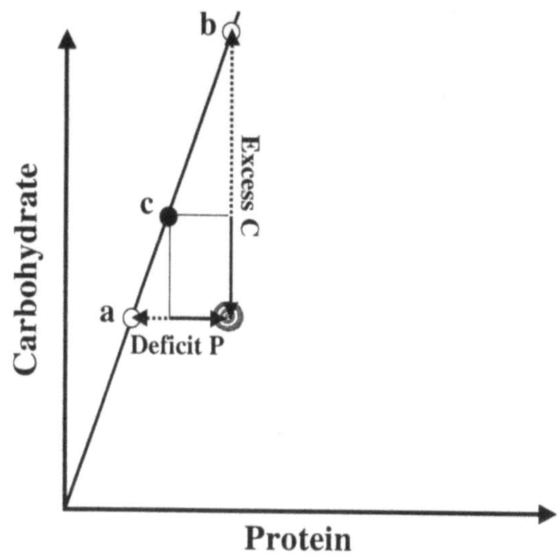


*Figure 8.* Nutritional compromises when feeding on imbalanced foods. An animal that feeds to point a satisfies its requirements for carbohydrate, but suffers a deficit of protein. By feeding to point b it ingests sufficient protein but an excess of carbohydrate, while at point c it experiences both an excess of carbohydrate and a deficit of protein, neither of which are as large as at the extremes.

generalists. On the contrary, one reason for including a large number of food types in a diet might be to avoid any changes (e.g., seasonal or developmental) in the composition of the host(s), thus regulating nutrient intake within a limited region of nutrient space. Food specialists, by contrast, might be nutrient generalists to the extent that changes in the composition of their hosts leave them little option but to deviate from the target across wide areas of nutrient space.

## Nutritional conflicts: when the intake target cannot be reached

*Rules of compromise.* Above, we have seen that animals are capable of achieving the intake target when they are fed nutritionally balanced foods, or are able to switch among nutritionally imbalanced, but complementary, foods (Figure 2a and c). However, animals might frequently have available neither balanced nor complementary foods, and thus be forced to ingest foods that are to some extent imbalanced. In this predicament, the animal confronts a situation of conflict among its requirements for different nutrients (Raubenheimer, 1992). As is illustrated in Figure 8a, it can satisfy its requirements for carbohydrate by feeding to point a, but in so doing will suffer a deficit of protein. At the other extreme, it could satisfy its requirement for protein (point b), but only if it ingests an excess of carbohydrate. Alternatively, it could feed to an intermediate point (e.g., point c) at which it suffers both an excess of one nutrient and a deficit of the other, but neither of which are as large as the extremes. The regulated point of intake in such circumstances – the *rule of compromise* – should reflect the relative costs and benefits of over-ingesting one of the nutrients and under-ingesting the other (Raubenheimer & Simpson, 1993, 1997; Simpson & Raubenheimer, 1993a). Therefore, like intake targets, rules of compromise are an important nutritional datum, which might differ both among species and within species at different growth stages and for different combinations of nutrients.

But how to measure such rules? Having determined the position of the intake target, the rule of compromise can simply be determined by confining an animal to a single nutritionally imbalanced food and recording its intake of the relevant nutrients, as depicted in Figure 9a. This, however, provides a very limited picture, as there is no guarantee that the rule will not differ for foods that are imbalanced to a greater or lesser extent, or indeed for those that are imbalanced in the opposite direction (excess P relative to C). A fuller picture can be obtained by confining insects to a range of foods, and measuring their intake across the relevant period (Figure 9b). The array

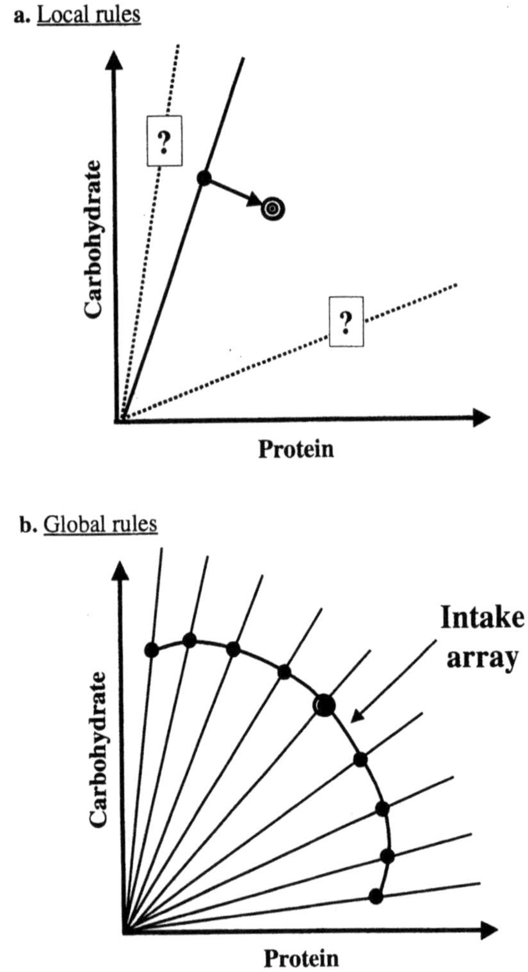


*Figure 9.* A single intake point provides information about the rule of compromise local to the nutritional rail (a), but says nothing about the rule on other foods. The intake array resulting from an experiment including rails representing a large area of nutrient space characterises the global rule of compromise (b).

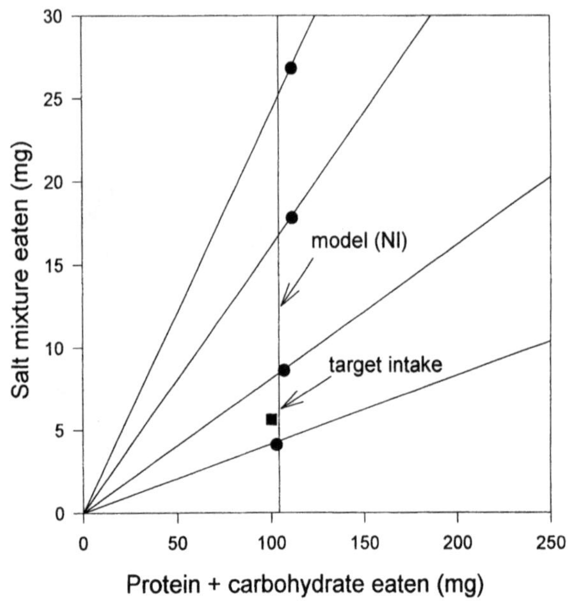


*Figure 10.* Intake array for 5th stadium *L. migratoria* made to trade off a mineral salt mixture against a compound axis of protein and digestible carbohydrate. The vertical line represents the prediction for the 'no interaction' rule of compromise.

of intake points resulting from such an experiment provides a comprehensive description of the rule of compromise for the nutrients under consideration.

*Interpreting intake arrays.* Among the intake arrays recorded in this manner for insects are, firstly, the response of *Locusta migratoria* to mineral salts when forced into conflict with a balanced complement of protein and carbohydrate [Figure 10; data from Trumper & Simpson (1993)]. Secondly, the intake arrays for protein and carbohydrate have been measured for the generalist locust *Schistocerca gregaria* (Figure 11a), and the grass feeder *Locusta migratoria* (Figure 11b) (Raubenheimer & Simpson, in prep.). Superimposed on the data are computer-generated models of the single rule that best fits the data in each case.

Intake arrays thus provide a visual representation of the patterns of nutrient intake when animals trade-off the over- and under-ingestion of nutritionally imbalanced foods. But how to interpret what the various rules mean? To illustrate, we will use the model data superimposed on the three intake arrays observed in locusts (Figures 10 and 11). For clarity we assume in the models a symmetrical intake target, but the interpretation would apply equally for an asymmetrical target. The model data are plotted in Figure 12.

Since intake arrays represent the rules adopted in trading off nutritional errors (excesses and deficits), they can best be interpreted by examining these errors, as illustrated in Figure 13. Plotted on the *y*-axis in these summary plots is the nutritional error (distance from the *X* and *Y* co-ordinates of the intake target) experienced by the animals on each nutritional rail (*X*-axis). The horizontal line at zero error depicts the target position (no error for either nutrient), while a negative error represents a deficit and a positive error an excess of a nutrient.

It can readily be seen in these terms (Figure 13a) what the array in Figure 12a (and Figure 10) means. These animals have regulated intake so as to satisfy without error their requirement for the combined com-

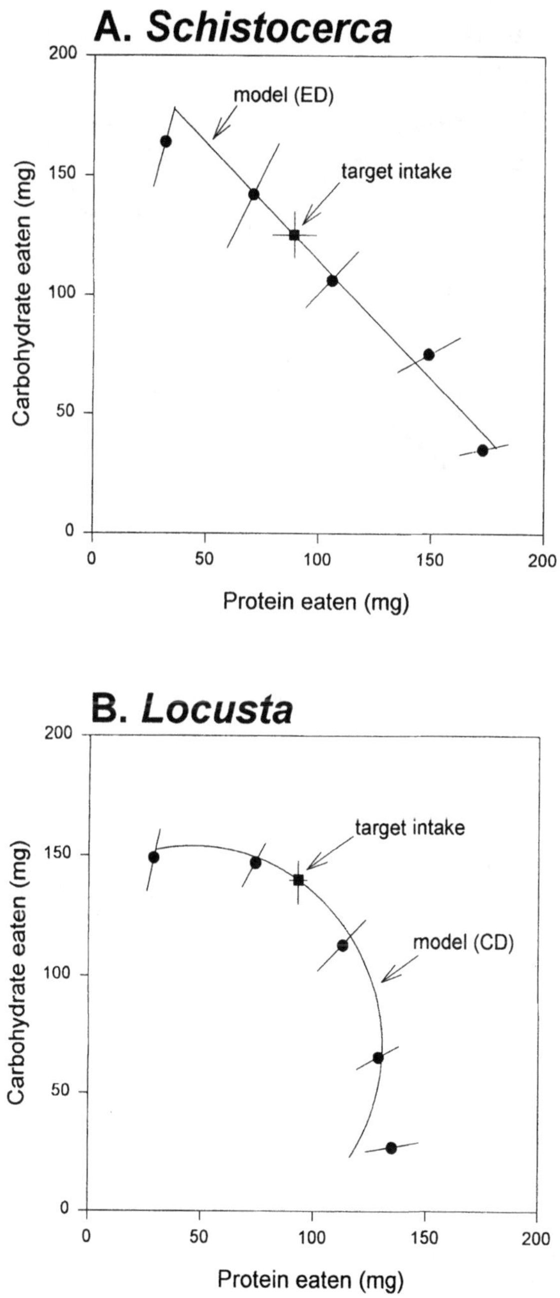


*Figure 11.* Protein-carbohydrate intake arrays for (A) the generalist feeder *S. gregaria* and (B) the grass feeder *L. migratoria* over the first 3 days of the 5th stadium. Filled circles represent means ± SE of 10 insects fed one of 5 foods varying in digestible carbohydrate and protein. The solid squares represents the selected intake points of locusts allowed to switch between foods 14:28 and 28:14 (%protein:%carbohydrate). The solid lines represent the prediction for the equal distance (A) and the closest distance (B) rules of compromise.

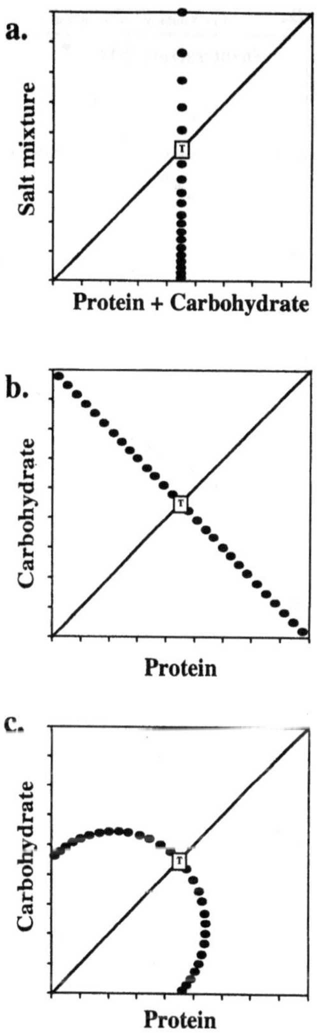


*Figure 12.* Model data for the no interaction (a), equal distance (b) and closest distance (c) rules of compromise.

plement of protein and carbohydrate, irrespective of the excess or shortfall of mineral salts ingested. Therefore, from the perspective of regulatory mechanisms, there has been no interaction between the two groups of nutrients, and this rule is accordingly termed the *no interaction rule* (NI).

Analysis of the equivalent plot for the intake array in Figure 12b (and Figure 11a) reveals that the rule of compromise displayed by these animals is to regulate intake such that the deficit incurred in one nutrient exactly matches the excess incurred in the other. In other words, these animals regulate to a point in nutrient space which is equidistant from the target in the two dimensions, and this rule is thus termed the *equal distance rule* (ED).

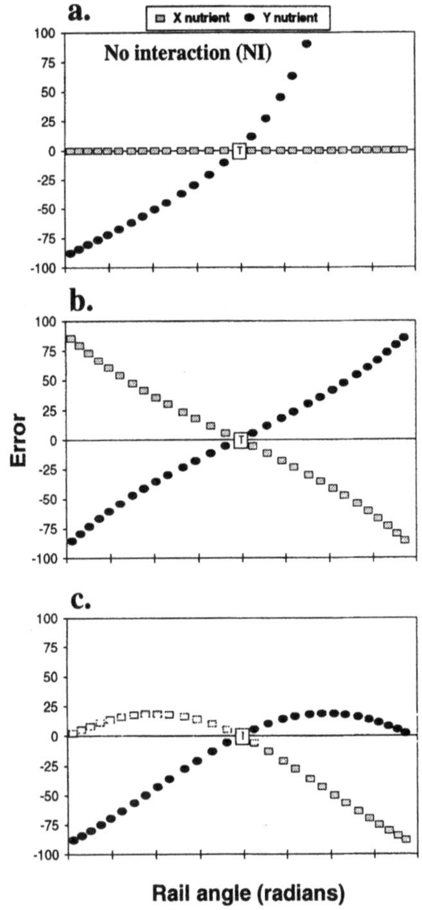


*Figure 13.* Plot of nutritional errors for three rules of compromise. A positive error on the *y*-axis represents the excess and a negative error the deficit of nutrients ingested by animals on each nutritional rail (*x*-axis). It is only animals on the target food (boxed *T*) that have no error.

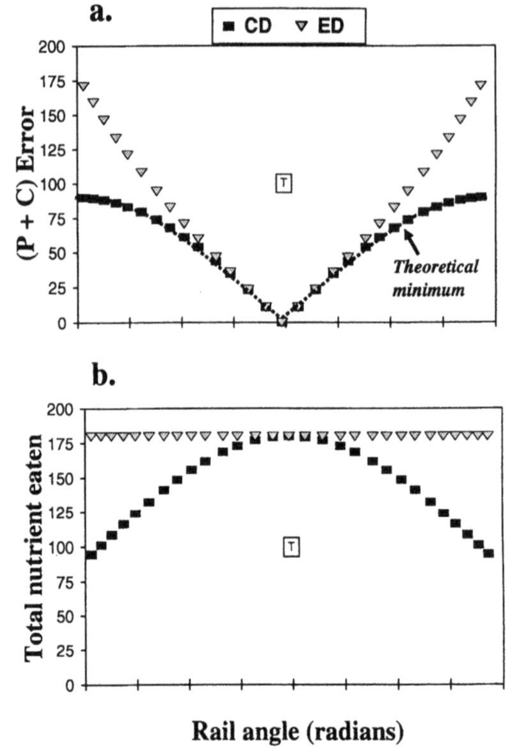


*Figure 14.* The sum of nutritional errors (a) and the total nutrients ingested (b) by hypothetical animals adopting the closest distance (CD) and equal distance (ED) rules of compromise. The dotted line in a represents the theoretical minimum possible for this variable.

Considering the error plot for the rule depicted in Figure 12c (and Figure 11b), a first striking point to note is that these animals are prepared to suffer a relatively large shortfall of the deficient nutrient (whether it be protein or carbohydrate) in order to avoid ingesting a relatively small surplus of the excessive one (Figure 13c). This result emphasises very strongly the importance of an integrative approach to nutrition, in contrast with the uni-dimensional approaches which consider the deficient nutrient to be the one that is primarily limiting (see above).

Beyond this, it is not easy to interpret in these terms what the pattern in Figure 13c means. To assist, a further summary plot is presented, which shows the relationship between rail angle and the sum of the absolute value of the two errors (protein error + carbohydrate error) incurred by animals fed each food (Figure 14). This is, in other words, a plot of rail vs. the two-dimensional distance from the target achieved by animals fed the various foods. Figure 14a shows that this value is considerably lower for the rule in Figure 12c than for the equal distance rule. In fact, the former represents the minimum possible two-dimensional distance that can be achieved from the intake target. Therefore, animals following this rule feed to the point in nutrient space which, for a given food rail, represents the closest possible geometrical (in this case two-dimensional) distance from the intake target. Accordingly, we have termed this rule the *closest distance* (CD) rule (Raubenheimer & Simpson 1993).

This analysis provides an interesting basis for comparing the nutritional strategies in protein and carbohydrate regulation of the generalist locust *S. gregaria* (Figure 11a) with that of the grass specialist, *L. migratoria* (Figure 11b). Figure 14a suggests that the grass feeder has evolved to minimise the total deviation in nutrient space from the intake target incurred when feeding on imbalanced foods. By contrast, the generalist incurs a considerably larger error; how, then, might they benefit from this strategy? Figure 14b suggests one potential explanation. This figure shows that the total amount of nutrient (protein + carbohydrate) eaten by the generalist is constant, and on imbalanced foods considerably greater than for the specialist. Perhaps this reflects the greater probability that a generalist will subsequently encounter an edible plant with a complementary imbalance. The ingested excess from the current food would then balance the deficit in the subsequent one, thus converting excess into useful nutrient. Overall, therefore, this comparison suggests that the grass specialist feeder *L. migratoria* has adopted a strategy of error minimisation, while the generalist *S. gregaria* is a nutrient maximiser.

## Post-ingestive regulation

Our data therefore demonstrate that, where available foods permit, locusts regulate homeostatically to a point in protein-carbohydrate nutrient space, the intake target. When constrained from reaching this point, they show a clearly defined strategy of compromise which differs for the generalist feeder *S. gregaria* and the grass specialist *L. migratoria*. In this section we demonstrate a further capability in their already impressive arsenal of homeostatic responses, post-ingestive regulation.

By definition, animals that reach the intake target ingest an optimal amount and balance of nutrients, and are thus able to allocate nutrients optimally to growth. We have termed the optimal levels of nutrients that can be allocated to growth the *growth target* (Raubenheimer & Simpson, 1993). Post-ingestive regulation enables animals feeding on imbalanced foods to reach their growth target, even though they are unable to reach the intake target.

This is illustrated for *L. migratoria* in Figure 15 (data from Raubenheimer & Simpson 1993). The figure shows, firstly, the protein-carbohydrate intake array measured across the duration of the 5th stadium (in contrast with Figure 11b which shows the equivalent array measured over the first 3 days of the stadium). The arc of the closest distance array of intake points observed on the earlier measurement is still to some extent visible for insects fed the more moderate (central) foods, represented in the figure by filled circles. However, the animals on the more extreme foods have now moved outwards from the closest distance intake array. This is most pronounced in the near-vertical

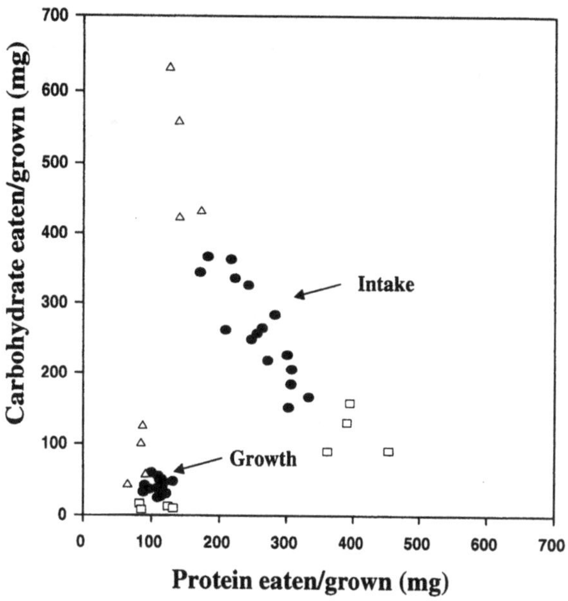


*Figure 15.* Protein-carbohydrate intake and growth arrays over the 5th stadium of *L. migratoria*. Filled circles represent these variables for insects fed the less extreme foods, while those fed foods with a relatively large excess of carbohydrate and proteins are represented by open triangles and squares respectively.

arm formed by animals fed foods containing excess carbohydrates (hollow triangles), but also apparent for those fed the excess protein foods (hollow squares). We will return to this change in shape of the intake array shortly.

Also shown in Figure 15 are the amounts of protein and carbohydrate allocated to growth by each group of insects represented in the intake array. This figure therefore provides a geometrical representation of a two-dimensional nutrient budget (Raubenheimer & Simpson, 1995), showing the amounts of ingested nutrient that were allocated to growth by each group of animals. The most striking result is how closely the growth points – and particularly those for animals fed the less extreme, central foods – clustered in comparison with the wide range of intake points, this cluster presumably encompassing the growth target. The extent of this regulation is shown in Figure 16, where the amount of nutrient ingested and allocated to growth are plotted separately for protein and carbohydrate against rail number. The vertical distance separating the growth and intake points for each rail represents the amount of nutrient ingested but not allocated to growth – in other words that which was eliminated from the system. The figure shows very clearly how, through regulating this latter quantity, growth was held constant in the face of fairly severe perturbations to

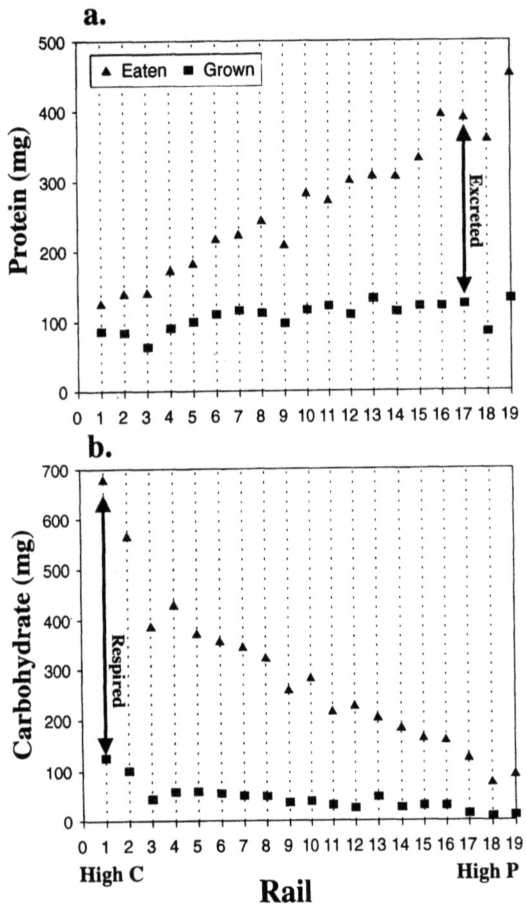


*Figure 16.* Growth and intake of protein and carbohydrate by *L. migratoria* fed one of 19 foods varying in the balance of these nutrients (*x* axis). The vertical distance between growth and intake points represents the nutrient eliminated from the animals. In the case of protein, this was via the faeces in the form of uric acid, while excess carbohydrate was respired.

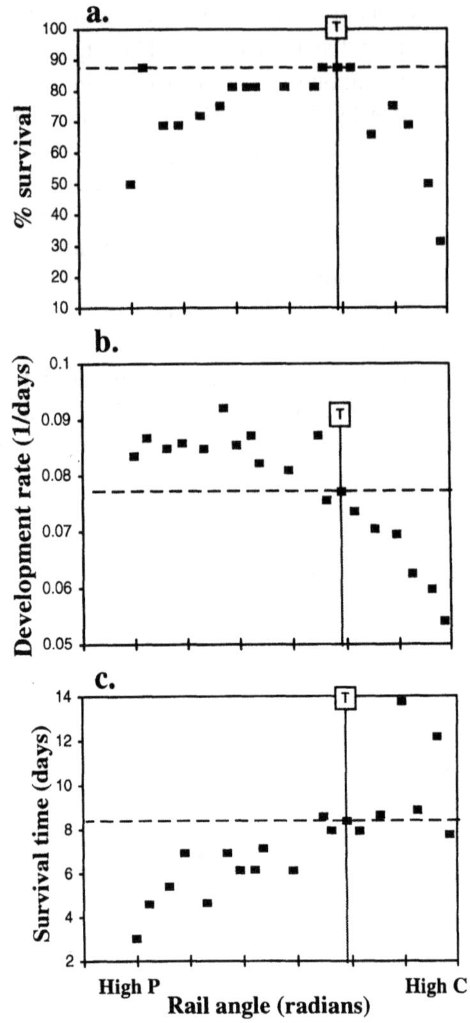


*Figure 17.* Performance of 5th stadium *L. migratoria* fed foods varying in protein-carbohydrate balance (*x*-axis). The target rail is indicated by the vertical lines.

nutrient intake. Experiments have shown that in these locusts excess ingested protein is de-aminated and the nitrogen excreted via the faeces as uric acid, while excess carbohydrate is metabolised through an increased metabolic rate and the carbon component eliminated as carbon dioxide (Zanotto et al., 1993, 1994, 1997).

A further interesting aspect of Figure 15, are the reasons the animals on the extreme foods broke away from the closest distance intake array. The formation of the near-vertical arm on the foods with the greatest excess of carbohydrate is due to these animals having extended the stadium duration and thus continued to feed for up to 5 days longer than those on less extreme foods (Raubenheimer & Simpson, 1993; see also Figure 17b). A noteworthy feature, is how these intake points aligned with the protein co-ordinate of the growth target, suggesting that the animals delayed the moult until some minimum level of protein-derived growth had been achieved. In so doing, however, they over-extended their ability to void the excess ingested of carbohydrate, resulting in them exceeding the growth target in the carbohydrate dimension. This can also clearly be seen in Figure 16b, as a breakdown in regulation of carbohydrate-derived growth.

Animals fed the foods with the greatest excess of proteins also moved outwards from the central array. Unlike those fed an excess of carbohydrate, this was not a result of extended stadium duration, since these animals moulted at a similar time or even sooner than those on the central foods (Raubenheimer & Simpson, 1993; Figure 17b). In fact, Figure 11b shows that even after 3 days on the foods, there was already a tendency for these animals to move outwards from the closest distance intake array, demonstrating a greater ability to ingest an excess of protein than would be predicted by the closest distance rule. One possible explanation for this is that the carbon-based breakdown products of excess ingested protein could be channelled into partly offsetting the deficit of carbohydrate ingested on these foods. This would not only contribute directly to carbohydrate requirements, but would also reduce the accumulation in the organism of excess protein, thus facilitating additional intake of carbohydrate.

**Performance measures**

Despite the impressive ability of locusts to achieve balanced growth across a wide range of nutrient intakes, it remains a fact that when given suitable complementary foods they regulate to a specific point in nutrient space. This gives rise to the question of what are the benefits of selecting this point, rather than some other point from which they could achieve similar growth. Such functional questions are addressed in our framework using an additional axis, the *performance axis*, which maps onto nutrient space the consequences to an animal of being at a given point at a given time.

In Figure 17 is presented three performance measures for *L. migratoria*. The first plot shows that the percentage survival through the fifth stadium was low for animals fed foods containing an excess of protein, then increased up to the target rail, and once again decreased on foods containing an excess of carbohydrate. Development rate (Figure 17b), by contrast, was highest for animals fed the excess-protein foods, then dropped towards and beyond the target rail. This demonstrates that these locusts could in fact achieve a higher development rate than they do under ad-libitum conditions, by selecting an intake point closer to the protein axis – why don't they?

While a comprehensive answer would require a detailed analysis of various performance measures, one possible reason is suggested in Figure 17c. This shows the survival time of locusts brought through the 5th (final) larval stadium on one of the foods then deprived of food as adults – it is, in other words, a measure of resistance to starvation. The figure shows that animals fed the high protein foods succumbed to starvation several days earlier than those fed foods containing relatively higher levels of carbohydrate. Therefore, increasing development rate by selecting diets with higher protein content would make these locusts vulnerable to food shortages. Quite possibly, the selected intake point represents a prioritisation of survival through the fifth stadium, together with a balanced trade-off between development rate and resistance of adults to starvation.

**Nutritional dimensionality**

Our data have demonstrated that for locusts, a large proportion of the variance in nutritional regulation and performance can be explained through the interaction of two nutrient groups, proteins and carbohydrates. Many foods are, of course, more heterogeneous than this, as are the nutrient requirements of animals. This gives rise to the important issues of how to select the number of dimensions to be included in a model, and how to expand the geometrical approach to consider more than two dimensions.

The number of relevant dimensions depends, obviously, on the biological details of the system in question, but also on the research approach being taken. At the one extreme are studies of insect behaviour on, for instance, whole plants in the wild, where the researcher has little or no control over the chemical composition of the foods. In such cases, a range of chemical variables might be measured and attempts made to map the animals' responses onto underlying nutritional patterns. Here, determining the number of relevant dimensions would be a standard statistical issue of identifying the nutrients that explain a good proportion of the variance in the animals' nutritional responses, while not over-saturating the model with redundant terms (Wright, 1998). At the other extreme are laboratory studies using synthetic foods, where the number of relevant dimensions can to some extent be determined by the experimenter, as discussed further below. Somewhere between the two is the use of genetically manipulated plants, where key nutrient groups are controlled within a complex plant physiological milieu (Wright, 1998).

Whether in a field-type study or a laboratory experiment using synthetic foods, the geometrical approach can readily be extended to analyse data of dimensionality exceeding two. One approach would be to use three-dimensional graphical representations. While this might be heuristically useful as a graphical illustration of increased nutritional dimensionality that is within the conceptual grasp of non-mathematicians, in practice it has limitations. Firstly, it does not readily extend to more than three dimensions, and secondly it is in our experience frequently difficult to interpret three-dimensional scatter plots visually. A second approach, and one that we have found useful in the study of three-dimensional systems (Simpson & Raubenheimer, 1997), is to reduce a multi-dimensional representation to a series of two-dimensional cross-sections. There are several advantages to this approach, including its simplicity and the fact that it can cope with any number of dimensions. However, since the number of cross-sections increases geometrically with the number of dimensions, there are advantages to restricting the number of dimensions included in a model.

One means of reducing the number of dimensions is to use *compound axes*, which represent more than one nutrient (Simpson & Raubenheimer, 1993, 1995). There are several contexts in which this approach might be applied. For example, including all amino acids separately in a model will yield a cumbersome explosion of parameters, so it may be preferable to represent them jointly on a single, compound, protein axis. This sort of coarse-graining might be more than a methodological convenience, but potentially a fair reflection of reality, since insects are themselves incapable of regulating separately the intake of all amino acids (Simpson et al., 1990; Simpson & Simpson, 1992). A second reason for compounding axes is where two or more nutrient groups are functionally interchangeable. For instance, many vertebrates channel both ingested carbohydrates and lipids into energy metabolism, and these nutrients can therefore appear on a single axis representing energy. Likewise, amino acids might all constitute sources of nitrogen, and may therefore be represented as a single axis. A third application of compound axes is where two or more nutrients are present in a food in approximately balanced proportions. Since in this case there is no conflict among the nutrients concerned – once the animal has satisfied its requirements for one, it has done so for all – they can reasonably be considered a single resource and tested jointly in pairwise models against other nutrients. Trumper & Simpson (1993) used this approach to investigate the regulation of salt intake in 5th stadium *Locusta migratoria* Figure 10.

This latter experiment, and several unpublished studies involving vitamins and other microconstituents of the synthetic foods we have used in our research into locust nutrition, explains our focus on a protein-carbohydrate two-dimensional model. Jointly, the evidence suggests that among the dry diet components protein and carbohydrate explain a substantial proportion of the variance in nutritional regulation in these insects. While this explains our interest in protein and carbohydrate, it does not justify our omission of the micronutrients from the models, an issue which does require explanation since it is known that many are indispensable components of an insect's diet (Dadd, 1960). We have dealt with this by ensuring, firstly, that all micronutrients are present in the foods at levels that support good growth and development. Secondly, the use of non-nutritional crystalline cellulose in the foods enables any manipulations of the levels of macronutrients to be compensated, such that the relative concentrations of all other components remain unchanged (Simpson & Abisgold, 1985; Raubenheimer & Simpson, 1993).

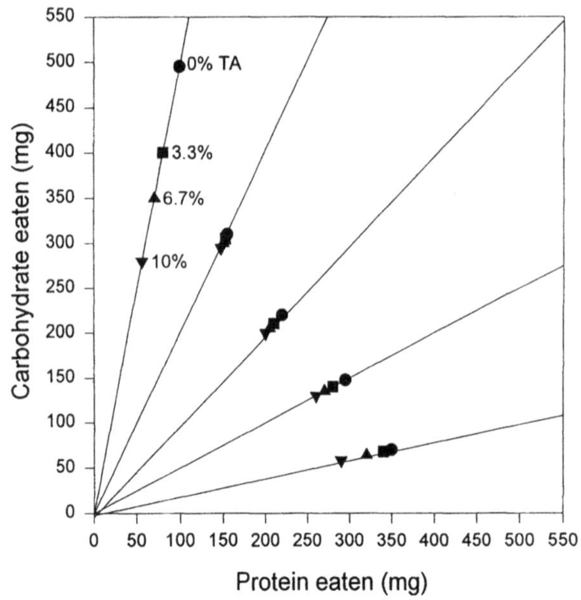


*Figure 18.* Protein-carbohydrate intake arrays across the 5th stadium for *L. migratoria* fed foods containing one of four levels of tannic acid.

There remains, however, an important macronutrient not represented in any of the models presented in this review: water (but see Raubenheimer & Gäde, 1993). The way we have dealt with this in the work presented above, illustrates a further approach to reducing dimensionality, namely by providing one or more nutrients separately from the rest. This allows the animals to regulate the separate nutrient(s) orthogonally, such that there is no conflict between them and the other nutrients.

## Allelochemicals

Finally, the geometrical approach is not restricted to nutrients, but can readily incorporate non-nutritional food components such as allelochemicals. An example of the influence of allelochemicals on intake arrays is taken from S. J. Simpson & D. Raubenheimer (unpubl.). This study showed that tannic acid reduced nutrient intake by *L. migratoria*, but only on foods containing a large excess of carbohydrate relative to protein (Figure 18). This contextual effect was echoed in performance (mortality). Interestingly, while there was little effect of tannic acid on consumption by insects fed excess-protein foods, tannic acid did increase mortality on these foods, suggesting a post-ingestive toxic effect. These data therefore emphasise a point that is central to our motives in developing the geometrical approach: The influence of a given factor (in this case an allelochemical) on a biological system can be heavily dependent on the context in which it occurs (in this case nutritional balance), and an understanding of the system requires systematic study of both the factor and the context. It requires, in other words, an integrative approach.

## References


Belovsky, G. E., 1990. How important are nutrient constraints in optimal foraging models or are spatial/temporal factors more important? In: R. N. Hughes (ed.), Behavioural Mechanisms of Food Selection, Nato ASI Series, Vol. 20. Springer-Verlag, Berlin, pp. 255–278.

Calow, P., 1976. Biological Machines. A Cybernetic Approach. Edward Arnold, London.

Chambers, P. G., S. J. Simpson & D. Raubenheimer, 1995. Behavioural mechanisms of nutrient balancing in *Locusta migratoria*. Animal Behaviour 50: 1513–1523.

Chambers, P. G., D. Raubenheimer & S. J. Simpson, 1997. The selection of nutritionally balanced foods by *Locusta migratoria*: the interaction between food nutrients and added flavours. Physiological Entomology 22: 199–206.

Chyb, S. & S.J. Simpson, 1990. Dietary selection in adult *Locusta migratoria*. Entomologia Experimentalis et Applicata 56: 47–60.

Dadd, R. H., 1960. The nutritional requirements of locusts-I Development of synthetic diets. Journal of Insect Physiology 4: 319–347.

Dawkins, M. S., 1989. The future of ethology: how many legs are we standing on? Perspectives in Ethology 8: 47–54.

Dewsbury, D. A., 1992. On the problems studied in ethology, comparative psychology, and animal behaviour. Ethology 92: 89–107.

Jutsum, A. R. & G. J. Goldsworthy, 1976. Fuels for flight in *Locusta*. Journal of Insect Physiology 22: 243–249.

Mattson, W. J., 1980. Herbivory in relation to nitrogen. Annual Review of Ecology and Systematics 11: 119–161.

McFarland, D. J. & A. Houston, 1981. Quantitative Ethology. Pitman Advanced Publishing Program, Boston.

McNeill, S. & T. R. E. Southwood, 1978. Role of nitrogen in the development of insect-plant relations. In: J. Harborne (ed.), Biochemical Aspects of Plant and Animal Coevolution. Academic Press, New York, pp. 77–98.

Pulliam, H. R., 1975. Diet optimization with nutrient constraints. American Naturalist 109: 765–768.

Raubenheimer, D., 1992. Tannic acid, protein and digestible carbohydrate: dietary imbalance and nutritional compensation in the African migratory locust. Ecology 73: 1012–1927.

Raubenheimer, D. & G. Gäde, 1994. Hunger-thirst interactions in the locust, *Locusta migratoria*. Journal of Insect Physiology 40: 631–639.

Raubenheimer, D. & S. J. Simpson, 1993. The geometry of feeding in the locust. Animal Behaviour 45: 953–964.

Raubenheimer, D. & S. J. Simpson, 1994. The analysis of nutrient budgets. Functional Ecology 8: 783–791.

Raubenheimer, D. & S. J. Simpson, 1995. Constructing nutrient budgets. Entomologia Experimentalis et Applicata 77: 99–104.

Raubenheimer, D. & S. J. Simpson, 1996. Meeting nutrient requirements: the roles of power and efficiency. Entomologia Experimentalis et Applicata 80: 65–68.

Raubenheimer, D. & S. J. Simpson, 1997. Integrative models of nutrient balancing: application to insects and vertebrates. Nutrition Research Reviews 10: 151–179.

Raubenheimer, D. & S. J. Simpson, 1998. Nutrient transfer functions: the site of integration between feeding behaviour and nutritional physiology. Chemoecology 8: 61–68.

Simpson, C. L., S. J. Simpson & J. D. Abisgold, 1990. An amino acid feedback and the control of locust feeding. Symposium Biologica Hungarica 39. 39–46.

Simpson, S. J. & J. D. Abisgold, 1985. Compensation by locusts for changes in dietary nutrients: behavioural mechanisms. Physiological Entomology 10: 443–452.

Simpson, S. J. & D. Raubenheimer, 1993a. A multi-level analysis of feeding behaviour: the geometry of nutritional decisions. Philosophical Transactions of the Royal Society 342: 381–402.

Simpson, S. J. & D. Raubenheimer, 1993b. The central role of the haemolymph in the regulation of feeding. Physiological Entomology 18: 395–403.

Simpson, S. J. & D. Raubenheimer, 1995. The geometric analysis of feeding and nutrition: a user's guide. Journal of Insect Physiology 41: 545–553.

Simpson, S. J. & D. Raubenheimer, 1996. Feeding behaviour, sensory physiology and nutritional feedback: an integrative model. Entomologia Experimentalis et Applicata 80: 55–64.

Simpson, S. J. & D. Raubenheimer, 1997. The geometric analysis of macronutrient selection in the rat. Appetite 28: 201–213.

Simpson, S. J. & C. L. Simpson, 1992. Mechanisms controlling modulation by haemolymph amino acids of gustatory responsiveness in the locust. Journal of Experimental Biology 168: 269–287.

Simpson, S. J., D. Raubenheimer & P. G. Chambers, 1995. The mechanisms of nutritional homeostasis. In: R. F. Chapman & G. de Boer (eds), Regulatory Mechanisms of Insect Feeding. Chapman and Hall, New York, pp. 251–278.

Stephens, D. W. & J. R. Krebs, 1986. Foraging Theory. Princeton University Press, Princeton.

Tinbergen, N., 1951. The Study of Instinct. Oxford University Press, Oxford.

Tinbergen, N. ,1963. On aims and methods of ethology. Zeitschrift für Tierpsychologie 20: 410–433.

Toates, F., 1986. Motivational Systems. Cambridge University Press, Cambridge.

Trumper, S. & S. J. Simpson, 1993. Regulation of salt intake by nymphs of *Locusta migratoria.* Journal of Insect Physiology 39: 857–864.

White, T. C. R., 1993. The Inadequate Environment: Nitrogen and the Abundance of Animals. Springer-Verlag, Berlin.

Wright, G. A., 1998. The Effect of Variation in Plant Chemistry on the Pattern of Feeding of two Generalist Insect Herbivores. D.Phil. thesis, Oxford University.

Zanotto, F. P., S. J. Simpson & D. Raubenheimer, 1993. The regulation of growth by locusts through post-ingestive compensation for variation in the levels of dietary protein and carbohydrate. Physiological Entomology 18: 425–434.

Zanotto, F. P., D. Raubenheimer & S. J. Simpson, 1994. Selective egestion of lysine by locusts fed nutritionally unbalanced foods. Journal of Insect Physiology 40: 259–265.

Zanotto, F. P., D. Raubenheimer & S. J. Simpson, 1996. The levels in the haemolymph of amino acids and sugars in locusts fed unbalanced diets. Journal of Comparative Physiology, B. 166: 223–229.

Zanotto, F. P., S. M. Gouveia, S. J. Simpson, D. Raubenheimer & P. C. Calder, 1997. Nutritional homeostasis in locusts: is there a mechanism for increased energy expenditure during carbohydrate overfeeding? Journal of Experimental Biology 200: 2473–2448.

*Entomologia Experimentalis et Applicata* **91:** 83–88, 1999.



# Absence of food-aversion learning by a polyphagous scarab, *Popillia japonica*, following intoxication by geranium, *Pelargonium* x *hortorum*

Daniel A. Potter & David W. Held
*Department of Entomology, S-225 Agriculture Science Bldg. N., University of Kentucky, Lexington, KY 40546-0091, USA*

Accepted: October 17, 1998

*Key words:* Japanese beetle, Scarabaeidae, food-aversion learning, host-plant selection, polyphagy

**Abstract**

It is commonly held that food-aversion learning should be more prevalent in polyphagous herbivores than in specialists. We tested the ability of *Popillia japonica*, a polyphagous scarab, to learn avoidance of a palatable but illness-inducing plant. Beetles that feed on flowers of geranium, *Pelargonium* × *hortorum*, became paralyzed, although most recovered within 24 h. In choice tests, naive beetles strongly preferred geranium petals over leaves of linden, *Tilia cordata*, a highly suitable host. Experienced beetles maintained this preference although it resulted in repeated bouts of paralysis. Fecundity was $>10$ times higher for beetles fed linden foliage for 2 wk than for those fed only geranium. Nevertheless, when a surplus of both foods was provided, the beetles fed mainly on geranium, resulting in greatly reduced fecundity. These results contradict the view that generalists should show propensity for food-aversion learning. Indeed, in this case, *P. japonica* continued to prefer the toxic plant, compromising its fitness.

## Introduction

It is commonly held that the ability to learn to associate a particular palatable food with temporary, post-ingestional illness, and to avoid future ingestion of that food, should be characteristic of polyphagous herbivores rather than specialists (Gelperin & Forsythe, 1975; Dethier & Yost, 1979; Dethier, 1980, 1988; Bernays & Lee, 1988; Lee & Bernays, 1990). The rationale is that selection for aversion learning ability presumably is stronger for generalists that may consume palatable but toxic plants, whereas the restricted diet of oligophagous or monophagous species normally prevents such mistakes. Limited empirical evidence seems to support this view. Generalists, including slugs (Gelperin, 1975), larvae of the arctiids *Diacrisia virginica* and *Estigmene congrua* (Dethier, 1980, 1988), and the locust *Schistocerca gregaria* (Bernays & Lee, 1988; Lee & Bernays, 1990) showed some capacity for food-aversion learning, whereas the oligophagous sphingid *Manduca sexta* did not alter its behavior after similar experience with an illness-inducing plant (Dethier & Yost, 1979).

The Japanese beetle, *Popillia japonica* Newman, feeds on nearly 300 plant species in ≈79 plant families in the United States (Fleming, 1972). The greatest damage is caused by the beetles' proclivity to aggregrate on particular hosts, a response that involves attraction to feeding-induced plant volatiles (Loughrin et al., 1996). Nevertheless, certain palatable plants reportedly are toxic to *P. japonica* (Fleming, 1972). Ballou (1929), for example, found that beetles feeding on cultivated geranium, *Pelargonium* × *hortorum*, soon became paralyzed and dropped from the plants. Inflorescences were more attractive to beetles and more toxic to them than the foliage, and the effect was enhanced when the plants were grown in full sunlight. Paralysis began with the hind legs and progressed cephalad. Most of the paralyzed beetles recovered within 24 h, although about a third of those under observation died within 4 days (Ballou, 1929).

Because of their polyphagy, longevity, and mobility, Japanese beetles frequently may encounter palatable but potentially toxic food plants. Such an insect presumably would benefit from food-aversion learning. We therefore tested the capacity of *P. japonica* to learn to avoid blooms of geranium after recovery from post-ingestive paralysis, and assessed the consequences for survival and fecundity when beetles had equal access to suitable and unsuitable food.

## Materials and methods

All experiments were conducted during July and August of 1997. Beetles were captured with Japanese beetle traps baited with food-type lures (2-phenylethyl propionate, eugenol, and geraniol; 1:2:1 ratio; Trécé Inc., Garwood, NJ). Traps were situated in a clover-dominated meadow in Versailles, KY, where no geraniums are present. Thus, the field-captured beetles presumably had no experience with geranium. Traps were emptied twice daily and beetles used in assays were held overnight without food. Only females were used for preference tests.

*Choice tests for food-aversion learning.* Naive or experienced beetles were offered a choice between fresh foliage of linden, *Tilia cordata* L., a highly preferred and suitable host (Fleming, 1972; Rowe & Potter, 1996), or freshly-harvested geranium flowers. Inflorescences of 'Orbit Red' zonal geranium, a seed-propagated cultivar, were collected from two mulched beds (≈200 plants each) that had been established on 1 May 1997. Geraniums were growing in full sunlight. Linden leaves were harvested from six trees of relatively uniform height (7–8 m) and trunk diameter (22–26 cm at 1 m height). All of the plants were on the University of Kentucky main campus, Lexington, KY. Leaves and flowers used were free of apparent disease symptoms. Plant materials were collected into sealable plastic bags and transported in coolers.

Preference test arenas were similar to those of Jermy et al. (1968). Leaf discs (22 mm diameter) were mounted on insect pins between bits of cellulose acetate film (4 $mm^2$) to prevent the disc from tearing away during feeding. Individual flowers were similarly pinned through the center of the five-petaled corolla. To ensure representative samples, each disc was cut from a separate linden leaf, and the geranium flowers were dissected from inflorescences harvested from eight different plants in each of the two beds. The tests were run in 474-ml containers (Solo Inc., Baltimore, MD) one-third filled with paraffin, with 15 replicates per comparison. A moistened filter paper was placed in the bottom before each assay. Two linden discs and two geranium flowers were alternately pinned around the perimeter of the arena, one female beetle was added, and the container then was topped with clear plastic wrap. This arrangement ensured that both treatments would be encountered before sustained feeding occurred. Beetles were placed in the arenas at ≈1200 h, and allowed to feed for 5 h. This interval was sufficient for all beetles that fed on geranium to become paralyzed. Containers were held in a growth chamber at 25 ± 2 °C under fluorescent lights. Individual discs and flowers were of similar area (means, 380 versus 287 $mm^2$, respectively), and neither food was >75% consumed during any assay.

After the 5-h feeding period, all beetles were removed from the arenas and held overnight, without food, in individual plastic containers (237 ml) with 2 cm of moist, sifted, autoclaved soil to allow them the opportunity to oviposit. Whether or not the beetle had become paralyzed was noted. Remaining plant tissues were then removed, and areas with feeding damage were traced onto clear acetate sheets. Traced areas were blackened with a marker, photocopied, and then measured with an electronic area meter. To account for differences in thickness of leaf discs and flowers, missing areas were converted to fresh weight (mg) eaten using a constant calculated from area/fresh weight standards. Data were analyzed within paired comparisons by Wilcoxon signed rank tests (Hollander & Wolfe, 1973).

To determine the capacity of *P. japonica* for food-aversion learning, the same group of beetles was re-tested on two additional, successive days. Naive beetles that fed on geranium became temporarily paralyzed, but they were sufficiently recovered by morning to crawl normally. All survivors were retested between 1200 and 1700 h during the afternoons of the second and third days. Thus, relative consumption of geranium or linden was measured for naive beetles, and for the same group after one or two 5-h exposures to geranium. Procedures for choice tests, confinement of beetles between trials, and data analysis were as described above.

Concurrent with the above tests, a separate, no-choice trial was run to verify the palatability of linden leaf discs. Leaves were collected from the same six trees used for the choice tests, and four leaf discs were pinned in the 474-ml arenas as previously described. There were 15 replicates, each with a single female captured and handled in the same manner as those used in the main test. The amount of leaf tissue consumed in 5 h was determined as described earlier.

*Consequences of non-learning for survival and fecundity.* We sought to determine if the beetles' proclivity to feed on geranium flowers, even in presence of more suitable food, would compromise their survival and fecundity. Beetles were collected with traps as described earlier. Groups of 10 females and 10 males were confined in clear plastic boxes (27.3 × 19.5 × 9.5 cm; TriState Plastics, Dixon, Ky.) containing 4 cm of finely sifted, moist, autoclaved soil for oviposition. Each group was fed either linden foliage only, geranium flowers only, or a 1:1 mixture of both foods for 14 d. Plant material, obtained from the same sources used for the choice tests, was collected into sealable plastic bags, transported in coolers, and provided to beetles within 1 h of harvest. There were six replicate boxes for each diet. Boxes for the linden-only treatment were provisioned with two shoots (≈25 cm long), each with 8 to 10 undamaged leaves. Each replicate received foliage from a different tree. Boxes for the geranium-only treatment were provisioned with two whole inflorescences from different plants. Mixed-food boxes received one linden shoot and one whole inflorescence. To maintain freshness, shoots and inflorescences were placed in florists' water pics before being placed on the soil surface. Leaves or inflorescences were replaced every 2 days, so that neither food was ever depleted. The boxes were held in a growth chamber at 25 ± 2 °C under fluorescent lights (L14:D10).

The beetles' status was recorded every 2 d, when the food plants were changed. We noted the numbers on leaves or flowers, as well as on or in the soil. All beetles were transferred to a holding cup which was then inverted and the beetles' response observed. Beetles were scored as normal, paralyzed, or dead. Normal beetles quickly righted themselves and crawled or initiated pre-flight movements. Paralyzed beetles were unable to crawl normally or to right themselves when inverted, but twitched the antennae, palps, or legs when these structures were touched with a probe. Beetles that failed to respond in this manner were scored as dead. Normal and paralyzed beetles were returned to their respective boxes, whereas dead beetles were removed and their sex determined from foretibial characters (Fleming, 1972). Egg counts were taken after 7 and 14 d. On each date, normal and paralyzed beetles were transferred to holding containers and any dead beetles were removed. The soil then was carefully examined for eggs, which were counted and removed. Beetles and plant material were then returned to the appropriate box. Weekly and total egg counts were analyzed by 2-way analysis of variance with means separated by LSD. Numbers of surviving beetles were compared among treatments by ANOVA for repeated measures. Nitrogen content of linden leaves and geranium blooms was compared by standard Kjeldahl analysis of lyophilized tissue samples.

*Choice test with conditioned beetles.* To determine if feeding preference was altered after prolonged exposure to geranium, linden, or the mixed-plant diet, we randomly selected 15 surviving, apparently normal females from the 2-week feeding trial on each plant diet. These beetles were then offered a choice between linden leaf discs or geranium flowers as described earlier. We compared the amount (mg) of geranium or linden consumed by beetles from each group, as well as the proportion that became paralyzed within the 5-h trial. Data for consumption of geranium, and incidence of paralysis, were compared among groups by 1-way ANOVA or a chi-square test for independence, respectively.

## Results

*Choice tests for food-aversion learning.* In the no-choice test, beetles consumed relatively large amounts of linden foliage (23.5 ± 2.1 mg per female in 5 h), confirming that linden is a highly palatable food. Nevertheless, naive, field-collected beetles overwhelmingly preferred geranium flowers over linden (Figure 1). All of the beetles fed on geranium and became paralyzed during the 5-h assay. Recovery was signalled by normal locomotion and a subsequent return of appetite as evidenced by the post-illness tests described below. All beetles recovered by the following morning, i.e. within 16 h.

Following recovery, the beetles continued to prefer toxic geranium flowers over linden foliage (Figure 1). As had occurred during the first exposure, 100% of them became paralyzed during the second, 5-h bout. The next day, following two successive bouts of poisoning, the recovered beetles still strongly preferred the illness-inducing geranium over linden foliage.

*Consequences of non-learning for survival and fecundity.* By the end of the second week, fecundity of

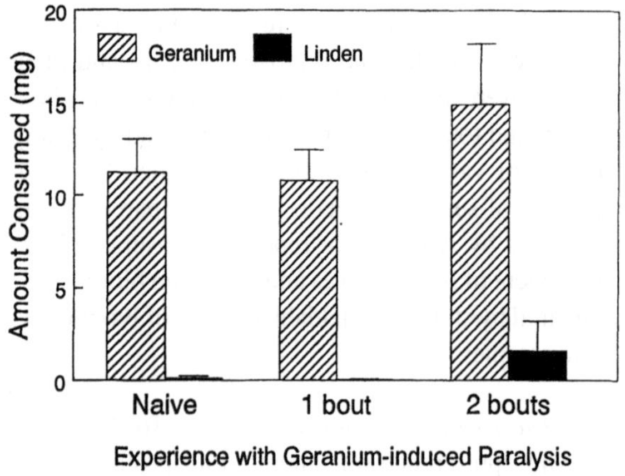


*Figure 1.* Amounts of toxic geranium or suitable linden consumed in choice tests by naive *P. japonica* and by the same beetles after one or two bouts of geranium-induced paralysis. Geranium was preferred each time (Wilcoxon signed rank tests, P < 0.01).

beetles that were confined with geranium had declined to <8% of that for beetles that fed only on linden leaves. Because beetles with access to both linden and geranium fed mainly on the latter, they, too, experienced repeated paralysis and had greatly reduced fecundity (Figure 2). Cumulative mortality (Figure 2) was higher on geranium than in the mixed-plant or linden-only treatments (ANOVA for repeated measures, $F_{2,15} = 34.32$, P < 0.001 for diet), with males being more severely affected than females ($F_{1,12} = 11.72$, P < 0.006 for sex). Survival after 14 d was higher for linden-fed males than for those having access to both foods (LSD, P < 0.05). Mortality of females was highest on the geranium-only diet (LSD, P < 0.05) but there was no difference between the linden-only and mixed diets (Figure 3). Nitrogen content was significantly higher in linden foliage than in geranium flowers (means ± SE: 1.99 ± 0.04 versus 1.05 ± 0.05, respectively; $t = 14.83$, df = 10, P < 0.001).

*Choice tests with conditioned beetles.* Beetles from all three groups continued to strongly prefer geranium over linden even after 2 wks on their respective diets (Figure 4). This was despite the fact that those that had been provisioned with geranium or the geranium/linden mix had experienced repeated bouts of paralysis. Notably, the amount of geranium consumed during the 5-h assay was much less for linden-conditioned beetles that had not previously experienced geranium than for those that had been fed geranium or the mixed diet ($F_{2,39} = 5.38$; P < 0.01). Nevertheless, these geranium-naive beetles

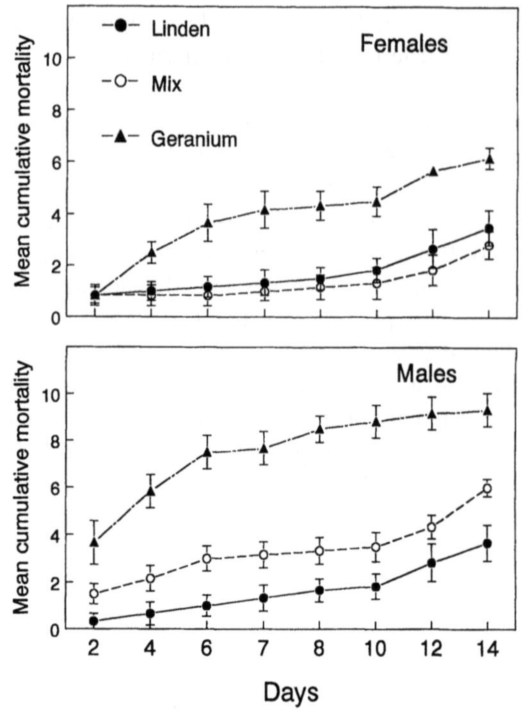


*Figure 2.* Cumulative mortality of male and female *P. japonica* fed either linden foliage, geranium flowers, or a mix of both foods ad libitum for 2 wk. Data are mean (±SE) number of dead beetles for six replicates; ten initial pairs of beetles per cage.

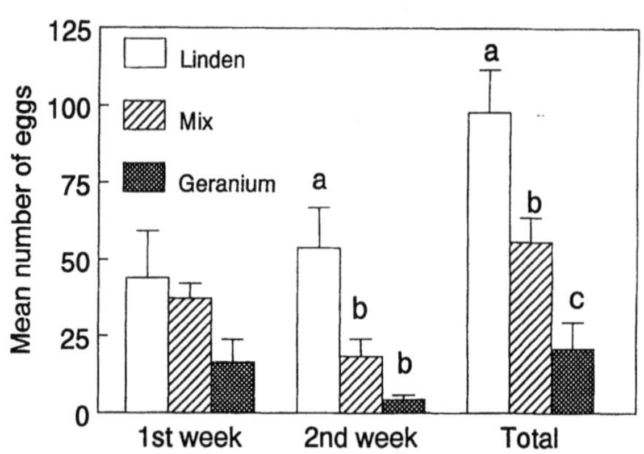


*Figure 3.* Fecundity of *P. japonica* fed either linden leaves, geranium blooms, or both foods ad libitum for 2 weeks. Data are means (±SE) for six replicates; ten initial pairs of beetles per cage. Within each set, means marked with different letters differ significantly (ANOVA; LSD; P < 0.01)

were somewhat more susceptible to paralysis. Most (80%) of the linden-conditioned beetles became paralyzed during the assay, whereas more than half of those conditioned with geranium or the mixed diet appeared normal after the assay (Figure 5).

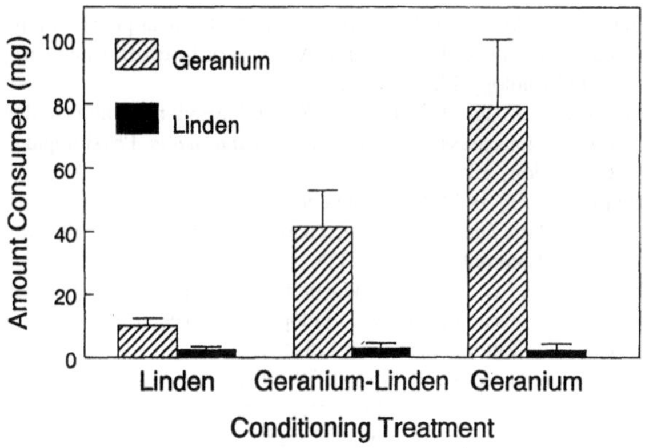


*Figure 4.* Relative amounts of geranium or linden consumed in choice tests by *P. japonica* that had been conditioned by feeding on linden, geranium, or a mix of both foods for 2 wk before the test. Geranium was preferred by all groups (Wilcoxon signed rank tests, P < 0.01).

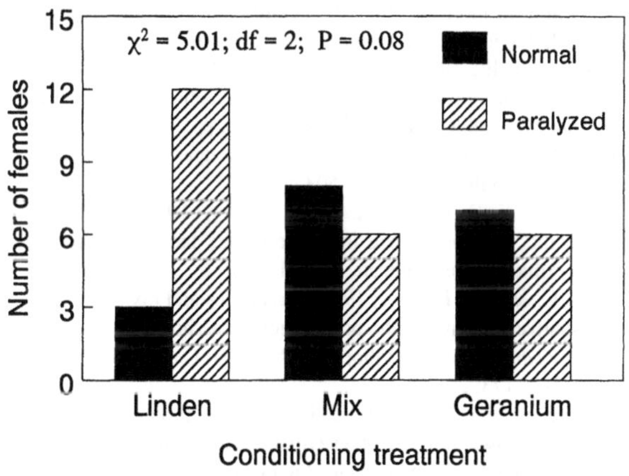


*Figure 5.* Number of beetles (out of 15) succumbing to geranium-induced paralysis during the 5-h choice test with *P. japonica* that had been conditioned by feeding on linden, geranium, or a mix of both foods for the preceding 2 weeks.

## Discussion

Japanese beetles are polyphagous, feeding to varying extents on nearly 300 plant species. Individual beetles live for 4 to 6 wk or longer, take discrete meals, and frequently fly from plant to plant (Fleming, 1972). Females usually spend one or more days feeding and mating on host plants before seeking sites with moist soil for oviposition. During her life a female may enter the soil 16 or more times (Fleming, 1972), alternating with bouts of feeding. Furthermore, *P. japonica* seems to be attracted to a wide array of plant species regardless of their acceptibility as hosts (Loughrin et al., 1997). Each beetle therefore is likely to encounter and sample many potential food plants varying both in surface gustatory cues and in their chemical defenses. Such foraging habits theoretically place a premium on food discrimination tactics, such as aversion learning (Gelperin & Forsythe, 1975; Dethier, 1980; Lee & Bernays, 1990).

Toxic geranium flowers were strongly preferred over suitable linden foliage in all choice tests. Contrary to expectation, however, beetles that had experienced geranium-induced paralysis did not avoid geranium upon subsequent encounters. Thus, contrary to the hypothesis that generalists should develop aversion to toxic but palatable plants, *P. japonica* did not exhibit food-aversion learning of the kind demonstrated by certain slugs, polyphagous caterpillars, or locusts (Gelperin, 1975; Dethier, 1980; Bernays & Lee, 1988). Even when a more suitable food was available, they continued feeding on intoxicating geranium to the extent that their survival and fecundity were compromised.

A limitation of using a choice test as a learning experience is that subjects that feed on both choices have no way of knowing which food had the noxious effect. In our tests, however, naive beetles fed almost exclusively on geranium, which seemingly would minimize this potential confounding effect. Moreover, beetles from the conditioning assay that had fed solely on geranium preferred the illness-inducing plant in the final choice test. Because the toxic effect appears to be neurological, affected beetles' ability to learn may also be compromised.

Naive field-collected or linden-conditioned beetles became paralyzed after consuming relatively small amounts (10 to 11 mg) of geranium in the choice tests. In contrast, geranium-conditioned beetles consumed more than seven times as much geranium during the same 5-h assay, with somewhat lower incidence of paralysis. This may reflect induction of polysubstrate monooxygenases or other detoxication enzymes, such as occurs in other insects (Lindroth, 1991). Indeed, generalists such as *P. japonica* may exhibit greater inducibility and range of responsiveness of enzymatic detoxication systems than specialists (Lindroth, 1991; Berenbaum, 1991).

Insects and vertebrates may acquire taste aversions more easily to foods that are less preferred (Bernays & Lee, 1988). Polyphagous locusts, for example, did not learn to associate highly palatable blackberry with an aversive stimulus, but did learn to avoid less palatable onion and spinach (Lee & Bernays, 1990). Possibly, the strong attractiveness and palatability of geranium provides such a high level of sensory stimulation for *P. japonica* that negative feedback from illness cannot override it. Japanese beetles reportedly will feed on other toxic, but palatable plants. For example, they readily consume the flowers of bottlebrush buckeye, *Aesculus parviflora*, and if enough is eaten, become paralyzed and eventually die (Hawley & Metzger, 1940). *Pelargonium* × *hortorum* is a complex hybrid of *Pelargonium*, a genus consisting of ≈280 species of herbs or shrubs native mostly to southern Africa (Baily Hortorium, 1976). Although *P. japonica* does not have a long evolutionary history with this plant, its polyphagous habits suggest that encounters with other, relatively uncommon toxic plants to which it is not adapted may occur within its native habitat in Japan.

The substance in geranium that causes paralysis of *P. japonica* is not known, but several anthocyanidins and flavonols are present in the flowers (Harborne, 1991; Mitchell et al., 1998). Ballou (1929) showed that the poisonous effect of geranium on Japanese beetles is much greater when feeding takes place on plants exposed to sunlight than on shaded plants. Interestingly, one characteristic of flavonoids is that their synthesis and accumulation are increased by light (Harborne, 1991).

We expect that mortality of geranium-intoxicated *P. japonica* would be greater in field situations than in our assays. Paralyzed beetles seemingly would be susceptible to desiccation. Those dropping off the plant would be vulnerable to ground-dwelling predators (e.g., ants, carabids), whereas beetles incapacitated on inflorescences might be taken by birds or other opportunistic predators.

## Acknowledgements

This study was supported by USDA-SRIPM grant 96-34103-3346. We thank K. F. Haynes for helpful discussions, and L. K. Rieske-Kinney and B. A. Webb for constructive reviews. Contribution no. 98-08-69 of the Kentucky Agricultural Experiment Station.

## References

Bailey Hortorium, 1976. Hortus III. MacMillan, New York, NY, 1290 pp.

Ballou, C. H., 1929. Effects of geranium on the Japanese beetle. Journal of Economic Entomology 22: 289–293.

Berenbaum, M. R., 1991. Comparative allelochemical processing in the Papilionidae (Lepidoptera). Archives of Insect Biochemistry and Physiology 17: 213–221.

Bernays, E. A., and J. C. Lee., 1988. Food aversion learning in the polyphagous grasshopper *Schistocerca americana*. Physiological Entomology 13: 131–137.

Dethier, V. G., 1980. Food-aversion learning in two polyphagous caterpillars, *Diacrisia virginica* and *Estigmene congrua*. Physiological Entomology 5: 321–325.

Dethier, V. G., 1988. Induction and aversion-learning in polyphagous arctiid larvae (Lepidoptera) in an ecological setting. Canadian Entomologist 120: 125–131.

Dethier, V. G. & M. T. Yost, 1979. Oligophagy and absence of food-aversion learning in tobaccco hornworms, *Manduca sexta*. Physiological Entomology 4: 125–130.

Fleming, W. E., 1972. Biology of the Japanese beetle. U.S. Department of Agriculture (technical bulletin 1383). Washington, D.C..

Gelperin, A., 1975. Rapid food-aversion learning by a terrestrial mollusk. Science 189: 567–570.

Gelperin, A. & D. Forsythe, 1975. Neuroethological studies of learning in mollusks. In: J. C. Fentress (ed), Simpler Networks and Behavior. Sinauer, Sunderland, MA, pp. 239–250.

Harborne, J. B., 1991. Flavonoids. In: G. A. Rosenthal & M. R. Berenbaum (eds), Herbivores, Their Interaction with Secondary Plant Metabolites. Academic, San Diego, pp. 389–429.

Hollander, M. & D. A. Wolfe, 1973. Nonparametric statistical methods. Wiley, New York, 503 pp.

Jermy, T., F. E. Hanson & V. G. Dethier, 1968. Induction of specific food preference in lepidopterous larvae. Entomologia experimentalis et applicata 11: 211–230.

Lee, J. C. & E. A. Bernays, 1990. Food tastes and toxic effects: associative learning by the polyphagous grasshopper *Schistocerca americana* (Drury) (Orthoptera: Acrididae). Animal Behaviour 39: 163–173.

Lindroth, R. L., 1991. Differential toxicity of plant allelochemicals to insects: roles of enzymatic detoxication systems. In: E. Bernays (ed.), Insect-Plant Interactions, Vol. 3., CRC Press, Boca Raton, FL, pp. 1–33.

Loughrin, J. H., D. A. Potter, T. R. Hamilton-Kemp & M. E. Byers, 1996. Role of feeding- induced plant volatiles in aggregative behavior of the Japanese beetle (Coleoptera: Scarabaeidae). Environmental Entomology 25: 1188–1191.

Loughrin, J. H., D. A. Potter, T. R. Hamilton-Kemp & M. E. Byers, 1997. Response of Japanese beetles (Coleoptera: Scarabaeidae) to leaf volatiles of susceptible and resistant maple species. Environmental Entomology 26: 334–342.

Mitchell, K. A., K. R. Markham & M. R. Boase. Pigment chemistry and color of *Pelargonium* flowers. Phytochemistry 47: 355–361.

Rowe, W. J. & D. A. Potter, 1996. Vertical stratification of feeding by Japanese beetles within linden tree canopies: selective foraging or height per se? Oecologia 108: 459–466.

*Entomologia Experimentalis et Applicata* **91:** 89–95, 1999.



# Examining the hierarchy threshold model in a no-choice feeding assay

Toni M. Withers
*CRC for Tropical Pest Management, Alan Fletcher Research Station, P.O. Box 36, Sherwood, QLD 4075, Australia; Present address: Forest Research, Private Bag 3020, Rotorua, New Zealand*

Accepted: October 18, 1998

*Key words:* Chrysomelidae, hierarchy threshold model, deprivation, discrimination, feeding behaviour

**Abstract**

The hierarchy threshold model of individual insect diet predicts the acceptance or rejection of individual hosts when encountered by insects. One assumption of the hierarchy threshold model is that 'post-discrimination phase' insects which accept lower ranked hosts will also accept hosts that are ranked higher. This assumption does not however suggest whether or not such insects behave differently when encountering these two hosts. This question is explored using *Zygogramma bicolorata* Pallister (Coleoptera: Chrysomelidae). Adult beetles that had just fed, or were one inter-meal interval after feeding, or were deprived for three or six days after feeding, were individually observed on a plant that either ranked high, *Parthenium hysterophorus* L., or low *Xanthium occidentale* Bertol. (Heliantheae: Ambrosiinae) in choice tests. Just-fed beetles were generally unresponsive to either host plant and did not feed. Beetles held for one inter-meal interval showed more feeding responses (shorter lag time between sampling and feeding, higher proportions of individuals feeding, and more time feeding) towards the higher ranked plant. As the beetles approached six days deprivation, behaviours of sampling, feeding and locomotion differed less between the two plants, but were still apparent. I conclude that food deprived *Z. bicolorata* beetles do still discriminate between the two plants. This result partially supports the hierarchy threshold model.

## Introduction

The hierarchy-threshold model of individual insect diet (Courtney et al., 1989) proposes that as insects encounter hosts, each has a certain acceptability to the insect based upon the balance of factors internal to the insect and sensory cues received from that plant. At any time when a host is encountered the behavioural threshold of responsiveness of the insect will determine whether that host is accepted or rejected. These authors also predict that when the behavioural threshold lowers, the highest ranked host remains acceptable (Courtney et al., 1989), but, lower ranked hosts also become acceptable (such as following deprivation from hosts). As is stated in Assumption Two (p. 57) of this model, '... an insect which accepts a host type will accept other host types which are ranked higher...'. I wanted to explore this aspect of acceptance of high and low ranked hosts, to examine whether acceptance of the two hosts by insects is identical during this post-discrimination phase, or if despite their acceptability, some vestige of discrimination remains. Should both hosts be acceptable but some discrimination remain, this was expected to be expressed in behaviour at the plant surface.

The leaf feeding beetle, *Z. bicolorata,* a biocontrol agent on parthenium weed (*P. hysterophorus*), was chosen for this study as it was known to be oligophagous on related Ambrosiinae hosts (Withers, 1998). A no-choice behavioural feeding assay was used to examine host acceptance behaviour. Treatments included first obtaining insects at four different levels of food deprivation, including satiated insects as a control. All individual insects were then observed during a single encounter with either a low or high-ranked host plant.

## Methods

*Insects.* The *Z. bicolorata* population used in these experiments originated from adults collected from parthenium weed in Mexico in 1980, and were reared on parthenium weed in glasshouses in Brisbane, Australia until 1983. Since this time *Z. bicolorata* has also been abundant in the field on the closely related ragweed (*Ambrosia artemisiifolia* L.) (Heliantheae: Ambrosiinae), as well as being an effective biological control agent against parthenium weed in both India and Australia (Dhileepan & McFadyen, 1997; Jayanth & Bali, 1994). Adult *Z. bicolorata* were collected from ragweed in Brisbane, and their offspring reared for one generation on potted ragweed plants. As adults eclosed from pupation sites in the soil of these pots, they were collected and held in cages with ragweed plants in a greenhouse maintained between 24–34 °C and 85–95% r.h. Adults were up to 3 weeks old (mean of 10 days old) at the time of experimentation. Adult *Z. bicolorata* spend their time resting and feeding, alone or in mating pairs, on their host plants. The mean inter-meal interval on a highly ranked host plant is approximately 2.3 h for females and 4 h for males (obtained by scan sampling 10 individuals at 2-min intervals for 11 h; T. Withers, unpubl.). Among plants tested, parthenium weed is both more acceptable in no-choice tests, and preferred over a related plant, *X. occidentale* (Noogoora burr, which is also a weed in Australia) (Withers, 1998; T. Withers, unpubl.).

*Z. bicolorata* adults were obtained at four different food deprivation levels before being individually tested on either parthenium weed or Noogoora burr:

(1) *Just-fed*; caged beetles feeding on ragweed plants were closely observed during each experimental day, and after completing a meal individuals were held in a small plastic container and tested as below within 5–15 min of completing feeding.

(2) *5 hours deprived;* caged beetles on ragweed plants were observed from 08:30, and after completing their first meal of the day, were removed and held together in a 5 × 10 × 20 cm plastic box, with a mesh insert in the lid, containing a moistened sand/bark substrate for sheltering. They were held for 5 h to ensure both males and females had passed through at least one inter-meal interval before testing as below on a host plant.

(3) *3 days deprived*; beetles were removed from ragweed plants at 11:00 h, 3 days before observations, and placed into a container as described above, until time of testing.

(4) *6 days deprived*; beetles were removed from ragweed plants at 11:00 h, 6 days before observations, and placed into a container as described, until the time of testing. Such deprived beetles remain fairly inactive during deprivation, female beetles resorb eggs, but are otherwise thought to be healthy. Adult *Z. bicolorata* can survive for at least 2 weeks without food under such conditions (K. Dhileepan, pers. comm.).

A total of 16 beetles were observed each day over 6 days of observations, corresponding to two beetles per plant-by-deprivation treatment daily ($n = 12$ per treatment). Observations on each day were carried out in a complete randomised block design, except with the order of observing beetles in deprivation treatments 1 and 2, which was dictated by the time at which they fed prior to the observation.

*Test procedure.* The arena was a cotton mesh cage with an open front (measuring 40 × 40 × 80 cm high) situated in a greenhouse (24–34 °C). Either a single 4-week-old Noogoora burr or a 6-week-old parthenium weed plant (both 8–12 leaf stage) was placed in the centre of the cage. A false, bark-covered floor was then placed around the base of the plant 15 cm above the true floor of the cage to create a more realistic physical environment. A fresh plant was used for each treatment, each day. Beetle behaviour was individually recorded using a portable computer programmed with 'The Observer for Windows 3.0' (Noldus, 1995). Observations began daily at 09:30 h, with an individual beetle coaxed with a camel hair brush from the holding container into a small plastic vial and from there allowed to walk onto a young leaf of the plant. Recording began when the beetle initiated walking on the leaf and ended 12 min later, or earlier if the beetle left the plant.

Walking, pausing, feeding, and ovipositing behaviours of each individual were recorded to the nearest second as exclusive continuous behavioural categories (Withers, 1998). Each instance of sample biting (intermittent squeezing of the leaf edge with mandibles usually removing $<2$ mm$^2$ of leaf into the buccal cavity) was recorded as an instantaneous event. Feeding behaviour was recorded when multiple instances of sample biting occurred in the same place on a leaf edge. The difficulty of ascertaining whether such a collection of sample bites did in fact constitute feeding initiation, lead me to impose a post-observation criterion upon the data for initiation of feeding. An arbitrary cut-off was imposed after observing a higher than expected number of feeding periods less than

12 s. Feeding was therefore only considered to have occurred when >12.5 s of continuously recorded feeding had taken place. In addition to behaviour, the location of the beetle on the plant and in the cage was simultaneously recorded. At the conclusion of observations, beetles were sexed.

*Data analysis.* A logistic regression model was used to test the effects of host plant and deprivation time on the proportion of beetles that took a sample bite, initiated feeding, and that remained on the plant within the 12 min observation period. This procedure gave a $\chi^2$ statistic with one degree of freedom for host plant and three degrees of freedom for deprivation time (McCullagh & Nelder, 1989). Latency from introduction to sampling biting was analysed only on those beetles that sampled the leaf. Measures including total or proportions of time spent performing a behaviour included all individuals. Measured parameters were tested for homogeneity of variances using a Bartlett's test (Sokal & Rohlf, 1981). Where heteroscedasticity remained, median values were compared between treatments using a Kruskal–Wallis test at $P < 0.05$. For parameters that were homogeneous or that could be made homogeneous using a $\log_{10} + 1$ transformation, data were tested with ANOVA's testing for host plant and deprivation effects using a Generalised Linear Model. When treatment effects existed, means separations were performed at $P < 0.05$ (Minitab version 11, Minitab, 1995).

## Results

Deprivation time significantly affected the proportion of beetles that sampled the plant ($\chi^2 = 23.1$, df = 3, $P < 0.001$), whereas host plant had no influence ($\chi^2 = 0.46$, df = 1, $P < 0.5$). None of the just-fed beetles sampled Noogoora burr and only two just-fed beetles sampled parthenium weed, compared to almost all of the more deprived beetles doing so. The latency from introduction of a beetle to the plant, to the time when it initiated its first sample bite of leaf was significantly influenced by deprivation of the beetle (Figure 1) ($F = 11.4$, df = 3, 68, $P < 0.0001$). Host plant had no influence on the latency to sample bite ($F = 61.1$, df = 1, 68, $P < 0.3$).

The proportion of the beetles that fed on the plant was affected by both host plant ($\chi^2 = 7.55$, df = 1, $P < 0.01$) and deprivation effects ($\chi^2 = 19.10$, df = 3, $P < 0.001$). A lower proportion of 5-hour and 3- and 6-day deprived beetles fed on Noogoora burr than on parthenium weed, but the proportion increased with time since the beetles last fed (Figure 2A). The lag time from when the first sample bite was taken to when the first feeding bout was initiated was also examined in those beetles that fed (Figure 2B). Deprivation time did not significantly influence this lag to feeding ($F = 0.7$, df = 3, 58, $P < 0.57$). Host plant showed a significant effect on this parameter ($F = 23.2$, df = 1, 58, $P < 0.001$), with the lag between sample biting and feeding consistently longer on Noogoora burr, than on parthenium weed.

For both plant types, the total number of sample bites (see Methods) on the plant during the observation period increased significantly with deprivation time (Table 1). Fewer sample bites were taken from parthenium weed, compared with the number of bites from Noogoora burr (Table 1). That sample biting was greatest on Noogoora burr by the most deprived beetles, indicates that *Z. bicolorata* initiated multiple instances of sampling and feeding on the less preferred plant.

Both host plant and deprivation effects significantly influenced both the measures of total time spent feeding during the observation period (host plant $F = 20.2$, df = 1, 95, $P < 0.0001$; deprivation $F = 33.7$, df = 3, 95, $P < 0.0001$) (Figure 3), and the proportion of the observation time spent feeding (host plant $F = 23.6$, df = 1, 95, $P < 0.0001$; deprivation $F = 30.4$, df = 3, 95, $P < 0.0001$). Total time spent feeding is shown instead of meal durations, as some meals had not concluded at the completion of the observation period and as such would not be true durations. In both cases there was a significantly greater time spent feeding by 5-h-deprived beetles on parthenium weed than on Noogoora burr. This difference became less marked with increasing deprivation, until by 6 days of deprivation, beetles showed a slight but insignificant difference in feeding behaviour between the two plants (Figure 3).

Activity levels of the beetles on the plants reflect both levels of responsiveness and host plant effects in more subtle ways. There was no significant effect of treatment on the proportion of time beetles spent pausing or resting on the plant (host plant $F = 0.4$, df = 1, 95, $P < 0.56$; deprivation $F = 2.1$, df = 3, 95, $P < 0.11$). The proportion of time that was spent walking, however, appears negatively correlated to the proportion of time spent feeding, and is significantly influenced by both host plant and deprivation (Table 1). Just-fed beetles on both plants, and 5-h-

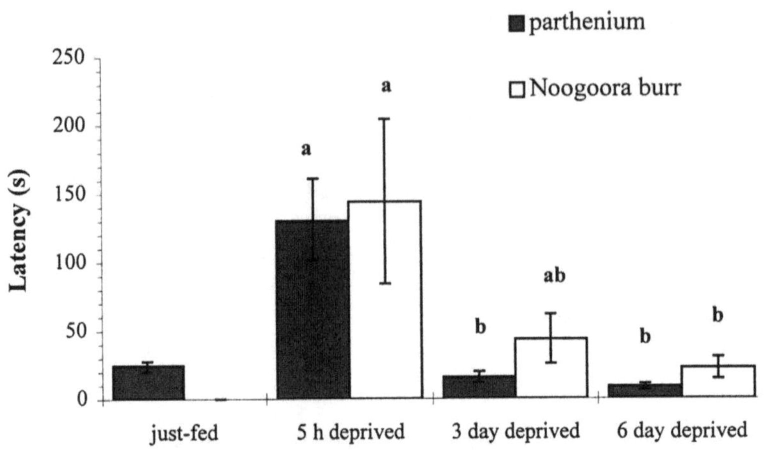


*Figure 1.* The influence of deprivation upon the mean (± SE) latency to sample bite by adult *Z. bicolorata* on either a high ranked (parthenium weed) or low ranked (Noogoora burr) host. Means sharing the same letter are not significantly different across all treatments. All sample sizes are 10–12 individuals, except for two just-fed beetles biting parthenium weed and no just-fed beetles biting Noogoora burr.

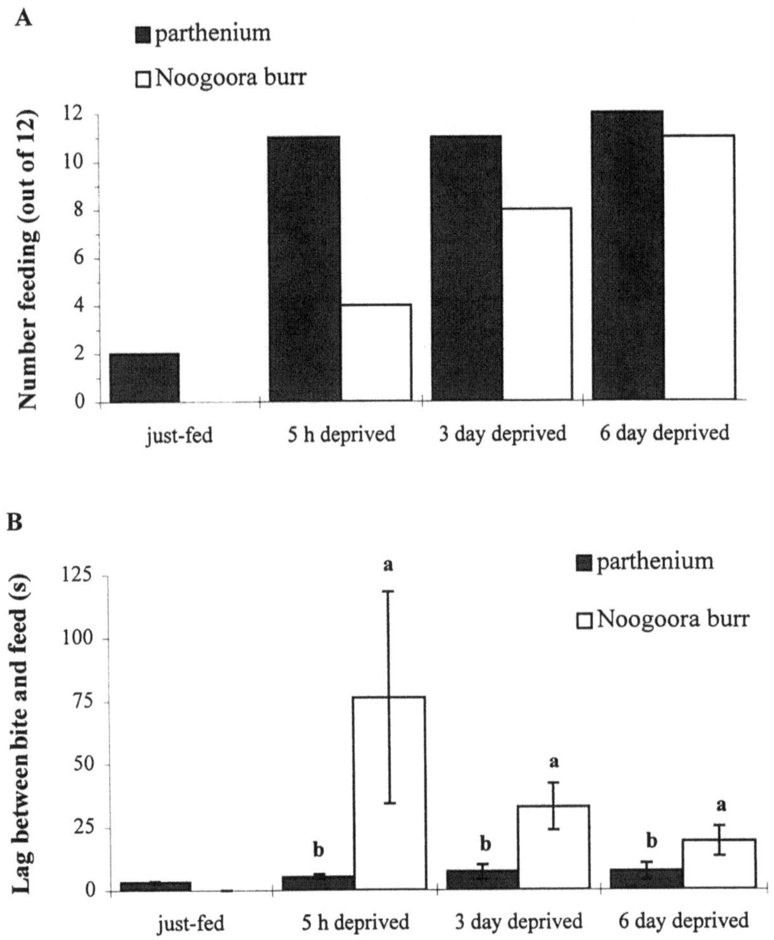


*Figure 2.* The influence of deprivation upon (A) the number of adult *Z. bicolorata* out of 12 that initiated feeding, and (B) the mean (± SE) lag between initiation of the first sample bite and the initiation of feeding by those that fed on either parthenium weed (high ranked) or Noogoora burr (low ranked). In (A) both host plant and deprivation effects were significant at $P < 0.05$, in (B) means sharing the same letter are not significantly different across all treatments.

*Table 1.* Mean (± SE) behaviours of adult *Z. bicolorata* under different deprivation treatments and on different host plants. Means that are accompanied by the same letter are not significantly different across all treatments

| | Total no. bites | | Time on first leaf (s) | | Prop. time spent walking | | Prop. leaving plant | |
|---|---|---|---|---|---|---|---|---|
| | parthenium | Noogoora burr | parthenium | Noogoora burr | parthenium | Noogoora burr | parthenium | Noogoora burr |
| Just-fed | 0.5 (0.4) | 0 | 330.7 (87)**cd** | 389.9 (88)**c** | 41.4 (7.3)**ab** | 42.3 (9.6)**ab** | 0.83 | 0.83 |
| 5 hrs deprived | 4.1 (0.8) | 15.1 (3.8) | 369.8 (90)**c** | 284.9 (59)**d** | 20.1 (3.4)**bc** | 56.4 (5.8)**a** | 0 | 0.83 |
| 3 days deprived | 3.5 (0.6) | 22.7 (5.9) | 544.7 (73)**b** | 547.4 (74)**b** | 13.2 (3.1)**c** | 32.2 (6.9)**b** | 0 | 0.92 |
| 6 days deprived | 3.2 (0.5) | 30.6 (9.1) | 650.2 (37)**a** | 587.1 (59)**ab** | 11.9 (3.3)**c** | 29.0 (7.4)**bc** | 0 | 0 |
| significance of treatments | $H = 53.5$ $P < 0.0001$ | | plant $F = 0.17$, $P < 0.68$ deprivation $F = 7.6$, $P < 0.0001$ | | plant $F = 10.5$, $P < 0.002$ deprivation $F = 6.8$, $P < 0.0001$ | | plant $\chi^2 = 1.4$, $P > 0.5$ deprivation $\chi^2 = 2.2$, $P > 0.5$ | |

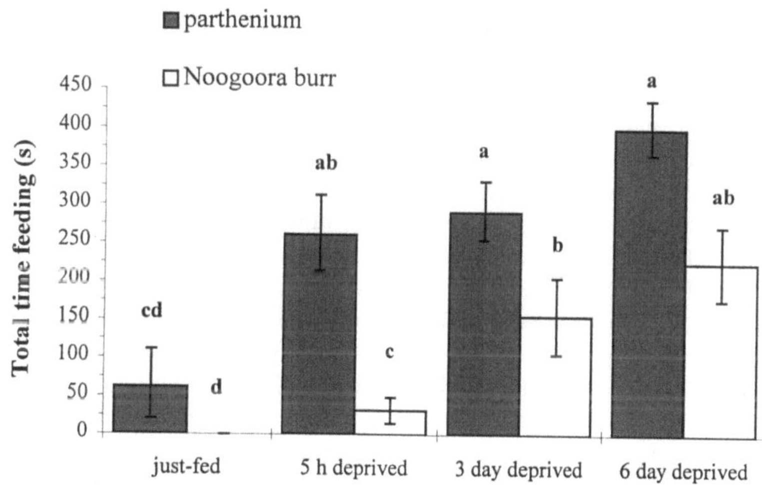


*Figure 3.* The influence of deprivation upon the mean (± SE) total time spent feeding by adult *Z. bicolorata* on either parthenium weed (high ranked) or Noogoora burr (low ranked). Means sharing the same letter are not significantly different across all treatments.

deprived beetles on Noogoora burr, showed the highest locomotory activity on the plant. Three- and 6-day deprived beetles on parthenium weed spent the least time walking. The beetles most likely to leave the plant before the end of the observation time appeared to be those that had just fed, but this was not significant (Table 1). The measure of time spent on the first leaf (if only one leaf was visited, 12 min is taken as the time on that leaf) showed a significant effect of deprivation treatment, but not of the host plant it was on (Table 1). In particular just-fed and 5-h-deprived beetles remained on the first leaf they contacted for less time than 3- and 6-day deprived beetles.

There was an insignificant effect of both host plant ($F = 3.1$, $df = 1, 95$, $P < 0.08$) and deprivation ($F = 2.4$, $df = 3, 95$, $P < 0.08$) on the proportion of time spent in oviposition. In the 5-h-deprived beetles on parthenium weed, on average 10% (SE = 6%) of their time was spent in oviposition, while just-fed beetles also on parthenium spent a mean of 1% (SE = 0.7%) of their time ovipositing. No oviposition occurred in any other treatment. There was no significant effect of beetle sex found on any parameters other than ovipositing when tested with a Generalised Linear Model (Minitab, 1995) with sex, plant, and deprivation level as factors.

## Discussion

The host-plant relationships of *Z. bicolorata* in Australia where a population exists on an alternative host plant, ragweed, enabled the comparison of adult beetle behaviour on two different host plants without the complicating factor of pre-test experience of either a significantly higher or lower ranked host plant (ragweed is similar in ranking to parthenium weed,

T. Withers, unpubl.). Furthermore preliminary experiments showed the handling process used in experiments did not artificially raise the response level of *Z. bicolorata*.

The four deprivation treatments permitted the comparison of behaviour between *Z. bicolorata* beetles at different behavioural 'threshold phases' as outlined in the models of host selection of Courtney et al. (1989), Courtney & Kibota (1990), and Singer et al. (1992). All just-fed *Z. bicolorata* (except for two individuals) tested within 15 min of feeding on ragweed were unresponsive to both host plants. This is a typical response from post-prandial insects (Simpson, 1995). These satiated beetles either rested on the underside of a leaf, walked actively about the plant, or left the plant in less than 12 min. Their behaviour did not differ greatly between the two host plants.

*Z. bicolorata* beetles that were 5-h-deprived (approximately one to two intermeal inter-vals) before being introduced to one of the host plants, showed many of the behaviours associated with 'discrimination phase' insects (Singer, 1982). These *Z. bicolorata* initiated feeding more rapidly and fed for significantly longer on parthenium weed, and usually rejected Noogoora burr after sample biting (Figure 2B & 3). In addition, 5-h-deprived females were the most likely to oviposit during observations onto the higher ranked host. The behaviour which apparently disagrees with the hierarchy threshold model for insects in this behavioural state was the lack of significant difference according to host plant in the proportion of insects taking a sample bite, and the latency to biting (Figure 1). This suggests that *Z. bicolorata* does not discriminate between Ambrosiinae host plants through initial contact, antennation and walking on the plant leaf. It was only after exhibiting sampling behaviours (breaking the leaf surface by biting) that the beetles began to show discriminating behaviour. Thus for *Z. bicolorata*, the stage of the host finding sequence that encompasses the critical behavioural decision to proceed from examining to consuming on potential hosts, appears to be when sensory cues received from the labial and maxillary palps, and sensory structures in the buccal cavity, are exposed to internal plant constituents, i.e., during sample biting (Miller & Strickler, 1984).

The 3-day deprived *Z. bicolorata* showed intermediate characteristics between discrimination and post-discrimination phase insects. In some behaviours they appeared to still discriminate actively, with the highest ranked host being accepted significantly more than the lower ranked (lag between sampling and feeding, and total time feeding), but in other behavioural measures they behaved as if in the post-discrimination phase, by responding similarly to both hosts in a way that approached the behaviour of the 6-day-deprived beetles (latency to sample bite, and time spent visiting the first leaf). The 6-day-deprived beetles showed many of the characteristics of post-discrimination phase insects; with latency to sample bite and proportions of insects sampling and feeding, total time spent feeding, as well as locomotory behaviours, not differing significantly between the two host plants. Yet some behaviours of post-discriminatory beetles consistently differed between the two host plants. These behavioural differences were that feeding on the lower ranked host occurred at multiple sites, with each meal preceded by a greater number of sample bites, and a resultant longer lag time between sampling and feeding.

The results from this experiment indicate that *Z. bicolorata* beetles increasingly accept lower ranked host plants after between 3 and 6 days of complete food deprivation. After 6 days of deprivation plant cues from the differently ranked hosts no longer had as strong an impact on behaviour as they did when the beetles were undeprived. This finding supports the predictions and assumptions of the hierarchy-threshold model of insect diet, but persisting behavioural differences suggest that although greater deprivation does cause differently ranked host plants to become more closely acceptable, that the insect does not lose the ability to discriminate between them.

Such a finding in which progressively deprived insects have come to feed on previously unacceptable host plants have also been shown in the acridids *Locusta migratoria* L. (Bernays et al., 1976) and *Chortoicetes terminifera* Walker (Bernays & Chapman, 1973), the psyllid *Cacopsylla pyricola* Foerster (Horton & Krysan, 1991), and many others. These results demonstrate how important internal factors are to the outcome of numerous types of insect feeding assays.

In conclusion, the temporal nature of changes in the internal levels of responsiveness are a vital part of all behavioural assays with insects. Without knowledge of the time since the last ingested meal, the nutritional quality of that ingested meal, and the mean inter-meal interval (Simpson, 1995) insects placed into feeding assays will differ vastly from each other in time-dependent internal state. When the acceptability of various plants as food or oviposition sites is being examined, such insects will also differ in their relative responses to the plants, with differences in acceptability greatest when insects are in the discrimination phase and smallest when insects are in the post-discrimination phase.

## Acknowledgements

Thanks to Anna Yeomans for invaluable technical assistance with rearing plants and insects, to Lindsay Barton Browne for stimulating discussions, and to Marion Harris and other referees for their helpful comments. This research was undertaken as part of the CRC for Tropical Pest Managements' host specificity research program, and publication was supported by the New Zealand Forest Research Institute Ltd.

## References

Bernays, E. A. & R. F. Chapman, 1973. The role of food plants in the survival and development of *Chortoicetes terminifera* (Walker) under drought conditions. Australian Journal of Zoology 21: 575–592.

Bernays, E. A., R. F. Chapman, J. Macdonald & J. E. R. Salter, 1976. The degree of oligophagy in *Locusta migratoria* (L.). Ecological Entomology 1: 223–230.

Courtney, S. P. & T. T. Kibota, 1990. Mother doesn't know best: selection of hosts by ovipositing insects. In: E. A. Bernays (ed.), Insect-Plant Interactions II. CRC Press, Boca Raton, Florida, pp. 161–188.

Courtney, S. P., G. K. Chen & A. Gardner, 1989. A general model for individual host selection. Oikos 55: 55–65.

Dhileepan, K. & R. E. McFadyen, 1997. Biological control of parthenium in Australia: progress and prospects. In: M. Mahadevappa & V. C. Patil (eds), First International Conference on Parthenium Management. University of Agricultural Sciences, Dharwad, India., pp. 40–44.

Horton, D. R. & J. L. Krysan, 1991. Host acceptance behavior of pear psylla (Homoptera: Psyllidae) affected by plant species, host deprivation, habituation, and egg-load. Annals of the Entomological Society of America 84: 612–627.

Jayanth, K. P. & G. Bali, 1994. Biological control of *Parthenium hysterophorus* by the beetle *Zygogramma bicolorata* in India. FAO Plant Protection Bulletin 42: 207–213.

McCullagh, P. & J. A. Nelder, 1989. Generalized Linear Models, 2nd ed. Chapman & Hall, London.

Miller, J. R. & K. L. Strickler, 1984. Finding and accepting host plants. In: W. J. Bell & R. T. Cardé (eds), Chemical Ecology of Insects. Sinauer, Sunderland, Mass. pp. 127–157.

Minitab, 1995. Minitab User's Guide. Minitab Inc., State College, PA.

Noldus, I. T., 1995. The Observer, Base Package for Windows. 3.0 ed., Noldus Information Technology b.v., Wageningen.

Simpson, S. J., 1995. Regulation of a meal: chewing insects. In: R. F. Chapman & G. de Boer (eds), Regulatory Mechanisms in Insect Feeding. Chapman and Hall, London, pp. 137–156.

Singer, M. C., 1982. Quantification of host preference by manipulation of oviposition behaviour in the butterfly *Euphydryas editha*. Oecologia 52: 224–229.

Singer, M. C., D. Vasco, C. Parmesan, C. D. Thomas & D. Ng, 1992. Distinguishing between 'preference' and 'motivation' in food choice: an example from insect oviposition. Animal Behaviour 44: 463–471.

Sokal, R. R. & F. J. Rohlf, 1981. Biometry. W. H. Freeman and Company, New York.

Withers, T. M., 1998. Influence of plant species on host acceptance behaviour of the biocontrol agent *Zygogramma bicolorata* (Coleoptera: Chrysomelidae). Biological Control 13: 55–62.

*Entomologia Experimentalis et Applicata* **91:** 97–103, 1999.



# Anomalous stylet punctures of phloem sieve elements by aphids

W. Fred Tjallingii[1] & Beata Gabryś[2]
[1]*Laboratory of Entomology, Agricultural University, POB 8031, 6709PD Wageningen, The Netherlands;*
[2]*Department of Agricultural Entomology, Agricultural University, Cybulskiego 32. 50–205 Wroclaw, Poland*

Accepted: October 10, 1998

*Key words:* EPG, potential drop, intracellular, sieve element, phloem, probing

## Abstract

A deviating type of potential drop (pd), i.e. the electrically recorded intracellular punctures during the pathway phase of plant penetration by aphids, is described and discussed. This so-called repetitive pd (R-pd) has been observed in 4 aphid species, two on herbs and two on sycamore. Apart from a regular repetition with short intervals, the most characteristic feature of the R-pd is the middle intracellular part, i.e. sub-phase 2, which is extended, making the individual R-pd 2–3 times longer than the normal pd. Differences between the normal and R-pd are described, as well as differences within each pd type between species. The R-pd's seem to reflect repeated punctures of phloem sieve elements. Their biological significance is unknown.

## Introduction

Electrical penetration graphs (EPGs) use an electrical circuit in which aphid and plant are incorporated (Tjallingii, 1988). Waveforms arise as soon as the stylets are inserted into the plant, thus completing the circuit. Three probing phases, each comprising one ore more waveforms have been distinguished. The technique has been used in many applied studies, such as host plant resistance (Van Helden & Tjallingii, 1993; Caillaud et al., 1995; Lei et al., 1998) and virus transmission (Prado & Tjallingii, 1994; Martin et al., 1997) as well as in fundamental studies (Tjallingii & Hogen Esch, 1993). Aphid EPGs show many brief intracellular punctures of plant cells by the aphid stylets. These punctures are known as potential drops (pd's), because the signal voltage suddenly drops by about 100 mV, from the extra- to the intracellular potential and *vice versa* (Tjallingii, 1985). Potential drops typically occur during the pathway phase of probing, i.e., before the phloem is reached. From the epidermis onwards, nearly every cell along the intercellular stylet route is punctured, the vascular bundle cells often more than once. The punctures appear to play a crucial role in the transmission of non-persistent plant viruses (Powell, 1991), concerning virus acquisition as well as inoculation. Recently, two out of three waveform sub-phases during the intracellular period of a pd (Powell et al., 1995; Collar et al., 1997) were correlated behaviourally, using viruses as markers (Martin et al., 1997).

In addition to the 'normal' pd, some aphid species appear to show a different type of pd, occurring in high numbers and in regular repetition and are therefore called repetitive potential drops (R-pd's). This is not a novel observation. McLean and Kinsey (1968) have observed them with their AC system, using *Acyrthosiphon pisum* (Harris), albeit without understanding their nature, and they have been mentioned earlier for the DC EPG system as well (Tjallingii, 1985).

Our aim was to study and describe the occurrence, waveform features, and histology at the stylet tips in a few aphid species.

## Materials and methods

For the EPG technique, the aphid and plant are made part of an electrical circuit by attaching one electrode of thin gold wire to the aphid's dorsum and inserting a second electrode, a metal rod, in the potting soil.

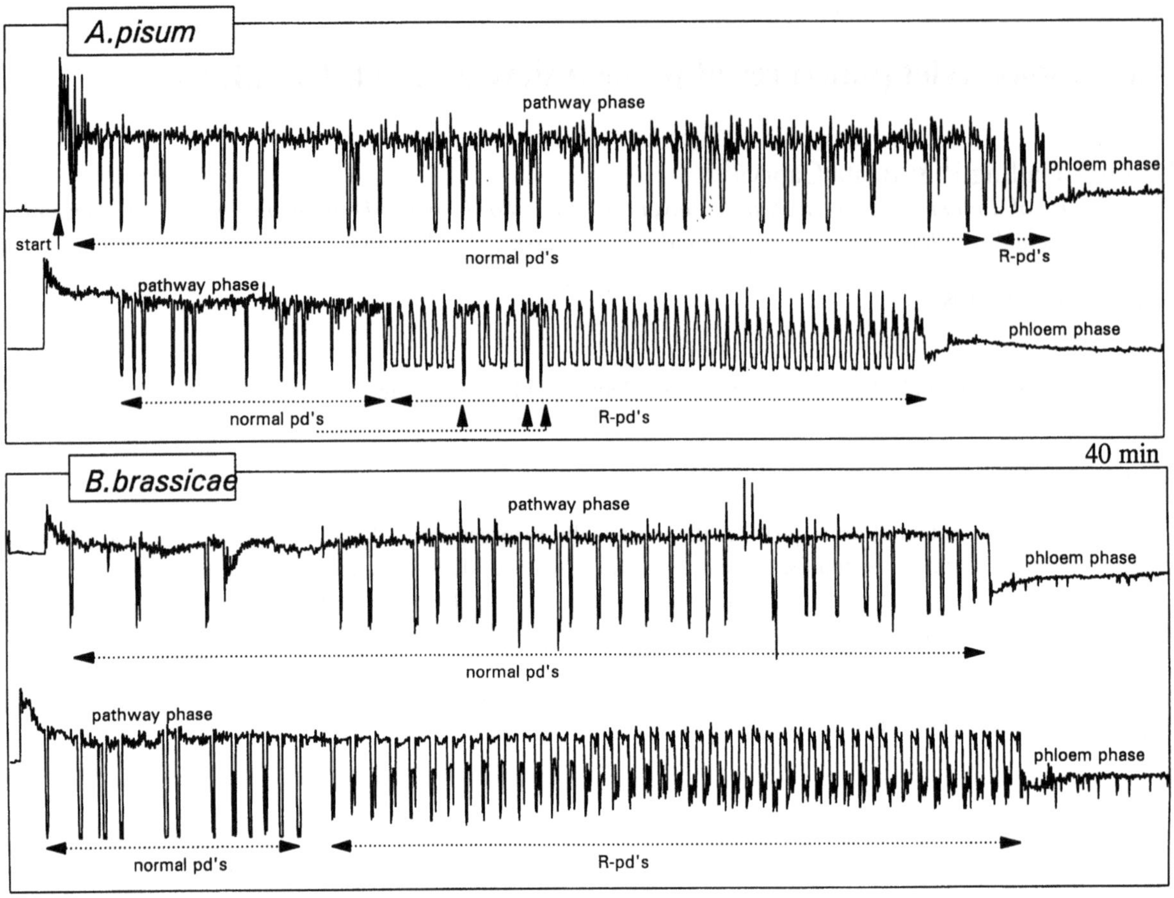


*Figure 1.* Sequences of normal and repetitive pd's in probes by *A. pisum* and *B. brassicae*. Start of probe, pathway and transition to phloem phase.

The circuit is completed when the aphid's stylets are inserted into the plant (Tjallingii, 1988). We recorded in the laboratory (ca. 20 °C) from adult apterous virginoparous aphids known to produce the R-pd's, i.e., *A. pisum* and *Brevicoryne brassicae* L., on 3–5-week-old *Vicia faba* L. and *Sinapis alba* L. potted plants, respectively. Aphids were reared at 18–20 °C and L16:D8, *B. brassicae* on Brussels sprouts, *Brassica oleracea* L., originating from a field-collected colony (Wageningen, 1996) and *A. pisum* on *V. faba* originating from a French colony (INSA-INRA, Villeurbane). Additionally, recordings were made of *Drepanosiphum platanoidis* (Shrank) and *Periphyllus acericola* (Walker) (alate virginoparae) collected on sycamore trees outside in May 1994. Alate adults were recorded in the laboratory on *Acer platanoides* L. seedlings, which had been reared in pots outside.

Stylectomy was carried out on *B. brassicae* after an R-pd sequence had been recorded on mustard. The leaf region probed was processed for electron microscopy, ultra thin sections and micrographs of the stylet tips in the sieve element were made according to methods described earlier (Tjallingii & Hogen Esch, 1993).

## Results and discussion

The two main species investigated showed long R-pd sequences (Figure 1), occasionally longer than one hour. Nevertheless, sequences of normal pd's were more common. The R-pd sequences mostly follow a period with normal pd's in a probe and generally they precede a phloem phase (waveforms E1 and E2), although sometimes they may follow a phloem phase or, occur between two phloem phases. The normal pd's of the two species last about 5 s, show three waveform sub-phases (Figure 2), and occur at very irregular intervals (see SD in Figure 3). R-pd's have a 2–3 times longer duration (*A. pisum* 12.6 s, *B. brassicae* 14.7 s) which is mainly caused by an increased sub-phase 2

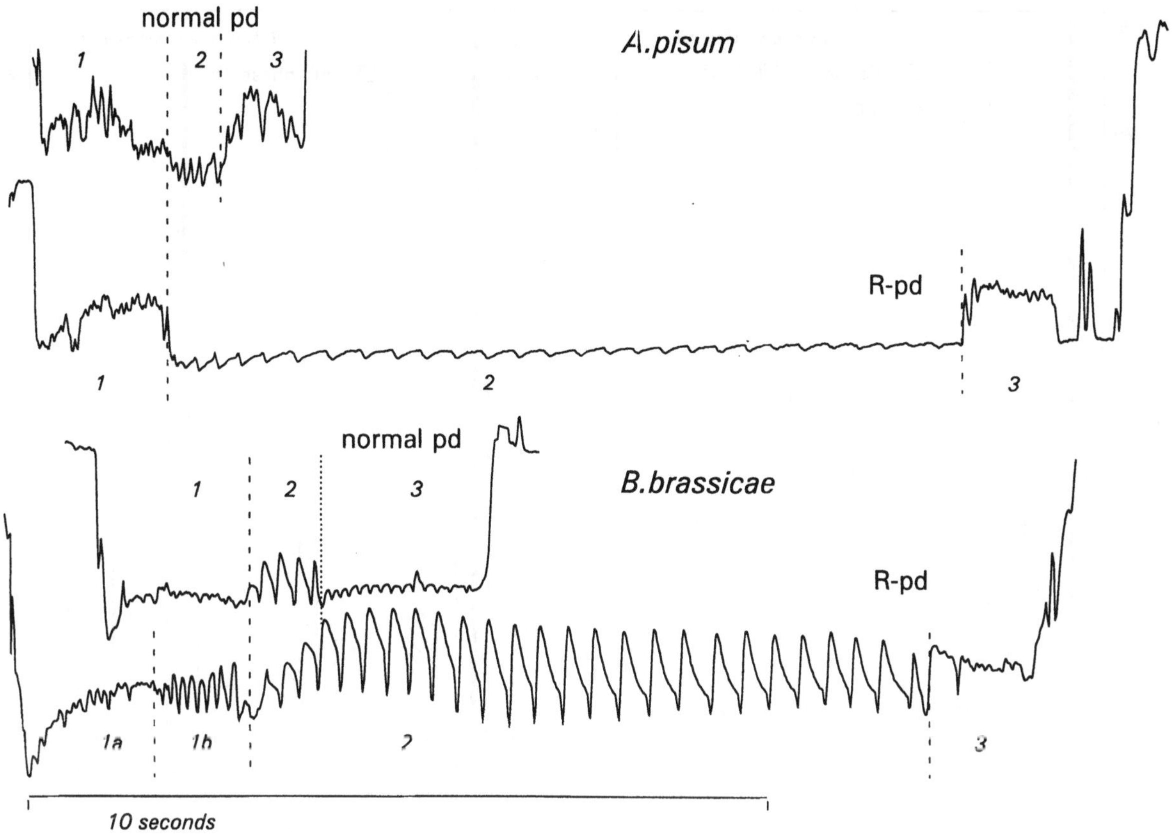


*Figure 2.* Single normal and repetitive pd's by *A. pisum* and *B. brassicae*. Waveform sub-phases separated by dotted lines and indicated by numbers.

(Figure 2). The most dominant feature, however, is the repetitive occurrence caused by the high constancy of the interval between individual R-pd's, which is also shorter than between normal pd's (Figure 3).

The magnitudes of the last normal pd's before phloem phase (Figure 1) were generally smaller than those of the preceding pd's. In *A. pisum* there is only one normal pd followed by three R-pd's. The potential drop as the stylets puncture the phloem is similar. This is consistent with Van Bel & Van Rijen's (1994) findings that the sieve element-companion cell complex has a lower transmembrane potential than the phloem parenchyma cells. Thus, these phloem preceding pd's with lower magnitudes are likely to reflect punctures of a sieve element or its companion cell. In the R-pd sequences, all R-pd magnitudes are smaller than the preceding or intermittent normal pd's. Thus, the R-pd's might also reflect sieve element or companion cell punctures. Differences between normal pd magnitudes, as also observed in Figure 1, would reflect punctures in different mesophyll cells (or plasmadesma interlinked groups of cells). So, the constancy of the pd magnitudes suggests that only one sieve element, or a sieve element and a companion cell, is punctured during an R-pd sequence.

Apart from the foregoing common R-pd features and differences between normal and repetitive pd's, there are a number of differences between pd's of the two species. In sequences with normal pd's by *A. pisum* the last pd's before phloem phase often have R-pd features, i.e., an extended sub-phase 2 and short and regular intervals (Figure 1), whereas in *B. brassicae* such changed last pd's do not occur. On the other hand, *B. brassicae* seems to have an extra sub-phase 1 in its R-pd's, with features that seem intermediate between sub-phase 1 and sub-phase 2 of other species, i.e. sub-phase 1b (Figures 2 and 3, stacked bars). Also, nearly all normal pd's by *B. brassicae* show the same high waveform amplitude during sub-phase 2 as the R-pd's do (Figure 2), whereas in *A. pisum* the sub-phase 2 has a low amplitude in both types of pd's.

Two other aphid species showed R-pd like phenomena, *P. acericola* and *D. platanoidis*, both occurring on sycamore trees (recently we also found R-pd's

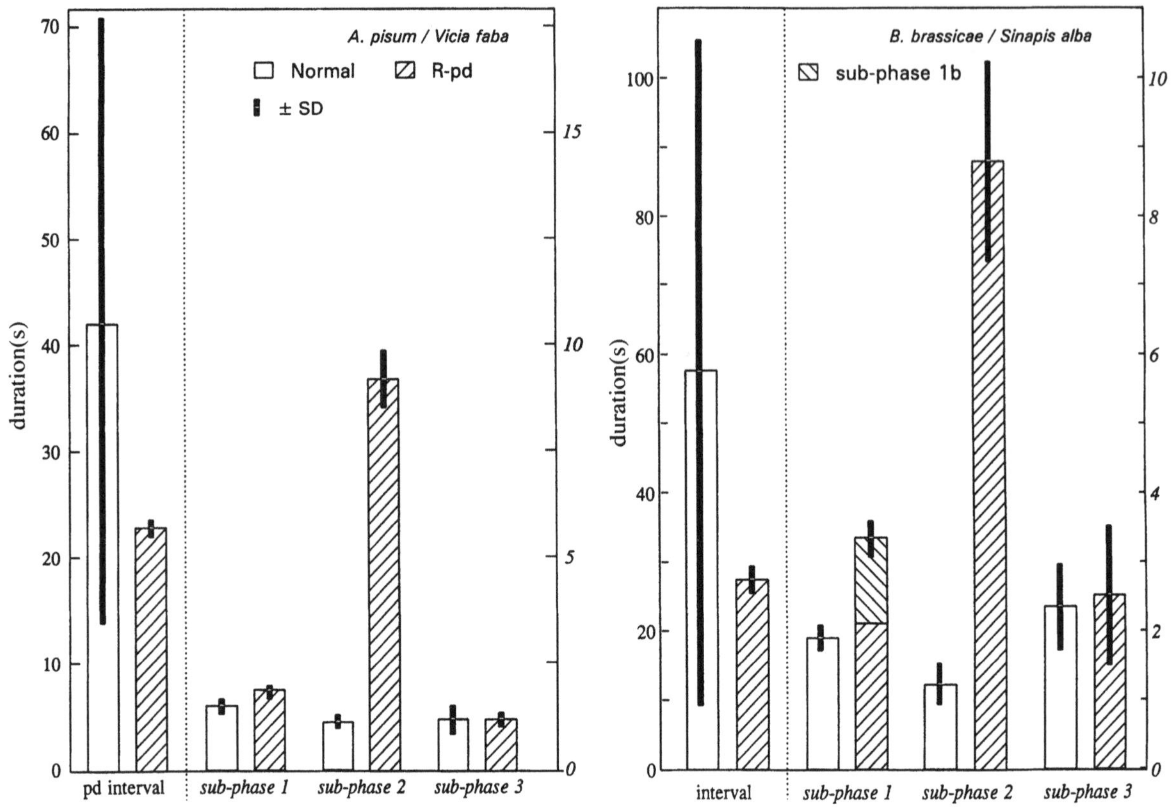


*Figure 3.* Durations (s) of normal pd and R-pd aspects in two aphids (mean ± SD). About 30 pd's from sequences, just before phloem phase; one R-pd and one normal pd sequence for each aphid. In the normal pd sequence, the few final pd's with R-pd features (*A. pisum*, especially) were excluded. Y axis (right side) with scale figures in italics refer to bars with labels in italics.

in *Tuberolachnus salignus* (Gmelin) on willow). Figures 4 and 5 show that R-pd's occur before phloem phases in these species. R-pd sequences were mostly shorter than in the two foregoing species. In *P. acericola*, R-pd's resemble those described for *A. pisum*, whereas in *D. platanoidis* a typical increase in the R-pd durations is shown (Figure 5) with time. The closer to the phloem phase the longer the R-pd duration becomes, up to several minutes ultimately, which is caused completely by an extended sub-phase 2 duration. R-pd's of both species show more or less the same additional waveform sub-phase 1b as described for *B. brassicae* above.

Sub-phase 2 features have a vague resemblance to E1 waveforms (phloem salivation). However, the similarity was not supported when the signals were observed in more detail and depicted side by side as shown for the two sycamore aphids in Figures 4 and 5. *Brevicoryne brassicae* and the two sycamore aphids always showed pd sub-phase waveforms immediately before the start of waveform E1 (phloem phase), whereas *A. pisum* did not (Figures 4 and 5).

Transmission electron micrographs (data not shown) of sections of the stylet tips, made after a sequence of R-pd's and a successive phloem phase (after stylectomy of *B. brassicae*) showed no special features of the punctured area as compared to micrographs with stylet tips in sieve elements when the phloem phase was preceded by normal pd's (Tjallingii & Hogen Esch, 1993). No increased numbers of punctures in the cell walls could be detected, neither in the sieve element nor in companion or other cells, earlier on the stylet route. This suggests that the repeated punctures of one sequence were all in the same place, using the same hole in the cell wall and seem to be restricted to repeated plasmalemma punctures only. The idea that there is no change in stylet tip position between successive R-pd's is also supported by the very short duration of the intervals between successive R-pd's in a sequence, allowing no pathway between the punctures.

We may conclude that the R-pd sequences in all four species seem to represent repeated plasmalemma punctures of one and the same sieve element, accord-

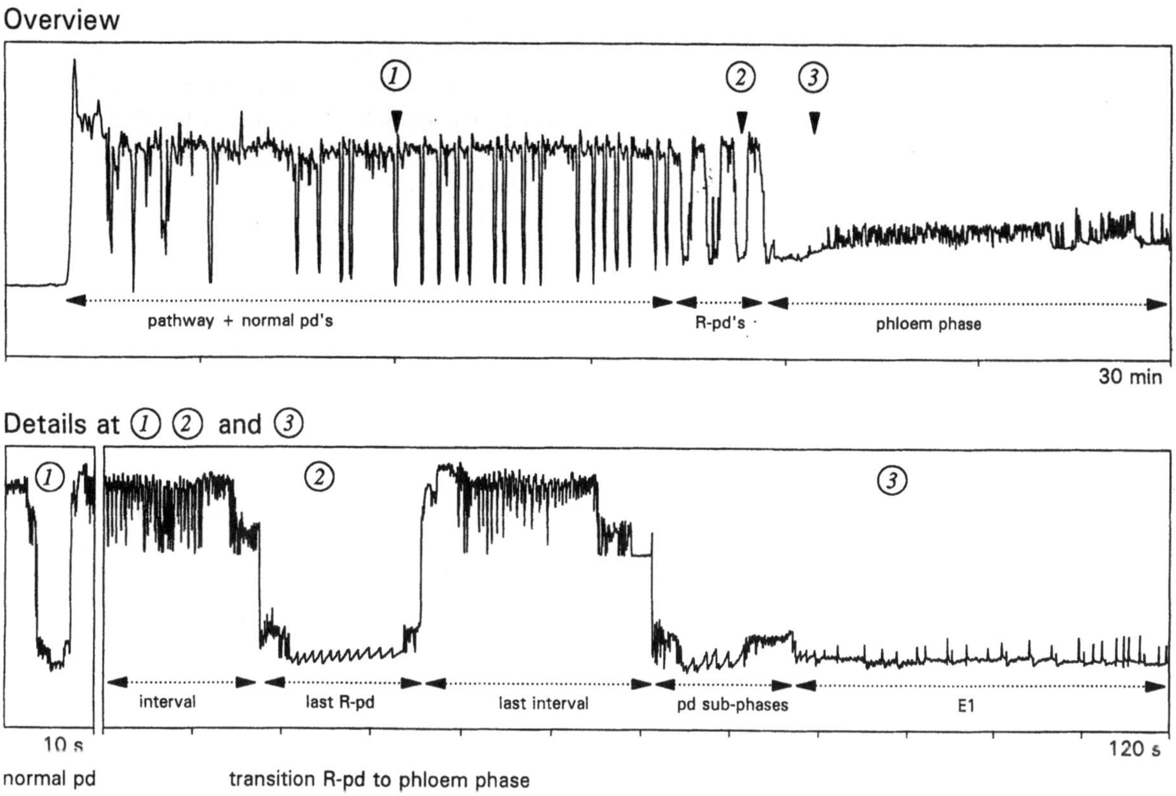


*Figure 4.* R-pd's before phloem phase by *P. acericola* on *A. platanoides* Encircled detail numbers refer to numbers in overview.

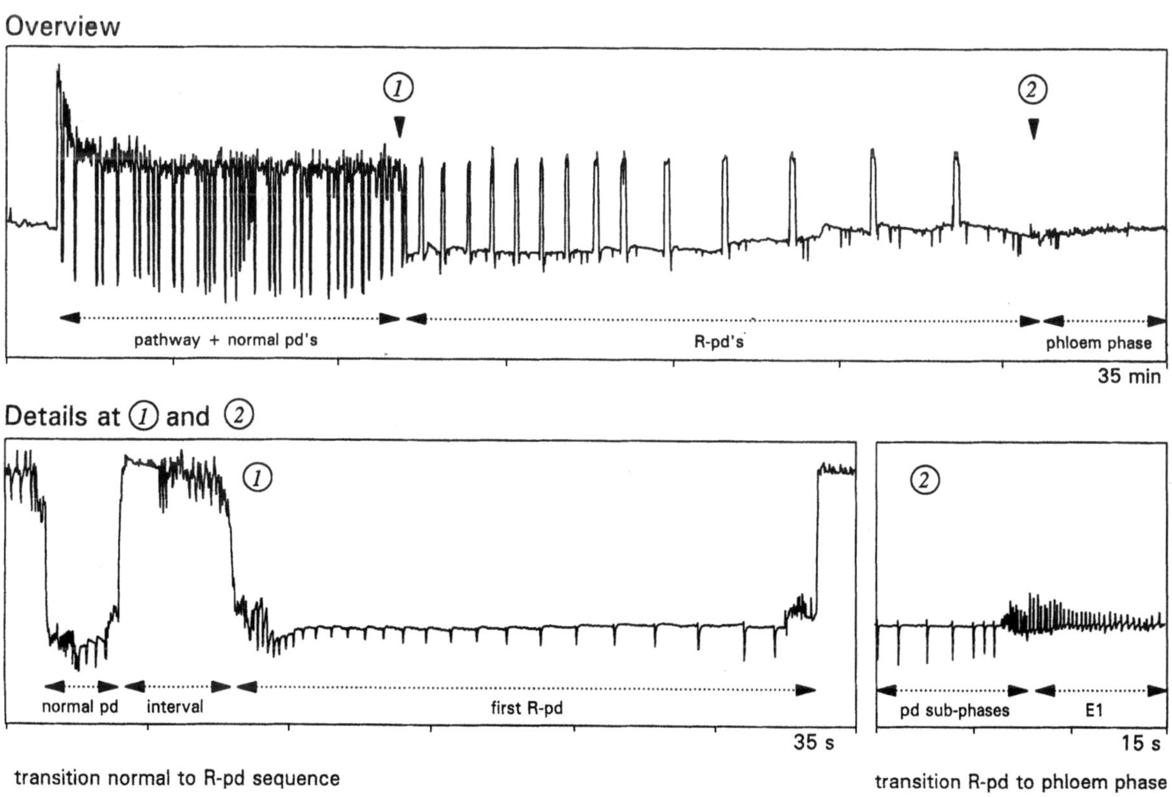


*Figure 5.* R-pd's before phloem phase by *D. platanoidis* on *A. platanoides.* Encircled detail numbers refer to numbers in overview.

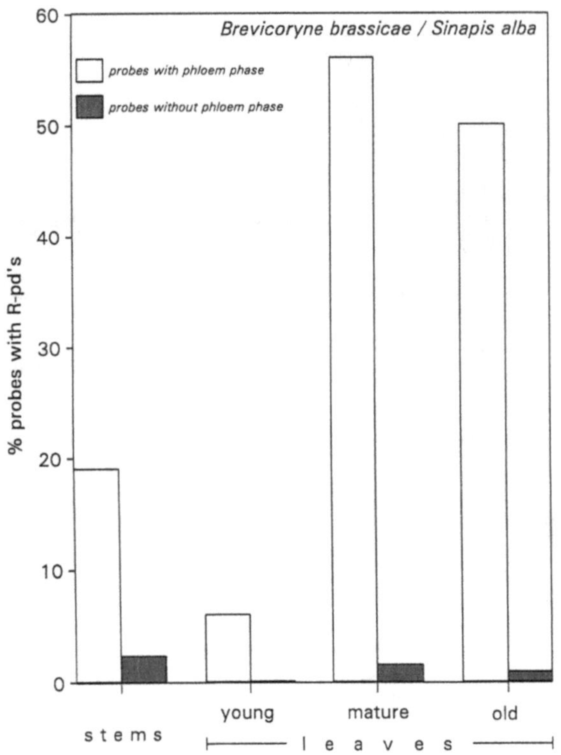


*Figure 6.* Occurrence of R-pd sequences by *B. brassicae* on different plant parts of white mustard.

ing to the EPG and micrograph data. Why sieve elements are punctured repeatedly by these aphid species remains unclear. As sub-phase 1 in each pd represents salivation and sub-phase 3 represents ingestion (Martin et al., 1997), these activities are repeated as often as the number of R-pd's in a series. However, now it would be interesting to know more about the functional aspects of sub-phase 2, as this is the part which is extended in R-pd's, especially in the case of *D. platanoidis.*

*Brevicoryne brassicae* on white mustard shows a clear preference for young stems to leaves, especially to senescent and old leaves (Gabrys et al., 1997). R-pd sequences occurred more on the less preferred mature and old leaves (Figure 6). Moreover, this figure shows a clear link between the R-pd sequences and the phloem phases in a probe. Although there were a few exceptions (solid bars), there is a high coincidence in the occurrence of both phenomena together. With the other species we have no detailed observations. In *A. pisum* most waveform transitions from pathway to phloem phase show a few R-pd's but long sequences are much less common and may depend on plant part, age, nutrition, etc. The sycamore aphids seem to show the described R-pd sequences in nearly every pathway to phloem phase transition.

We need further studies, and presumably additional techniques, to clarify the relationships between stylet activities and the overall function of pd's and R-pd's in aphid-plant relationships. There seems no reason to suppose that their occurrence forms a prerequisite for phloem feeding. Whiteflies, homopterans that also feed predominantly on phloem, do not show pd's, at least not as a regular activity during pathway phases of probing (Janssen et al., 1989).

## Acknowledgements

We thank Du Yong-jun for data collection and Thie Hogen Esch for the TEM work and Dr. Yvan Rahbé for providing the *A. pisum* clone.

## References

Bel, A. J. E. van, & H. V. M. van Rijen, 1994. Microelectrode-recorded development of the symplasmic autonomy of the sieve element/companion cell complex in the stem phloem of *Lupinus luteus* L. Planta 192: 165–175.

Caillaud, C. M., J. S. Pierre, B. Chaubet & J. P. di Pietro, 1995. Analysis of wheat resistance to the cereal aphid *Sitobion avenae* using electrical penetration graphs and flow charts combined with correspondence analysis. Entomologia Experimentalis et Applicata 75: 9–18.

Collar, J. L., C.Avilla & A. Fereres, 1997. New correlations between aphid stylet paths and non-persistent virus transmission. Environmental Entomology 26: 537–544.

Gabrys, B., W. F. Tjallingii & T. A. van Beek, 1997. Analysis of EPG recorded probing by cabbage aphid on host plant parts with different glucosinolate contents. Journal of Chemical Ecology 23: 1661–1673.

Helden, M. van, & W. F. Tjallingii, 1993. Tissue localisation of lettuce resistance to the aphid *Nasonovia ribisnigri* using electrical penetration graphs. Entomologia Experimentalis et Applicata 68: 269–278.

Janssen, J. A. M., W. F. Tjallingii & J. C. van Lenteren, 1989. Electrical recording and ultrastructure of stylet penetration by the greenhouse whitefly. Entomologia Experimentalis et Applicata 52: 69–81.

Lei, H., W. F. Tjallingii & J. C. van Lenteren, 1998. Probing and feeding characteristics of greenhouse whitefly in association with host-plant acceptance and whitefly strains. Entomologia Experimentalis et Applicata 88: 73–80.

Martin, B., J. L. Collar, W. F. Tjallingii & A. Fereres, 1997. Intracellular ingestion and salivation by aphids may cause the acquisition and inoculation of non-persistently transmitted plant viruses. Journal of General Virology 78: 2701–2705.

McLean, D. L. & M. G.Kinsey, 1968. Probing behavior of the pea aphid, *Acyrthosiphon pisum.* III. Effect of temperature on certain probing activities. Annals of the Entomological Society of America 61: 927–933.

Powell, G., 1991. Cell membrane punctures during epidermal penetrations by aphids: consequences for the transmission of two potyviruses. Annals of Applied Biology 119: 313–321.

Powell, G., T. Pirone & J. Hardie, 1995. Aphid stylet activities during potyvirus acquisition from plants and an *in vitro* system that correlate with subsequent transmission. European Journal of Plant Pathology 101: 411–420.

Prado, E., & W. F. Tjallingii, 1994. Aphid activities during sieve element punctures. Entomologia Experimentalis et Applicata 72: 157–165.

Tjallingii, W. F., 1985. Membrane potentials as an indication for plant cell penetration by aphid stylets. Entomologia Experimentalis et Applicata 38: 187–193.

Tjallingii, W. F., 1988. Electrical recording of stylet penetration activities. In: A. K. Minks & P. Harrewijn (eds), Aphids, their Biology, Natural Enemies and Control. Elsevier, Amsterdam, pp. 95–108.

Tjallingii, W. F. & Th. Hogen Esch, 1993. Fine structure of the stylet route in plant tissues by some aphids. Physiological Entomology 18: 317–328.

*Entomologia Experimentalis et Applicata* **91:** 105–109, 1999.




# Acceptability of different species of Brassicaceae as hosts for the cabbage aphid

Beata Gabryś & Magdalena Pawluk
*Department of Agricultural Entomology, Agricultural University. Cybulskiego 32, 50-205 Wrocław, Poland*

Accepted: October 15, 1998

*Key words:* aphid, EPG, probing behaviour, plant resistance


## Abstract

The probing and feeding behaviour of the cabbage aphid, *Brevicoryne brassicae* (L.), (Homoptera, Aphididae) was studied on several plant species that represented various levels of acceptability: *Sinapis alba* L. (a permanent host plant), *Capsella bursa-pastoris* (L.) Med., *Thlaspi arvense* L., *Lunaria annua* L., *Erysimum cheiranthoides* L. (accidental host plants), *Vicia faba* L. (a non-host plant), using the electrical penetration graph technique (EPG). *B. brassicae* on *V. faba* did not show any patterns related to penetration of phloem vessels. Stylet penetration was deterred on *L. annua* and *E. cheiranthoides* where non-penetration prevailed, the periods of sap ingestion were short or did not occur, the percentage of time spent in the phloem was consistently low (5–6%) and E1 salivation predominated. The pathway activities were not suppressed on *C. bursa-pastoris* and *T. arvense* and the aphids spent an average of 3 h in the phloem during the 8-h experiment. However, a considerable delay between finding and accepting the phloem and a substantial proportion of E1 salivation (20–30% of all phloem activities) indicated a deterrent factor in the sieve elements of these plants. Aphid probing and sap ingestion were rarely interrupted on *S. alba*. The results of this study suggest that the deterrent agents vary in activity and may hinder stylet penetration at different levels (epidermis, parenchymatous tissues and/or phloem elements), depending on the plant species.


## Introduction

The family Brassicaceae (= Cruciferae) comprises 350 genera and circa 3000 species which occur mainly in the Northern Hemisphere (Hegi, 1986). The brassicaceous plants are very common and play an important role as hosts for many phytophagous insects. The cabbage aphid, *Brevicoryne brassicae* (L.) (Homoptera, Aphididae), is widely distributed in temperate regions of the world where it may severely damage brassicaceous field crops by feeding and transmitting about 20 plant viruses (Blackman & Eastop, 1984). In Poland and Central Europe, the cabbage aphid may use at least 97 species of Brassicaceae as permanent or temporal hosts (Lipa et al., 1977). According to Markkula (1953), the true hosts of *B. brassicae* may be divided into three groups depending on their ability to support the aphid population:

- permanent host-plants, which may support the cabbage aphid population throughout the vegetation period;
- temporal host-plants, which may support only 2–3 generations;
- accidental hosts, on which the aphid may develop less than one generation.

The acceptability of a host plant for an aphid is often measured by its quality as a food source, especially in studies on plant resistance mechanisms (Montllor & Tjallingii, 1989). Valuable information on resistance factors and their localisation in plant tissues which may affect feeding is provided by monitoring of aphid probing behaviour using the electrical penetration graph (EPG) technique (Tjallingii & Mayoral, 1992). The analysis of parameters derived from EPG (frequency, duration and sequence of different waveforms) may reflect behavioural responses of aphids to differences in plant suitability (Mayoral et al., 1996).

This study was carried out to investigate the probing behaviour of *B. brassicae* on several plant species that represent various levels of acceptability: *Sinapis alba* L. (a permanent host plant), *Capsella bursa-pastoris* (L.) Med., *Thlaspi arvense* L., *Lunaria annua* L., *Erysimum cheiranthoides* L. (accidental host plants), *Vicia faba* L. (a non-host plant).

## Materials and methods

*Aphids.* *Brevicoryne brassicae* was collected from an oilseed rape field in Pawłowice near Wrocław. The aphids were reared on cabbage, *Brassica oleracea* var. *capitata* L., in the laboratory at about 20 °C, 70% r.h. and L16:D8 photoperiod.

*Plants.* *Sinapis alba*, *Lunaria annua* and *Vicia faba* were grown from commercially available seeds in plastic pots, in a standardised soil mixture. *Capsella bursa-pastoris*, *Thlaspi arvense* and *Erysimum cheiranthoides* were collected from natural sites, transferred to the laboratory and re-planted in plastic pots. All plants were maintained in a growth chamber at about 20 °C, 70% r.h. and L16:8D photoperiod. Plants collected in the field were conditioned in the laboratory for 3–5 days prior to the experiments. All plants, except *L. annua*, were used at the beginning of the flowering stage; the field-collected plants had a few young pods. *L. annua* was used at the 4-leaf stage. All experiments were carried out in May–June.

*EPG recording.* The probing behaviour of 2-4-days-old adult apterous aphids was recorded using the Giga-4 DC-amplifier (Wageningen Agricultural University, The Netherlands) which had $10^9$ Ohm input resistance and an adjustable plant voltage. The EPG recording was performed inside a Faraday cage in the laboratory at about 20 °C during the day for 8 h continuously, one aphid per plant, and the signals were stored on the hard disk of a PC. EPGs were recorded from aphids on stems of *S. alba* (17 individuals), *V. faba* (16), *C. bursa-pastoris* (13), *T. arvense* (14) and *E. cheiranthoides* (13), and on leaves of *L. annua* (15).

*EPG analysis.* The acquisition and analysis of the EPG signals were done with STYLET 2.0 computer program (Tjallingii & Mayoral, 1992). The waveform patterns were identified according to Tjallingii (1990). The aphid behaviour during the experiments was divided into probing and non-probing periods. The probing, i.e., the aphid stylet penetration in plant tissues was further divided into the pathway phase (A, B, C, and F waveforms) and the phloem phase (E1 – salivation and E2 – sap ingestion). A number of parameters were derived from the EPGs. Total time of probing, proportion of phloem patterns E1 and E2, number of probes, proportion of short (<3 min) probes before the 1st phloem phase, delay between finding and accepting the sieve elements, are good indicators of plant suitability or interference of probing by chemical or physical factors in certain plant tissues (Mayoral et al., 1996). Probes before the first phloem phase equal to or less than 3 min long would not go beyond about one cell layer (Gabryś et al., 1997). The data were statistically analysed using the analysis of variance (Kruskal–Wallis test; $P = 0.05$).

## Results and discussion

All aphids showed distinguishable patterns of probing. The waveforms representing pathway activities (A, B, C and F), phloem salivation (E1) and phloem sap ingestion (E2) were recognised and pattern F (stylet penetration difficulties during pathway) occurred sporadically. However, the electronic monitoring of the feeding behaviour showed important differences in acceptance of brassicaceous plants by the cabbage aphid.

*Sinapis alba* was undoubtedly the most favourable host plant used in this study where probing and sap ingestion were rarely interrupted. The first encountered sieve element was accepted for sustained (i.e., longer than 10 min) sap ingestion which lasted for almost 3 h on average, and phloem salivation (E1) was less than 3% of all activities in sieve elements (Table 1). The proportion of E (=E1+E2) patterns gradually increased to reach 70% of all probing activities at the end of the experiment (Figure 1).

*Capsella bursa-pastoris* and *Thlaspi arvense* appeared acceptable to *B. brassicae* because the pathway activities were not suppressed and the aphids spent 3 h on average in phloem activities during the 8-h experiment. However, a considerable delay between finding and accepting the phloem (time between 1st E and 1st E2>10 min) and a substantial proportion of phloem salivation (20–30% of time spent in the phloem) (Table 1) indicate a deterrent factor in the phloem elements of these plants (Tjallingii, 1994), which may impede aphids settling (Markkula, 1953;

*Table 1.* Probing behaviour of *Brevicoryne brassicae* on host and non-host plants (mean values per aphid ±s.d.)

| Plant species<br>EPG parameter | | *Sinapis alba* | *Capsella bursa-pastoris* | *Thlapsi arvense* | *Lunaria annua* | *Erysimum cheiranthoides* | *Vicia faba* |
|---|---|---|---|---|---|---|---|
| Total penetration time A+B+C+F+E | min | 445 ± 27.1b | 334.8 ± 120.9cd | 392.5 ± 77.2d | 276.4 ± 128.3c | 175 ± 95.5a | 343.5 ± 59.8c |
| Number of probes | # | 6.0 ± 6.1b | 13.7 ± 13.7cd | 8.6 ± 4.8bc | 15.9 ± 15.8c | 12.0 ± 9.5c | 26.3 ± 11.8a |
| Probes <3 min before 1st E1 | % | 30.6 | 46.9 | 48.4 | 58.5 | 64.5 | – |
| Total phloem pattern in time E1+E2 | min | 252.2 ± 111.1c | 147.1 ± 165.4a | 180.4 ± 112.6c | 17.5 ± 16.1b | 35.0 ± 46.8b | – |
| Total sap ingestion time E2 | min | 247.9 ± 112.3c | 140.4 ± 166.4b | 160 ± 120.0bc | 9.7 ± 13.6a | 30.5 ± 46.8a | – |
| E/E+A+B+C+F | % | 55.6 ± 22.4d | 34.7 ± 35.7bc | 43.6 ± 23.1cd | 5.5 ± 5.1a | 18.0 ± 24.8ab | – |
| E1/E1+E2 | % | 2.4 ± 3.3a | 26.4 ± 35.0b | 19.8 ± 17.0b | 53.8 ± 28.9c | 34.9 ± 37.0bc | – |
| Time from 1st E to 1st E2>10 min | min | 0.0a | 80.3 ± 89.2b | 135.9 ± 114.1b | – | 104.1 ± 101.5b | – |
| Aphids showing E1 | % | 100.0 | 76.9 | 100.0 | 80.0 | 69.2 | – |
| Aphids showing E2>10 min | % | 100.0 | 61.5 | 85.7 | 20.0 | 46.2 | – |

Different letters show significant difference (Kruskal–Wallis test; P = 0.05); E=E1+E+2; E2>10 min = sustained sap ingestion.

Gadomski, 1992). Prolonged E1 salivation has often been reported characteristic of aphid behaviour on resistant plant cultivars (van Helden & Tjallingii, 1993; Klingler et al., 1998; Wilkinson & Douglas, 1998). It has been established that aphid saliva might be used for neutralising plant defence mechanisms located in sieve elements (Miles, 1990). Less salivation into sieve elements was reported for *Aphis fabae* Scopoli on a previously colonised leaf of *V. faba* (Prado & Tjallingii, 1997). Salivary enzymes (polyphenol oxidase and peroxidase) of the bird cherry-oat aphid, *Rhopalosiphum padi* L., are involved in detoxification of plant phenolic compounds (Urbanska & Leszczynski, 1992).

On *Lunaria annua* and *Erysimum cheiranthoides* aphid probing was clearly deterred, apparently by factors present both in the peripheral tissues and in the phloem. For most of the study time the stylets did not penetrate the plant, short (<3 min) probes prevailed, the sap ingestion periods were short or did not occur, the percentage of time spent on phloem activities was consistently low (5–6%) (Figure 1) and E1 salivation predominated. 60% of aphids on *L. annua* and 20% on *E. cheiranthoides* did not accept sieve elements after periods of E1 salivation (Table 1). The lack of feeding was probably the reason for the severely reduced fecundity of *B. brassicae* on these plants (Markkula, 1953).

On *Vicia faba*, *B. brassicae* did not show any patterns related to penetration of phloem vessels. However, aphids were not deterred from penetrating the leaf surface and the total penetration time was comparable to that on brassicaceous plants. It is likely that repeated probing on this non-host plant was partly due to the effect of tethering: aphids might have walked away after a short probe had they been free to move (Powell et al., 1993; Gabryś et al., 1997). Nevertheless, it appears that host acceptance by the cabbage aphid is not mediated by factors occurring at the leaf surface. The average duration of a probe in stems of *V. faba* (about 15 min) suggests that aphid stylets penetrated beyond epidermis and outer parenchyma allowing the sampling from phloem elements during brief cell punctures along pathway (Tjallingii, 1994).

*B. brassicae* recognises its host plants by using chemical cues present in plant tissues. The glucosinolates, which are the feeding stimulants, occur in all plants of the Brassicaceae family (Fenwick et al., 1983). However, atypical secondary compounds also exist in brassicaceous plants and might be responsible for variation in acceptability of different species to specialised phytophages (Usher & Feeny, 1983). Indeed, in the case of *Pieris brassicae* L. and *Pieris rapae* L. (Lepidoptera, Pieridae), the balance between stimulatory compounds (the glucosinolates) and the deterrent ones (cardenolides) determined the reaction of these species to *E. cheiranthoides* (Huang et al., 1993). It is likely that similar factors are responsible for the different acceptability of brassicaceous plants to the cabbage aphid. Our data suggest that if these chemicals are involved they differ in the deterrent activity and may impede stylet penetration at the level of epidermis, parenchymatous tissues, as well as in the phloem elements, depending on the plant species.

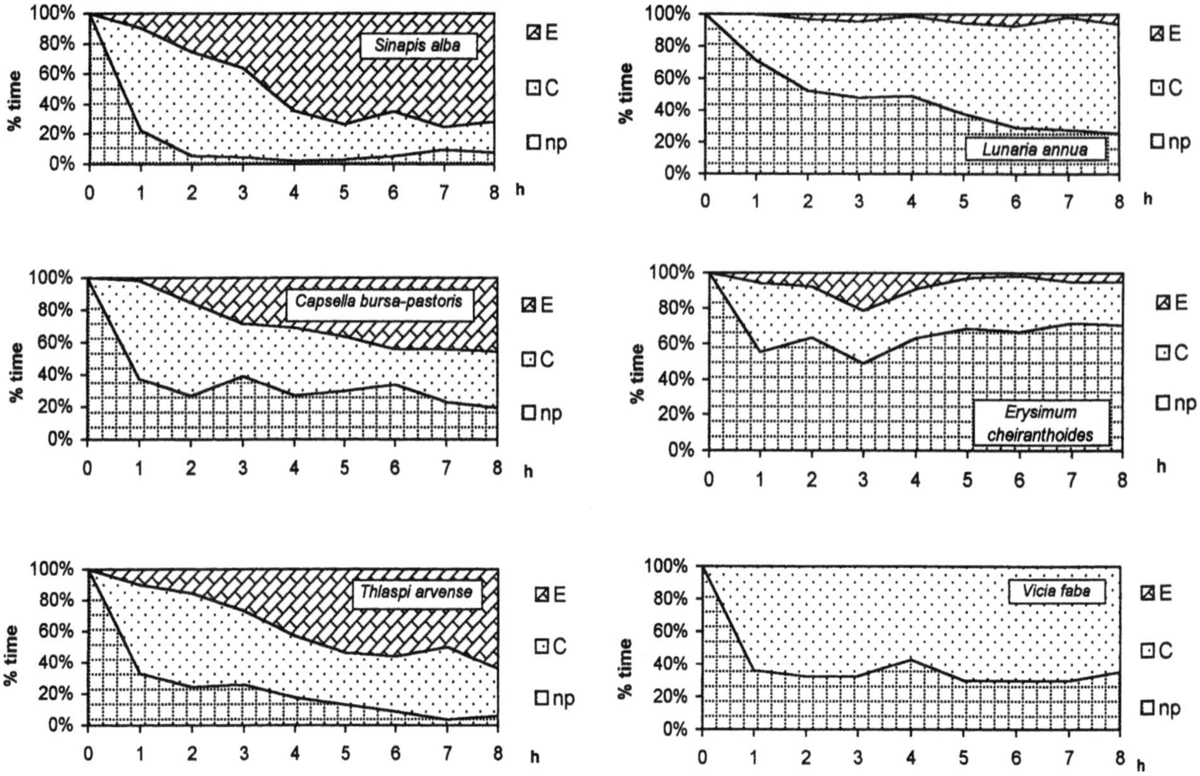


*Figure 1.* Probing behaviour of *Brevicoryne brassicae* on host and non-host plants during 8-h EPG experiments. E=E1+E2; C=A+B+C+F; np=non-probing.

## References


Blackman, R. L. & V. F. Eastop, 1984. Aphids on the World Crops. An Identification Guide. John Wiley & Sons, New York.

Fenwick, G. R., R. K. Heaney & W. J. Mullin, 1983. Glucosinolates and their breakdown products in food and food plants. CRC Critical Review in Food Science and Nutrition 18: 123–202.

Gabryś, B., W. F. Tjallingii & T. A. van Beek, 1997. Analysis of EPG recorded probing by cabbage aphid on host plant parts with different glucosinolate contents. Journal of Chemical Ecology 23: 1661–1673.

Gadomski, H., 1992. Oviposition of *Brevicoryne brassicae* (L.) (Homoptera: Aphididae) on cruciferous plants. Aphids and Other Homopterous Insects. PAS, Warsaw. 3: 33–38.

Helden, M. van & W. F. Tjallingii, 1993. Tissue localisation of lettuce resistance to the aphid *Nasonovia ribisnigri* using electrical penetration graphs. Entomologia Experimentalis et Applicata 68: 269–278.

Hegi, G., 1986. Illustrierte Flora von Mitteleuropa. Verlag Paul Parey, Berlin.

Huang, H., J. A. A. Renwick & K. Sadchev-Gupta, 1993. A chemical basis for differential acceptance of *Erysimum cheiranthoides* by two *Pieris* species. Journal of Chemical Ecology 19: 195–210.

Klingler, J., G. Powell, G. A. Thompson & R. Isaacs, 1998. Phloem specific aphid resistance in *Cucumis melo* line AR 5: effects on feeding behaviour and performance of *Aphis gossypii*. Entomologia Experimentalis et Applicata 86: 79–88.

Lipa, J. J., A. Studziński & D. Małachowska, 1977. Insects and Mites Associated with Cultivated and Weedy Cruciferous Plants (Cruciferae) in Poland and Central Europe. Polish Scientific Publishers, Warszawa-Poznan.

Markkula, M., 1953. Biologisch-ökologische Untersuchungen über die Kohlblattlaus, *Brevicoryne brassicae* (L.) (Hem., Aphididae). Annales Zoologici Societatis Zoologicae Botanicae Fennicae 'Vanamo' 15: 1–133.

Mayoral, A. M., W. F. Tjallingii & P. Castanera, 1996. Probing behaviour of *Diuraphis noxia* on five cereal species with different hydroxyamic acid levels. Entomologia Experimentalis et Applicata 78: 341–348.

Miles, P., 1990. Aphid salivary secretions and their involvement in plant toxicoses. In: R. K. Campbell & R. D. Eikenbary (eds), Aphid-Plant Genotype Interactions. Elsevier Science Publishers, Amsterdam, pp. 131–147.

Montllor, C. B. & W. F. Tjallingii, 1989. Stylet penetration by two aphid species on susceptible and resistant lettuce. Entomologia Experimentalis et Applicata 52: 103–111.

Powell, G., J. Hardie & J. A. Pickett, 1993. Effects of the antifeedant polygodial on plant penetration by aphids, assessed by video and electrical recording. Entomologia Experimentalis et Applicata 68: 193–200.

Prado, E. & W. F. Tjallingii, 1997. Effects of previous plant infestation on sieve element acceptance by two aphids. Entomologia Experimentalis et Applicata 82: 189–200.

Tjallingii, W. F., 1990. Continuous recording of stylet penetration activities by aphids. In: R. K. Campbell & R. D. Eikenbary, (eds), Aphid-Plant Genotype Interactions. Elsevier Science Publisher., B. V. Amsterdam, pp. 88–89.

Tjallingii, W. F., 1994. Sieve element acceptance by aphids. European Journal of Entomology 91: 47–52.

Tjallingii, W. F. & A. M. Mayoral, 1992. Criteria for host plant acceptance by aphids. In: S. B. J. Menken, J. H. Visser & P. Harrewijn (eds), Proceedings 8th Int. Symposium on Insect-Plant Relationships. Kluwer Academic Publishers, Dordrecht, pp. 280–282.

Urbanska, A. & B. Leszczynski, 1992. Biochemical adaptations of cereal aphids to host plants. In: S. B. J. Menken, J. H. Visser & P. Harrewijn (eds), Proc. 8th Int. Symposium Insect-Plant Relationships. Kluwer Academic Publishers, Dordrecht, pp. 277–279.

Usher, B. F. & P. Feeny, 1983. Atypical secondary compounds in the family Cruciferae: tests for toxicity to *Pieris rapae*, an adapted crucifer-feeding insect. Entomologia Experimentalis et Applicata 34: 257–262.

Wilkinson, T. L. & A. E. Douglas, 1998. Plant penetration by pea aphids (*Acyrthosiphon pisum*) of different plant range. Entomologia Experimentalis et Applicata 87: 43–50.

Entomologia Experimentalis et Applicata **91:** 111–114, 1999.




# Salivation into sieve elements in relation to plant chemistry: the case of the aphid *Sitobion fragariae* and the wheat, *Triticum aestivum*

Claudio C. Ramírez & Hermann M. Niemeyer
*Laboratorio de Química Ecológica, Departamento de Ciencias Ecológicas, Facultad de Ciencias, Universidad de Chile, Casilla 653, Santiago, Chile*

Accepted: October 15, 1998

*Key words:* sieve element, salivation, aphid, plant resistance, wheat, *Sitobion fragariae*, *Triticum aestivum*

**Abstract**

Extended sieve element salivation (E1 waveform in the electrical penetration graph) is a characteristic activity during early sieve element punctures, particularly in resistant plants. In order to explore a chemically-mediated mechanism of resistance associated with sieve element salivation, we compared the pattern of feeding behaviour of the aphid, *Sitobion fragariae* (Walker), on two cultivars of the wheat *Triticum aestivum* L., with different concentrations of hydroxamic acids (Hx). During 24 h of electronic monitoring, aphids dedicated over 50% of the total time to phloem ingestion from the sieve elements. Total time allocated to E1 in the experiment, time to first E1 within the experiment, time allocated to E1 before a sustained phloem ingestion (E2) and the contribution of sieve element salivation to the phloem phase (E1/[E1+E2]) were significantly higher in the high-Hx cultivar. The increased salivation in plants with higher contents of Hx suggests the existence, at least in this system, of a chemically-mediated sieve element constraint.

## Introduction

Feeding behaviour of cereal aphids has been intensively studied in relation to plant resistance. Comparison of feeding behaviour in susceptible/resistant plants as well as in host/non-host plants has been used to explore the plant factors accounting for the resistance (Spiller, 1988; Girma et al., 1992; Caillaud et al., 1995; Mayoral et al., 1996). Recently, studies using the electrical penetration graph (EPG) technique have reported increased salivation into sieve elements (E1 pattern) (see review in Prado, 1997), as well as delay in sieve element ingestion (Givovich & Niemeyer, 1995; Caillaud et al., 1995) of aphids feeding on resistant plants. In order to explore a chemically-mediated mechanism associated with sieve element salivation, we studied the feeding behaviour of the aphid *Sitobion fragariae* (Walker) on two cultivars of wheat, *T. aestivum*, varying in their content of hydroxamic acids (Hx), with special emphasis on EPG parameters related to salivation in sieve elements. Several investigations on Hx have addressed their role in the resistance of wheat against aphids (Niemeyer, 1990; Nicol et al., 1992; Givovich et al., 1994; Mayoral et al., 1996, Nicol & Wratten, 1997).

## Materials and methods

*Insects.* A clonal colony of *S. fragariae,* derived from an individual collected in central Chile, was established in the laboratory on oat (*Avena sativa* L. cv. Nahuén) at 20 ± 2 °C and L16:D8 photoperiod.

*Plants.* The first leaves of 7-day-old seedlings were used from cultivars Naofén and Millaleu of the wheat *T. aestivum.* Chemical analysis showed that the concentration of Hx (determined as the glucoside of DIMBOA following the procedure described by Weibull & Niemeyer, 1995) was significantly higher in cv. Naofén than in cv. Millaleu (1.94±0.31 mmol/kg fresh weight, $n = 8$, and 1.09 ± 0.30 mmol/kg fresh weight, $n = 8$, respectively; $F(1, 14) = 29.21$, $n = 16$, $P < 0.001$).

*Experimental set-up.* Each experimental aphid was monitored over 24 h using the EPG technique (Tjallingii, 1978). All signals were recorded on a PC hard disk and analysed with the EPGview software (Flores et al., 1998, unpubl.). Detailed assessment of aphid activities could be achieved as during aphid probing the location of the stylet tips and their activities produce recognisable patterns of voltage changes in the recorded signal (Tjallingii & Hogen Esch, 1993; Prado & Tjallingii, 1994).

*Parameters used to study probing behaviour.* Aphid feeding behaviour was continuously monitored for 24 h on cv. Millaleu (low-Hx) and cv. Naofén (high-Hx) seedlings. A total of 23 recordings were performed for each cultivar. The following wave forms were recorded and recognised: non-probing (NP), pathways activities (C), salivation into a sieve element (E1), ingestion from the sieve elements (E2), xylem ingestion (G), and difficulties during stylet penetration (F). Timing the occurrence of each pattern also allowed the evaluation of the following parameters related to activities in sieve elements: (1) proportion of the time allocated to each activity, (2) time to first E1 within the experiment, (3) total duration of E1 periods within the experiment, (4) number of E1 periods, (5) mean duration of E1 periods, (6) duration of E1 periods before the first sustained E2 (i.e., E2 for more than 8 minutes) in the experiment, (7) duration of E1 periods before the first sustained E2 within a probe (a probe, here, is a continuous period without withdrawal of the stylets from the plant), (8) mean duration of the E1 period preceding the first sustained E2, (9) contribution of total E1 to total phloem phase (E1/[E1+E2]), and (10) time to the first sustained E2 in the experiment.

## Results and discussion

The major aphid activity during 24 h of electronic monitoring was phloem ingestion, to which aphids dedicated over 50% of the total time (Figure 1). The number of aphids showing phloem phase was not significantly different between cultivars (20 out of 23 and 21 out of 23 in low and high-Hx cultivar, respectively; $Z = 0.94$, $P = 0.34$, test for two proportions [Zar, 1996]). Similarly, the proportion of time devoted to phloem ingestion did not show differences associated with Hx variation, a result comparable with those reported for other cereal aphid species (Girma et al., 1992; Givovich & Niemeyer, 1996; Mayoral et al., 1996). Surprisingly, xylem ingestion (G pattern) was significantly higher in the low-Hx cultivar, although the number of aphids showing g pattern was not significantly different (12 out of 20 and 8 out of 21 in low and high-Hx cultivar, respectively; $Z = 1.09$, $P = 0.27$, test for two proportions [Zar, 1996]). These results are in contradiction with other studies in other cereal aphid species (Givovich & Niemeyer, 1995). It is likely that in the present system, xylem ingestion was reduced simply as a consequence of the higher proportion time devoted to E1 in high-Hx cultivar (Figure 1).

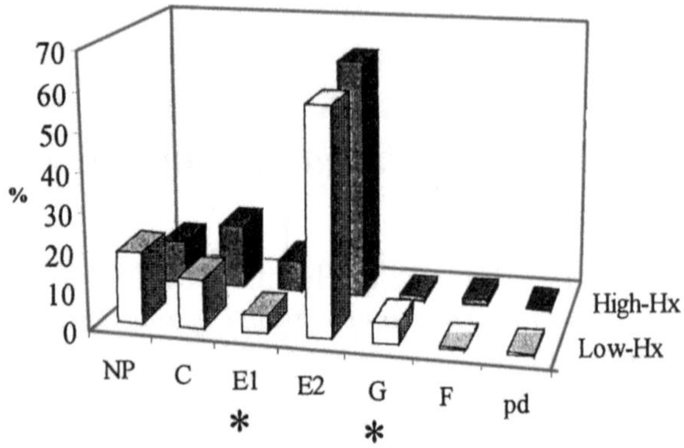


*Figure 1.* Proportion of the time (%) allocated to each activity during 24 h of EPG-recorded probing behaviour. Asterisks indicate significant differences $P < 0.05$, Mann–Whitney U test. NP: non-probing; C: pathways activities; E1: salivation into a sieve element; E2: ingestion from the sieve elements; G: xylem ingestion; F: difficulties during stylet penetration.

Focusing on sieve element salivation, the high-Hx cultivar showed significantly different results to the low-Hx cultivar in the following E1-related parameters (see Table 1): time to first E1 within the experiment, total duration of E1 periods, duration of the E1 periods before the first sustained E2 and contribution of sieve element salivation to the phloem phase. Clearly, Hx affect the process of salivation into the sieve elements.

Since Hx are mainly present in meristematic, epidermal, mesophyll and vascular tissue (Epstein et al., 1986, Argandoña et al., 1987), it is reasonable to speculate that Hx are perceived by the aphid during cell punctures during the pathway to sieve elements. The deterrent effect of Hx would cause a delay in the time needed to attain a first sieve element salivation (Niemeyer & Pérez, 1995; Ramírez et al., 1999). Brief cell punctures in the sieve elements may also occur (Tjallingii & Hogen Esch, 1993) and can contribute in such a delay. On the other hand, the increased duration of salivation periods seems more likely to be the result

*Table 1.* EPG parameters of feeding behaviour related to salivation and ingestion in the sieve elements over the 24-h recording period. Probabilities were calculated using a Mann–Whitney test

| EPG parameter | Low Hx $n = 20$ (X ± SE) | High Hx $n = 21$ (X ± SE) | P | |
|---|---|---|---|---|
| 2. Time to first E1 (min) | 69.8 ± 10.4 | 140.5 ± 36.2 | 0.033 | |
| 3. Total duration of E1 periods (min per 24 h)* | 44.5 ± 10.2 | 102.2 ± 16.0 | 0.002 | |
| 4. Number of E1 periods (#) | 5.1 ± 0.9 | 8.4 ± 1.6 | 0.260 | ns |
| 5. Mean duration of E1 periods (min) | 9.9 ± 1.1 | 28.4 ± 9.1 | 0.170 | ns |
| 6. Duration of E1 before first sustained E2 (min) | 30.3 ± 7.1 | 54.7 ± 9.7 | 0.041 | |
| 7. Duration of E1 before first sustained E2 within a probe (min) | 25.7 ± 5.8 | 46.6 ± 9.3 | 0.077 | ns |
| 8. Mean duration of the E1 preceding first sustained E2 (min) | 11.1 ± 1.6 | 30.2 ± 9.0 | 0.123 | ns |
| 9. Contribution of E1 to phloem phase (%) | 11.2 ± 4.4 | 19.3 ± 4.9 | 0.018 | |
| 10. Time to first sustained E2 (min) | 228.6 ± 46.5 | 291.2 ± 54.5 | 0.251 | ns |

*Figure 1 shows these values as the proportion of total time allocated to different probing activities.

of Hx in the phloem sap (Givovich et al., 1994; but see Caillaud & Niemeyer, 1996). However, a direct causal link between the Hx content and E1 duration must be made with caution because the Hx content of extracts of macerated whole leaves does not correlate strongly with the content in the phloem sap (Givovich et al., 1994).

In conclusion we have shown that Hx are associated with a delay in the time to start the process of salivation in the sieve elements and with a increase in the process of salivation itself, suggesting that these compounds may be acting both at the level of the epidermis/mesophyll and the phloem.

## Acknowledgements

We gratefully acknowledge funding from Universidad de Chile (graduate research grant PG-047-96 to C.C.R.), FONDECYT (grant 1961035 to H.M.N.), and the International Program in the Chemical Sciences (IPICS). C.C.R. acknowledges CONICYT for a graduate fellowship during part of his PhD studies. C.C.R. was supported during most of this work by the Presidential Chair in Science awarded to H.M.N.

## References

Argandoña, V. H., G. E. Zúñiga & L. J. Corcuera, 1987. Distribution of gramine and hydroxamic acid in barley and wheat leaves. Phytochemistry 26: 1917–1918.

Caillaud, C. M. & H. M. Niemeyer, 1996. Possible involvement of the phloem sealing system in the acceptance of a plant as host by an aphid. Experientia 52: 927–931.

Caillaud, C. M., J. S. Pierre, B. Chaubet & J. P. Di Pietro, 1995. Analysis of wheat resistance to the cereal aphid *Sitobion avenae* using electrical penetration graphs and flow charts combined with correspondence analysis. Entomologia Experimentalis et Applicata 75: 9–18.

Epstein, W. W, C. N. Rowsemitt, P. J. Berger & N. C. Negus, 1986. Dynamics of 6-methoxybenzoxazolimone in winter wheat. Journal of Chemical Ecology 12: 2011–2020.

Girma, M., G. D. Wilde & J. C. Reese, 1992. Russian wheat aphid (Homoptera: Aphididae) feeding on host and non-host plants. European Journal of Entomology 85: 395–401.

Givovich, A. & H. M. Niemeyer, 1995. Comparison of the effect of hydroxamic acids from wheat on five species of cereal aphids. Entomologia Experimentalis et Applicata 74: 115–119.

Givovich, A. & H. M. Niemeyer, 1996. Role of hydroxamic acids in the resistance of wheat to the Russian wheat aphid, *Diuraphis noxia* (Mordvilko) (Hom., Aphididae). Journal of Applied Entomology 120: 537–539.

Givovich, A. J., Sandström, H. M. Niemeyer & J. Pettersson, 1994. Presence of a hydroxamic acid glucoside in wheat phloem sap, and its consequences for performance of *Rhopalosiphum padi* (L.) (Homoptera: Aphididae). Journal of Chemical Ecology 20: 1923–1930.

Mayoral, A. M., W. F. Tjallingii & P. Castañera, 1996. Probing behaviour of *Diuraphis noxia* on five cereal species with different hydroxamic acid levels. Entomologia Experimentalis et Applicata 78: 341–348.

Nicol, D. & S. D. Wratten, 1977. The effect of hydroxamic acid concentration at late growth stages of wheat on the performance of the aphid *Sitobion avenae*. Annals of Applied Biology 130: 387–396.

Nicol, D., S. V. Copaja, S. D. Wratten & H. M. Niemeyer, 1992. A screen of worldwide wheat cultivars for hydroxamic acid levels and aphid antixenosis. Annals of Applied Biology 121: 11–18.

Niemeyer, H. M., 1990. The role of secondary plant compounds in aphid-host interactions. In: R. K. Campbell & R. D. Eikenbary (eds), Aphid: Plant Genotype Interactions. Elsevier Amsterdam, pp. 187–205.

Niemeyer, H. M. & F. J. Pérez, 1995. Potential of hydroxamic acids in the control of cereal pests, diseases and weeds. In: K.M.M. Inderjit, Dakshini & F. A. Einhellig (eds), Allelopathy: Organisms, Processes, and Applications. American Chemical Society Symposium Series, Washington, DC, pp. 260–270.

Prado, E., 1997. Aphid-plant interaction at phloem level, a behavioural study. PhD Thesis. Wageningen Agricultural University.

Prado, E. & W. F. Tjallingii, 1994. Aphid activities during sieve element punctures. Entomologia Experimentalis et Applicata 72: 157–165.

Prado, E. & W. F. Tjallingii, 1997. Effects of previous plant infestation on sieve element acceptance by two aphids. Entomologia Experimentalis et Applicata 82: 189–200.

Ramirez, C.C., P.P. Caballero & H.M. Niemeyer, 1999. Effect of previous exposure to hydroxamic acids in probing behavior of aphid *Sitobion fragariae* on wheat seedlings. Journal of Chemical Ecology 25: 771–779.

Spiller, N. J., 1988. Electronic recording of plant penetration by the cereal aphids *Rhopalosiphum padi* and *Metopolophium dirhodum* on resistant and susceptible wheat seedlings. Annals of Applied Biology 112: 471–478.

Tjallingii, W. F., 1978. Electronic recording of penetration behaviour by aphids. Entomologia Experimentalis et Applicata 24: 721–730.

Tjallingii, W. F. & T. Hogen Esch, 1993. Fine structure of aphid stylet routes in plant tissues in correlation with EPG signals. Physiological Entomology 18: 317–328.

Weibull, J. & H. M. Niemeyer, 1995. Changes in dihydroxymethoxybenzoxazinone glycoside content in wheat plants infected by three plant pathogenic fungi. Physiological and Molecular Plant Pathology 47: 201–212.

Zar, J. H. 1996. Biostatistical Analysis. Third edition. Prentice Hall. New Jersey.

Entomologia Experimentalis et Applicata **91:** 115–123, 1999.



# Aphid responses to non-host epicuticular lipids

Glen Powell[1], Sangita P. Maniar[2], John A. Pickett[2] & Jim Hardie[1]
[1]*Aphid Biology Group, Department of Biology, Imperial College at Silwood Park, Ascot, Berks, SL5 7PY, UK;*
[2]*Department of Biological and Ecological Chemistry, IACR-Rothamsted, Harpenden, Herts, AL5 2JQ, UK*

Accepted: October 20, 1998

*Key words:* aphid behaviour, *Aphis fabae*, 1-hexacosanol, host selection, plant waxes

**Abstract**

Initiation of stylet penetration by aphids depends on the sensory assessment of a number of plant surface features including colour, texture and phytochemicals (volatile and non-volatile). Video recording behaviour of the black bean aphid, *Aphis fabae,* showed that these insects rapidly inserted their stylets following contact with host plants (beans), but were reluctant to penetrate non-hosts (oats). However, when epicuticular waxes were stripped from oats, using cellulose acetate, aphids penetrated the plant surface significantly earlier than on oats with the wax layer intact. Chloroform extraction of epicuticular lipids, followed by coupled gas chromatography-mass spectrometry, revealed a complex blend of wax components on beans, whereas one compound (1-hexacosanol) predominated on oats. Epicuticular lipids were applied to artificial (glass) substrates in order to investigate their behavioural activity. Initiation of a stylet penetration attempt by *A. fabae* was delayed when the oat extract or pure 1-hexacosanol were applied, but the bean extract had no behavioural effect. The results suggest that epicuticular lipids play an important role in the early stages of host-plant selection by *A. fabae.*

## Introduction

Stimuli detected by phytophagous insects at the plant surface are often important cues for host plant selection, particularly during the first moments of plant contact (Städler, 1986; Woodhead & Chapman, 1986). For aphids, the initial decision to insert the stylets depends on substrate colour and texture (Moericke, 1950; Ibbotson & Kennedy, 1959; Pelletier, 1990), but phytochemicals detected at the plant surface are also important. Volatile phytochemicals may influence the initiation of stylet penetration (Storer et al., 1996), and trichome exudates have been reported to disrupt settling behaviour (Goffreda et al., 1988; Neal et al., 1990; Rodriguez et al., 1993; Musetti & Neal, 1997). Insects at the plant surface are also exposed to lipid components of the epicuticular wax layer, and these compounds often play important roles as mediators of insect-plant interactions (Eigenbrode & Espelie, 1995; Eigenbrode, 1996). Much of the experimental evidence to support this claim is correlative, for example depending on differences in insect behaviour or performance on normal vs 'glossy' plant genotypes (Eigenbrode & Espelie, 1995). However, direct evidence that epicuticular lipids function as allelochemicals has been obtained, especially by extraction and deposition onto artificial surfaces for examination of behavioural responses. Such studies have included insects from the orders Orthoptera (e.g., Woodhead, 1983), Diptera (e.g., Foster & Harris, 1992) and Lepidoptera (e.g., Varela & Bernays, 1988; Eigenbrode et al., 1991).

Little is known about how epicuticular lipids affect responses of Homoptera, but there are some indications that plant waxes play a role in host selection by aphids. Observations of the strawberry aphid (*Chaetosiphon fragaefolii* Cockerell) on artificial (glass filter paper) surfaces indicated that organic solvent extracts of host-plant leaves stimulated probing, but the active compounds were not identified (Shanks & Finnigan, 1970). Settling by pea aphids (*Acyrthosiphon pisum* Harris) was enhanced on non-host plants which were treated with host-plant wax (Klingauf, 1975; Klingauf et al., 1978). In the present study, the behaviour of two aphid species (the black bean aphid, *Aphis fabae* Scopoli, and the grain aphid, *Sitobion avenae* F.) was compared on host and non-host plants. *Aphis fabae* showed clear behavioural differences before stylet insertion: the insects readily penetrated hosts (broad beans, *Vicia faba* L.), but not non-hosts (oats, *Avena sativa* L.). The role of epicuticular waxes in this discrimination was investigated by assessing insect responses on plants with the wax layer removed, and by extracting lipids for chemical analysis and deposition onto glass surfaces for further behavioural assays.

## Materials and methods

*Insects.* The aphids used were from clonal laboratory cultures of *A. fabae* (Kennedy & Booth, 1951) and *S. avenae* (Powell et al., 1997), reared at 15 °C and L16:D8 photoperiod. Recently-moulted adult apterae were collected and confined to plastic Petri dishes for 1–2 h before use in experiments.

*Plants.* 'Dula' oats and 'Tick' beans were chosen as experimental host plants for *S. avenae* and *A. fabae* respectively, as these were the plants used for rearing the insects. The reciprocal aphid/plant combination was used to investigate responses to non-host plants. Small (approximately 3 cm high) seedlings were used in behavioural experiments: oats at growth stage 10 (Tottman & Makepeace, 1979) and beans at the hook stage.

*Removal of epicuticular wax layer.* Cellulose acetate was dissolved in acetone (10% w/v) and used to strip wax from the plant surface (Baker et al., 1983). This viscous solution was painted onto the whole seedling (stem and all exposed leaf surfaces). After evaporation of the solvent, the cellulose acetate film was easily removed using fine forceps.

*Extraction of epicuticular lipids and deposition onto glass substrates.* Oat and bean seedlings were cut at soil level and immediately dipped in chloroform for 10 s, taking care not to immerse the cut ends. Extracts were analysed using a coupled gas chromatograph and mass spectrometer (GC-MS) as described below. Samples for behavioural tests were passed through glass filter paper, concentrated by rotary evaporation, and stored at 4 °C before use. Since aphids will attempt to penetrate a glass surface (Moericke, 1950), glass tubes provided a convenient substrate for the application of extracts for behavioural tests. The tubes were cut from the base of a Pasteur pipette (Volac D810, John Poulten Ltd, internal diameter = 5 mm; external diameter = 6 mm; length of cut portion = 50 mm) and coated with plant lipids by brief (1 s) immersion in the concentrated extract. The fresh plant material was weighed immediately before dipping. The volume of extract applied to each tube was estimated from the difference before and after dipping, although this would have included evaporative loss. These figures were used to estimate the fresh weight of plant material (in gram leaf equivalents, gle) represented by the amount of extract on each glass tube. Control tubes were immersed in the chloroform solvent only. After chemical treatment and solvent evaporation, a green or white paper tube (50 mm long, 5 mm diameter) was pushed inside each glass tube. This arrangement allowed independent manipulation of substrate surface chemistry (lipid extract or control) and colour (white or green).

*Video recording of behaviour.* Oat or bean seedlings were transplanted into a pot of damp sand for video monitoring of insect behaviour on plants. Similarly, artificial test substrate tubes were positioned vertically by pushing into damp sand, leaving the top 30 mm exposed for each experiment. A small mirror was placed behind each plant or tube, so that the reflected image of the insect could be monitored when the aphid walked out of direct view. Insects were placed at the base of each seedling/tube using a glass pipette tip aspirator (pooter) and allowed to walk up. Behaviour of each insect was video-recorded using a Cohu video camera (4710 series), until the first stylet penetration attempt. Aphids which walked off onto the sand were returned to the base of the experimental substrate using the pooter. Each insect was allowed a total substrate contact time of 5 min (this did not include periods spent off the plant), and if no attempt to penetrate was made, the aphid was removed. Each insect and substrate was used only once, and 20 aphids were tested with each type of substrate. A vision mixer (Electrocraft VMC89) was used to monitor 2 insects simultaneously, using 2 separate cameras, and a time base displayed 0.1 s intervals. Video tapes were later analysed and antennal movements used to score the start of a stylet penetration attempt (Hardie et al., 1992). The time between initial substrate contact and first stylet penetration attempt was calculated. This was scored as 300 s if the aphid did not attempt to pen-

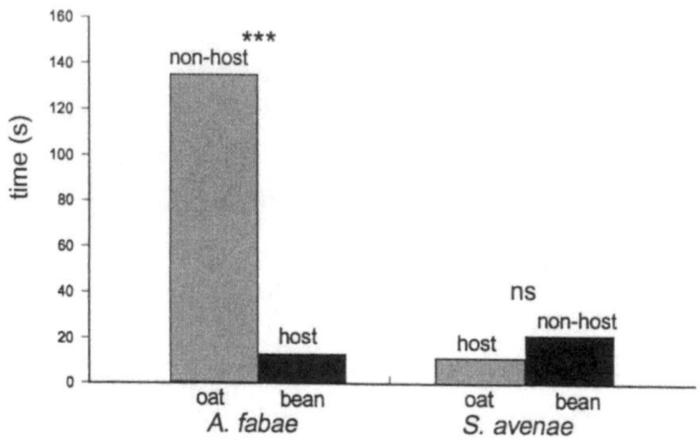


*Figure 1.* Median time to first stylet penetration by *A. fabae* and *S. avenae* on host and non-host plants (***P < 0.001; ns, no significant difference).

etrate the test substrate during the 5 min access period. The Mann–Whitney U-test was used to compare time to first probing attempt from different experimental groups.

*Spectral reflectance* of plant and paper surfaces was measured using a Bentham spectrophotometer (DM150EC).

*GC-MS analysis of extracts.* Coupled GC-MS was performed using a Hewlett-Packard 5880A GC with an HP-5 capillary column (30 m × 0.32 mm ID, 0.25 μm film thickness) directly coupled to either a VG 70-250 MS or a VG Autospec (Fisons Instruments) (electron impact, 70 eV, 250 °C). The sample was introduced via a cold (30 °C) on-column injector. The GC was temperature-programmed with an initial 5 min at 30 °C, then a rise of 10 °C per min to a final isothermal period at 300 °C.

*Scanning electron microscopy.* Plant (intact epicuticular layer and stripped) and glass specimens were mounted onto a cryo-stage using aqueous Carbon Dag, and frozen in liquid nitrogen slush (approximately −210 °C) in a cryo-workstation (Emscope SP2000). Specimens were then transferred under vacuum to a Philips 501(B) microscope, and frost removed by heating the stage to −65 °C while monitoring the sublimation visually. Specimens were sputter coated with gold in the cryo-workstation (10–15 mA for 2.5 min, 0.2 Torr) before transfer back to the microscope for examination and photography.

## Results and discussion

Video recording of behaviour on bean and oat seedlings revealed that *A. fabae* showed clear differences on the host vs non-host plants. The insects inserted stylets very quickly after initial contact with beans, but penetrated oats much later (Figure 1; P < 0.001). In contrast, *S. avenae* readily penetrated on both plant species, showing that the unsuitability of beans as a host for this aphid is not detected before stylet penetration. Although regarded as a grass specialist, *S. avenae* has been reported infesting legumes and may even be capable of limited phloem feeding and survival on such 'non-hosts' (Nault, 1975).

The role of plant surface factors in the ability of *A. fabae* to discriminate beans from oats was investigated by removing the epicuticular waxes from these plants using the cellulose acetate technique. Scanning electron microscopy showed that the technique removed the epicuticular wax layer from oats very effectively (Figure 2, A vs B), leaving an undamaged but wax-free plant surface suitable for testing insect behaviour. Wax crystals cover the cuticle of intact, unstripped oat plants (Baum & Hadland, 1975), but were absent on stripped plants. The stripping technique was also assumed to be effective with beans, but the epicuticular wax on these plants had an amorphous appearance (Figure 2C) and it was not possible to visualize differences between stripped and control seedlings.

When *A. fabae* were placed on bean plants with the wax removed, they initiated stylet insertion as early as on intact plants (Figure 3), indicating that no stimulatory factors had been removed from the host-plant surface. However, on stripped oat plants, aphids initiated stylet penetration significantly earlier than on intact oats (P < 0.001; Figure 3), and indeed showed similar initial behaviour to those insects on the bean host (P > 0.05). This result suggests that the presence of epicuticular waxes on oat leaf surfaces inhibits stylet penetration by *A. fabae*.

Several properties of the oat wax may be detected by the aphids. Surface reflectance of both plant species was affected by the wax removal (Figures 4A, B and C), being notably increased after stripping the adaxial oat surface, but decreased on wax-free beans. We therefore cannot exclude the possibility that the responses to oat wax by *A. fabae* occurs as a visual response. Furthermore, the crystalline structure of the oat wax (Figure 2A) may be detected by aphid mechanosensory sensilla. A further possibility is that

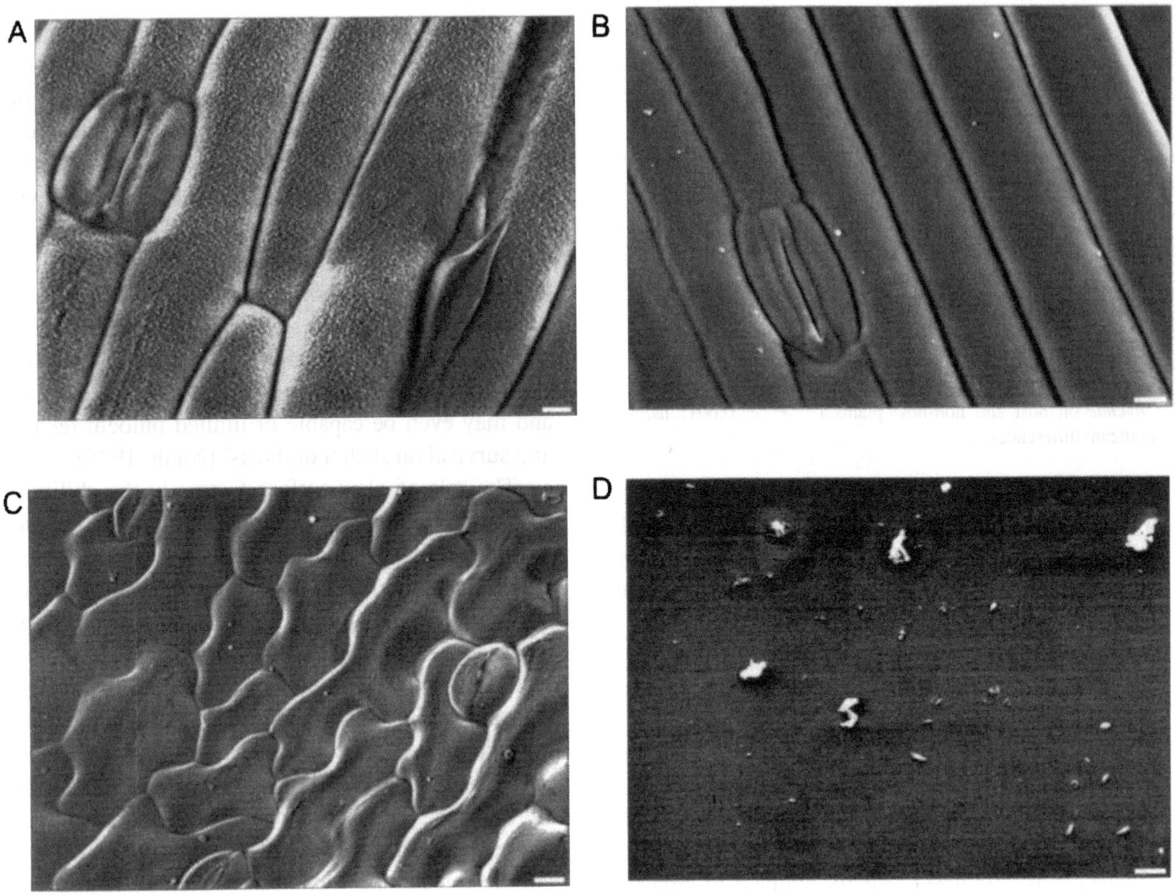


*Figure 2.* Scanning electron micrographs: the abaxial surface of oat seedlings, before (A) and after (B) stripping off the epicuticular wax layer using cellulose acetate. Crystalline epicuticular waxes were clearly visible on unstripped oat leaves (A), but were not observed on the stripped oat surface (B), an unstripped bean seedling (C, area near tip shown) or a glass surface treated with oat lipid extract (D) (scale bar = 10 $\mu$m).

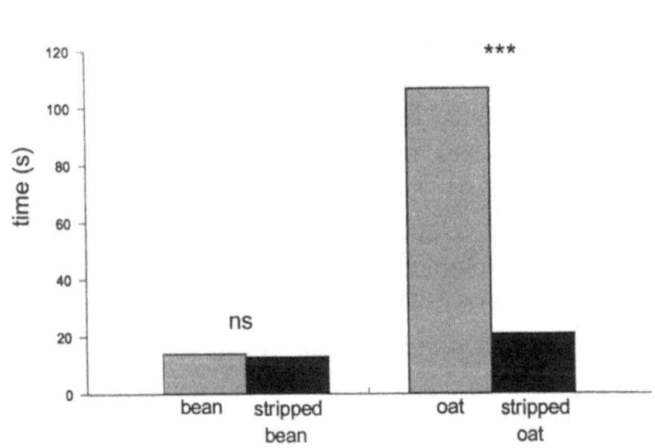


*Figure 3.* Median time to first stylet penetration by *A. fabae* on intact and cellulose acetate-stripped oat and bean seedlings (***P < 0.001; ns, no significant difference).

epicuticular lipids or other wax components are detected by a chemosensory mechanism, and function as allelochemicals. A system for assessing aphid behaviour on artificial surfaces was therefore developed, in order to manipulate substrate properties and identify factors which affect the probing response.

Aphids were presented with different combinations of substrate colour (white or green) and surface texture (glass or paper). The green paper had a maximum reflectance at around 550 nm (Figure 4D), and elicited early probing attempts whether the aphids walked on the paper itself or on a glass tube covering the paper (Figure 5). Direct contact with white paper did not stimulate probing, probably due to a high reflectance across a broad wavelength range (Figure 4E), but stylet penetration attempts were initiated rapidly on a glass-covered white surface (Figure 5). The smooth texture of the glass apparently elicits a probing response, overcoming the visual inhibitory effect of the white surface. Thus stylet penetration may be evoked independently by an appropriate surface reflectance and by substrate texture. As walking and probing are antagonistic (Ibbotson & Kennedy, 1959), any factor

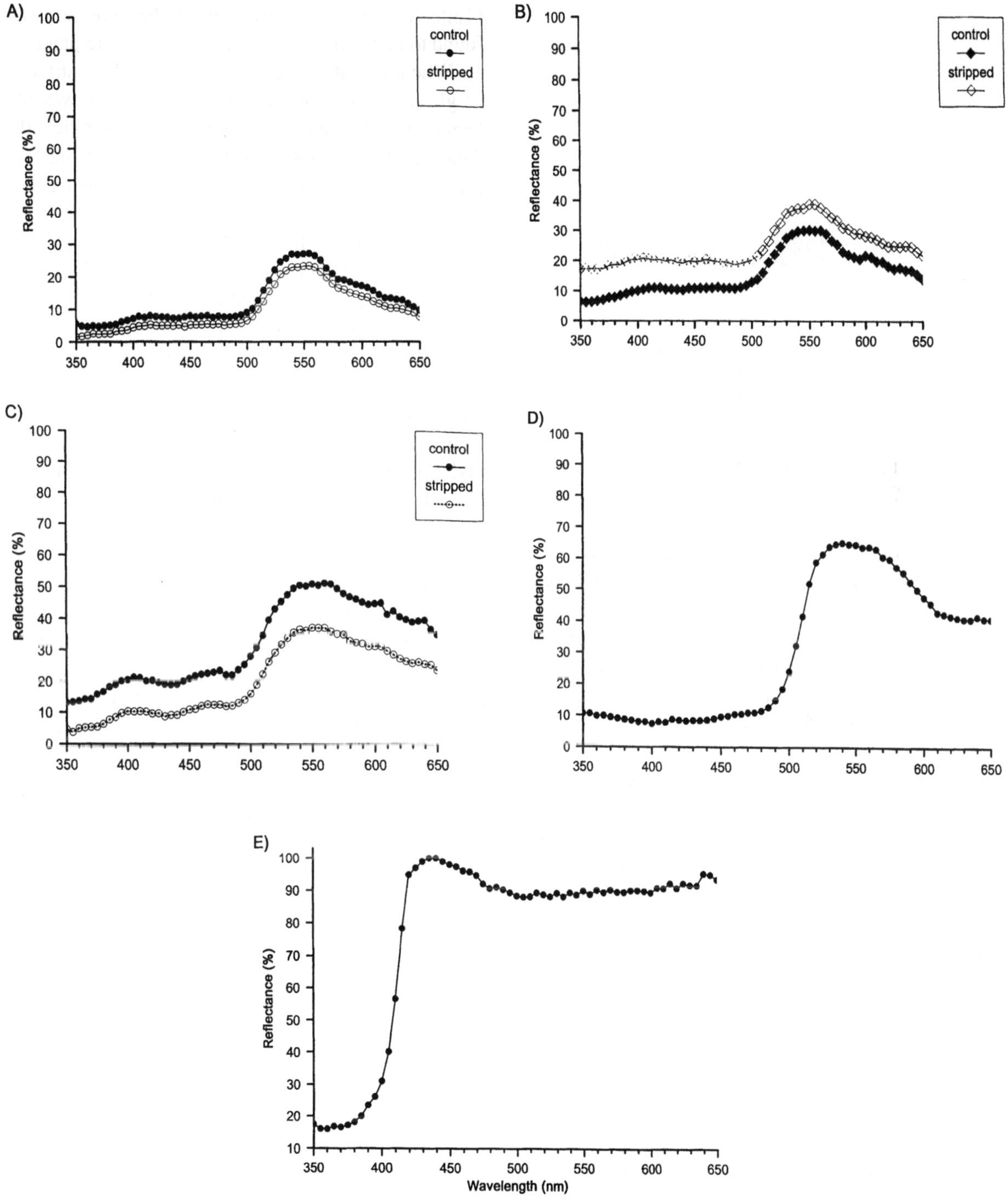


*Figure 4.* Spectral reflectance of plant (cellulose acetate-stripped and control) and paper surfaces; (A) oat seedling (abaxial leaf surface); (B) oat seedling (adaxial leaf surface); (C) bean seedling (stipule area near tip); (D) green paper; (E) white paper.

which inhibits walking will tend to elicit probing. The initiation of stylet penetration attempts on glass may therefore occur indirectly, as a result of locomotory inhibition. On a smooth glass surface, aphid adhesion is effected by the tarsal pulvilli, the claws contributing no additional purchase (Lees & Hardie, 1988). The aphids in this study appeared to walk well on glass tubes and rarely lost their foothold. It seems likely that the smooth glass surface stimulates probing directly, rather than as a consequence of locomotory inhibition.

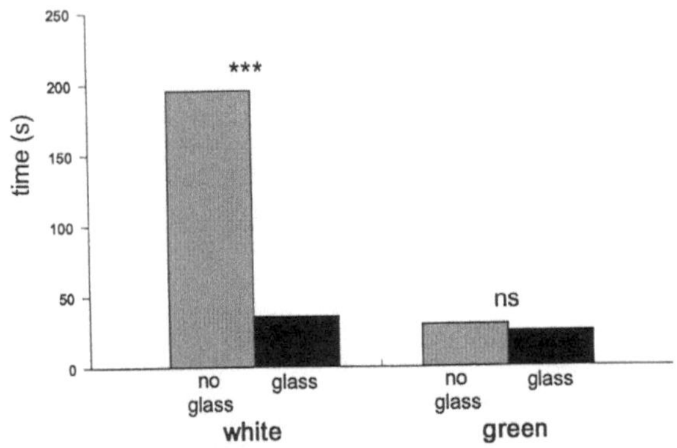


*Figure 5.* Median time to first stylet penetration attempt by *A. fabae* walking on white or green paper tubes, with and without a glass covering (***P < 0.001; ns, no significant difference).

Glass tubes therefore provided a convenient substrate for testing the behavioural effects of non-host plant extracts.

Chloroform extraction of epicuticular lipids, followed by GC-MS, revealed that oat and bean seedlings have very different epicuticular lipid compositions (Figure 6). The bean extract (Figure 6A) comprised a complex blend of compounds, including hydrocarbons, acetates and aldehydes. $C_{23}$ hydrocarbons were the most abundant compounds. Identification of the hydrocarbons was confirmed by peak enhancement with synthetic compounds (Aldrich), by GC-MS from the characteristic mass spectra fragmentation patterns. The broad bean plants investigated in this study had a different wax composition from those investigated by Klingauf (1975), where the $C_{31}$ *n*-alkane was most abundant, but this is probably due to different plant ages, genotypes and growing conditions. Only two lipid components were found on the surface of oats (Figure 6B): 1-hexacosanol (estimated to be ~ 88% of the total lipids by GC-MS) and a minor component, 1-hexacosanal (~8%). These compounds were identified by mass spectrometry from their fragmentation patterns. Although the position of the hydroxy group of the hexacosanol has not yet been unequivocally established, the presence of the corresponding aldehyde, hexacosanal, which would be formed directly by oxidation of the primary alcohol, strongly suggests 1-hexacosanol. Furthermore, nuclear magnetic resonance (NMR) analysis of the total oat epicuticular lipids showed the major component (>80%) to be a primary alcohol. The GC-MS and NMR analysis of the extracted lipids from young oat seedlings agree well with an earlier analysis of oat wax composition (Tulloch & Hoffman, 1973), where hexacosanol was found to be the major lipid present on mature plants.

When the oat extract was applied to glass surfaces (2 gle per tube), the behaviour of *A. fabae* was affected. Stylet penetration attempts were significantly delayed when the aphids were exposed to the oat extract over a white surface (Figure 7A), showing that the stimulatory effect of the glass was inhibited by chloroform-soluble factors. A very similar result was obtained using a green substrate, indicating that probing behaviour was inhibited by non-host plant lipids even when an additional positive (visual) stimulus was present. This inhibition is unlikely to be caused by the physical presence of wax on the glass surface; application of the bean extract to glass-covered white paper caused no behavioural response (median time to first stylet penetration attempt = 26 s on bean wax-treated glass tubes vs 21 s on solvent-treated control tubes, P > 0.05).

Scanning electron microscopy of glass surfaces showed that the oat crystals present on leaves (Figure 2A) were not reconstituted on glass following application of the extract (Figure 2D). These results suggest that the behavioural effect of the oat epicuticular lipids does not occur as an exclusively mechanosensory response to the crystalline wax structure, but probably involves a chemosensory component. The sensory organs involved are unknown. Anderson & Bromley (1987) have speculated that sensilla on the aphid tarsi may have a chemosensory function but there is no published ultrastructural or electrophysiological evidence to support this. However, contact chemosensory sensilla are known to occur at the aphid antennal tips (Bromley et al., 1980), and to contact the substrate during walking (Powell et al., 1995).

1-Hexacosanol is likely to be the behaviourally active oat lipid, as the other oat compound, hexacosanal, was less abundant and also occurred on the surface of bean seedlings (Figure 6A). 1-Hexacosanol occurs in the epicuticular waxes of many plants, including hops (*Humulus lupulus* L.; Gülz et al., 1993) and mulberry (*Morus alba* L.; Mori, 1982), and is a phagostimulant for larvae of the silkworm moth, *Bombyx mori* L. (Mori, 1982) and adult cottonwood leaf beetles, *Chrysomela scripta* F. (Lin *et al.*, 1998). The synthetic compound was obtained commercially (Aldrich, purity = 98%) and applied to glass tubes at a concentration of 0.5% in chloroform. The behavioural effects of pure 1-hexacosanol were similar to the response produced by the oat extract, with delayed stylet penetration on both white and green substrates (Fig-

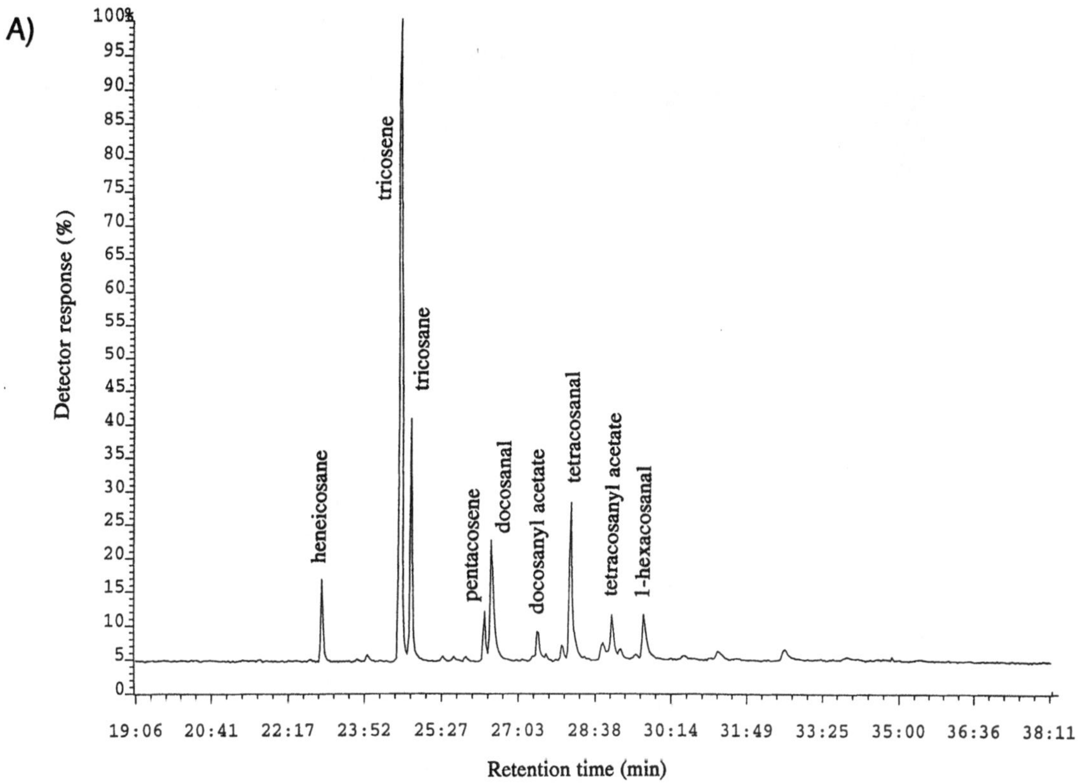


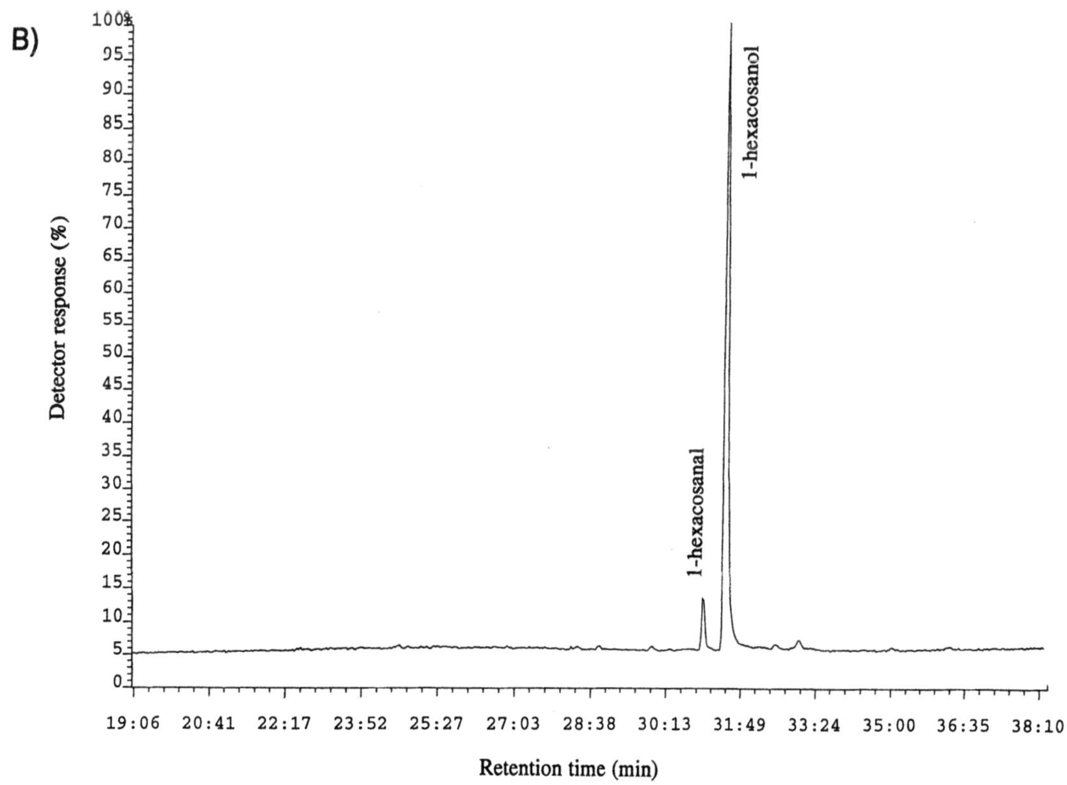


*Figure 6.* Gas chromatographs of chloroform extracts of (A) bean and (B) oat seedlings.

A)

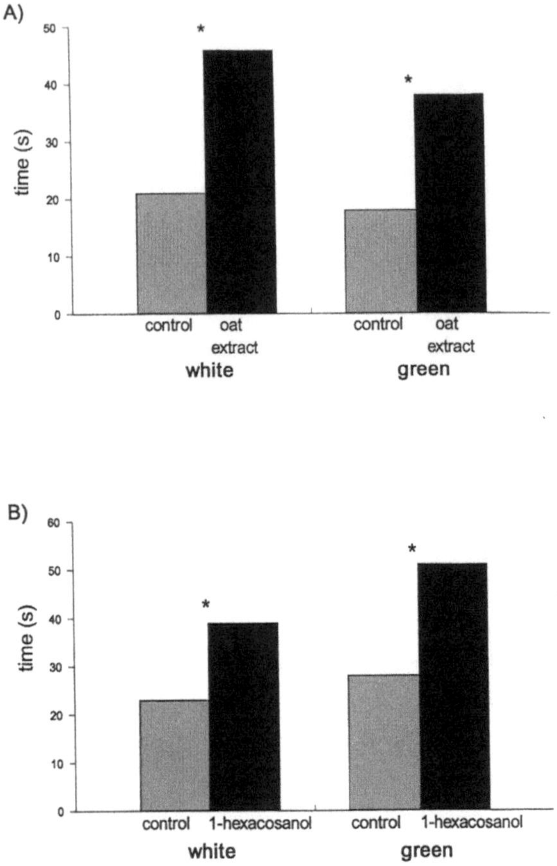


B)

*Figure 7.* Median time to first stylet penetration attempt by *A. fabae* on glass tubes covered with (A) oat leaf extract (2 gle per tube) or (B) 0.5% synthetic 1-hexacosanol vs chloroform solvent controls. The experiments were repeated using white and green paper under the glass (*P < 0.05).

ure 7B). Thus while we were unable to find evidence that bean epicuticular lipids stimulate stylet penetration by *A. fabae*, the most dominant component of oat wax contributed to an early change in behaviour on this non-host plant.

## Acknowledgements

We thank Ian Morris for help with electron microscopy. The work at Imperial College was supported by the Biotechnology and Biological Sciences Research Council of the UK. The BBSRC also provided grant-aided support to IACR-Rothamsted, and the work at IACR was in part supported by the UK Ministry of Agriculture, Fisheries and Food.

## References

Anderson, M. & A. K. Bromley, 1987. Sensory system. In: A. K. Minks & P. Harrewijn (eds), Aphids: Their Biology, Natural Enemies and Control, Vol. 2A. Elsevier, Amsterdam, pp. 153–162.

Baker, E. A., G. M. Hunt & P. J. G. Stevens, 1983. Studies of plant cuticle and spray droplet interactions: a fresh approach. Pesticide Science 14: 645–658.

Baum, B. R. & V. E. Hadland, 1975. The epicuticular waxes of glumes of *Avena*: a scanning electron microscopic study of the morphological patterns in all the species. Canadian Journal of Botany 53: 1712–1718.

Bromley, A. K., J. A. Dunn & M. Anderson, 1980. Ultrastructure of the antennal sensilla of aphids. II. Trichoid, chordontal and campaniform sensilla. Cell and Tissue Research 205: 493–511.

Eigenbrode, S. D., 1996. Plant surface waxes and insect behaviour. In: G. Kerstiens (ed.), Plant Cuticles. An Integrated Functional Approach. Bios Scientific Publishers, Oxford, pp. 201–221.

Eigenbrode, S. D. & K. E. Espelie, 1995. Effects of plant epicuticular lipids on insect herbivores. Annual Review of Entomology 40: 171–194.

Eigenbrode, S. D., K. E. Espelie & A. M. Shelton, 1991. Behavior of neonate diamondback moth larvae (*Plutella xylostella* L.) on leaves and on extracted leaf waxes of resistant and susceptible cabbages. Journal of Chemical Ecology 17: 1691–1704.

Foster, S. P. & M. O. Harris, 1992. Foliar chemicals of wheat and related grasses influencing oviposition by hessian fly, *Mayetiola destructor* (Say) (Diptera: Cecidomyiidae). Journal of Chemical Ecology 18: 1965–1980.

Goffreda, J. C., M. A. Mutschler & W. M. Tingey, 1988. Feeding behavior of potato aphid affected by glandular trichomes of wild tomato. Entomologia Experimentalis et Applicata 48: 101–107.

Gülz, P. G., E. Müller, T. Herrmann & P. Lösel, 1993. Epicuticular leaf waxes of the hop (*Humulus lupulus*). Chemical composition and surface structures. Zeitschrift für Naturforschung 48c: 689–696.

Hardie, J., M. Holyoak, N. J. Taylor & D. C. Griffiths, 1992. The combination of electronic monitoring and video-assisted observations of plant penetration by aphids and behavioural effects of polygodial. Entomologia Experimentalis et Applicata 62: 233–239.

Ibbotson, A. & J. S. Kennedy, 1959. Interaction between walking and probing in *Aphis fabae*. Journal of Experimental Biology 36: 377–390.

Kennedy, J. S. & C. O. Booth, 1951. Host alternation in *Aphis fabae* Scop. I. Feeding preferences and fecundity in relation to the age and kind of leaves. Annals of Applied Biology 38: 25–64.

Klingauf, F., 1975. Die Frühphase in der Wirtswahl von Blattläusen. Mededelingen van de Faculteit Landbouwwetenschappen Rijksuniversiteit Gent 40: 351–363.

Klingauf, F., K. Nöcker-Wenzel & U. Röttger, 1978. Die Rolle peripherer Pflanzenwachse für den Befall durch phytophage Insekten. Zeitschrift für Pflanzenkrankheiten und Pflanzenschutz 85: 228–237.

Lees, A. D. & J. Hardie, 1988. The organs of adhesion in the aphid *Megoura viciae*. Journal of Experimental Biology 136: 209–228.

Lin, S., B.F. Binder & E.R. Hart, 1998. Insect feeding stimulants from the leaf surface of *Populus*. Journal of Chemical Ecology 24: 1781–1790.

Moericke, V., 1950. Über das Farbsehen der Pfirsichblattlaus (*Myzodes persicae* Sulz.). Zeitschrift für Tierpsychologie 7: 265–274.

Mori, M., 1982. n-Hexacosanol and n-octacosanol: feeding stimulants for larvae of the silkworm, *Bombyx mori.* Journal of Insect Physiology 28: 969–973.

Musetti, L. & J. J. Neal, 1997. Resistance to the pink potato aphid, *Macrosiphum euphorbiae,* in two accessions of *Lycopersicon hirsutum* f. *glabratum.* Entomologia Experimentalis et Applicata 84: 137–146.

Nault, L. R., 1975. Tests for transmission of pea enation mosaic virus by oligophagous mustard and grain aphids. Phytopathology 65: 496–497.

Neal, J. J., W. M. Tingey, & J. C. Steffens, 1990. Sucrose esters of carboxylic acids in glandular trichomes of *Solanum berthaultii* deter settling and probing by green peach aphid. Journal of Chemical Ecology 16: 487–497.

Pelletier, Y., 1990. The role of the color of the substratum on the initiation of the probing behaviour in *Myzus persicae* (Sulzer) and *Macrosiphum euphorbiae* (Thomas) (Homoptera: Aphididae). Canadian Journal of Zoology 68: 694–698.

Powell, G., J. Hardie & J. A. Pickett, 1995. Behavioural evidence for detection of the repellent polygodial by aphid antennal tip sensilla. Physiological Entomology 20: 141–146.

Powell, G., J. Hardie & J. A. Pickett, 1997. Laboratory evaluation of antifeedant compounds for inhibiting settling by cereal aphids. Entomologia Experimentalis et Applicata 84: 189–193.

Rodriguez, A. E., W. M. Tingey & M. A. Mutschler, 1993. Acylsugars of *Lycopersicon pennellii* deter settling and feeding of the green peach aphid (Homoptera: Aphididae). Journal of Economic Entomology 86: 34–39.

Shanks, C. H. & B. Finnigan, 1970. Probing behavior of the strawberry aphid. Annals of the Entomological Society of America 63: 734–737.

Städler, E. 1986. Oviposition and feeding stimuli in leaf surface waxes. In: B. E. Juniper & T. R. E. Southwood (eds), Insects and the Plant Surface. Edward Arnold, London, pp. 105–121.

Storer, J. R., G. Powell & J. Hardie, 1996. Settling responses of aphids in air permeated with non-host plant volatiles. Entomologia Experimentalis et Applicata 80: 76–78.

Tottman, D. R. & R. J. Makepeace, 1979. An explanation of the decimal code for the growth stages of cereals, with illustrations. Annals of Applied Biology 93: 221–234.

Tulloch, A. P. & L. L. Hoffman, 1973. Leaf wax of oats. Lipids 8: 617–622.

Varela, L. G. & E. A Bernays, 1988. Behavior of newly hatched potato tuber moth larvae, *Phthorimaea operculella* Zell. (Lepidoptera: Gelechiidae), in relation to their host plants. Journal of Insect Behavior 1: 261–275.

Woodhead, S., 1983. Surface chemistry of *Sorghum bicolor* and its importance in feeding by *Locusta migratoria.* Physiological Entomology 8: 345–352.

Woodhead, S. & R. F. Chapman, 1986. Insect behaviour and the chemistry of plant surface waxes. In: B. E. Juniper & T. R. E. Southwood (eds), Insects and the Plant Surface. Edward Arnold, London, pp. 123–135.

Entomologia Experimentalis et Applicata **91**: 125–130, 1999.



# Effects of *Brassica oleracea* waxblooms on predation and attachment by *Hippodamia convergens*

Sanford D. Eigenbrode & Nelson N. Kabalo
*Department of Plant, Soil and Entomological Sciences, University of Idaho, Moscow, ID 83844-2339, USA*

Accepted: October 15, 1998

*Key words:* Plant surface waxes, predation, tritrophic interactions, *Plutella xylostella*, adhesion, attachment

## Abstract

Four mutations that reduce waxbloom in *Brassica oleracea* L. were examined for their effects on predation, mobility, and adhesion to the plant surface by the general predator *Hippodamia convergens* (Guérin-Menéville) (Coleoptera: Coccinellidae). The mutations reduce waxbloom to different degrees, but all produce a 'glossy' phenotype. Plants tested were inbred lines, near isogenic lines, or segregating $F_2$ populations, depending on the mutation. In an experiment on caged leaves, predation of *Plutella xylostella* L. larvae by *H. convergens* adult females was significantly greater on glossy types as compared with 'normal-wax' or wild-type counterparts. Although the trend was the same for each mutation, individual comparisons between glossy and normal-wax lines or segregants were only significant for two of them, those producing mutant alleles $gl_a$ and $gl_d$. Individual *H. convergens* were observed to spend more time walking on leaf edges and less time walking on leaf surfaces of normal-wax plants than glossy plants. *Hippodamia convergens* also obtained better adhesion to the surfaces of glossy plants than to normal-wax plants when tested using a centrifugal device. Two of the mutations produced similarly strong effects on predation, behaviour, and adhesion by *H. convergens*. These two are the same previously determined by us to provide the strongest similar effects on another generalist predator, *Chrysoperla plorabunda* (Fitch). The results indicate that waxbloom variation in nature could affect herbivore populations through its effects on generalist predators.

## Introduction

Plant surface waxes often form conspicuous crystalline 'waxblooms' over the underlying amorphous epicuticular wax layer (Martin & Juniper, 1970). The minute crystals (≈ 1–30 μm) can easily be abraded from the surface and this interferes with attachment by insects (Stork, 1980). Through this mechanism, waxblooms can provide protection against insect herbivory or other kinds of depredation (Stork, 1980; Bodnaryk, 1992; Juniper, 1995). Waxblooms also potentially affect attachment by predatory insects and thus influence trophic interactions among insects on the plants (Eigenbrode et al., 1995, 1996, 1998, 1999).

Genetic variation in waxbloom provides an effective tool for examination of how waxbloom can influence insect herbivory and predation. In *Brassica oleracea* L., several mutations occur that reduce waxbloom (Macey & Barber, 1970; Eigenbrode et al., 1991b). The mutations reduce crystal densities to varying degrees and alter crystal shape. The changes in morphology are associated with differences in composition of the waxes (Macey & Barber, 1970; Eigenbrode et al., 1991a, b). In a recent study (Eigenbrode et al., 1999) we examined the effects of four such waxbloom mutations on the behaviour, effectiveness and adhesion to the plant surface by the generalist predator *Chrysoperla plorabunda* (Fitch) (Neuroptera: Chrysopidae). All the mutant types enhanced predation and adhesion by this insect feeding on *Plutella xylostella* L. larvae, but to different degrees.

Here we report the results of a set of experiments to examine the effects of the same four waxbloom mutations on walking behaviour, adhesion, and effectiveness of a coccinellid generalist predator, *Hippodamia convergens* (Guérin-Menéville). The objective was to determine whether effects similar to those observed for *C. plorabunda* extend to a representative of the Coccinellidae, one of the most important taxa of generalist predators in natural systems and biological control.

## Materials and methods

*Plants.* *Brassica oleracea* expressing four recessive mutations that reduce surface waxbloom, and their wild-type counterparts, were used in the experiments. In this paper, those with reduced waxbloom will be referred to as 'glossy' and the wild-types will be referred to as 'normal-wax'. These glossy mutations were described and assigned gene labels by Eigenbrode et al. (1991b). All have been associated with resistance to Lepidoptera, aphids, or both in the field (Dickson & Eckenrode, 1975, 1980; Stoner, 1990; Eigenbrode et al., 1991b). The mutant allele $gl_a$ was tested in glossy broccoli NY 1406 (obtained from M. H. Dickson, Cornell University) and compared with the cabbage hybrid 'Round-Up'. The mutant allele $gl_d$ was tested in the inbred cauliflower line 'Glossy Andes' (obtained from K. A. Stoner, Connecticut Agricultural Experiment Station) and compared with its parental variety 'Andes'.

*Plants.* Two of the glossy mutations were tested in segregating $F_2$ populations obtained from K. A. Stoner. The KCR4 × 'Packman' $F_2$ population segregates for $gl_{a-1}$ and the Broc5 × 'Packman' $F_2$ population segregates for $gl_c$. 'Packman' is a standard broccoli variety with a normal-wax bloom. KCR4 is an inbred glossy line of kale and Broc5 is an inbred glossy line of broccoli. Glossy segregants were compared with normal-wax types in each of these populations. The KCR4 × 'Packman' $F_2$ included three phenotypes based on their appearance to the unaided eye and under scanning electron microscopy (Eigenbrode et al., 1999): a normal-wax, a glossy wax, and an intermediate wax phenotype. Each of these three phenotypes was tested separately. To simplify presentation in this paper, *B. oleracea* varieties, inbred lines, or segregating phenotypes will be referred to generically as wax types.

All plants were grown in the greenhouse under supplemental metal halide lighting (L18:D6) (500 $\mu$mol photosynthetically active radiation $m^{-2}$), in 10-cm pots with greenhouse potting soil (Sunshine Mix 1, SunGro Horticulture, Bellevue, WA). Plants were used for experiments six to 10 weeks after germination.

*Insects.* *Plutella xylostella* larvae were from a colony maintained on artificial diet (Shelton et al., 1991) at the University of Idaho and derived from the 'Geneva 88' population (A. M. Shelton, Cornell University). *Hippodamia convergens* adults were obtained commercially (Nature's Control, Medford, OR). Females that had been starved for 24 h were used in all bioassays.

*Predation by* H. convergens *on caged leaves.* Individual leaves (5th to 7th fully expanded) on intact plants of each *B. oleracea* type were enclosed in plastic cages (20 cm × 40 cm × 10 cm) screened on two sides with 'no-see-um' netting (REI, Spokane, WA, USA). The petioles passed through foam gaskets, sealed with 'Tack'n Stick® ' adhesive (American Glue Corp.). Each leaf was infested with 30 *P. xylostella* eggs, with visible developing embryos, collected from the colony on aluminum foil. The foil strip with eggs was pinned to the abaxial side of the leaf. After 48 h, the foil strip was removed and the number of eggs that had hatched was determined by counting empty chorions. Two female adult *H. convergens* were then introduced into one of the cages on each plant and the other cage served as a predator-free control. Forty-eight hours after infestation, the number of surviving *P. xylostella* was counted on each caged leaf. The proportion of *P. xylostella* surviving was calculated by dividing the number of surviving larvae by the number of eggs that had hatched on each leaf. There were 6 replicates of each of the 9 wax types × 2 predator treatments, for a total of 108 caged leaves.

The effect of wax phenotype, predator, and their interaction on proportion of *P. xylostella* surviving were examined with a two-way analysis of variance (ANOVA) on the transformed proportions (arcsin $\sqrt{x}$). For this analysis, wax phenotypes were classified as normal-wax, glossy, or intermediate. A 1-way ANOVA was also calculated, with each treatment combination (line or wax type within a population × predator absent or present) considered a class level. Planned contrasts compared the effect of adding *H. convergens* versus controls on *P. xylostella* survival within each wax type. An estimate of the reduction in *P. xylostella* survival attributable to predation on each wax type was obtained by computing Abbot's (1925) formula for effectiveness: (*P. xylostella* survival on predator-free control − *P. xylostella* survival with

*H. convergens*)/(Predator-free control + *P. xylostella* survival with *H. convergens*) × 100.

Hippodamia convergens *behaviour.* Individual *H. convergens* adult females were observed on each *B. oleracea* wax type for 5 min. Eighteen animals were observed on the KRC4 × 'Pacman' $F_2$ intermediate and 24 to 32 on each of the other test entries. The beetle's movement on the plant was continuously recorded using The Observer (Noldus Information Technologies, Wageningen). The distribution of walking among leaf upper surfaces, leaf lower surfaces, and leaf edges was compared with multivariate analysis of variance (MANOVA) (Harris, 1985) with percentage of time allocated to each location as dependent variables and wax type as the independent variable. Significance of the model was assessed on the basis of Wilks' Lambda statistic. Individual ANOVAs were compared for walking on each leaf part (upper surface, lower surface, edge) if the overall MANOVA was significant. Planned contrasts compared walking in each location on glossy and normal-wax paired lines or segregants within an $F_2$.

Hippodamia convergens *adhesion.* Adhesive force (mN) produced by *H. convergens* adult females on the leaf surfaces of each *B. oleracea* wax type was measured with a centrifugal device (Eigenbrode et al., 1999). Twelve insects were tested on the upper leaf surface of each of the wax types. Each insect was tested on a fresh leaf segment and there were two leaf segments from each of six plants from each wax type. Analysis of raw data was performed with ANOVA.

Statistical analyses of the experiments were performed using the procedures MEANS and GLM with LSMEANS in SAS (SAS Institute, Cary, NC).

## Results and discussion

*Predation by* H. convergens. Based on two-way ANOVA, waxbloom class (glossy, normal-wax, intermediate) (df = 2, 99; F = 1.57; P = 0.0115) and *H. convergens* (df = 1, 99; F = 7.24; P = 0.0084) significantly affected *P. xylostella* survival on caged leaves. The interaction between *H. convergens* and waxbloom class was not significant (with $\alpha = 0.05$) (df = 2, 99; F = 2.88; P = 0.0611). Suppression of *P. xylostella* survival by *H. convergens* was consistently greater on glossy types within pairs of lines or $F_2$ populations (Table 1). Although the effect of glossy wax on *H. convergens* predation of *P. xylostella* is not as great as previously reported between two cabbage genotypes differing in expression of $gl_a$ (Eigenbrode et al., 1995), here we have shown that four different glossy mutations in *B. oleracea* can enhance predation by this insect. Together with our previous study of *C. plorabunda* (Fitch) predation on these *B. oleracea* wax types (Eigenbrode et al., 1999), this result indicates that different reduced waxbloom mutations can enhance predation by at least two active predators in different orders. The greatest increase in *H. convergens* predation occurred on plants expressing $gl_a$ (NY 1406) or $gl_d$ (Glossy Andes), which produce the strongest reductions in wax crystal density and wax load ($\mu$g cm$^{-2}$) (Eigenbrode et al., 1991b; 1999; Eigenbrode unpubl.). These two characteristics are apparently most important for affecting predator effectiveness. The trend within the KCR4 × 'Packman' $F_2$ towards increasing predation from normal-wax, through intermediate, to glossy types (Table 1), also is consistent with wax phenotype affecting predation. The intermediate phenotype has a crystal density intermediate between normal-wax and glossy types in the KCR4 × 'Packman' $F_2$ (Eigenbrode et al., 1999).

*Plutella xylostella* survival was similar on controls without *H. convergens,* except for 'Round-Up', on which survival was reduced. This hybrid variety matured faster than the other test entries and the leaves were tougher, apparently producing this effect. Reduced predation on this line relative to NY 1406 may have been caused by lower prey populations as well as wax phenotype.

Hippodamia convergens *behaviour.* *H. convergens* adult females spent ≥ 75% of the time walking during 5-min observations on all wax types. This walking was distributed differently among parts of leaves of the *B. oleracea* wax types (Table 2). The insects consistently and significantly spent more time walking on upper or lower leaf surfaces of glossy types than normal-wax types, and more time walking on leaf edges of normal-wax types than glossy types (Table 2). The proportion of time *H. convergens* spent walking on upper surfaces and edges of intermediate KCR4 × 'Packman' $F_2$ was similar to the proportion of time spent walking on glossy segregants in this cross. Walking on lower surfaces was similar on the intermediate and normal-wax segregants in the KCR4 × 'Packman' $F_2$ population.

These differences in the distribution of walking by *H. convergens* can explain the effectiveness of

*Table 1.* Survival (% ± SEM) of neonate *Plutella xylostella* on caged leaves of *Brassica oleracea* with different surface waxbloom with and without the predator *Hippodamia convergens*

| Lines varieties or population | Wax class[a] | Predator-free control | With *H. convergens* | Reduction attributed to *H. convergens*[b] | P[c] |
|---|---|---|---|---|---|
| NY 1406 | G | 69.9 ± 5.7 | 39.9 ± 7.1 | 27.3 | 0.0175 |
| 'Round -Up' | N | 32.5 ± 12.5 | 26.0 ± 8.2 | 11.1 | 0.5473 |
| KCR4 × | G | 61.3 ± 12.5 | 40.0 ± 6.2 | 21.0 | 0.0715 |
| 'Packman' F2 | I | 64.8 ± 10.0 | 55.7 ± 10.0 | 7.6 | 0.3954 |
| | N | 61.1 ± 6.3 | 65.1 ± 11.8 | −3.2 | 0.4993 |
| 'Glossy Andes' | G | 47.0 ± 6.3 | 20.5 ± 6.3 | 39.3 | 0.0157 |
| 'Andes' | N | 55.7 ± 5.1 | 48.8 ± 6.8 | 6.6 | 0.5703 |
| Broc5 × | G | 51.8 ± 9.0 | 31.8 ± 2.0 | 23.9 | 0.1288 |
| 'Packman' F2 | N | 62.5 ± 6.4 | 48.5 ± 5.1 | 12.6 | 0.2475 |

ANOVA df 17, 90; $F = 3.82$; $P = 0.0001$.
[a]G = glossy, reduced waxbloom; N = normal, wild-type waxbloom; I = intermediate waxbloom.
[b]Mortality attributed to predation was estimated by Abbot's (1925) formula: (*P. xylostella* survival on predator-free control − *P. xylostella* survival with *H. convergens*) / (Predator-free control + *P. xylostella* survival with *H. convergens*) × 100.
[c]P value for planned contrast comparing *P. xylostella* survival on Control with *P. xylostella* survival with *H. convergens*.

*Table 2.* Percentage of time (± SEM) *Hippodamia convergens* spent walking on leaf parts of *Brassica oleracea* with different surface waxbloom

| Lines varieties or population | Wax class[a] | *n* | Upper surfaces[b] | Lower surfaces[b] | Edges[b] |
|---|---|---|---|---|---|
| NY 1406 | G | 28 | 42.2 ± 3.6*** | 6.1 ± 1.0** | 47.9 ± 3.0*** |
| 'Round -Up' | N | 28 | 13.8 ± 3.0 | 0.0 ± 0.0 | 76.0 ± 4.5 |
| KCR4 × | G | 24 | 14.2 ± 3.5*** | 2.3 ± 1.0* | 54.2 ± 5.7* |
| 'Packman' F2 | I | 18 | 25.6 ± 3.5*** | 0.9 ± 0.9* | 54.3 ± 5.2** |
| | N | 28 | 5.7 ± 1.9 | 0.8 ± 0.6 | 68.7 ± 4.3 |
| 'Glossy Andes' | G | 30 | 13.8 ± 2.2** | 9.4 ± 2.4** | 52.1 ± 3.8* |
| 'Andes' | N | 32 | 4.9 ± 1.7 | 4.3 ± 1.6 | 65.5 ± 4.3 |
| Broc5 × | G | 27 | 15.4 ± 2.4** | 3.9 ± 1.6* | 52.4 ± 4.6* |
| 'Packman' F2 | N | 30 | 2.8 ± 1.0 | 0.7 ± 0.5 | 66.8 ± 4.4 |
| ANOVA | | | | | |
| df | | | 8, 244 | 8, 244 | 8, 244 |
| F | | | 23.39 | 6.05 | 4.57 |
| P | | | <0.0001 | <0.0001 | <0.0001 |

MANOVA Wilk's Lambda, df = 24, 702.5 $F = 10.68$; $P < 0.0001$.
[a]G = glossy, reduced waxbloom; N = normal-wax, wild-type waxbloom; I = intermediate waxbloom.
[b]Asterisks indicate P value for planned contrast between reduced waxbloom (G or I) and normal waxbloom type (N) within each group; * =≤ 0.05, ** =≤ 0.01, *** =≤ 0.001.

*Table 3.* Adhesive force (mN ± SEM) produced by *Hippodamia convergens* on *Brassica oleracea* with different surface waxbloom

| Lines varieties or population | Wax class[a] | mNewtons[b] |
|---|---|---|
| NY 1406 | G | 16.5 ± 3.7*** |
| 'Round -Up' | N | 1.9 ± 0.1 |
| KCR4 × | G | 17.3 ± 3.5*** |
| 'Packman' F2 | I | 1.9 ± 0.2 |
| | N | 1.2 ± 0.1 |
| 'Glossy Andes' | G | 36.5 ± 5.3*** |
| 'Andes' | N | 1.0 ± 0.1 |
| Broc5 × | G | 13.4 ± 0.2** |
| 'Packman' F2 | N | 1.0 ± 0.1 |
| ANOVA | | |
| df | | 8, 99 |
| F | | 24.47 |
| P | | < 0.0001 |

[a]G = glossy, reduced waxbloom; N = normal, wild-type waxbloom; I = intermediate waxbloom.
[b]Asterisks indicate P value for planned contrast between reduced waxbloom (G or I) and normal waxbloom type (N) within each group; ** =≤ 0.01, *** =≤ 0.001.

this predator on the *B. oleracea* lines tested. *Plutella xylostella* tend to feed or mine on the leaf surface, especially on lower leaf surfaces (Salinas, 1984; Eigenbrode, 1990). The greater time these predators spend walking over leaf surfaces of glossy plants should increase their rates of encounter with *P. xylostella*.

Hippodamia convergens *adhesion.* Adhesive forces generated by *H. convergens* differed significantly among the *B. oleracea* wax types, ranging from nearly 10-fold to 30-fold higher on glossy versus normal-wax types within pairs of lines or $F_2$ segregants (Table 3). Adhesive force produced by *H. convergens* on the KCR4 × 'Packman' $F_2$ intermediate wax type was not different from normal-wax segregants in this population.

Differences in adhesion to the plant surface provide a mechanistic explanation for *H. convergens* walking more on leaf surfaces of glossy *B. oleracea*. On normal-wax plants, increased walking on the leaf edge must compensate for poor adhesion to the leaf surface. Walking and feeding on leaf edges is a strategy employed by some chrysomelid species for coping with poor adhesion to the leaf surface (Edwards & Wanjura, 1990). Insects straddling a leaf edge can probably use muscular force to increase molecular or frictional adhesion between opposing tarsi and the leaf surface.

In this experiment, *H. convergens* adhesion to the upper leaf surface was measured. Adhesion to the lower surface may help explain behavioural differences because time walking on the lower and upper surface did not correspond. For example, *H. convergens* allocated walking time similarly to the upper surface of KCR4 × 'Packman' $F_2$ intermediate and glossy segregants, but walking on the lower surface of intermediates was similar to that on normal-wax segregants.

## Conclusions

The three experiments presented here support the hypothesis that genetic reduction in waxbloom of *B. oleracea* improves adhesion, mobility (indicated by the insects walking over more of the leaf surface), and predatory effectiveness of *H. convergens* foraging on the plants. This result is similar to that obtained with first instar *C. plorabunda* on the same set of *B. oleracea* types (Eigenbrode et al., 1999). This neuropteran predator obtained better adhesion, greater mobility, and was more effective as a predator on the glossy types. The general agreement between the results with two predators suggests reduced waxbloom may benefit other active generalist predator species. However, effects of waxbloom on behaviour of the two predator species differ in detail. In contrast to *H. convergens*, *C. plorabunda* spent more total time walking on glossy types and most of this walking was on leaf surfaces. *Chrysoperla plorabunda* spent little time walking on leaf edges and this did not differ substantially among the *B. oleracea* types (Eigenbrode et al., 1999).

Previously reported observations of *H. convergens* and *C. plorabunda* on a normal-wax and glossy cabbage (expressing $gl_a$) (Eigenbrode et al., 1996) also differ in detail from those we report here. In that study, differences were observed in grooming behaviour and time spent walking on glossy and normal-wax plants by both predator species. The genetic background in which the glossy gene was expressed, and the environmental conditions for plant production (L13:D11 natural lighting versus L16:D8 supplemented lighting in the present study; 25-cm diam versus 15-cm diam pots in the present study) may account for these differences. Nonetheless, predation (Eigenbrode et al.,

1995) by *H. convergens* and *C. plorabunda* on the glossy type were similar to those reported here and in Eigenbrode et al. (1999).

All the work examining this phenomenon has been done in laboratory or greenhouses and its importance in the field is uncertain. However, the results suggest that improvement in predator effectiveness for biological control might be achieved by deploying *B. oleracea* varieties with reduced waxbloom. The gene $gl_a$ apparently provides the greatest potential benefit of this type. This strategy would have tradeoffs because glossy types may be more susceptible to attack by *Phyllotreta* flea beetles. At least this was shown to be the case for other reduced wax mutations of *B. napus* (Bodnaryk, 1992). In natural systems, intra- and interspecific variation in waxbloom (e.g., Martin & Juniper, 1970; Mulroy, 1976; Tipton & White, 1995) could affect active generalist predators and have consequences for regulation of herbivore populations.


## Acknowledgements

K. A. Stoner, Connecticut Agricultural Experiment Station, provided the $F_2$ populations. M. Rohde contributed to the development of several of the methods used. S. Castle conducted the adhesion bioassays. The research was funded by USDA CSREES NRI Grant #95-37302-1836. This is paper #98722 of the Idaho Agricultural Experiment Station.


## References


Abbot, W. S., 1925. A method of computing the effectiveness of an insecticide. Journal of Economic Entomology 18: 265–267.

Bodnaryk, R. P., 1992. Leaf epicuticular wax, an antixenotic factor in Brassicaceae that affects the rate and pattern of feeding in flea beetles, *Phyllotreta cruciferae* (Goeze). Canadian Journal of Plant Science 72: 1295–1303.

Dickson, M. H. & C. J. Eckenrode, 1975. Variation in *Brassica oleracea* resistance to cabbage looper and imported cabbage worm in the greenhouse and field. Journal of Economic Entomology 68: 757–760.

Dickson, M. H. & C. J. Eckenrode, 1980. Breeding for resistance in cabbage and cauliflower to cabbage looper, imported cabbageworm and diamondback moth. Journal of the American Society for Horticultural Science 105: 782–785.

Edwards, P. B. & W. J. Wanjura. 1991. Physical attributes of eucalypt leaves and the host range of chrysomelid beetles. In: A. Szentesi & T. Jermy. (eds.), Insects-Plants '89. Symposia Biologica Hungarica 39: 227–236.

Eigenbrode, S. D. 1990. Mechanisms of resistance to diamondback moth in *Brassica*. PhD Thesis, Cornell University, Ithaca, NY.

Eigenbrode, S. D., K. E. Espelie & A. M. Shelton, 1991a. Behavior of neonate diamondback moth larvae on leaves and leaf waxes of resistant and susceptible cabbages. Journal of Chemical Ecology 17: 1691–1704.

Eigenbrode, S. D., S. Moodie & T. Castagnola, 1995. Generalist predators mediate resistance to a phytophagous pest in cabbage with glossy leaf wax. Entomologia Experimentalis et Applicata 77: 335–342.

Eigenbrode, S. D., N. N. Kabalo & K. A. Stoner, 1999. Predation, behavior, and attachment by *Chrysoperla plorabunda* larvae on *Brassica oleracea* with different surface waxblooms. Entomologia Experimentalis et Applicata (in press).

Eigenbrode, S. D., C. White & M. Rohde, 1998. Behavior and effectiveness of adult *Hippodamia convergens* (Coleoptera: Coccinellidae) as a predator of *Acyrthosiphon pisum* on a glossy-wax mutant of *Pisum sativum*. Environmental Entomology 27: 902–909.

Eigenbrode, S. D., T. Castagnola, M.-B. Roux & L. Steljes, 1996. Mobility of three generalist predators is greater on cabbage with glossy leaf wax than on cabbage with a wax bloom. Entomologia Experimentalis et Applicata 81: 335–343.

Eigenbrode, S. D., K. A. Stoner, A. M. Shelton & W. C. Kain, 1991b. Characteristics of leaf waxes of *Brassica oleracea* associated with resistance to diamondback moth. Journal of Economic Entomology 83: 1609–1618.

Harris, J. J., 1985. A Primer of Multivariate Statistics. Academic Press, New York.

Juniper, B. E. 1995. Waxes on plant surfaces and their interactions with insects. In: R. J. Hamilton (ed.), Waxes: Chemistry, Molecular Biology and Functions. Dundee, The Oily Press, pp. 157–174.

Macey, M. J. K. & H. N. Barber, 1970. Chemical genetics of wax formation on leaves of *Brassica oleracea*. Phytochemistry 9: 13–23.

Martin, J. T. & B. E. Juniper, 1970. The Cuticles of Plants. Edward Arnold, London.

Mulroy, T. W. 1976. The adaptive significance of glaucescence in *Dudleya* (Crassulaceae). Ph.D. Thesis, University of California, Irvine, Irvine, CA.

Salinas, P. J., 1984. Studies on the behaviour of the larvae of *Plutella xylostella* (Linnaeus) (Lepidoptera: Plutellidae), a world pest of cruciferous crops: normal and 'spacing' behaviour. Turrialba 34: 77–84.

Shelton, A. M., R. J. Cooley, W. T. Kroening & S. D. Eigenbrode, 1991. Comparative analysis of two rearing procedures for diamondback moth, *Plutella xylostella*. Journal of Entomological Science 26: 17–26.

Stoner, K. A., 1990. Glossy leaf wax and host-plant resistance to insects in *Brassica oleracea* L. under natural infestation. Environmental Entomology 19: 730–739.

Stork, N. E., 1980. Role of waxblooms in preventing attachment to brassicas by the mustard beetle, *Phaedon cochleariae*. Entomologia Experimentalis et Applicata 28: 100–107.

Tipton, J. T. & M. White, 1995. Differences in leaf cuticle structure and efficacy among Eastern redbud and Mexican redbud phenotypes. Journal of the American Society for Horticultural Science 120: 59–64.

*Entomologia Experimentalis et Applicata* **91:** 131–142, 1999.



# Are herbivore-induced plant volatiles reliable indicators of herbivore identity to foraging carnivorous arthropods?

Marcel Dicke
*Laboratory of Entomology, Wageningen Agricultural University, P.O. Box 8031, NL-6700 EH Wageningen, The Netherlands*

Accepted: October 17, 1998

*Key words:* tritrophic interactions, volatile infochemicals, predators, parasitoids, Hymenoptera, Phytoseiidae, learning, specific hunger, specificity, headspace composition, behavioural ecology

## Abstract

Plants that are damaged by herbivorous arthropods provide carnivorous enemies of the herbivores with important information. They emit an induced volatile blend that is highly detectable to the carnivores from a distance. Such detectable signals that indicate herbivore presence are important for the carnivores because herbivores themselves are under strong selection not to expose themselves. In addition, carnivores would benefit from a specificity of the induced plant volatiles. Whether herbivore-induced plant volatiles are reliable indicators of herbivore identity, however, has not been resolved unambiguously. Some studies support the reliability of herbivore-induced plant volatiles, while others do not. Different approaches have been used such as chemical analysis, behavioural analysis or a combination of the two. Based on the total of chemical studies one might conclude that in most cases herbivore-induced plant volatiles are not very specific for the herbivore that damages the plant. However, arthropod chemosensors are much more sensitive than the detectors of analytical instruments. Therefore, chemical analyses are not suitable to demonstrate whether or not herbivore-induced plant volatiles are reliable indicators of herbivore identity to carnivores. Behavioural studies should provide this information. In analysing carnivore behaviour it should be realised, however, that arthropod behaviour can be highly variable. Arthropod foraging decisions are affected by external and internal factors such as (a) abiotic environmental factors, (b) presence of competitors or enemies, (c) deprivation of food or oviposition sites, (d) specific deprivation of certain nutrients or (e) learning. In this paper their effect on discrimination of carnivores between volatile blends emitted by plants infested by different herbivores is reviewed. This provides testable hypotheses of why discrimination was not found in some studies. The ability of carnivores to discriminate is likely to be more common than is clear to date, which should invoke functional studies of the conditions that influence the occurrence of this discrimination.

## Introduction

In the past decade it has been well documented that carnivorous arthropods can exploit herbivore-induced plant volatiles to locate their herbivorous victim (for reviews see Vet & Dicke, 1992; Turlings et al., 1993a, 1995; Takabayashi & Dicke, 1996). These volatiles are produced by plants in response to herbivory (Dicke et al., 1990a; Turlings et al., 1990a; Paré & Tumlinson, 1997) and are considered a solution to the so-called 'reliability-detectability problem' of foraging carnivores (Vet et al., 1991; Vet & Dicke, 1992). Herbivores are under strong selection not to emit cues that allow carnivores to find them. Plants, having a much larger biomass than herbivorous arthropods, emit larger quantities of volatiles than herbivores. Plant volatiles produced in response to herbivore damage may contain information on the identity of the herbivore. Therefore, herbivore-induced plant volatiles may be both detectable and reliable indicators of herbivore presence and identity (Vet et al., 1991; Vet & Dicke, 1992).

It is clear that herbivore-induced plant volatiles are highly detectable to carnivores. Carnivores are readily attracted to herbivore-induced plant volatiles, while they usually are not attracted to herbivores isolated from a plant (Vet & Dicke, 1992). Herbivore products such as faeces can attract carnivores, but this attraction is generally much weaker than that of herbivore-induced plant volatiles (Vet & Dicke, 1992). Carnivores as small as 0.8 mm can detect herbivore-induced plant volatiles at distances of at least several metres (i.e., a few thousand body lengths away) from the source (Sabelis & van der Weel, 1993). Under field conditions herbivore-induced plant volatiles lead to the attraction of considerable numbers of carnivore individuals (Drukker et al., 1995; Shimoda et al., 1997).

In contrast to the consensus on the detectability of herbivore-induced plant volatiles, their reliability in indicating the herbivore species that infests a plant has not been demonstrated unequivocally and is subject to debate. Studies that support a specificity of herbivore-induced plant volatiles (e.g., Sabelis & van de Baan, 1983; Takabayashi et al., 1995; Du et al., 1996; Powell et al., 1998; de Moraes et al., 1998) as well as data that do not support such a specificity (e.g., Turlings et al., 1993b, 1998; Mattiacci & Dicke, 1995; Röse et al., 1998) have been presented for different plant-herbivore-carnivore systems. It may very well be that the degree of reliability is system-specific. However, different studies have based their conclusions on different experimental approaches. Some conclusions are based on chemical analyses of headspace composition, others are based on the behavioural response of carnivores and sometimes a combination of the two is used. Each of these two approaches has its own advantages and disadavantages as will be discussed in this paper. I will critically evaluate current evidence on specificity of herbivore-induced plant volatiles as indicators of the type of herbivore that feeds on a plant.

## Variation in chemical composition of blends of herbivore-induced plant volatiles

The chemical composition of herbivore-induced plant volatiles has been studied for a growing number of plant-herbivore systems (see for reviews Dicke et al., 1990b; Turlings et al., 1993a, 1995; Takabayashi et al., 1994; Takabayashi & Dicke, 1996; Dicke, 1999). In total more than 12 plant species and more than 12 herbivore species were used in chemical analyses to date (Dicke, 1999).

Because the induced volatiles are plant products, it is no surprise that chemical analyses show that plant species differ largely in the composition of volatiles that they release in response to herbivory. These are qualitative as well as quantitative differences. For instance, corn plants infested with beet armyworm larvae emit linalool, (3*E*)-4,8-dimethyl-1,3,7-nonatriene, (*trans*)-$\alpha$-bergamotene and (*E*)-$\beta$-farnesene as major compounds, while these are absent in the headspace from soybean plants infested by this herbivore (Turlings et al., 1993a).

Plant species also vary in the degree to which their volatile blends differ between mechanically damaged plants and herbivore-damaged plants. At one extreme the blends of plants of these treatments are qualitatively very different. Examples are e.g., Lima bean and corn that emit large amounts of several terpenoids in response to herbivory but not in response to mechanical damage (Dicke et al., 1990a; Turlings et al., 1990a). At the other extreme, only small qualitative differences exist between the blends emitted by mechanically damaged and herbivore-damaged plants of a species. An example is cabbage (Mattiacci et al., 1994; Agelopoulos & Keller, 1994b; Geervliet et al., 1997).

Within a plant species, there is variation in headspace composition among genotypes, which is mostly restricted to quantitative differences in the contribution of individual components to blend composition (Takabayashi et al., 1991; Loughrin et al., 1995). For example, in the headspace of apple leaves of cultivar Cox Orange Pippin that are infested by two-spotted spider mites, the terpenoids (*E*)-$\beta$-ocimene and (*E,E*)-$\alpha$-farnesene represent ca. 5% of the total volatile blend, while this is ca. 25 and 55% respectively in the headspace of apple leaves of cultivar Summer Red infested by the same herbivore species (Takabayashi et al., 1991).

Chemical analysis of variation in headspace composition among plants of the same species that are infested by different herbivores has been performed for seven plant species: apple (Takabayashi et al., 1991), corn (Turlings et al., 1993a, 1998; Takabayashi et al., 1995; de Moraes et al., 1998), cabbage (Figure 1) (Blaakmeer et al., 1994; Agelopoulos & Keller, 1994b; Geervliet et al., 1997), nasturtium (Geervliet et al., 1997), broad bean (Du et al., 1998), tobacco and cotton (de Moraes et al., 1998). In all cases quantitative differences among volatile blends have been reported (e.g., Figure 1). In only two studies were substantial qualitative differences observed. Broad bean

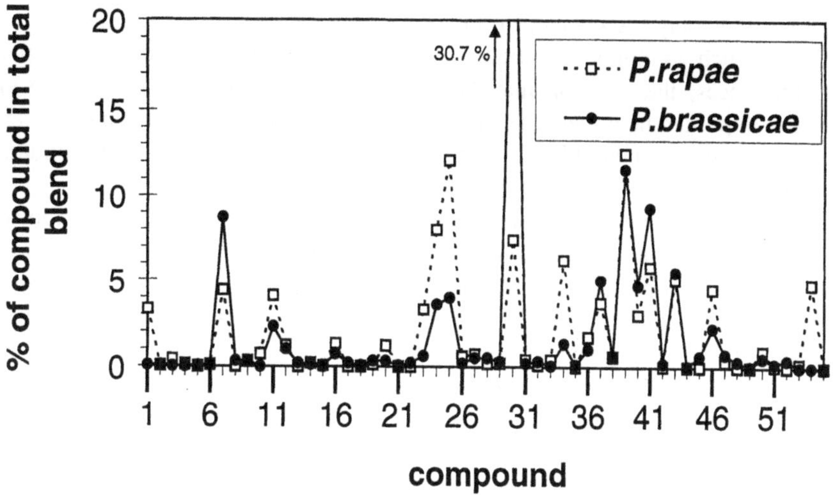


*Figure 1.* Composition of headspace of white cabbage plants infested with caterpillars of *P. brassicae* or *P. rapae*. Compound numbers: 1= 1-butanol; 2= 1-penten-3-ol; 3= 3-methyl-2-pentanol, 4= 1-pentanol; 5= 2-penten-1-ol; 6= 1-hexanol; 7= (Z)-3-hexen-1-ol; 8= 1-heptanol; 9= 1-octanol; 10= 2-pentanone; 11= 3-methyl-2-pentanone; 12= 1-penten-3-one; 13= 2-heptanone; 14= 3-octanone; 15= 1-octen-3-one; 16= 6-methyl-5-hepten-2-one; 17= 2-methyl-2-cyclopenten-1-one; 18= 2-nonanone; 19= 3,5,5-trimethyl-3-cyclohexen-1-one; 20= hexanal; 21= 2-pentenal; 22= heptanal; 23= octanal; 24= nonanal; 25= decanal; 26= undecanal; 27= dodecanal; 28= 3-methyl-3-buten-1-yl acetate; 29= hexyl acetate; 30= (Z)-3-hexen-1-yl acetate; 31= methyl salicylate; 32= 3-methylbutanenitrile; 33= 2-methylbutanenitrile; 34= benzonitrile; 35= benzyl cyanide; 36= $\alpha$-pinene; 37= $\alpha$-thujene; 38= $\beta$-pinene; 39= sabinene; 40= myrcene; 41= limonene; 42= $\beta$-phellandrene; 43=1,8-cineole; 44= $\gamma$-terpinene; 45= p-cymene; 46= (3E)-4,8-dimethyl-1,3,7-nonatriene; 47=trans-sabinene hydrate; 48= linalool; 49= (3E,7E)-4,8,12-trimethyl-1,3,7,11-tridecatetraene; 50= dimethyl disulfide; 51= dimethyl trisulfide; 52= methyl(iso)thiocyanate; 53= butanoic acid; 54= hexanoic acid; 55= indole Based on data from Geervliet et al. (1997).

plants infested by the pea aphid *Acyrthosiphon pisum* emit a blend that comprises 6-methyl-5-hepten-2-one, while conspecific plants infested by the black bean aphid *Aphis fabae* do not. Otherwise the two blends are qualitatively similar, but quantitative differences exist (Du et al., 1998). An even more striking difference was found for maize plants infested with either young (1st/2nd or 3rd larval instars) or old (6th larval instar) caterpillars. Plants infested with young caterpillars emitted large amounts of induced volatiles that were not found in the headspace of mechanically damaged corn leaves. In contrast, in the headspace of plants infested with old caterpillars almost none of these compounds was recorded. The headspace of plants infested with old caterpillars to a large extent resembled the headspace of mechanically damaged plants (Takabayashi et al., 1995). So far, this large difference in headspace composition between plants infested by different types of herbivores seems to be an exception. However, the number of studies addressing this issue is still small.

In conclusion, chemical analyses indicate that differences between blends of herbivore-induced plant volatiles are largest between plant species and lowest between plants of one species that are infested by different herbivores. Arthropods are well known for a well-developed sense of smell (Mustaparta, 1984). Therefore, it is interesting to investigate to what extent carnivorous arthropods can discriminate between blends with varying degrees of differences.

## Biological importance of chemical differences in odour blends

When chemical analyses show large differences between odour blends, such as qualitative differences in major blend components it can be expected that carnivorous arthropods can easily discriminate among these blends. Especially, if the chemical differences are associated with biologically relevant differences for the carnivore. For instance, carnivores are strongly attracted by herbivore-infested plants but not by mechanically damaged plants in those cases where plants emit a qualitatively very different blend in response to herbivores as compared to mechanical damage (Turlings et al., 1990a; Shimoda et al., 1997; Powell et al., 1998). Furthermore, carnivores can discriminate between plants of different species that are infested by herbivores of the same species (Turlings et al., 1990b; Dicke et al., 1990b; McCall et al., 1993; Takabayashi et al., 1994).

When chemical analyses only show small differences between blends, such as minor quantitative and/or qualitative differences, the situation may be more complex. This is important here, because such differences seem to be common among blends from plants of the same species that are infested by different herbivore species. Whether small differences in blends are biologically relevant is dependent on the sensitivity of the animals to perceive them. For instance, the presence/absence of minor peaks that are consistently associated with a biologically relevant parameter, may provide the animal with the opportunity to discriminate between odour blends. Whether this difference is detected through a chemical analysis depends on the sensitivity of the analytical equipment. A behavioural analysis should ultimately reveal whether herbivore-induced plant volatiles are reliable indicators of herbivore identity. In reaching conclusions from behavioural analyses, however, one should be aware that behavioural decisions of arthropods are flexible and can be influenced by various internal and external factors (Stephens & Krebs, 1986). Therefore, if the animal does not discriminate between odour sources, the conclusion should be that either the animal cannot discriminate or that the animal is able to discriminate but does not do so, because the right conditions for a discrimination to be functional have not been met (Vet et al., 1998). In the next section I will review the factors that may influence behavioural decisions of carnivorous arthropods to induced plant volatiles.

## Variation in behavioural response of carnivorous arthropods to herbivore-induced plant volatiles

*Biotic environmental conditions.* Foraging decisions can be affected by the level of competition to be expected or actually experienced (Visser et al., 1990). The response of carnivores to induced plant volatiles can be affected by conspecific or heterospecific competitors. The predatory mite *Phytoseiulus persimilis* prefers the volatiles from bean leaves infested with the spider mite *Tetranychus urticae* over those from bean leaves infested with spider mites plus conspecific predators (Janssen et al., 1997). In a windtunnel, the parasitoid *Cotesia glomerata* preferentially flies towards a cabbage plant infested with *Pieris rapae* larvae when the alternative is a similar plant on which the parasitoid *Cotesia rubecula* is parasitising the *Pieris* larvae (Geervliet, 1997, p. 154). The presence of an enemy of the carnivore can affect the response to volatiles from herbivore-infested plants in a similar way (Höller et al., 1994). It remains unknown whether an experience with competitors or enemies just prior to testing affects carnivore behaviour.

*Deprivation.* In predators, the food content of the gut is an important indicator of motivational state (Sabelis, 1981; Mols, 1988). This is reflected in foraging decisions that are based on induced plant volatiles. For instance, the predatory mite *Typhlodromus pyri* is attracted to volatiles from a larger number of plant-prey combinations with increasing degrees of starvation (Dicke, 1988).

In many parasitoid species, adult females have to search for food as well as hosts. When starved, the females may die and that would end reproduction, but when only searching for food they do not contribute their genes to the next generation. Parasitoids can locate nectar sources through flower volatiles (Wäckers, 1994) and their preference for flower volatiles vs. volatiles from host-infested plants is affected by starvation (Lewis & Takasu, 1990; Wäckers, 1994). Egg-load is also known to affect foraging decisions of parasitoids (Minkenberg et al., 1992), but no reports on the effect of egg-load on responses to herbivore-induced plant volatiles are known to me.

*Specific deprivation.* Specific deprivation reflects a deficiency in a particular nutrient or group of nutrients. For example, flies that have been reared under protein-deficient conditions forage in a way that maximises protein intake (Dethier, 1976) and protein-related volatiles play a role in this (Pospisil, 1958). Similar examples also exist for responses of carnivorous arthropods to induced plant volatiles. The response of the polyphagous predatory mite *Amblyseius andersoni* (= *A. potentillae*) is affected by carotenoid availability in its diet. Carotenoids are essential for diapause induction and thus for survival in winter (van Zon et al., 1981). When predators are reared on one of two diets that only differ in the presence of a single carotenoid, carotenoid deficiency results in attraction to volatiles from a wider range of herbivore-plant combinations (Figure 2) (Dicke et al., 1986; Dicke & Groeneveld, 1986). The herbivores involved are inferior prey to the predators but they can all relieve the carotenoid deficiency. The deficiency does not lead to indiscriminate attraction, however: a plant that is infested by a herbivore that cannot serve as food, e.g., larvae of the leafroller *Adoxophyes orana,* does not lead to attraction (Figure 2). The effect of carotenoid deprivation on foraging decisions has been observed in other species of predatory mites as well (Dicke, 1988; Dicke et al., 1988, 1989).

*Learning.* The response of many carnivorous arthropods is affected by previous foraging experiences (Vet & Dicke, 1992; Turlings et al., 1993a; Vet et al., 1995; Dicke & Vet, 1999). Experience with a host may increase the responsiveness to volatiles emitted by plants with a low host density (Geervliet et al., 1998a). Moreover, discrimination by parasitoids between two plant species infested with their hosts (Geervliet et al., 1998a) or between conspecific plants infested by different herbivore species (Turlings et al., 1993a) can be affected as well. Associative learning enables carnivorous arthropods to adjust their foraging decisions to the prevailing circumstances so as to maximise reproductive success. However, the adaptive value of learning is expected to vary among carnivore species depending on their diet breadth at both the herbivore and the plant level (Vet & Dicke, 1992; Vet et al., 1995). This is supported by a comparative investigation of the closely related parasitoids *C. rubecula* and *C. glomerata*, of which the former is specialised on caterpillars of *P. rapae* on crucifers, while *C. glomerata* is a generalist, accepting several pierid species as hosts on a range of cruciferous and rosaceous plants. The behaviour of the specialist *C. rubecula* is not affected by previous foraging experience in contrast to the behaviour of the generalist *C. glomerata* (Geervliet et al., 1998a,b). Although it is expected that the behaviour of generalists is more flexible than that of specialists, it remains unclear why the specialist *C. rubecula* does not discriminate between plants with its host *P. rapae* and plants with the non-host *P. brassicae* (see Geervliet et al., 1998b for more elaborate discussion).

Most research on the effect of experience on carnivore foraging behaviour has considered rewarding experiences such as the finding of a host or a host's products (Turlings et al., 1993a; Vet et al., 1995). However, unrewarding experiences can play a role as well (Papaj et al., 1994; Vet et al., 1998). The discrimination between odour sources was significantly improved in the parasitoid *Leptopilina heterotoma* when the females had a rewarding experience in relation to one of two odour sources in addition to an unrewarding experience in relation to the other odour source (Vet et al., 1998). Especially when differences between odour sources were small, e.g., a quantitative difference in just one blend component, the wasps only discriminated after they had received a combination of a rewarding and an unrewarding experience. In the case of such a double experience, the information the wasps had on the difference in profitability of the two odour sources was more complete than after a rewarding experience at one of the odour sources only. The conclusion drawn by Vet et al. (1998) was that a non-discrimination between odour sources is not a lack of ability to discriminate but a functional decision by the parasitoid. This conclusion is of great importance in the context of discrimination between odour blends from plants infested by different herbivores, that often show only minor quantitative differences.

## Do carnivorous arthropods discriminate between induced volatiles from plants infested with different herbivore types?

Several studies have shown carnivorous arthropods to discriminate between plants infested with different herbivores (Table 1). Among these are examples of naive parasitoids that are able to discriminate between qualitatively different blends (Takabayashi et al., 1995; Du et al., 1996; Powell et al., 1998) or predatory mites that had never encountered prey in association with the, quantitatively different, odour blends they are offered (Sabelis & van de Baan, 1983; Sabelis & Dicke, 1985; Dicke & Groeneveld, 1986). With smaller differences between blends, the recording of discrimination sometimes depended on previous foraging experience (Turlings et al., 1993a; McCall et al., 1993; Geervliet et al., 1998a).

In contrast, other studies have provided examples where a carnivorous arthropod did not discriminate among volatiles from plants infested with different herbivore species (Table 2). It may be possible that some species have not been selected for the ability to discriminate, which may depend on their diet breadth. For instance, selection on the ability to discriminate between plants infested with different herbivore species is expected to be stronger for carnivores that can exploit only one herbivore species than for carnivores that can exploit several herbivore species that all feed on the same host plant (Vet & Dicke, 1992).

Alternatively, it may be that the carnivores tested did not have the appropriate condition to demonstrate discrimination (Table 2). Females of the parasitoid *Microplitis croceipes* that were reared on hosts on artificial diet and that had never encountered a host or

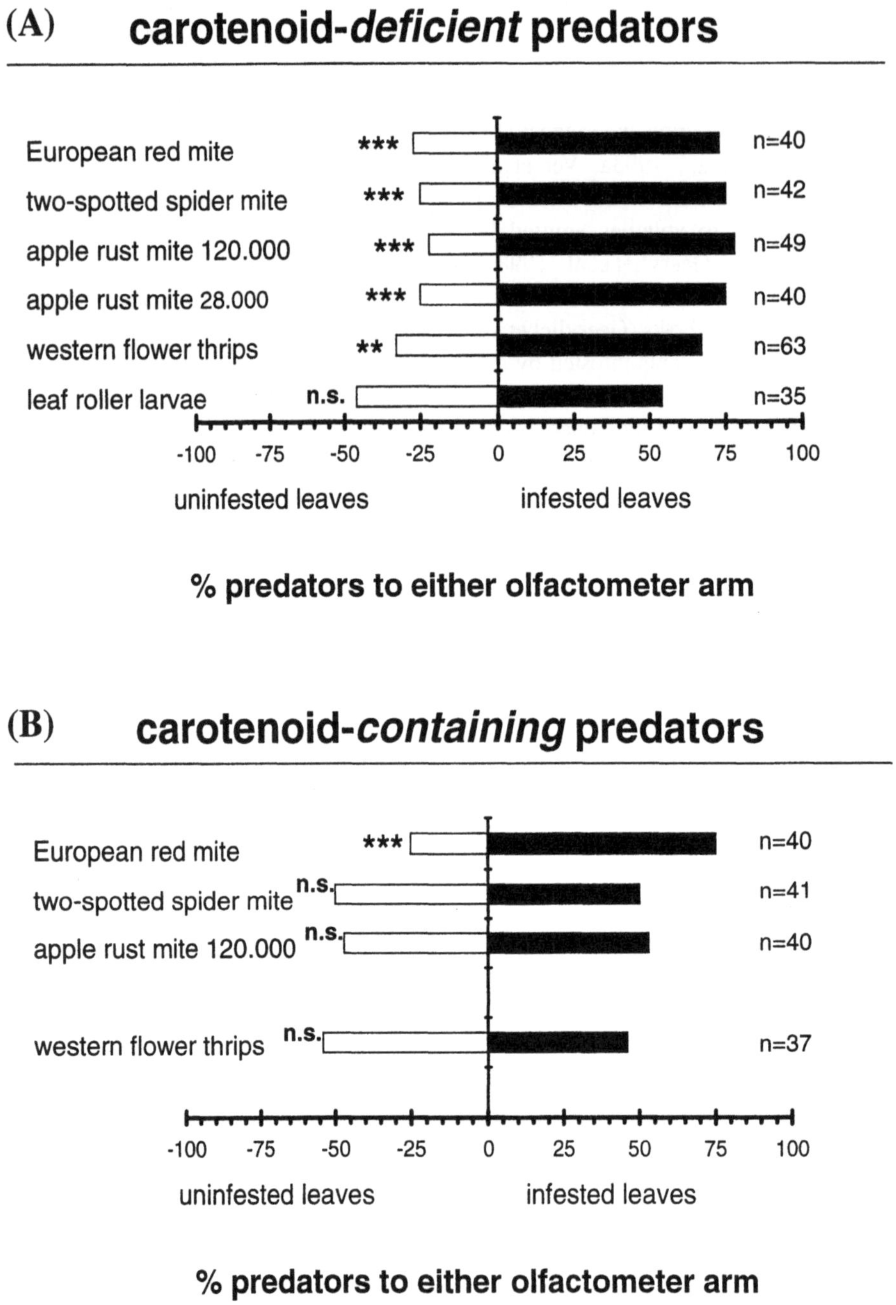


*Figure 2.* Behavioural response of 24-h starved adult females of the predatory mite *Amblyseius andersoni* (= *A. potentillae*) to volatiles from plant leaves infested by herbivores vs. volatiles from uninfested plant leaves in a Y-tube olfactometer. (A) Predators had been reared on a diet of *Vicia faba* pollen on plastic plates. (B) Predators had been reared on a diet of *Vicia faba* pollen mixed with $\beta$-carotene on plastic plates. Odour sources: European red mites (*Panonychus ulmi*) on apple leaves, two-spotted spider mites (*T. urticae*) on Lima bean leaves, apple rust mites (*Aculus schlechtendali*) on apple leaves, Western flower thrips (*Frankliniella pallida*) on Lima bean leaves, leaf rollers (*Adoxophyes orana*) on Lima bean leaves. For each odour source combination the uninfested leaves were from the same plant species as the infested leaves. Numbers next to herbivore names represent the number of herbivores on infested leaves. After data from Dicke & Groeneveld (1986).

*Table 1.* Behavioural studies showing that carnivorous arthropods respond differently to volatiles emitted by plants infested with different herbivore species

| Carnivore species | Plant species | Herbivore species | Specific condition of carnivore | Reference |
|---|---|---|---|---|
| *Amblyseius andersoni*<br>*Amblyseius finlandicus*<br>*Phytoseiulus persimilis*<br>(Acari: Phytoseiidae) | apple | *Panonychus ulmi*<br>*Tetranychus urticae*<br>(Acari: Tetranychidae) | • no experience with odours from apple leaves infested by either spider mite species<br>• *A. andersoni* and *A. finlandicus* starved for 24 h<br>• *P. persimilis* satiated or starved for 24 h<br>• behaviour functional in terms of predator reproductive success | Sabelis & van de Baan, 1983; Sabelis & Dicke, 1985 |
| *Amblyseius andersoni*<br>*Amblyseius finlandicus*<br>(Acari: Phytoseiidae) | apple | *Tetranychus viennensis*<br>*Tetranychus urticae*<br>(Acari: Tetranychidae) | • no experience with odours from apple leaves infested by either spider mite species<br>• starved for 24 h<br>• behaviour functional in terms of predator reproductive success | Sabelis & van de Baan, 1983 |
| *Amblyseius andersoni*<br>(Acari: Phytoseiidae) | Lima bean | *Tetranychus urticae*<br>(Acari: Tetranychidae)<br>*Adoxophyes orana*<br>(Lepidoptera: Tortricidae) | • reared on broad bean pollen on plastic plate, so no experience with odours from bean leaves infested by either herbivore<br>• behaviour functional in terms of carotenoid intake | Dicke & Groeneveld, 1986 |
| *Amblyseius andersoni*<br>(Acari: Phytoseiidae) | apple | *Panonychus ulmi*<br>(Acari: Tetranychidae)<br>*Aculus schlechtendali*<br>(Acari: Eriophyiidae) | • reared on broad bean pollen plus carotene on plastic plate, so no experience with odours from apple leaves infested by either herbivore | Dicke & Groeneveld, 1986 |
| *Typhlodromus pyri*<br>(Acari: Phytoseiidae) | apple | *Panonychus ulmi*<br>(Acari: Tetranychidae)<br>*Aculus schlechtendali*<br>(Acari: Eriophyiidae) | • reared on broad bean pollen on plastic plate, so no experience with odours from apple leaves infested by either herbivore<br>• predators starved for 20 h<br>• behaviour functional in terms of carotenoid intake (Dicke et al., 1990c) | Dicke, 1988 |
| *Cotesia marginiventris*<br>(Hymenoptera: Braconidae) | maize | *Spodoptera exigua*<br>*Spodoptera frugiperda*<br>(Lepidoptera: Noctuidae) | • discrimination after oviposition experience on one of the two herbivores | Turlings et al., 1993a |
| *Microplitis croceipes*<br>(Hymenoptera: Braconidae) | cotton,cowpea | *Helicoverpa zea*<br>*Spodoptera exigua*<br>(Lepidoptera: Noctuidae) | • two oviposition experiences on *H. zea* on the host plant species tested | McCall et al., 1993 |
| *Microplitis croceipes*<br>(Hymenoptera: Braconidae) | cowpea | *Helicoverpa zea*<br>*Trichoplusia ni*<br>(Lepidoptera: Noctuidae) | • complete information, i.e. an oviposition experience on the host *H. zea* on cowpea followed by an oviposition on the non-host *T. ni* on cowpea | McCall et al., 1993 |
| *Cotesia rubecula*<br>(Hymenoptera: Braconidae) | cabbage | *Pieris rapae*<br>(Lepidoptera: Pieridae)<br>*Plutella xylostella*<br>(Plutellidae) | • experience with volatiles from *Pieris rapae*-damaged cabbage plant<br>• discrimination is correlated to reproductive success of the parasitoids on each of the two herbivore species | Agelopoulos & Keller, 1994a |
| *Cotesia rubecula*<br>(Hymenoptera: Braconidae) | cabbage | *Pieris rapae*<br>(Lepidoptera: Pieridae)<br>snail *Helix aspera* | • experience with volatiles from *Pieris rapae*-damaged cabbage plant<br>• exhibited discrimination is correlated to reproductive success of the parasitoids on each of the two herbivore species | Agelopoulos & Keller, 1994a |

*Table 1.* Continued

| Carnivore species | Plant species | Herbivore species | Specific condition of carnivore | Reference |
|---|---|---|---|---|
| *Cotesia kariyai* (Hymenoptera: Braconidae) | maize | *Pseudaletia separata* 1st-4th vs. 5th-6th larval instars (Lepidoptera: Noctuidae) | • naive: no oviposition experience, nor any experience with maize plants | Takabayashi et al., 1995 |
| *Cotesia glomerata* (Hymenoptera: Braconidae) | Brussels sprouts | *Pieris brassicae*, *Pieris rapae* (Lepidoptera: Pieridae) | • oviposition experience on each herbivore on Brussels sprouts plant, or contact experience with damage done by each herbivore species on Brussels sprouts plant | Geervliet et al., 1998b |
| *Aphelinus ervi* (Hymenoptera: Aphidiidae) | broad bean | *Acyrthosiphon pisum* *Aphis fabae* (Homoptera: Aphididae) | • naive: no oviposition experience, nor any experience with broad bean plants | Powell et al., 1998 |
| *Cardiochiles nigriceps* (Hymenoptera: Braconidae) | tobacco, cotton | *Heliothis virescens* *Helicoverpa zea* (Lepidoptera: Noctuidae) | • field population, so parasitoids are likely to have had many oviposition experiences | de Moraes et al., 1998 |

its products in association with the odour sources offered, were attracted to systemically induced volatiles produced by cotton plants infested by a non-host herbivore (Röse et al., 1998). Possibly, the systemic volatiles are not specific for the herbivore species that damages the plant, as the authors conclude, although no experiments were done in which the parasitoids were offered a choice between plants with volatiles systemically induced by a host and a non-host species. However, alternatively the parasitoids may need exposure to plant cues present in or on the host upon emergence (Corbet, 1985) and/or oviposition experience(s) on herbivore-damaged cotton before this discrimination can be achieved (Turlings et al., 1993a).

The parasitoid *C. glomerata* did not discriminate between cabbage plants infested with young (1st or 2nd larval instar) or old (5th larval instar) *P. brassicae* larvae (Mattiacci & Dicke, 1995). The parasitoids tested were either naive, or had contact experience with feeding damage done by old larvae, by young larvae or a combination of two different experiences, with damage done by young or by old larvae. Encountering old larvae bears a large risk, since they heavily defend themselves through spitting, kicking and biting. In this study none of the parasitoids used had ever experienced an encounter with an old larva. Possibly a negative experience with a 5th instar caterpillar in combination with a rewarding experience with 1st instars results in discrimination between damage done by old and young larvae. After all, for *C. glomerata* to discriminate between cabbage plants with different host species, a double experience is needed (Geervliet et al., 1998a). Alternatively, it may be that cabbage does not emit a blend that is specific for the herbivore instar.

The parasitoids *Cotesia marginiventris* and *M. croceipes* did not discriminate between maize seedlings that had been incubated with their stem in a solution of regurgitant of either beet armyworm (*Spodoptera exigua*) or the grasshopper *Schistocerca americana* (Turlings et al., 1993b). The wasps had received a 20-s contact experience with corn seedlings fed upon by *S. exigua* caterpillars (*C. marginiventris* wasps) or by *Helicoverpa zea* caterpillars (*M. croceipes* wasps) just before they were used in the windtunnel bioassay. Thus, *M. croceipes* had no experience with corn damaged by *Spodoptera frugiperda* or *S. americana*. The experience of *M. croceipes* with host-damaged corn seedlings just prior to the experiment may have resulted in a generalisation (see Vet et al., 1998) that led to equal attraction to seedlings treated with regurgitate from two different herbivores. However, for *C. marginiventris* this cannot explain the absence of discrimination between the two odour sources (Turlings et al., 1993b). It is interesting in this context that a 30-s contact experience of *C. marginiventris* with corn damaged by *S. frugiperda* or *S. exigua* resulted in significantly different responses when the wasps were subsequently offered the odour of corn infested by either of these two *Spodoptera* species (Turlings et al., 1993a). Possibly, the actual feeding by caterpillars results in different odour blends as was recorded through chemical analyses (Turlings et al., 1993a)

*Table 2.* Behavioural studies in which a carnivorous arthropod did *not* respond differently to volatiles emitted by plants infested with different herbivore species

| Carnivore species | Plant species | Herbivore species | Potential cause for non-discrimination | Reference |
|---|---|---|---|---|
| *Cotesia marginiventris* *Microplitis croceipes* (Hymenoptera: Braconidae) | maize | regurgitant from *Spodoptera frugiperda* (Lepidoptera: Noctuidae) *Schistocerca americana* (Orthoptera) | • plants treated with regurgitant instead of plants damaged by feeding herbivores; feeding can result in specific blends (Turlings et al., 1993a) <br> • for *M. croceipes* generalization (see Vet et al., 1998) due to experience with an odour source that is different from the two odour sources offered | Turlings et al., 1993b |
| *Microplitis croceipes* (Hymenoptera: Braconidae) | cotton cowpea | *Helicoverpa zea*, *Spodoptera exigua* (Lepidoptera: Noctuidae) | • lack of experience of parasitoids; after two oviposition experiences on *H. zea* on cotton parasitoids did discriminate | McCall et al., 1993 |
| *Microplitis croceipes* (Hymenoptera: Braconidae) | cowpea | *Helicoverpa zea Trichoplusia ni* (Lepidoptera: Noctuidae) | • experience of parasitoids, especially complete information; after oviposition experience on the host *H. zea* on cowpea, followed by oviposition experience on the non-host *T. ni* on cowpea parasitoids did discriminate | McCall et al., 1993 |
| *Cotesia glomerata* (Hymenoptera: Braconidae) | Brussels sprouts | *Pieris brassicae, Pieris rapae, Pieris napi* (Lepidoptera: Pieridae) *Mamestra brassicae* (Lepidoptera: Noctuidae) | • wasps were naive; after oviposition experience on *Pieris brassicae* and *P. rapae* on Brussels sprouts plant they were able to discriminate between Brussels sprouts plants infested with either of these two herbivore species (Geervliet et al., 1998b) | Geervliet et al., 1996 |
| *Cotesia rubecula* (Hymenoptera: Braconidae) | Brussels sprouts | *Pieris brassicae, Pieris rapae,* (Lepidoptera: Pieridae) | • the closely related *C. glomerata* does discriminate in same setup with same odour sources after 2 oviposition experiences; *C. rubecula* may need more experiences. | Geervliet et al., 1996, 1998b |
| *Cotesia glomerata* (Hymenoptera: Braconidae) | Brussels sprouts | *Pieris brassicae* 1st/2nd and 5th larval instars (Lepidoptera: Pieridae) | • parasitoids were either naive, or had single or double experience with feeding damage by 2nd or 5th instar lavae, or had an oviposition experience in a 2nd instar larva plus a contact experience with L2-damaged cabbage <br> • possibly a negative experience with a 5th larval instar in combination with a positive experience with a 2nd larval instar does have an effect | Mattiacci & Dicke, 1995 |

while application of regurgitant through the stem does not (Turlings et al., 1993a).

## Epilogue

The above evaluation of studies on the specificity of herbivore-induced plant volatiles as indicators of the herbivore species that feeds on a plant, shows that carnivorous arthropods are highly sensitive to the volatiles. Chemical analyses may provide information on the degree of variation among odour blends. But since they cannot provide information on the ability of an arthropod to discriminate between the blends, the importance of the recorded differences to carnivores cannot be concluded on.

The recording of a behavioural discrimination by the arthropod is the final proof that is needed to demonstrate that plant volatile blends are specific for the herbivore(s) that feed on the plant. This has been done in several studies (Table 1).

The ability to discriminate between blends of herbivore-induced plant volatiles has been studied mainly for the very first phase in carnivore foraging behaviour such as the period before the parasitoids have encountered three herbivores. This may not be representative for most foraging decisions in nature, where previous foraging experiences, both positive and negative ones, can be integrated. It will be interesting to expand the study of carnivore responses to herbivore-induced plant volatiles to longer observations of foraging behaviour that include an accumulation of foraging experiences.

The discrimination of carnivores among plant volatiles induced by different herbivores has been demonstrated for several predators and parasitoid species. A range of factors that may affect the discrimination by carnivores has been reviewed. An important factor that has not been mentioned is genetic variation among individuals of a carnivore population. Very little is known on genetic variation in responses to herbivore-induced plant volatiles (but see Prevost & Lewis, 1990), and nothing is known on genetic variation in the ability to discriminate among plant volatiles induced by different herbivore species. In contrast, a lot of knowledge is available on context-specificity of carnivore discrimination among plant volatiles induced by different herbivore species. Factors involved are e.g., previous experiences, starvation and specific hunger. This context-specificity brings up an important question: why does carnivore discrimination vary with these factors? This question should be answered through a functional approach in terms of fitness that aims at understanding the costs and benefits of foraging decisions such as to discriminate or not to discriminate (Robertson et al., 1995; Vet, 1999). This evolutionary approach takes variation in discrimination among herbivore-induced plant volatiles as a starting point rather than as an end point. It considers the role of natural selection in shaping foraging decisions of carnivorous arthropods, while taking into account constraints of information availability as well as physiological, developmental and phylogenetic constraints of the behavioural response. This approach is likely to make important contributions not only to our understanding of three-trophic level interactions at the individual level, but also to our understanding of multitrophic interactions and population and community processes (Vet, 1999; Janssen et al., 1998; Dicke & Vet, 1999).

## Acknowledgements

The manuscript benefited from comments by Louise E.M. Vet on a previous version. MD was partly funded by the Uyttenboogaart-Eliasen Foundation.

## References

Agelopoulos, N. G. & M. A. Keller, 1994a. Plant-natural enemy association in the tritrophic system, *Cotesia rubecula-Pieris rapae*-Brassicaceae (Crucifera): I. Sources of infochemicals. Journal of Chemical Ecology 20: 1725–1734.

Agelopoulos, N. G. & M. A. Keller, 1994b. Plant-natural enemy association in the tritrophic system, *Cotesia rubecula-Pieris rapae*-Brassicaceae (Crucifera): III. Collection and identification of plant and frass volatiles. Journal of Chemical Ecology 20: 1955–1967.

Blaakmeer, A., J. B. F. Geervliet, J. J. A. van Loon, M. A. Posthumus, T. A. van Beek & A. E. de Groot, 1994. Comparative headspace analysis of cabbage plants damaged by two species of *Pieris* caterpillars: consequences for in-flight host location by *Cotesia* parasitoids. Entomologia Experimentalis et Applicata 73: 175–182.

Corbet, S. A., 1985. Insect chemosensory responses: a chemical legacy hypothesis. Ecological Entomology 10: 143–153.

Dethier, V. G., 1976. The Hungry Fly: A Physiological Study of the Behavior Associated with Feeding. Harvard University Press. Cambridge, Massachusetts, 489 pp.

Dicke, M., 1988. Prey preference of the phytoseiid mite *Typhlodromus pyri*: 1. Response to volatile kairomones. Experimental and Applied Acarology 4: 1–13.

Dicke, M., 1999. Evolution of induced indirect defence of plants. In: R. Tollrian & C. D. Harvell (eds), The Ecology and Evolution of Inducible Defenses. Princeton University Press, Princeton, New Jersey, pp. 62–68.

Dicke, M. & A. Groeneveld, 1986. Hierarchical structure in kairomone preference of the predatory mite *Amblyseius potentillae*: dietary component indispensable for diapause induction affects prey location behaviour. Ecological Entomology 11: 131–138.

Dicke, M. & L. E. M. Vet, 1999. Plant-carnivore interactions: evolutionary and ecological consequences for plant, herbivore and carnivore. In: H. Olff, V. K. Brown & R. H. Drent (eds), Herbivores: Between Plants and Predators. Blackwell Science, Oxford, UK, pp. 483–520.

Dicke, M., M. W. Sabelis & A. Groeneveld, 1986. Vitamin A deficiency modifies response of predatory mite *Amblyseius potentillae* to volatile kairomone of two-spotted spider mite, *Tetranychus urticae*. Journal of Chemical Ecology 12: 1389–1396.

Dicke, M., M. W. Sabelis & M. de Jong, 1988. Analysis of prey preference of phytoseiid mites as determined with an olfactometer, predation models and electrophoresis. Experimental and Applied Acarology 5: 225–241.

Dicke, M., M. de Jong, M. P. T. Alers, F. C. T. Stelder, R. Wunderink & J. Post, 1989. Quality control of mass-reared arthropods: Nutritional effects on performance of predatory mites. Journal of Applied Entomology 108: 462–475.

Dicke, M., T. A. van Beek, M. A. Posthumus, N. Ben Dom, H. van Bokhoven & A. E. de Groot, 1990a. Isolation and identification of volatile kairomone that affects acarine predator-prey interactions. Involvement of host plant in its production. Journal of Chemical Ecology 16: 381–396.

Dicke, M., M. W. Sabelis, J. Takabayashi, J. Bruin, & M. A. Posthumus, 1990b. Plant strategies of manipulating predator-prey interactions through allelochemicals: prospects for application in pest control. Journal of Chemical Ecology 16: 3091–3118.

Dicke, M., M. W. Sabelis, M. de Jong & M. P. T. Alers, 1990c. Do phytoseiid mites select the best prey species in terms of reproductive success? Experimental and Applied Acarology 8: 161–173.

Drukker, B., P. Scutareanu & M. W. Sabelis, 1995. Do anthocorid predators respond to synomones from *Psylla*-infested pear trees under field conditions? Entomologia Experimentalis et Applicata 77: 193–203.

Du, Y.-J., G. M. Poppy & W. Powell, 1996. Relative importance of semiochemicals from first and second trophic levels in host foraging behavior of *Aphidius ervi*. Journal of Chemical Ecology 22: 1591–1605.

Du, Y., G. M. Poppy, W. Powell, J. A. Pickett, L. J. Wadhams & C. M. Woodcock, 1998. Identification of semiochemicals released during aphid feeding that attract the parasitoid *Aphidius ervi*. Journal of Chemical Ecology 24: 1355–1368.

Geervliet, J. B. F., 1997. Infochemical use by insect parasitoids in a tritrophic context: comparison of a generalist and a specialist. PhD-thesis Wageningen Agricultural University, 200 pp.

Geervliet, J. B. F., L. E. M. Vet & M. Dicke, 1996. Innate responses of the parasitoids *Cotesia glomerata* and *C. rubecula* (Hymenoptera: Braconidae) to volatiles from different plant-herbivore complexes. Journal of Insect Behavior 9: 525–538.

Geervliet, J. B. F., M. A. Posthumus, L. E. M. Vet, M. & Dicke, 1997. Comparative analysis of headspace volatiles from different caterpillar-infested and uninfested food plants of *Pieris* species. Journal of Chemical Ecology 23: 2935–2954.

Geervliet, J. B. F., S. Ariëns, M. Dicke & L. E. M. Vet, 1998a. Long-distance assessment of patch profitability through volatile infochemicals by the parasitoids *Cotesia glomerata* and *C. rubecula* (Hymenoptera: Braconidae). Biological Control 11: 113–121.

Geervliet, J. B. F., A. I. Vreugdenhil, L. E. M. Vet & M. Dicke, 1998b. Learning to discriminate between infochemicals from different plant-host complexes by the parasitoids *Cotesia glomerata* and *C. rubecula* (Hymenoptera: Braconidae). Entomologia Experimentalis et Applicata 86: 241–252.

Höller, C., S. G. Micha, S. Schulz, W. Francke & J. A. Pickett, 1994. Enemy-induced dispersal in a parasitic wasp. Experientia 50: 182–185.

Janssen, A., J. Bruin, G. Jacobs, R. Schraag & M. W. Sabelis, 1997. Predators use volatiles to avoid prey patches with conspecifics. Journal of Animal Ecology 66: 223–232.

Janssen, A., A. Pallini, M. Venzon & M. W. Sabelis, 1998. Behaviour and indirect food web interactions among plant inhabiting arthropods. Experimental and Applied Acarology 22: 497–521.

Lewis, W. J. & K. Takasu, 1990. Use of learned odours by a parasitic wasp in accordance with host and food needs. Nature 348: 635–636.

Loughrin, J. H., A. Manukian, R. R. Heath, & J. H. Tumlinson, 1995. Volatiles emitted by different cotton varieties damaged by feeding beet armyworm larvae. Journal of Chemical Ecology 21: 1217–1227.

Mattiacci, L. & M. Dicke, 1995. Host-age discrimination during host location by *Cotesia glomerata*, a larval parasitoid of *Pieris brassicae*. Entomologia Experimentalis et Applicata 76: 37–48.

Mattiacci, L., M. Dicke & M. A. Posthumus, 1994. Induction of parasitoid attracting synomone in brussels sprouts plants by feeding of *Pieris brassicae* larvae: role of mechanical damage and herbivore elicitor. Journal of Chemical Ecology 20: 2229–2247.

McCall, P. J., T. C. J. Turlings, W. J. Lewis, & J. H. Tumlinson, 1993. Role of plant volatiles in host location by the specialist parasitoid *Microplitis croceipes* Cresson (Braconidae: Hymenoptera). Journal of Insect Behavior 6: 625–639.

Minkenberg, O. P. J. M., M. Tatar & J. A. Rosenheim, 1992. Egg load as a major source of variability in insect foraging and oviposition behavior. Oikos 65: 134–142.

Mols, P. J. M., 1988. Simulation of hunger, feeding and egg production in the carabid beetle *Pterostichus coerulescens* L. (= *Poecilus versicolor* Sturm). Wageningen Agricultural University Papers 88(3): 1–99.

Moraes, C. M. de, W. J. Lewis, P. W. Paré, H. T. Alborn & J. H. Tumlinson, 1998. Herbivore-infested plants selectively attract parasitoids. Nature 393: 570–573.

Mustaparta, H., 1984. Olfaction. In: W. J. Bell & R. T. Cardé (eds), Chemical Ecology of Insects, Chapman & Hall, London, pp. 483–520.

Papaj, D. R., H. Snellen, K. Swaans & L. E. M. Vet, 1994. Unrewarding experiences and their effect on foraging in the parasitic wasp *Leptopilina heterotoma* (Hymenoptera: Eucoilidae). Journal of Insect Behavior 7: 465–481.

Paré, P. W. & J. H. Tumlinson, 1997. Induced synthesis of plant volatiles. Nature 385: 30–31.

Pospisil, J., 1958. Neketere problemy cichu u saprofilnich much (Some problems of the smell of the Saprophilic flies). Acta Societatis Entomologicae Cechosloveniae 55: 316–333.

Powell, W., F. Pennacchio, G. M. Poppy & E. Tremblay, 1998. Strategies involved in the location of hosts by the parasitoid *Aphidius ervi* Haliday (Hymenoptera: Braconidae: Aphidiinae). Biological Control 11: 104–112.

Prévost, G. & W. J. Lewis, 1990. Genetic differences in the response of *Microplitis croceipes* to volatile semiochemicals. Journal of Insect Behavior 3: 277–287.

Robertson, I. C., B. D. Roitberg, I. Williamson & S. E. Senger, 1995. Contextual chemical ecology: an evolutionary approach to the chemical ecology of insects. American Entomologist 41: 237–239.

Röse, U. S. R., W. J. Lewis & J. H. Tumlinson, 1998. Specificity of systemically released cotton volatiles as attractants for specialist and generalist parasitic wasps. Journal of Chemical Ecology 24: 303–319.

Sabelis, M. W., 1981. Biological control of two-spotted spider mites using phytoseiid predators. Part I: modelling the predator-prey interaction at the individual level. PhD thesis, Wageningen Agricultural University, 242 pp.

Sabelis, M. W. & H. E. van de Baan, 1983. Location of distant spider mite colonies by phytoseiid predators: demonstration of specific kairomones emitted by *Tetranychus urticae* and *Panonychus ulmi*. Entomologia Experimentalis et Applicata 33: 303–314.

Sabelis, M. W. & M. Dicke, M., 1985. Long-range dispersal and searching behaviour. In: W. Helle & M. W. Sabelis, (eds), Spider Mites. Their Biology, Natural Enemies and Control. World Crop Pests Vol. 1b. Elsevier, Amsterdam, pp. 141–160.

Sabelis, M. W. & J. J. van der Weel, 1993. Anemotactic responses of the predatory mite, *Phytoseiulus persimilis* Athias-Henriot, and their role in prey finding. Experimental and Applied Acarology 17: 521–529.

Shimoda, T., J. Takabayashi, W. Ashihara & A. Takafuji, 1997. Response of the predatory insect *Scolothrips takahashii* toward herbivore-induced plant synomone under both laboratory and field conditions. Journal of Chemical Ecology 23: 2033–2048.

Stephens, D. W. & J. R. Krebs, 1986. Foraging Theory. Princeton University Press, Princeton.

Takabayashi, J. & M. Dicke, 1996. Plant-carnivore mutualism through herbivore-induced carnivore attractants. Trends in Plant Science 1: 109–113.

Takabayashi, J., M. Dicke & M. A. Posthumus, 1991. Variation in composition of predator-attracting allelochemicals emitted by herbivore-infested plants: relative influence of plant and herbivore. Chemoecology 2: 1–6.

Takabayashi, J., M. Dicke & M. A. Posthumus, 1994. Volatile herbivore-induced terpenoids in plant-mite interactions: Variation caused by biotic and abiotic factors. Journal of Chemical Ecology 20: 1329–1354.

Takabayashi, J., S. Takahashi, M. Dicke & M. A. Posthumus, 1995. Developmental stage of herbivore *Pseudaletia separata* affects production of herbivore-induced synomone by corn plants. Journal of Chemical Ecology 21: 273–287.

Turlings, T. C. J., J. H. Tumlinson & W. J. Lewis, 1990a. Exploitation of herbivore-induced plant odors by host-seeking parasitic wasps. Science 250: 1251–1253.

Turlings, T. C. J., J. W. A. Scheepmaker, L. E. M. Vet, J. H. Tumlinson & W. J. Lewis, 1990b. How contact foraging experiences affect preferences for host-related odors in the larval parasitoid *Cotesia marginiventris* (Cresson) (Hymenoptera:Braconidae). Journal of Chemical Ecology 16: 1577–1589.

Turlings, T. C. J., F. L. Wäckers, L. E. M. Vet, W. J. Lewis & J. H. Tumlinson, 1993a. Learning of host-finding cues by Hymenopterous parasitoids. In: D. R. Papaj & A. C. Lewis (eds), Insect Learning. Chapman & Hall, New York, pp. 51–78.

Turlings, T. C. J., P. J. McCall, H. T. Alborn & J. H. Tumlinson, 1993b. An elicitor in caterpillar oral secretions that induces corn seedlings to emit chemical signals attractive to parasitic wasps. Journal of Chemical Ecology 19: 411–425.

Turlings, T. C. J., J. H. Loughrin, P. J. McCall, U. S. R. Röse, W. J. Lewis & J. H. Tumlinson, 1995. How caterpillar-damaged plants protect themselves by attracting parasitic wasps. Proceedings of the National Academy of Sciences of the USA 92: 4169–4174.

Turlings, T. C. J., M. Bernasconi, R. Bertossa, F. Bigler, G. Caloz & S. Dorn, 1998. The induction of volatile emissions in maize by three herbivore species with different feeding habits: possible consequences for their natural enemies. Biological Control 11: 122–129.

Vet, L. E. M., 1999. From chemical to population ecology: infochemical use in an evolutionary context. Journal of Chemical Ecology 25: 31–49.

Vet, L. E. M. & M. Dicke, 1992. Ecology of infochemical use by natural enemies in a tritrophic context. Annual Review of Entomology 37: 141–172.

Vet, L. E. M., F. L. Wäckers & M. Dicke, 1991. How to hunt for hiding hosts: the reliability-detectability problem in foraging parasitoids. Netherlands Journal of Zoology 41: 202–213.

Vet, L. E. M., W. J. Lewis, & R. T. Cardé, 1995. Parasitoid foraging and learning. In: R. T. Cardé & W. J. Bell (eds), Chemical Ecology of Insects 2. Chapman & Hall, New York, pp. 65–101.

Vet, L. E. M., A. G. de Jong, E. Franchi & D. R. Papaj, 1998. The effect of complete versus incomplete information on odour discrimination in a parasitic wasp. Animal Behaviour 55: 1271–1279.

Visser, M. E., J. J. M. van Alphen & H. W. Nell, 1990. Adaptive superparasitism and patch time allocation in solitary parasitoids: the influence of the number of parasitoids depleting a patch. Behaviour 114: 21–36.

Wäckers, F. L., 1994. The effect of food deprivation on the innate visual and olfactory preferences in the parasitoid *Cotesia rubecula*. Journal of Insect Physiology 40: 641–649.

Zon, A. Q. van, W. P. J. Overmeer & A. Veerman, 1981. Carotenoids function in photoperiodic induction of diapause in a predacious mite. Science 213: 1131–1133.

*Entomologia Experimentalis et Applicata* **91:** 143–148, 1999.



# Active defence of herbivorous hosts against parasitism: Adult parasitoid mortality risk involved in attacking a concealed stemboring host

R. P. J. Potting[1,3], N. E. Vermeulen[1,2] & D. E. Conlong[2]

[1]*Department of Entomology, Wageningen Agricultural University, P.O. BOX 8031, 6700 EH Wageningen, The Netherlands;* [2]*Department of Entomology, SASA Experimental Station, Private Bag X02, Mount Edgecombe, South Africa;* [3]*Present address: IACR-Rothamsted, Entomology and Nematology Department, AL5 2JQ Harpenden, Herts, UK*

Accepted: October 18, 1998

*Key words:* parasitoid, *Cotesia sesamiae*, host selection, mortality risk, stemborer, *Eldana saccharina*, *Sesamia calamistis*, defence behaviour, Hymenoptera, Braconidae

## Abstract

Phytophagous insects have several defence strategies to defend themselves against attack by parasitic wasps. Larval lepidopteran hosts can defend themselves actively to prevent oviposition by the parasitoid. Among the aggressive kinds of behaviour exhibited by hosts against parasitoids are violent wriggling, biting and spitting. The behaviour of the braconid parasitoid *Cotesia sesamiae* attacking stemboring larvae inside their feeding tunnel in the plant stem was investigated in maize and sugarcane stem pieces and transparent artificial tunnels. Attacking a defending stemborer host inside the confined space of a tunnel was shown to be risky for the female parasitoid. A considerable proportion (25%) of female wasps were killed in their attempt to attack the spitting and biting host. No difference was found in the behaviour of *C. sesamiae* attacking the suitable host *Sesamia calamistis* or the unsuitable host *Eldana saccharina*. The consequences of this high mortality risk involved in each host attack is discussed in relation to the ecology of the parasitoid.

## Introduction

Phytophagous insects have several strategies to defend themselves against attack by parasitic wasps. Besides immunological defence mechanisms after parasitization (Strand & Pech, 1995) adult and larval hosts can also defend themselves actively to prevent oviposition by the parasitoid (for reviews see Gross, 1993; Godfray, 1994). Among the aggressive behaviours exhibited by hosts against parasitoids are violent wriggling, biting and spitting (Askew, 1971; Pasteels et al., 1983: Lederhouse, 1990). There are only a few studies that investigated the direct mortality risk involved in attacking a defending host, especially in cases where the host tries to kill the adult parasitoid. Takasu & Overholt (1997, 1998) and Potting et al. (1997a) investigated for two parasitoid species the mortality risk involved in attacking a stemboring larva inside its tunnel. They found that at each oviposition attempt the endoparasitoid *Cotesia flavipes* (Hymenoptera: Braconidae) and the exoparasitoid *Goniozus indicus* (Hymenoptera: Bethylidae) had a high probability (30–70%) of being killed by defending stemborer larvae.

The braconid *Cotesia sesamiae* (Cameron) (Hymenoptera: Braconidae) is a gregarious larval endoparasitoid of stemborers. *Cotesia sesamiae* is native to Africa where it attacks several stemborer species including *Chilo partellus* (Swinhoe) (Lepidoptera: Pyralidae), *Sesamia calamistis* Hampson and *Busseola fusca* Fuller (Lepidoptera: Noctuidae) (Ullyett, 1935; Mohyuddin, 1971; Walker, 1994). *Cotesia sesamiae* is closely related to the well studied braconid *Cotesia flavipes* (Cameron) (Polaszek & Walker, 1991; Potting, 1997) and the behavioural ecology of *C. sesamiae* has been investigated by Ullyett (1935), Mohyuddin (1971), Ngi-Song & Overholt (1997) and Mbapila & Overholt (1997).

In Africa *Eldana saccharina* (Walker) (Lepidoptera: Pyralidae) is the main stemborer pest in sugarcane. Although there is one unconfirmed record of *C. sesamiae* from *E. saccharina* in East Africa (Mohyuddin & Greathead, 1970), *C. sesamiae* has never been recorded from *E. saccharina* in South Africa and Kenya, despite intensive sampling programs in sugarcane fields (Conlong, 1994; Overholt, pers. comm.). Host suitability studies in the laboratory have shown that in South Africa and Kenya *C. sesamiae* is unable to parasitize *E. saccharina*, because the parasitoid's eggs are successfully encapsulated by *E. saccharina* (Conlong, unpubl.; Overholt, pers. comm.). Within its geographical range, *S. calamistis* is a pest attacking maize and sorghum. In South Africa *S. calamistis* is an important secondary pest in young (3–5 months) sugarcane. An important natural enemy of *S. calamistis* in maize and sorghum is *C. sesamiae*. In South Africa *C. sesamiae* also attacks *S. calamistis* in sugarcane. For instance, in 1995 the parasitization rate of *S. calamistis* by *C. sesamiae* in sugarcane fields at the North coast of Kwazulu Natal, South Africa was 19.9% ($n$ = 948) (Vermeulen & Conlong, unpubl.). Thus in the sugarcane habitat *C. sesamiae* may encounter a suitable host (*S. calamistis*) and an unsuitable host (*E. saccharina*). In the present study we examined the mortality risk involved in attacking a host inside the stem and examined whether *C. sesamiae* can discriminate between these two host species and if so, in which stage of the hosts searching process. We observed the oviposition behaviour of *C. sesamiae* females towards *E. saccharina* or *S. calamistis* larvae inside their tunnel and examined host acceptance, oviposition success and mortality of the wasps. We hypothesize that if *C. sesamiae* has evolved the ability to recognize the unsuitable host *E. saccharina* it should do ideally so before contacting the offensive host by recognizing the species identity by the larval frass at the entrance of the tunnel.

## Materials and methods

*Insects.* Larvae of *S. calamistis* and *E. saccharina* were reared on an artificial diet as described by Graham & Conlong (1988). For the experiments fourth to fifth instar larvae were used. All *Cotesia sesamiae* parasitoids used in the experiments were field collected from parasitized *S. calamistis* from sugarcane (North coast Kwazulu Natal, South Africa). Emerged parasitoids were kept in vials (25–27 °C, 70% r.h.) and fed with honey. For the experiments 1–2 days old mated female *C. sesamiae* were used, that were naive with respect to oviposition or contact with host related cues.

*Response to frass.* The response to frass was tested in a Petri dish setup. Frass was collected from 4–5th instar larvae of *E. saccharina* and *S. calamistis* feeding on stems of maize or sugarcane. The material to be tested (0.5 gr of water saturated frass) was placed at the centre of a circular filter paper (Whatman #3.9 cm) that was placed in a glass Petri dish. The treatment zone was defined as the area 6 cm around the test material, marked with a thin pencil line. The residence time of individual *C. sesamiae* inside the treatment zone was measured. In the first series the response to frass from *S. calamistis* on maize and *E. saccharina* on sugarcane was compared. In the second series the response to frass from *S. calamistis* on maize and sugarcane was compared. All experiments (including the ones described below) were carried out in a temperature controlled room (25–27 °C, 70% r.h.). For each frass type 50 parasitoids were tested.

*Behaviour on stemborer infested stem.* The foraging behaviour of *C. sesamiae* females was observed on maize stems infested with *S. calamistis* and on sugarcane stems infested with *E. saccharina*. Infested stems (25 cm) were obtained by drilling a hole in the stem and introducing one fourth instar larva of *S. calamistis* or *E. saccharina* into the hole. The larvae were introduced into the stem 48 h prior to the experiments to allow the larva to create a typical stemborer tunnel. A female wasp was released on the stem 1 cm near the tunnel entrance and the behaviour was observed. An observation ended after 15 min or when the wasp left the host patch (i.e., when it crossed the marking lines 5 cm above and below the tunnel entrance). To rule out any day effects *S. calamistis* infested maize stems and *E. saccharina* infested sugarcane stems were tested alternately on the same day. For each replicate a new artificially infested stem was used. After each replicate the stem was opened to check if wasps that did not come out of the tunnel were dead or alive. The larva was maintained on the same diet for 10 days to check if it was parasitized (i.e., cocoon formation). Larvae without parasitoid cocoon formation were dissected to check if the eggs were encapsulated.

*Behaviour inside tunnel.* To record the behaviour of the parasitoid inside a stemborer tunnel an artificial transparent tunnel with a feeding larva was created by using small glass tubes. Glass tubes (7 × 0.5 cm) were filled with maize or sugarcane and one fourth instar larva of *S. calamistis* (maize) or *E. saccharina* (sugarcane) was introduced into the tube. The larva was allowed to feed for 48 h to create a tunnel. See Potting et al. (1997b) for a figure of a similar setup. The behaviour of individual female *C. sesamiae* was observed for 5 min after they were placed near the opening of the glass tube at the tail end of the larva. It was recorded whether the wasp contacted the larva and whether this resulted in oviposition.

## Results

*Response to frass.* Female *C. sesamiae* responded equally to frass from *E. saccharina* and *S. calamistis* in the Petri dish setup. Although there was no significant difference in the residence time on frass from *E. saccharina* on sugarcane compared to that on frass from *S. calamistis* on maize, there was a strong trend for a longer residence time on the latter (Figure 1a, *t*-test, $t = 1.75$, df = 98, P = 0.08). There was also no significant difference in the residence time of *C. sesamiae* on frass from the same host (*S. calamistis*) feeding on maize or sugarcane (Figure 1b, *t*-test, $t = 0.64$, df = 98, P = 0.52).

*Behaviour on stemborer infested stem.* The behaviour of female *C. sesamiae* on an *E. saccharina* infested sugarcane stem or a *S. calamistis* infested maize stem is summarized in Table 1. There was no difference in the willingness of the females to enter the different tunnel types. After release on the stem nearly all females entered the stemborer tunnel irrespective the stemborer species. However, the total time spent on a *S. calamistis* infested maize stem was significantly longer than on an *E. saccharina* infested sugarcane stem. This was due to the significantly longer time spent inside a *S. calamistis* tunnel compared to an *E. saccharina* tunnel. The wasps had the tendency to re-enter a *S. calamistis* tunnel more often than an *E. saccharina* tunnel. Around one quarter of the wasps entering both tunnel types were found dead inside the tunnel after being killed by the *E. saccharina* or *S. calamistis* host larva. None of the *E. saccharina* larvae was successfully parasitized (i.e., formation of parasitoid cocoons) by *C. sesamiae*, whereas 52% of the *S. calamistis* larvae were successfully parasitized by females entering a *S. calamistis* tunnel.

*Behaviour inside tunnel.* The behaviour of female *C. sesamiae* inside a stemborer tunnel was investigated with artificial transparent tunnels. The results of this experiment are summarized in Table 2. All the parasitoids tested contacted the host inside the tunnel. However, there was a significant difference in the acceptance level of the different host species. 82.3% of the contacts with a *S. calamistis* larva resulted in oviposition in this host, whereas only 7.4% of the contacts with an *E. saccharina* larva resulted in oviposition. The mortality rate of the female parasitoids was significantly higher in an *E. saccharina* tunnel (16.7%) compared to the mortality rate in a *S. calamistis* tunnel (4.8%). None of the *E. saccharina* larvae was successfully parasitized by *C. sesamiae*, whereas 55% of the *S. calamistis* larvae were successfully parasitized.

## Discussion

Attacking a concealed stemborer larva in the confined space of a stemborer tunnel is risky for a female parasitoid. At each host attack the female has a considerable risk of being killed by its aggressive host. In the present study we found that 25% of female *C. sesamiae* entering a stemborer tunnel in a maize or a sugarcane stem were killed be the defending host, irrespective of the host species (Table 1). In similar experiments Potting et al. (1997a) demonstrated that 30–40% of *C. flavipes* females attacking a *C. partellus* stemborer larva in its tunnel were killed by the spitting and biting larva. In the artificial transparent tunnels the mortality risk for the female wasps was lower (5–19%). However, in this experiment wasps were allowed to approach the larval host from its tail end, thus reducing the probability of being killed at first host contact. In a similar glass tube setup Takasu & Overholt (1997) showed that female *C. flavipes* approaching their larval host *Chilo partellus* towards the head had a much higher probability (40–70% mortality) of getting killed (i.e., bitten to death) than females approaching their host from the tails end (10–20% mortality). In addition to the direct mortality inflicted by the biting host Takasu & Overholt (1997) showed that 45% of the parasitoids that contacted the oral exudate of their host died within 24 h.

Despite the fact that a large proportion of the wasps is killed at each host attack it does not always prevent the defending host of being parasitized by the parasitoid. A female *C. sesamiae* needs only a few seconds to inject around 40 eggs into the host. In the

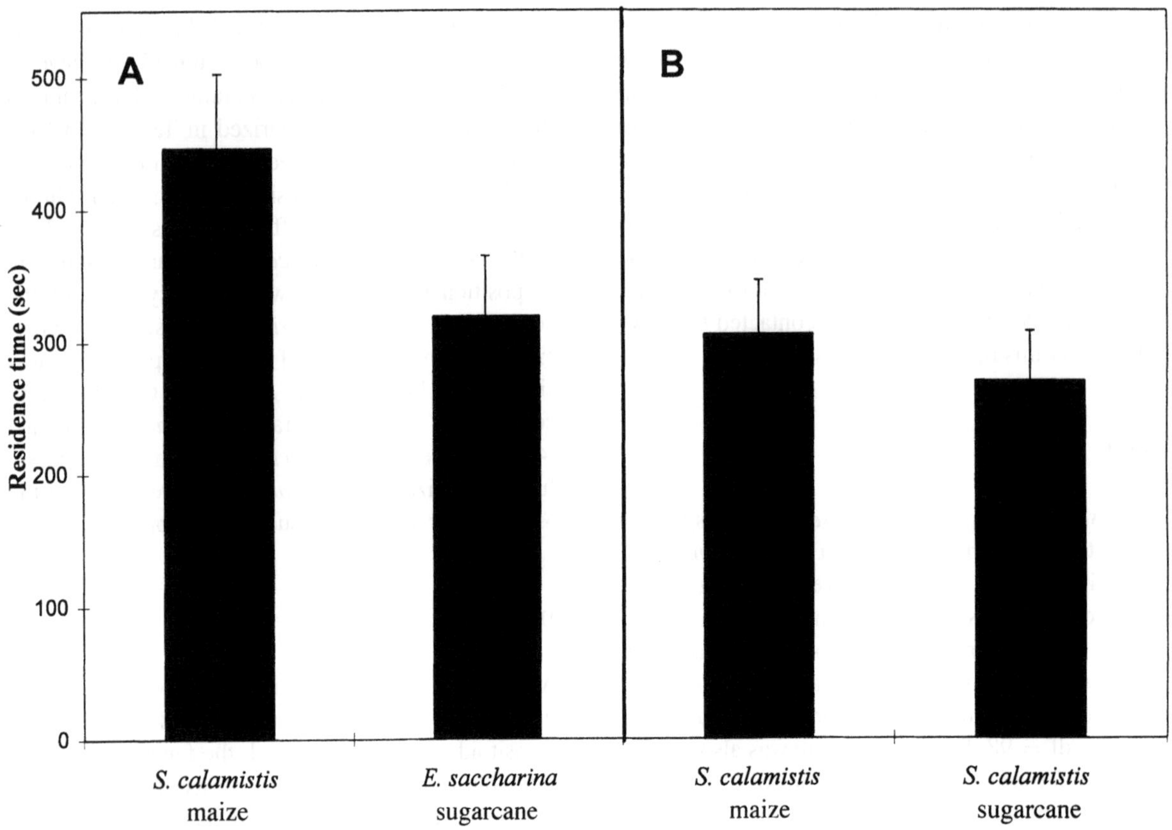


*Figure 1.* Response of *C. sesamiae* females to larval frass of *E. saccharina* feeding on sugarcane and *S. calamistis* feeding on sugarcane or maize. Indicated are average residence times (± SE) in treated patch. (A) Response to frass from different plant-host-complexes. (B) Response to frass from different plants with the same host.

stem pieces experiment 57% ($n = 7$) of the wasps that were killed by their *S. calamistis* host had successfully parasitized its host. In the glass tube experiment 31% ($n = 3$) of the killed wasps had successfully parasitized its host.

Foraging theory assumes that animals assess and select resources according to their relative impact on fitness (Stephens & Krebs, 1986). For instance, parasitoids generally prefer those host species that maximize larval performance and reproduction (Godfray, 1994). For several parasitoid species it has been demonstrated that female wasps can recognize the identity of the host species they are searching for by examination of the frass that is left behind by the larvae (e.g., Agelopoulos et al., 1995; Alborn et al., 1995). Agelopoulos et al. (1995) showed that the chemical profile emitted by frass of two pierid larvae feeding on the same cabbage variety was species specific and a parasitoid of these two herbivores (*Cotesia rubecula*) used this information to discriminate between these two host species. Alborn et al. (1995) identified for *Microplitis croceipes* the chemical substance in host frass used by this parasitoid to discriminate between frass from suitable and unsuitable host species.

Stemboring larvae maintain relatively clean tunnels and they push their frass out of the tunnel where it accumulates around the exit hole. Parasitoids of stemborers use this frass to find the tunnels of their host larvae (e.g., van Leerdam et al., 1985; Potting et al., 1997a; Ngi-Song & Overholt, 1997). However, female *C. sesamiae* do not recognize the identity of the host species inside a stemborer tunnel by examination of the frass. The results of our experiments indicate that *C. sesamiae* does not discriminate between a suitable host *S. calamistis* and an unsuitable host *E. saccharina* before host contact. There was no difference in the response to frass from these two host species (Figure 1) and in the willingness to enter a tunnel with one of these two host species (Tables 1 and 2). However, there was a significant difference in the host acceptance after contacting the host. In the glass tube

*Table 1.* Behaviour of *C. sesamiae* females on sugarcane stems infested with *E. saccharina* or maize stems infested with *S. calamistis*. Indicated are means (± SE, sample size)

| Parameter | *Eldana saccharina* in sugarcane stem | | *Sesamiae calamistis* in maize stem | | P* |
|---|---|---|---|---|---|
| Mean residence time on stem (s)[a] | 120.7 | (±26.9, 27) | 410.5 | (±73.5, 23) | 0.0005 |
| Percentage entering tunnel | 100% | (–, 36) | 100% | (–, 31) | n.s. |
| Total time spent inside tunnel (s)[a] | 63.0 | (±18.4, 27) | 358.4 | (±78.0, 23) | <0.0001 |
| Number of times entering | 1.1 | (± 0.06, 36) | 1.4 | (±0.15, 31) | n.s. |
| Percentage parasitized | 0 | (–, 36) | 51.6% | (–, 31) | <0.0001 |
| Percentage killed by larva[b] | 25% | (–, 36) | 25.8% | (–, 31) | n.s. |

[a]Excluding wasps that were killed inside stem.
[b]Of those wasps that entered tunnel.
*Means (± SE) tested with Mann–Whitney U-test (P = 0.05); ratios with G test (P = 0.05, Yates correction).

*Table 2.* Behaviour of *C. sesamiae* females in artificial transparent tunnel with one fourth instar *E. saccharina* larva or one fourth instar *S. calamistis* larva (see Table 1)

| Parameter | *Eldana saccharina* in sugarcane | | *Sesamiae calamistis* in maize | | P* |
|---|---|---|---|---|---|
| Percentage contacting host | 100% | (–, 54) | 100% | (–, 62) | n.s. |
| Percentage ovipositing in host | 7.4% | (–, 54) | 82.3% | (–, 62) | <0.0001 |
| Percentage parasitised | 0 | (–, 4) | 54.9% | (–, 51) | 0.03 |
| Percentage wasps killed by larva | 18.5% | (–, 54) | 4.8% | (–, 62) | 0.04 |

[a]Of those wasps that oviposited in host.
*Means (± SE) tested with Mann–Whitney U-test (P = 0.05); ratios with G test (P = 0.05, Yates correction).

experiment all the *E. saccharina* hosts were contacted by the female parasitoids, but only a fraction (7%) was accepted for oviposition, whereas a high proportion (82%) of the contacted *S. calamistis* larvae were accepted for oviposition. Thus, a *C. sesamiae* female can recognize the unsuitable host *E. saccharina*, but only after it has made contact with the defending host.

Previous studies have shown that *C. flavipes* and *C. sesamiae* generally do not discriminate between different stemborer species in the different stages of the host location process. In olfactometric tests it was shown that *C. flavipes* does not discriminate between volatiles emanating from maize plants infested by different stemborer species (Potting et al., 1993; Ngi-Song et al., 1996). Furthermore, both *C. sesamiae* and *C. flavipes* were unable to discriminate between volatiles from larval frass produced by four different stemborer species fed on maize (Ngi-Song & Overholt, 1997) and *C. flavipes* readily entered a stemborer tunnel with an unsuitable host (Potting et al., 1997a). For *C. flavipes* no difference was found in the acceptability rate of four different stemborer species (Ngi-Song et al., 1995). We found no evidence for our hypothesis that *C. sesamiae* should recognize a tunnel with a unsuitable host in such a way that it can avoid a superfluous mortality risk. Several studies indicated an imperfect degree of concordance between host selection and insect fitness (Fox & Lalonde, 1993). An imperfect host selection could arise from an inability of the insect to assess host suitability. There may be constraints in recognizing an unsuitable host before actual host contact. A stemborer tunnel made by an unsuitable host may be indistinguishable from a tunnel with a suitable host species. Females selected to avoid the unsuitable host are also selected to avoid the suitable host. Another explananation for the imperfect host selection might be that the selection pressure to avoid the unsuitable *E. saccharina* in sugarcane might not be high. *Cotesia sesamiae* generally attacks stemboring hosts in maize, sorghum and wild grasses and might follow *S. calamistis* when it attacks young sugarcane. In the sugarcane habitat it may also encounter *E. saccharina*. However, the association between *E. saccharina* and *C. sesamiae* might be relatively new. In South Africa, *E. saccharina* established in sugarcane around 1970.

## Acknowledgements

We thank W. A. Overholt (ICIPE, Kenya) for his useful comments and two anonymous referees for their constructive comments on a previous version of the manuscript. This work was supported by the Netherlands Foundation for the Advancement of Tropical Research (WOTRO, W84-325).

## References

Agelopoulos, N. G., M. Dicke & M. A. Posthumus, 1995. Role of volatile infochemicals emitted by feces of larvae in host searching behavior of parasitoid *Cotesia rubecula* (Hymenoptera: Braconidae): a behavioral and chemical study. Journal of Chemical Ecology 21: 1789–1811.

Alborn, H. T., W. J. Lewis & J. H. Tumlinson, 1995. Host specific recognition kairomone for the parasitoid *Microplitis croceipes* (Cresson). Journal of Chemical Ecology 21: 1697–1708.

Askew, R. R., 1971. Parasitic Insects. Heinemann, London.

Conlong, D. E., 1994. A review and perspectives for the biological control of the African sugarcane stalkborer *Eldana saccharina* Walker (Lepidoptera: Pyralidae). Agriculture Ecosystems and Environment 48: 9–17.

Fox, C. W. & R. G. Lalonde, 1993. Host confusion and the evolution of insect diet breadth. Oikos 67: 577–581.

Godfray, H. C. J., 1994. Parasitoids, Behavioral and Evolutionary Ecology. Princeton University Press, New Jersey.

Graham, D. Y. & D. E. Conlong, 1988. Improved laboratory rearing of *Eldana saccharina* (Lepidoptera: Pyralidae) and its indigenous parasitoid *Goniozus natalensis* (Hymenoptera: Bethylidae). Proceedings of South African Sugar Technology Association 62: 116–119.

Gross, P., 1993. Insect behavioral and morphological defenses against parasitoids. Annual Review of Entomology 38: 251–273.

Lederhouse, R. C., 1990. Avoiding the hunt: primary defenses of lepidopteran caterpillars. In: D. L. Evans & J. O. Schmidt (eds), Insect Defences. State University of New York Press, New York, pp. 175–190.

Leerdam, M. B. van, J. W. Smith Jr. & T. W. Fuchs, 1985. Frass-mediated, host finding behavior of *Cotesia flavipes*, a braconid parasite of *Diatraea saccharalis* (Lepidoptera: Pyralidae). Annals of the Entomological Society of America 78: 647–650.

Mbapila, J. C. & W. A. Overholt, 1997. Oviposition, development and searching behaviour of *Cotesia flavipes* Cameron (Hymenoptera: Braconidae) in relation to aestivating and non-aestivating *Chilo* spp. (Lepidoptera: Pyralidae) larvae in maize. African Entomology 5: 35–39.

Mohyuddin, A. I., 1971. Comparative biology and ecology of *Apanteles flavipes* (Cam.) and *A. sesamiae* Cam. as parasites of graminaceous borers. Bulletin of Entomological Research 61: 33–39.

Mohyuddin, A. I. & D. J. Greathead, 1970. An annotated list of the parasites of graminaceous stem borers in East Africa, with a discussion of their potential in biological control. Entomophaga 15: 241–274.

Ngi-Song, A. J. & W. A. Overholt, 1997. Host location and host acceptance by *Cotesia flavipes* Cameron and *C. sesamiae* Cameron (Hymenoptera: Braconidae), parasitoids of African gramineous stemborers: role of frass and other host cues. Biological Control 9: 136–142.

Ngi-Song, A. J., W. A. Overholt & J. N. Ayerty, 1995. Suitability of African gramineous stemborers for development of *Cotesia flavipes* and *C. sesamiae* (Hymenoptera: Braconidae). Environmental Entomology 24: 978–984.

Ngi-Song, A. J., W. A. Overholt, P. G. N. Njagi, M. Dicke, J. N. Ayerty & W. Lwande, 1996. Volatile infochemicals in host and host habitat location by *Cotesia flavipes* and *C. sesamiae* (Cameron) (Hymenoptera: Braconidae), two larval parasitoids of gramineous stemborers. Journal of Chemical Ecology 22: 307–323.

Pasteels, J. M., J. C. Grégoire & M. Rowell-Rahier, 1983. The chemical ecology of defense in arthropods. Annual Review of Entomology 28: 263–289.

Polaszek, A. & A. K. Walker, 1991. The *Cotesia flavipes* species-complex: parasitoids of cereal stemborers in the tropics. Redia 74: 335–341.

Potting, R. P. J., 1997. Evolutionary and applied aspects of the behavioural ecology of the stemborer parasitoid *Cotesia flavipes*. Insect Science and its Application 17: 109–118.

Potting, R. P. J., F. O. Danso, W. A. Overholt & A. J. Ngi-Song, 1993. Host selection in *Cotesia flavipes*, parasitoid of tropical stemborers. Proceedings of Experimental and Applied Entomology, NEV, Amsterdam 4: 47–52.

Potting, R. P. J., W. A. Overholt, F. O. Danso & K. Takasu, 1997a. Foraging behavior and life history of the stemborer parasitoid *Cotesia flavipes*. Journal of Insect Behavior 10: 13–29.

Potting, R. P. J., H. M. Snellen & L. E. M. Vet, 1997b. Fitness consequences of superparasitism and mechanism of host discrimination in the stemborer parasitoid *Cotesia flavipes*. Entomologia Experimentalis et Applicata 82: 341–348.

Stephens, D. W. & J. R. Krebs, 1986. Foraging Theory. Princeton University Press, Princeton.

Strand, M. R. & L. L. Pech, 1995. Immunological basis for compatiblity in parasitoid-host relationships. Annual Review of Entomology 40: 31–56.

Takasu, K. & W. A. Overholt, 1997. Aggressive behavior of *Chilo partellus* (Lepidoptera: Pyralidae) larvae against parasitization by the parasitoid *Cotesia flavipes* (Hymenoptera: Braconidae). Insect Science and its Application 17: 131–136.

Takasu, K. & W. A. Overholt, 1998. Brood guarding behavior and life history characteristics of *Goniozus indicus* Ashmead (Hymenoptera: Bethylidae), a larval ectoparasitoid of lepidopteran stemborers. Applied Entomology and Zoology 33: 121–126.

Ullyett, G. C., 1935. Notes on *Apanteles sesamiae* Cam., a parasite of the maize stem borer (*Busseola fusca*, Fuller) in South Africa. Bulletin of Entomological Research 26: 253–262.

Walker, A. K., 1994. Species of Microgastrinae (Hymenoptera: Braconidae) parasitizing lepidopterous cereal stem borers in Africa. Bulletin of Entomological Research 84: 421–431.

Entomologia Experimentalis et Applicata **91:** 149–154, 1999.



# Induction of cotton extrafloral nectar production in response to herbivory does not require a herbivore-specific elicitor

F. L. Wäckers & R. Wunderlin
*Institute for Plant Sciences, Applied Entomology, ETH, Clausiusstrasse 25, NW, 8092 Zürich*

Accepted: November 30, 1998

*Key words: Spodoptera littoralis*, cotton leafworm, *Gossypium herbaceum*, regurgitate, indirect defense, induction; elicitor, specificity, regurgitant

## Abstract

The production of extrafloral nectar is thought to represent an indirect plant defense, as it allows plants to recruit natural enemies which can protect the plant against herbivore attacks. In previous work we demonstrated that plants may show a strong increase in extrafloral nectar secretion in response to herbivory. Here we address the question of whether this induction is herbivory-specific, or simply a general response to tissue damage. We compared the level of induction in *Gossypium herbaceum* (L.) (Malvaceae) following either herbivory by *Spodoptera littoralis* (Boisduval) (Lepidoptera; Noctuidae) or mechanical damage with and without the addition of *S. littoralis* regurgitate. Both herbivore feeding and mechanical damage significantly raised nectar production. No difference in volume or pattern of nectar secretion was found between natural and mechanical damage, nor between artificially damaged plants treated with regurgitate or water. Our findings indicate that the induction of extrafloral nectar secretion constitutes a general response by the plant to tissue damage, rather than representing a herbivory-specific mechanism. The costs and benefits of such a non-specific strategy for the plant are discussed.

## Introduction

To defend themselves against herbivory, plants are thought to have developed a broad arsenal of morphological and chemical adaptations. These adaptations can be divided into direct and indirect defenses. Direct defenses, by definition, have a direct negative impact on herbivores and include thorns, spines (urticating), hairs and trichomes, sclerotizations, waxes, canal secretions (latex, resins), and a great diversity of secondary plant metabolites.

Indirect defenses, on the other hand, refer to plant characteristics which have a positive impact on the natural enemies of herbivores, whose predatory and parasitic potential can provide indirect protection to plants. Plant attributes which have been interpreted to represent such indirect defenses include: herbivory induced plant volatiles (used by predators and parasitoids to locate feeding herbivores) (Dicke & Sabelis, 1988; Turlings et al., 1990; Mattiacci et al., 1994; Drukker et al., 1995; Agrawal, 1998); domatia (morphological adaptations which serve ants or other predators as shelter and nesting sites) (Bakker & Klein, 1993; Whitman, 1996; Walter, 1996); and nutritional supplements (food bodies, extrafloral nectaries) (Koptur, 1989; Whitman, 1996).

With respect to the dynamics of defense induction, Rhoades (1979) and Tuomi et al. (1990) distinguished between constitutive defenses, which are pre-formed and not affected by the presence of (feeding) herbivores, and inducible defenses which are elicited by, or increase in response to herbivory. Many plant metabolites with presumed roles in direct plant defense have been shown to be inducible (for a recent review see Karban & Baldwin, 1997). Indirect defenses as well, can be inducible. In an increasing number of plant species it has been demonstrated that herbivore feeding induces the emission of volatiles which attract predators or parasitoids (e.g., Dicke and Sabelis, 1988; Turlings et al., 1990). The overall quantity of volatiles emitted by plants typically increases following herbivory. This increase is often combined with qualitative changes in the odor profile (Dicke & Vet, 1998).

Extrafloral nectar production provides another example of how plants can actively respond to herbivory in a way that increases the plant's attractiveness to natural enemies. The response can entail an increase in volume of nectar production, as has been demonstrated in *Vicia sativa* (L.) (2.5-fold increase) (Koptur, 1989), *Ricinus communis* (L.) (3-fold), and *Gossypium herbaceum* (L.) (12-fold) (Wäckers et al., unpubl.). In addition to this quantitative response, herbivory can also affect qualitative aspects of extrafloral nectar composition. Nectar of herbivore-damaged *Impatiens sultani* contained 5.6 times higher amino-acid concentration compared with an undamaged control, even though nectar quantity remained unaffected in this species (Smith et al., 1990). These herbivore-induced responses are thought to enhance indirect defense, since ants respond positively to both an increase in nectar volume and amino acid concentration (Del-Claro & Oliveira, 1993; Lanza et al., 1993). By mainly restricting induced effects to the damaged leaf (Wäckers et al., unpubl.) plants may be able to guide predators and parasitoids to the site under attack.

Despite the fact that induction of direct and indirect defenses seems to be a widespread phenomenon, relatively little is known regarding the elicitors involved in defense induction. To address the question of elicitor specificity, several studies have compared plant responses to either actual herbivory or mechanical damage (Karban & Baldwin, 1997).

The most comprehensive evidence for the existence of herbivore-specific elicitors has been furnished with respect to herbivore-induced plant volatiles. Turlings et al. (1990) demonstrated that the emission of plant volatiles following herbivory is specifically triggered by the interactions between mechanical damage and herbivore regurgitate. While mechanical damage or *Spodoptera exigua* (Hübner) regurgitate alone failed to elicit herbivory-specific volatiles, these volatiles were emitted when regurgitate was applied onto the mechanically damaged leaf. Recently, Alborn et al. (1997) managed to isolate and identify the elicitor from *S. exigua* larvae. The same elicitor has been found in a preliminary analysis of *S. littoralis* regurgitate (T. C. J. Turlings, pers. comm.). It remains to be seen whether this elicitor is also involved in other herbivore-plant systems for which feeding induction of volatile production has been demonstrated (recently reviewed by Dicke & Vet, 1998). An interesting analogous system has been described for the elm leaf beetle. Meiners & Hilker (1997) demonstrated that elm leaves emit volatiles in response to oviposition by the beetle onto previously damaged leaf material. In this system, the elicitor seems to stem from the cement with which the eggs are attached to the damaged leaves (T. Meiners, pers. comm.).

It has not yet been investigated whether the induction of extrafloral nectar production also requires a herbivore-specific elicitor. The few studies which have addressed the induction of extrafloral nectar production have worked with either actual herbivory or mechanical damage. The fact that in several systems mechanical damage elicited no nectar induction (Koptur, 1989) may imply that the method of mechanical damage is not a suitable mimic of herbivory.

To investigate whether the induction of nectar secretion requires a herbivory-specific elicitor, we conducted a set of experiments analogous to those conducted by Turlings et al. (1990), subjecting plants to either herbivory, or mechanical damage with and without regurgitate. We chose cotton as a system, based on the fact that this plant species has previously shown the highest degree of nectar increase in response to *S. littoralis* feeding.

## Materials and methods

### *Plants*

*Gossypium herbaceum* (L.) (Malvaceae) seedlings were grown individually in small pots (5 cm diameter) in a climate chamber (r.h. 80%; L14:D10; T: 26 °C (day)/10 °C (night)). Two weeks following germination, seedlings were transferred into 9 × 9.5 cm pots which were placed into a greenhouse (L16:D8, $T = 16 - 30$ °C, and r.h. = 60–80%). In the experiments, we used 2-month-old cotton plants. At this point, the plants had produced five-six leaves, on the underside of which a single nectary is located on the mid-vein. Only plants free of any visible damage were used in the experiments.

### *Clip-on cages*

These were constructed from a pair of plexiglas rings of 3.5 cm diameter attached by a metal clip. The edges contacting the plant surface were coated with foam rubber. The outsides were covered with screen mesh.

*Insects*

*Spodoptera littoralis* larvae which had been reared on an artificial diet were kindly provided by Novartis, Basel, Switzerland. Starting from the 2nd instar, larvae were fed cotton leaves. Before being used in the experiment, third-instar larvae were starved for 24 h to ensure that larvae started feeding as soon as they were placed on the cotton plants.

*Regurgitate*

Caterpillar regurgitate was obtained from fourth and fifth instar *S. littoralis*, which had been reared on cotton plants for at least 3 days prior to regurgitate collection. Regurgitate was collected with a 100 $\mu$l capillary tube connected to a vacuum pump. Regurgitation was induced by softly squeezing caterpillars behind the head with a pair of soft forceps. See Turlings et al. (1993) for further details of the collection technique. The collected regurgitate was divided into batches of 50 $\mu$l and stored in Eppendorf tubes at −60 °C.

*Experimental procedure*

Four days before treatment, groups of five plants were placed into screen cages (46 × 66.5 × 70 cm) under constant climatic conditions (L16:D8/5200 Lux; $T$ = 25 °C; r.h.= 60%) and subsequently watered every second day (100 ml water per plant). On day 0, cages of plants were randomly allocated to one of three treatments:

*Herbivory.* Plants were exposed to three early third instar *S. littoralis* larvae, introduced into a single clip-on cage. The cage was attached to the fourth leaf between nectary and leaf tip, covering approximately 25% of the total leaf area. After 24 h, larvae had consumed all leaf material within the cage, at which point larvae and cages were removed from the plants.

*Mechanical damage + regurgitate.* Mechanical damage was imposed by rubbing the upper epidermis with sandpaper. The location and area of damage was identical to the area consumed by *S. littoralis* in the herbivory treatment. Immediately following damage the damaged area was wetted with 10 $\mu$l caterpillar regurgitate (the approximate equivalent to the volume collected per larva).

*Mechanical damage + distilled water.* Plants were treated as described above. Instead of regurgitate, 10 $\mu$l distilled water was applied to the damaged area.

Nectar from the single nectary per leaf was collected every other day, starting 2 days before the treatment. On day 0, nectar was collected just before the treatment. Recordings continued until day 6, to allow time for potential induction effects to subside. Nectar was collected using 5 $\mu$l micropipettes with 1 $\mu$l divisions and the volume collected was calculated from nectar column length within the constant-bore pipettes. Treatments were conducted with 20 plants each.

Since data were not normally distributed (Normality Test, StatView), they were analyzed using non-parametric statistics (Kruskal–Wallis).

## Results

With respect to sample date, all three damage treatments showed a strong increase in nectar production within 2 days of damage infliction (Figure 1). In comparison to the pre-treatment level (day 0), nectar production increased by a factor 10 (regurgitate; $P < 0.001$), 12 (herbivory; $P < 0.001$), and 17 (water; $P < 0.001$). Four days following treatment, the induced increase in nectar production had somewhat receded, to an equivalent of 5-(regurgitate), 6-(herbivory), and 8-(water) times the pre-treatment level. Six days following treatment, induction had subsided and nectar production had returned to the pre-treatment level.

A non-parametric two-way analysis of variance showed a significant effect of sampling date ($P < 0.001$), but no effect of damage type on nectar production ($P = 0.24$). This demonstrates that induction of cotton extrafloral nectar secretion does not require a herbivore-specific elicitor.

The spatial distribution of nectar production following mechanical damage (day 2) was not affected by the addition of regurgitate (Figure 2) ($P = 0.62$). The induced increase in nectar secretion was mainly restricted to the damaged leaf, while there was an additional systemic increase in the youngest leaves. This distribution is comparable with the distribution found following herbivore induction (Wäckers et al., unpubl.). Since the two types of mechanical damage resulted in a similar increase in nectar volume and comparable effects on the spatial pattern of nectar production, it can be concluded that, under the given conditions, treatment of mechanical damage

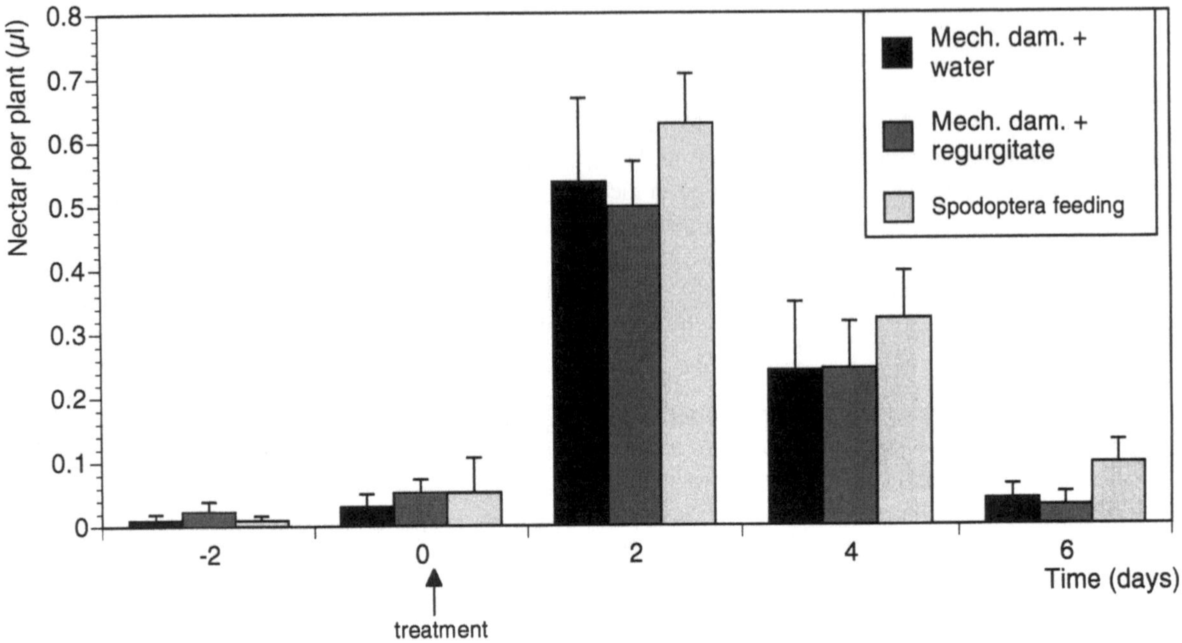


*Figure 1.* Extrafloral nectar production by *G. herbaceum* plants in response to *S. littoralis*-damage, mechanical damage treated with *S. littoralis* regurgitate, or mechanical damage treated with water. Nectar was collected at 48-h intervals. Damage was inflicted following the second collection (day 0). Bars show average nectar volume per plant and standard errors.

with *S. littoralis* regurgitate has no measurable effect on either the level, the timing, or the distribution of nectar induction.

## Discussion

Cotton has been intensively studied with regard to both direct and indirect defenses (Karban & Baldwin, 1997). Various studies have shown that damage induces changes in cotton plants which reduces their suitability to herbivores. Such induced resistance has been found in response to both mechanical damage (Croxford et al., 1989) and actual herbivory (Alborn et al., 1996). A comparative study was conducted by Karban (1985), which showed that mechanical damage was as effective in inducing resistance to the two-spotted spider mite as was previous spider mite infestation. Separate studies have addressed the effect of either herbivory or mechanical damage on the subsequent suitability of cotton plants to *Spodoptera* spp. Croxford et al. (1989) showed that *S. littoralis* larvae avoided feeding in the vicinity of mechanical damage. Palatability decreased both with time following damage (over 7 days) and with increasing proximity to the damage site. Alborn et al. (1996) reported that both herbivory by *S. littoralis* and *S. exigua* resulted in localized as well as systemic induction of feeding deterrency. Karban (1988) reported similar responses by *S. exigua* to mite-damaged cotton. Although these studies failed to establish whether the mechanisms underlying induction correspond, they did show that both mechanical damage and herbivory can give similar results in bioassays. This indicates that the specificity of direct defense induction in cotton may be limited.

In contrast to the seemingly rather general induction of direct defenses, the induction of volatile emission by cotton plants has been demonstrated to be more specific. Not only do herbivore damaged plants show a clearly higher rate of volatile emission over mechanically damaged plants (McCall et al., 1994), herbivore regurgitate also induces *de novo* synthesis of various terpenoids (Paré & Tumlinson, 1997) resulting in a quantitative as well as qualitative response to herbivory. However, the specificity of the plant response is not restricted to the differentiation between mechanical damage and herbivory. The composition of the induced volatile blend also varies distinctly between (even closely related) herbivore species (De Moraes et al., 1998).

Unlike the emission of herbivore induced volatiles, the induction of extrafloral nectar secretion in our experiments is elicited by tissue damage, irrespective of

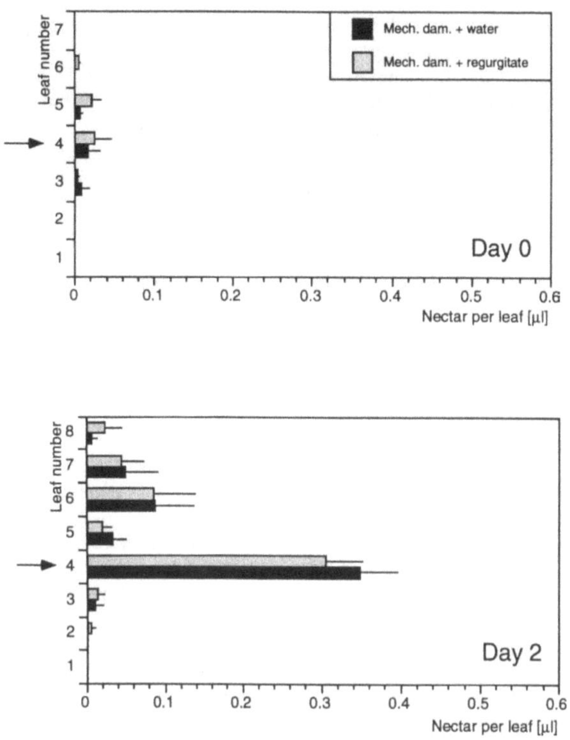


*Figure 2.* Distribution of nectar production within *G. herbaceum* plants before (day 0) and 48 hours following mechanical damage treated with either *S. littoralis* regurgitate, or water (day 2). Bars and standard errors represent nectar production by individual cotton leaves over a 48-h period. Mechanical damage was inflicted to leaf four following the nectar measurement on day 0.

whether this damage is mechanical or caused by actual herbivory. The addition of *S. littoralis* regurgitate had no significant effect on either the level, the timing, or the distribution of nectar secretion. The marked difference in induction specificity between the two indirect defense systems may reflect differences in the costs and benefits of such specificity.

In addition to the general defensive costs (biosynthetic costs, as well as costs for transport, storage, sequestration, and maintenance) (Simms, 1992; Lerdau & Gershenzon, 1997), specificity may entail particular further cost factors for the plant. A herbivory-specific response requires mechanisms by which plants can distinguish herbivory from other sources of damage. When the plant response is specifically triggered by a herbivore characteristic substance, this may require a receptor for identification of the elicitor. In systems in which the plant shows different responses to a range of herbivores (species specificity), more than one receptor may be involved. However, given the fact that in cotton specific mechanisms for herbivore recognition are in place (De Moraes et al., 1998), it seems unlikely that this cost factor would explain the difference in specificity between nectar secretion and volatile emission.

With respect to the benefits of a herbivory-specific response, specificity may help the plant to economize on its defenses, as well as improve its defense efficacy. The potential for cost saving is restricted by the fact that the defensive 'hardware' (plant structures involved in biosynthesis, transport, storage and sequestration of defensive substances) has to be in place irrespective of specificity. Defense specificity may, however, allow plants to economize on the cost of actually producing the defensive substance by restricting defense to true incidences of herbivory and avoiding defensive investment in 'false alarms' (e.g. mechanical damage due to treading, -wind, -sand blasts, -hail, and other adverse environmental conditions). The potential economic benefit of specificity consequently depends on the frequency of these false alarms, as well as the costs involved in expressing the defense. However, given the fact that the estimated cost for induced nectar production lies within the range of what plants have been estimated to invest in induced volatiles (Dicke & Vet, 1998; Wäckers et al., unpubl.) the biosynthetic costs do not explain why within one plant system (cotton), volatile induction shows a high level of specificity (Paré & Tumlinson, 1997; De Moraes et al., 1998), while induction of extrafloral nectar secretion is unspecific.

Instead, the explanation for the high level of specificity in the case of volatile emissions may lie in the fact that these volatiles serve parasitoids and predators as signals to locate herbivores. The effectiveness of this communication between plants and their potential guards depends on the honesty (reliability) of the signal (Vet et al., 1991). When volatile emissions are not reliably associated with herbivore presence, this indirect defense will break down (Godfray, 1995).

Extrafloral nectar, rather than serving as a signal to indicate the location of a reward, constitutes a reward in itself. The response by ants is determined by quantity and quality of the secreted nectar (Del-Claro & Oliveira, 1993; Lanza et al., 1993), not necessarily by the degree in which nectar secretion correlates with herbivore presence. An increase in nectar production following mechanical damage consequently has no negative implications for the efficacy of this indirect defense mechanism. An unspecific response to tissue damage might be the optimal solution for the induction of extrafloral nectar secretion, since it provides plants with the benefits of defense induction, without imposing possible additional costs for defense specificity.

## Acknowledgements

We thank M. Dicke, S. Wratten and an anonymous reviewer for helpful comments, T. Webb and A. Michel for corrections, P. Frick and J. Gutwald for technical support, and S. Dorn for providing the necessary infrastructure.

## References


Agrawal, A. A., 1998. Leaf damage and associated cues induce aggressive ant recruitment in a neotropical ant-plant. Ecology 79: 2100–2112.

Alborn, H. T., U. S. R. Röse & H. J. McAuslane, 1996. Systemic induction of feeding deterrents in cotton plants by feeding of *Spodoptera* spp. larvae. Journal of Chemical Ecology 22: 919–932.

Alborn, H. T., T. C. J. Turlings, T. H. Jones, G. Stenhagen, J. H. Loughrin & J. H. Tumlinson, 1997. An elicitor of plant volatiles from beet armyworm oral secretion. Science 276: 945–949.

Bakker, F. M. & M. E. Klein, 1992. Transtrophic interactions in cassava. Experimental & Applied Acarology 14: 293–311.

Croxford, A. C., P. J. Edwards & S. D. Wratten, 1989. Temporal and spatial variation in palatability of soybean and cotton leaves following wounding. Oecologia 79: 520–525.

De Moraes, C. M., W. J. Lewis, P. W. Paré, H. T. Alborn & J. H. Tumlinson, 1998. Herbivore-infested plants selectively attract parasitoids. Nature 393: 570–573.

Del-Claro, K. & P. S. Oliveira, 1993. Ant-Homoptera interaction: do alternative sugar sources distract tending ants? Oikos 68: 202–206.

Dicke, M. & M. W. Sabelis, 1988. How plants obtain predatory mites as bodyguards. Netherlands Journal of Zoology 38: 148–165.

Dicke, M. & L. E. M. Vet, 1998. Plant-carnivore interactions: evolutionary and ecological consequences for plant, herbivore and carnivore. In: H. Olff, V. K. Brown & R. H. Drent (eds), Herbivores: Between Plants and Predators. Blackwell Science, Oxford, pp. 483–520.

Drukker, B., P. Scutareanu & M. W. Sabelis, 1995. Do anthocorid predators respond to synomones from *Psylla*- infested pear trees under field conditions? Entomologia Experimentalis et Applicata 77: 193–203.

Godfray H. C. J., 1995. Communication between the first and the third trophic levels: an analysis using biological signalling theory. Oikos 72: 367–374.

Karban, R., 1985. Resistance against spider mites in cotton induced by mechanical abrasion. Entomologia Experimentalis et Applicata 37: 137–142.

Karban, R., 1988. Resistance to beet armyworms (*Spodoptera exigua*) induced by exposure to spider mites (*Tetranychus turkestani*) in cotton. American Midland Naturalist 119: 77–82.

Karban, R. & I. T. Baldwin, 1997. Induced Responses to Herbivory. The University of Chicago Press, Chicago, 319 pp.

Koptur, S., 1989. Is extrafloral nectar production an inducible defence? In: J. Bock & Y. Linhart (eds), Evolutionary Ecology of Plants. Westview Press, pp. 323–339.

Lanza, J., E. L. Vargo, S. Pulim, & Y. Z. Chang, 1993. Preferences of the fire ants *Solenopsis invicta* and *S. geminata* (Hymenoptera: Formicidae) for amino acid and sugar components of extrafloral nectars. Environmental Entomology 22: 411–417.

Lerdau, M. & J. Gershenzon, 1997. Allocation theory and chemical defense. In: F. A. Bazzaz & J. Graze (eds), Plant Resource Allocation. Academic Press, San Diego, pp. 265–277.

Mattiacci, L., M. Dicke & M. A. Posthumus, 1994. Induction of parasitoid attracting synomone in Brussels sprouts plants by feeding of *Pieris brassicae* larvae: role of mechanical damage and herbivore elicitor. Journal of Chemical Ecology 20: 2229–2247.

McCall, P. J., T. C. J. Turlings, J. Loughrin, A. T. Proveaux, & J. H. Tumlinson, 1994. Herbivore-induced volatile emissions from cotton (*Gossypium hirsutum* L.) seedlings. Journal of Chemical Ecology 20: 3039–3050.

Meiners, T. & M. Hilker, 1997. Host location in *Oomyzus gallerucae* (Hymenoptera: Eulophidae), an egg parasitoid of the elm leaf beetle *Xanthogaleruca luteola* (Coleoptera: Chrysomelidae). Oecologia 112: 87–93.

Paré, P. W. & J. H. Tumlinson, 1997. Induced synthesis of plant volatiles. Nature 385: 30–31.

Rhoades, D. F., 1979. Evolution of plant chemical defense against herbivores. In: G. A. Rosenthal & D. H. Janzen (eds), Herbivores, their Interaction with Secondary Plant Metabolites. Academic Press, New York, pp. 4–54.

Simms, E. L. 1992. Costs of plant resistance to herbivores. In: R. S. Fritz & E. L. Simms (eds), Plant Resistance to Herbivores and Pathogens. University of Chicago Press, Chicago, pp. 392–425.

Smith, L. L., J. Lanza & G. C. Smith, 1990. Amino acid concentrations in extrafloral nectar of *Impatiens sultani* increase after simulated herbivory. Ecology 71: 107–115.

Tuomi, J., P. Niemelä & S. Sirén, 1990. The Panglossian paradigm and delayed inducible accumulation of folia phenolics in mountain birch. Oikos 59: 399–410.

Turlings, T. C. J., J. H. Tumlinson & W. J. Lewis, 1990. Exploitation of herbivore-induced plant odors by host-seeking parasitic wasps. Science 250: 1251–1253.

Turlings, T. C. J., P. J. McCall, H. T. Alborn & J. H. Tumlinson, 1993. An elicitor in caterpillar oral secretions that induces corn seedlings to emit chemical signals attractive to parasitic wasps. Journal of Chemical Ecology 19: 411–425.

Vet, L. E. M., F. L. Wäckers & M. Dicke, 1991. How to hunt for hiding hosts: the reliability-detectability problem in foraging parasitoids. Netherlands Journal of Zoology 41: 202–213.

Walter, D. E., 1996. Living on leaves: Mites tomenta and leaf domatia. Annual Review of Entomology 41: 101–114.

Whitman, D. W., 1996. Plant bodyguards: mutualistic interactions between plants and the third trophic level. In: T. N. Ananthakrishnan (eds), Functional Dynamics of Phytophagous Insects. Oxford & IBH Publishing, New Delhi, pp. 207–248.

Entomologia Experimentalis et Applicata **91:** 155–161, 1999.



# Flowers in tri-trophic systems: mechanisms allowing selective exploitation by insect natural enemies for conservation biological control

L. R. Baggen[1], G. M. Gurr[1,*] & A. Meats[2]
[1]*Orange Agricultural College, The University of Sydney, Orange, New South Wales, 2800, Australia;* [2]*School of Biological Sciences, The University of Sydney, New South Wales, 2006, Australia;* **Author for correspondence*

Accepted: November 30, 1998

*Key words:* nectar, habitat manipulation, selectivity, *Copidosoma koehleri*, *Phthorimaea operculella*

## Abstract

Many insects have coevolved with certain angiosperm taxa to act as pollinators. However, the nectar and pollen from such flowers is also widely fed upon by other insects, including entomophagous species. Conservation biological control seeks to maximise the impact of these natural enemies on crop pests by enhancing availability of nectar and pollen-rich plants in agroecosystems. A risk with this approach is that pests may also benefit from the food resource. We show that the flowers of some plants (viz., buckwheat, *Fagopyron esculentum* Moench and dill, *Anethum graveolens* L.), and the extrafloral nectaries of faba bean (*Vicia faba* L.) benefit both *Copidosoma koehleri* Blanchard (Hymenoptera: Encyrtidae) and its host, the potato pest, *Phthorimaea operculella* Zeller (Lepidoptera: Gelechiidae). In contrast, phacelia (*Phacelia tanacetifolia* Benth) and nasturtium (*Tropaeoleum majus* L.) benefited only the parasitoid. When adult moths of *P. operculella* were caged with flowers of phacelia or nasturtium, longevity of males and females, egg laying life, fecundity, average oviposition rate, and number of eggs in ovaries at death were no greater than in the control treatment with access to shoots without flowers plus water. All the foregoing measures were increased compared to the control when the moths were allowed access to dill, buckwheat or faba bean extrafloral nectaries. Such 'selectivity' has the potential to make the use of floral resources in conservation biological control more strategic. We present morphometric and observational evidence to illustrate how such mechanisms may operate.

## Introduction

Conservation biological control is the least studied of all biological control approaches (Dent, 1995). Other than withholding the pesticides which may adversely affect natural enemy populations, conservation biological control also makes use of habitat manipulations to favour predators and parasites. These include provision of shelter, alternative hosts or prey, or food plants from which nectar and pollen may be obtained. The latter technique has been relatively widely used (e.g., Hickman & Wratten, 1996; Matthews-Gehringer et al., 1994; Topham & Beardsley, 1975). A number of workers have sought to maximise the benefit from providing food plants by screening to quantify the benefit to natural enemies of access to different flower types (Idris & Grafius, 1995; Patt et al., 1997; Orr & Plesants, 1996). Generally, plants with exposed nectaries such as are common in the family Apiaceae (formerly Umbelliferae) are recommended. However, applying this method to improve biological control of potato moth (*Phthorimaea operculella*) by its parasitoid *Copidosoma koehleri* proved only partially successful because pest adults apparently fed from the nectaries of the coriander (*Coriandrum sativum* L.) and/or faba bean (*Vicia faba* L.) planted alongside the potatoes (Baggen & Gurr, 1998). This observation led to laboratory screening of flower types similar to that carried out by other workers but also included tests to determine which flowers were also fed upon by the pest. Borage (*Borago officinalis* L.) was found to benefit only the parasitoid and the term 'selective food plant' proposed (Baggen & Gurr, 1998).

The primary aim of this study was to screen additional flower types to identify a broader range of selective food plants for use in habitat manipulation for *P. operculella*. The secondary aim was to make measurements and observations of both insects and flowers in order to determine the mechanism(s) by which selectivity may operate.

## Materials and methods

A laboratory colony of *P. operculella* was established using founder members caught from potato crops on the New South Wales tablelands close to Orange, Australia. Adults and eggs were harvested from the colony using methods based on Fenemore (1977). Adults were used in screening experiments and eggs to maintain a separate colony of *C. koehleri*. The latter was established using founders recovered as parasitised hosts from potato crops in the same area, and maintained using techniques based on Platner & Oatman (1972).

Plants of dill (*Anethum graveolens* L.), faba bean (*Vicia faba* L.), phacelia (*Phacelia tanacetifolia* Benth), nasturtium (*Tropaeoleum majus* L.), and buckwheat (*Fagopyron esculentum* Moench) were grown from seeds in a glasshouse. Unlike the method of Baggen & Gurr (1998), which required the synchronous blooming of all plant species to be tested and the availability of large numbers of the insects, the present approach was to test each plant species as it bloomed. Each plant was tested with a control consisting of a shoot without flowers to ensure comparable levels of humidity inside cages. In the case of faba bean, where extrafloral nectaries rather than flowers were to be tested, screening was possible prior to blooming and the control consisted of a shoot without extrafloral nectaries. Shoots were removed from plants and placed into individual 50 ml plastic vials containing water, the opening of which was sealed with a foam rubber plug. Through this plug, a cotton wick provided insects with a water source. Vials were fixed to the bases of flight cages with plastic modeling clay. Cages, constructed from a cylinder of clear acetate sheeting, were 120 mm tall and 90 mm in diameter. The base of each cylinder was a plastic Petri dish and the top was covered with fine nylon mesh held in place with a rubber band.

Adult longevity of *C. koehleri* was recorded when parasitoids were caged with nectary bearing shoots (test) or with shoots without nectaries (control) of buckwheat, faba bean, phacelia, and nasturtium. For each test and control plant seven replicates were conducted. To minimise variability between insects, screening any given flower against the control used females from the same brood. As brood sizes varied from 42–70, the number of females placed into each cage varied from 3 to 5 according to availability. Tests were conducted at a constant 25 °C and 14 h photoperiod. Surviving insects were subsequently counted at 24 h intervals.

For *P. operculella*, adult longevity and several fecundity parameters (see Table 1) were recorded when moths were caged with test and control shoots of buckwheat, faba bean, phacelia and dill. For each test and control 12 replicates were conducted. Each cage contained either a nectary bearing or nectary-free plant shoot and water source, plus a pair of adult insects. Insect size was standardised by selecting pupae from the colony with weights of 12–14 mg. These were sexed according to Atherton's (1936) description and held in individual vials. Within 24 h of eclosion, moths were randomly allocated to cages to give a male and female within each. In cages where fewer than eight eggs were deposited during the first four days of the experiment it was assumed that mating had not occurred or that the pair was infertile. Such cages were excluded from further examination as were those in which any moth died within two days. From the remaining cages enough were left to give ten replicates. Tests were conducted at a constant 25 °C and 14 h photoperiod. Surviving insects were subsequently counted at 24 h intervals. Every third day, the insects within these and *C. koehleri*-containing cages were disturbed until they alighted on either the acetate cylinder or top cover. This allowed the cage to be lifted from its base to replace the plant material and water. Once all *P. operculella* had died, females were dissected and the numbers of unlaid eggs remaining in ovaries counted.

Measurements of *C. koehleri* head widths and *P. operculella* proboscis lengths were made using an ocular micrometer fitted to a microscope. Twenty subjects were measured for each species. Proboscis length was measured for adults killed with Kahle's solution (Upton, 1991) and for which the proboscis was extended with fine forceps. Observations and measurements of corolla depth and width were made with the same microscopy equipment. This was carried out for the plants listed above and for the species tested previously by Baggen & Gurr (1998). Flowers of these (borage, coriander and faba bean) were obtained from plants grown as described above.

*Table 1.* Comparison of longevity, fecundity and egg production in *Phthorimaea operculella* caged with flowers/ extrafloral nectaries or control treatment with foliage. Values are means of 10 replicates ±s.e.

| Nectary type | Adult male longevity (days) | Adult female longevity (days) | Egg laying life (days) | Oviposition rate (eggs/day) | Eggs in ovaries at death |
|---|---|---|---|---|---|
| Buckwheat flowers (test) | 16.4 ± 1.1** | 13.4 ± 1.6* | 11.4 ± 1.2* | 14.3 ± 1.8 | 4.6 ± 2.7 |
| Buckwheat foliage (control) | 10.2 ± 1.0 | 9.5 ± 0.7 | 6.4 ± 0.7 | 12.6 ± 1.8 | 1.8 ± 0.8 |
| Faba bean extrafloral nectaries (test) | 13.4 ± 1.1* | 16.2 ± 0.9** | 14.2 ± 1.2** | 20.4 ± 3.7 | 10.7 ± 3.6* |
| Faba bean foliage (control) | 9.1 ± 0.7 | 9.6 ± 0.7 | 6.4 ± 1.2 | 13.0 ± 2.2 | 0.8 ± 0.4 |
| Phacelia flowers (test) | 10.0 ± 1.1 | 11.1 ± 0.9 | 8.6 ± 1.0 | 11.8 ± 1.8 | 2.4 ± 0.8 |
| Phacelia foliage (control) | 10.2 ± 0.5 | 10.5 ± 1.0 | 8.6 ± 1.1 | 11.0 ± 1.0 | 2.1 ± 0.8 |
| Dill flowers (test) | 20.1 ± 1.1** | 20.0 ± 1.1** | 16.6 ± 1.1** | 15.8 ± 1.2* | 5.2 ± 2.0* |
| Dill foliage (control) | 11.0 ± 0.9 | 12.8 ± 0.4 | 11.4 ± 0.3 | 10.5 ± 1.5 | 0.4 ± 0.2 |

Treatment mean differs from control: *P < 0.05; **P < 0.01.

Normality of insect data was confirmed using the Kolmogorov–Smirnov test and comparisons between each flower type and its control made with one-tailed Student's *t*-test. One way analysis of variance was used to compare corolla dimensions across flower types.

## Results

Access to the flowers of buckwheat, dill or the extrafloral nectaries of faba bean led to increased lifetime fecundity in *P. operculella* (Figure 1). Table 1 shows that this was coincident with increased male and female adult longevity and egg laying life ($P < 0.05$). Dill flowers also increased oviposition rate and numbers of eggs in ovaries at death, and faba bean extrafloral nectaries increased the latter. In contrast, access to flowers of phacelia did not afford a statistically significant benefit to *P. operculella* for any parameter of performance.

Access to any of the plant material with nectar sources led to a significant ($P < 0.0017$) increase in longevity of adult *C. koehleri* when compared to the water plus foliage control (Figure 2).

Corolla depth varied markedly from zero in dill to over 30 mm in nasturtium (Table 2). Size of corolla openings was smallest in borage and greatest in buckwheat. Though the corolla diameter of phacelia was relatively large (5.05 mm) access to the nectaries was occluded by stamen appendages leaving openings of just 0.15 mm. Below the stamen appendages, outward pointing hairs were present on the style and ovary.

Head widths of male and female *C. koehleri* were 0.43 ± 0.004 mm and 0.44 ± 0.003 mm respectively. Proboscis lengths in male and female *P. operculella* were 1.75±0.03 mm and 1.73±0.03 mm, respectively.

## Discussion

Baggen & Gurr (1998) showed that when adult longevity of *C. koehleri* is increased by the provision of food this translates to increased parasitism of the pest, *P. operculella*. Of the flowers tested in this study, only those of phacelia were beneficial to *C. koehleri* yet conferred no benefit to *P. operculella*. The fact that this species is a 'selective food plant' has important practical implications since it is relatively widely used in habitat manipulation (e.g., Hickman & Wratten, 1995; Holland et al., 1994; White et al., 1995). Its selection by these researchers is the result of its agronomic tractability combined with the production of copious pollen which is beneficial to hoverflies. Phacelia is also known to be a rich source of nectar for honey bees (Williams & Christian, 1991) and some work has shown it to be accessed by small hymenopteran parasitoids (see references in Jervis et al., 1993).

The selectivity of phacelia observed in the present study suggests that it may be well suited for use in future habitat manipulation attempts to enhance biological control of *P. operculella* by *C. koehleri*. The stamen appendages within the corollae of phacelia would prevent *P. operculella* adults crawling deep within the flower, but morphometric measurements suggest that the fully extended proboscis should, just, be able to reach the nectaries. However, the outward pointing hairs observed on the style and ovary are likely to impede the insertion of the proboscis. Like

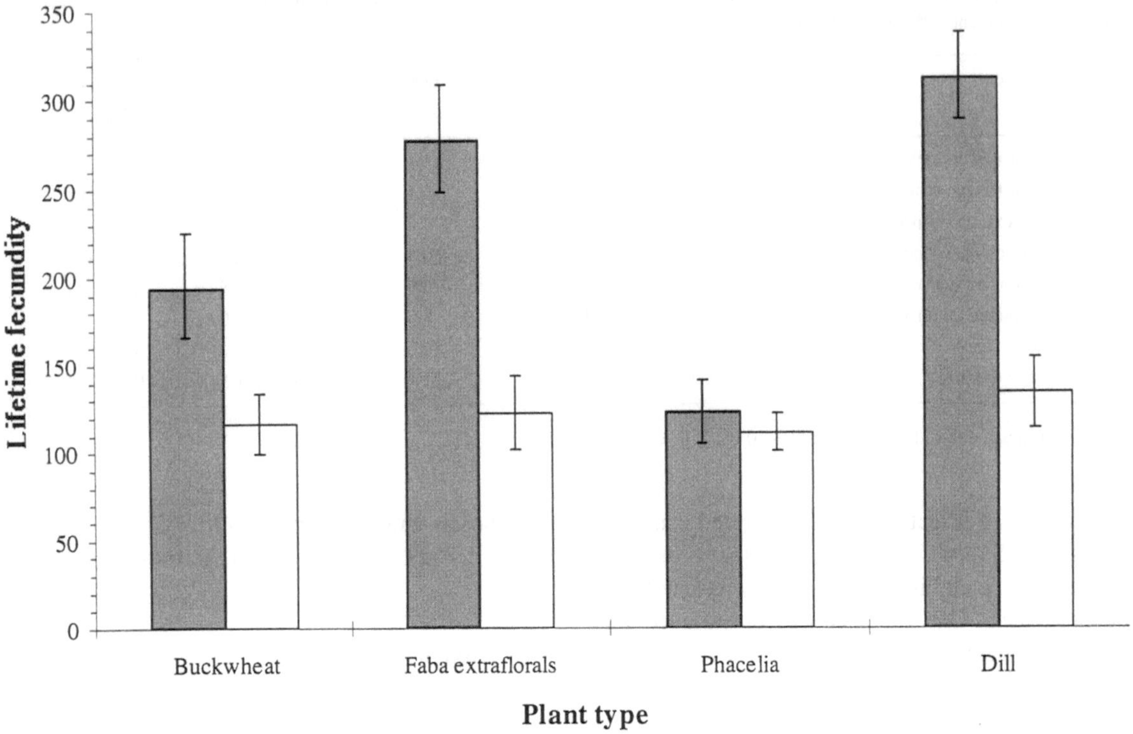


*Figure 1.* Comparison of *Phthorimaea operculella* lifetime fecundity when caged with nectary bearing plant shoots (solid bars), or with shoots without nectaries (unshaded bars) for a range of plant species.

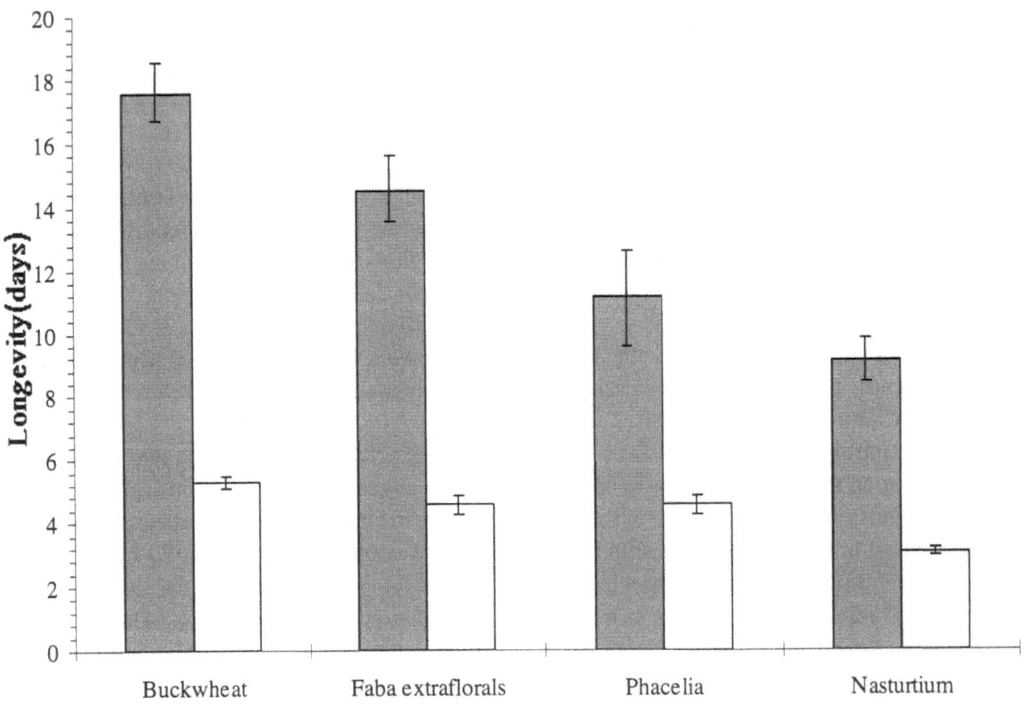


*Figure 2.* Comparison of *Copidosoma koehleri* adult longevity when caged with nectary bearing plant shoots (solid bars), or with shoots without nectaries (unshaded bars) for a range of plant species.

*Table 2.* Comparison of corolla depth and opening diameter for a range of plant types. ($n = 20$)

| Nectary type | Corolla depth (mm) | Corolla aperture (mm) | Insect(s) able to feed |
|---|---|---|---|
| Buckwheat flower | $0.54 \pm 0.01^{f}$ | $6.59 \pm 0.23^{a}$ | *C. koehleri* & *P. operculella*[1] |
| Dill flower | $0.00 \pm 0.00^{g}$ | $2.16 \pm 0.06^{c}$ | *C. koehleri* & *P. operculella*[2] |
| Faba bean extrafloral nectaries | $0.00 \pm 0.00^{g}$ | na | *C. koehleri* & *P. operculella*[1] |
| Faba bean flower | $18.40 \pm 0.26^{b}$ | $0.15 \pm 0.02^{f}$ | neither[3] |
| Nasturtium flower | $31.80 \pm 0.24^{a}$ | $1.75^{4} \pm 0.03^{d}$ | *C. koehleri*[2] |
| Phacelia flower | $7.07 \pm 0.15^{c}$ | $5.05 \pm 0.18^{b}$ | *C. koehleri*[1] |
| | $1.70^{5} \pm 0.05^{e}$ | $0.15^{6} \pm 0.01^{f}$ | *C. koehleri*[1] |
| Coriander | $0.00 \pm 0.00^{g}$ | $0.58 \pm 0.02^{e}$ | *C. koehleri* & *P. operculella*[3] |
| Borage | $2.85 \pm 0.05^{d}$ | $0.10 \pm 0.01^{f}$ | *C. koehleri*[3] |

[1]Inferred from the present study. [2]Inferred from the present study in conjunction with Baggen & Gurr (1998). [3] From Baggen & Gurr (1998). [4]Diameter where nectar first evident. [5]Depth from stamen appendages to nectaries. [6]Size of gaps in stamen appendages. Means followed by different letters are significantly different ($P = 0.05$).

the moth, *C. koehleri* would be unable to access nectar within phacelia flowers, but small volumes of nectar were seen on the outside of the stamen appendage area. Here it would be accessible to *C. koehleri* but present in volumes too small to significantly benefit larger *P. operculella* (compare Table 1). Its presence is likely to be the result of nectar rising up stamen appendages by capillary action, an effect described for the flowers of cardamom (Belevadi et al., 1997). *Copidosoma koehleri* were also seen to apply mouthparts to the anthers of phacelia (a behaviour not observed by Jervis et al. (1993) in a study of many more plant and parasitoid species) and thus are likely to supplement available nectar with pollen, a food source not available to the suctorial-feeding moth.

The floral nectaries of buckwheat and dill, and the extrafloral nectaries of faba bean, were freely fed upon by *P. operculella* as a result of the very shallow or non-existent corolla tube. As expected, *C. koehleri* also derived significant benefit from these nectar sources. The present study also showed nasturtium to be fed upon by the parasitoid. This is important because Baggen & Gurr (1998) found neither longevity or fecundity of *P. operculella* to be increased when allowed access to flowers of this plant. Thus, nasturtium is another example of a selective food plant. In this instance it is considered most likely that the selectivity of these flowers results from inability of *P. opercullella* to crawl deep within the long narrow corolla tube to access nectaries. The backward pointing scales of this insect may make escape from such a location difficult. In contrast, *C. koehleri* were observed to penetrate deep within corollas, as were ants (Formicidae) and earwigs (Dermaptera).

Baggen & Gurr (1998) identified borage as the first example of a selective food plant but did not determine the mechanism by which this plant afforded access to *C. koehleri* alone. The present study shows that flowers of this species had the narrowest of corolla openings through which it would have been impossible for *P. operculella* to crawl and a subsequent corolla depth greater than the length of the moth's proboscis. As in the case of phacelia, it is unclear from morphometric evidence alone how *C. koehleri* accesses nectar. However, Crane et al. (1984) describes borage as a morning producer of nectar, and its presence was noted on the exterior of these flowers, its escape being facilitated by their pendulous orientation. Equivalent build-up of nectar, resulting in its availability to hymenopterans has previously been noted by Orr & Plesants (1996). However, because moths are inactive during the day (Atherton, 1936) this nectar is likely to have crystallised or have become too viscous for the moth to feed upon by the time adults become active at dusk (Traynier, 1983). Daniel et al. (1989) showed that nectar sucrose concentrations of 35–45% were optimal for insects with a small cibarum, with energy intake declining sharply at greater concentrations (a result of the viscosity of the nectar). In contrast, hymenopteran adults with mandibulate mouthparts, such as *C. koehelri*, are able to feed on such foods.

Mandibulate mouthparts also allow certain hymenopteran species to chew holes in plant tissue to access nectar which would otherwise be unavailable (e.g., *Diadegma insulare* (Cresson) (Hymenoptera: Ichneumonidae), Idris & Grafius (1997)), though this behaviour was not observed for *C. koehleri* in the present study. Disruption of flower integrity by such destructive feeding, which may also include the dislocation of stamens, may subsequently allow feeding by other insects which otherwise would have been unable to access nectar. Thus, under field conditions 'selectivity' of plants, when based on nectar access alone, may not always be a constant.

However, a wide variety of other factors may operate in concert with flower architecture to maintain selectivity. These include flower colour and chemical factors such as the balance of sugars. Hexose-rich nectars, for example are generally unattractive to lepidopterans, whilst extrafloral nectar tends to be sucrose-rich and highly attractive to this order (Rogers, 1985). Toxic or distasteful phagodeterrents, such as non-protein amino acids, glycosides, alkaloids, phenolics and minerals (Kevan & Baker, 1984), may also be important.

Further studies are required to fully elucidate the nature of mechanisms for selective food plants, especially since the present findings are based on glasshouse-grown plants tested under non-choice conditions. Recent field trials with selective (borage) and non-selective (buckwheat) plants have shown that both are visited by *C. koehleri* but only the latter are fed upon by *P. operculella* (Baggen, unpubl.). Irrespective of such field verification, it is clear from the present study that in addition to comparisons of feeding structures with length of corolla tube other parameters, as outlined above, need to be considered when assessing selectivity.

In the present study a simple pest: natural enemy system has been considered because *P. operculella* is the key potato pest in inland New South Wales and *C. koehleri* the major natural enemy. For habitat manipulation to make a significant contribution towards pest management in potato growing areas with a more complex suite of pests and natural enemies, or towards the management of other nectar-feeding pest species (e.g., most lepidopterans), it will be important that the phenomenon of selectivity is better understood and that a wider range of selective food plants is identified. This will allow benefit to be extended to natural enemies without facilitating the 'co-location' (*sensu* Ehrlich, 1984) of those resources required by the pest in the modified habitat.

## Acknowledgements

This project was supported by a University of Sydney Postgraduate Award to LRB and a Department of Industry Science and Tourism grant to GMG.

## References

Atherton, D. O., 1936. Leaf miner and stem borer of tobacco in North Queensland. Queensland Agriculture Journal 45: 12–31.

Baggen, L. R. & G. M. Gurr, 1998. The influence of food on *Copidosoma koehleri*, and the use of flowering plants as a habitat management tool to enhance biological control of potato moth, *Phthorimaea operculella*. Biological Control 11: 9–17.

Belevadi, V. V. Venkateshalu & H. R. Vivek, 1997. Significance of style in cardamom corolla tubes for honey bees. Current Science 73: 287–290.

Crane, E., P. Walker & R. Day, 1984. Directory of Important Honey Sources. International Bee Research Association, London.

Daniel, T. L., J. G. Kingsolver & E. Meyhofer, 1989. Mechanical determinants of nectar feeding energetics in butterflies: muscle mechanics, feeding geometry, and functional significance. Oecologia 79: 66–75.

Dent, D., 1995. Integrated Pest Management. Chapman & Hall, London.

Ehrlich, P. R., 1984. The structure and dynamics of butterfly populations. In: R. I. Wright & P. R. Ackery (eds), The Biology of Butterflies. Academic Press, London, pp. 25–40.

Fenemore, P. G., 1977. Oviposition of the potato tuber moth, *Phthorimaea operculella* Zell. (Lepidoptera: Gelechiidae); fecundity in relation to mated state, age, and pupal weight, New Zealand Journal of Zoology 4: 187–191.

Hickman, J. M. & S. D. Wratten, 1996. Use of *Phacelia tanacetifolia* (Hydrophyllaceae) as a pollen source to enhance hoverfly (Diptera: Syrphidae) populations in cereal fields. Journal of Economic Entomology 89: 832–840.

Holland, J. M., S. R. Thomas & S. Courts, 1994. *Phacelia tanacetifolia* flower strips as a component of integrated farming. In: N. Boatman (ed), Field Margins: Integrating Agriculture and Conservation. British Crop Protection Council Monograph No. 58, Farnham, pp. 215–220.

Idris, A. B. & E. Grafius, 1995. Wildflowers as nectar sources for *Diadegma insulare* (Hymenoptera: Ichneumonidae), a parasitoid of diamondback moth (Lepidoptera: Yponomeutidae). Environmental Entomology 24: 1726–1735.

Idris, A. B. & E. Grafius, 1997. Nectar collecting behaviour of *Diadegma insulare* (Hymenoptera: Ichneumonidae), a parasitoid of diamondback moth (Lepidoptera: Plutellidae). Environmental Entomology 26: 114–120.

Jervis, M. A., N. A. Kidd, M. G. Fitton, T. Huddleston & H. A. Dawah, 1993. Flower-visiting by hymenopteran parasitoids. Journal of Natural History 27: 67–105.

Kevan, P. G. & H. G. Baker, 1984. Insects on flowers. In: C. B. Huffaker & R. L. Rabb (eds), Ecological Entomology. John Wiley, New York, pp. 607–631.

Matthews-Gehringer, D., E. Lachowski, S. Edwards, S. Wolfgang, T. Schettini, R. L. Bugg & C. M. Felland, 1994. Herbaceous Ground Covers in a Pennsylvania Apple Orchard: Establishment and Arthropod Assessment 1989–1991, HO-94/02, Rodale Institute Research Center, Kutztown.

Orr, D. B. & J. M. Pleasants, 1996. The potential of native prairie plant species to enhance the effectiveness of the *Ostrinia nubilalis* parasitoid *Macrocentrus grandii*. Journal of the Kansas Entomological Society 69: 133–143.

Patt, J. M., G. C. Hamilton & J. H. Lashomb, 1997. Foraging success of parasitoid wasps on flowers: interplay of insect morphology, floral architecture and searching behaviour. Entomologia Experimentalis et Applicata 83: 21–30.

Platner, G. R. & E. R. Oatman, 1972. Techniques for culturing and mass producing parasites of the potato tuberworm. Journal of Economic Entomology 65: 1336–1338.

Rogers, C. E. 1985. Extrafloral nectar: entomological implications. Bulletin of the Entomological Society of America 31: 15–20.

Topham, M. & J. W. Beardsley, 1975. An influence of nectar source plants on the New Guinea sugar cane weevil parasite, *Lixophaga sphenophori* (Villeneuve). Proceedings of the Hawaiian Entomological Society 22: 145–155.

Traynier, R. M. M., 1983. Influence of plants and adult food on the fecundity of potato moth, *Phthorimaea operculella*. Entomologia Experimentalis et Applicata 33: 145–154.

Upton, M. S., 1991. Methods for Collecting, Preserving, and Studying Insects and Allied Forms. 4th Edition. Miscellaneous Publication No. 3, The Australian Entomological Society, Brisbane.

White, A. J., S. D. Wratten, N. A. Berry & U. Weigmann, 1995. Habitat manipulation to enhance biological control of Brassica pests by hover flies (Diptera: Syrphidae). Journal of Economic Entomology 88: 1171–1176.

Williams, I. H. & D. G. Christian, 1991. Observations on *Phacelia tanacetifolia* Bentham (Hydrophyllacae) as food plant for honey bees and bumble bees. Journal of Apicultural Research 30: 3–12.

Entomologia Experimentalis et Applicata **91:** 163–167, 1999.



# Wound-induced increases in the glucosinolate content of oilseed rape and their effect on subsequent herbivory by a crucifer specialist

Elspeth Bartlet, Guy Kiddle, Ingrid Williams & Roger Wallsgrove
*IACR Rothamsted Experimental Station, Harpenden, Herts, AL5 2JQ, UK*

Accepted: November 1, 1998

*Key words: Brassica napus, Psylliodes chrysocephala*, glucosinolates, jasmonic acid, induction

## Abstract

Damage to the oilseed rape plant (*Brassica napus* L.) by the cabbage stem flea beetle, *Psylliodes chrysocephala* L. (Coleoptera: Chrysomelidae) induces systemic changes to the glucosinolate profile, most noticeably an increase in the concentration of indole glucosinolates. When jasmonic acid was applied to the cotyledons of the plant, a similar effect was observed. Feeding tests with artificial substrates compared a glucosinolate fraction from jasmonic acid-treated plants with a similar fraction from untreated plants. In these tests, alterations to the glucosinolate profile increased the feeding of a crucifer-specialist feeder (*P. chrysocephala*). However, in whole plant tests, *P. chrysocephala* did not feed more on the jasmonic acid treated plants than on the controls. This implies that other aspects of the damage response are being induced by the jasmonic acid treatment and having a negative effect on subsequent herbivory.

## Introduction

When an oilseed rape plant is damaged, either mechanically or by herbivores, a systemic increase in indole glucosinolate and decrease in aliphatic glucosinolate concentration occurs. The net result is usually an increase in total glucosinolates (Koritsas et al., 1991; Birch et al., 1992). Glucosinolates act as feeding deterrents for polyphagous herbivores and feeding stimulants for crucifer-specialists. For example, amongst oilseed rape pests, glucosinolates deter feeding by the polyphagous grey field slug (*Deroceras reticulatum* Müller) and stimulate feeding of the oligophagous cabbage stem flea beetle (*Psylliodes chrysocephala* L.) (Glen et al., 1990; Bartlet et al., 1994). Indole glucosinolates are known to be the most potent class of glucosinolates in stimulating several crucifer-feeding insects (Larsen et al., 1985; Städler et al., 1995; Roessingh et al. 1997; Isidoro et al. 1998). Thus, changes to the glucosinolate profile of oilseed rape following herbivory would be predicted to make the plant more palatable to crucifer specialists. Koritsas et al. (1991) even suggested that, by altering the indolyl glucosinolates, insects such as *P. chrysocephala* could be altering the metabolism of oilseed rape for their own benefit. However, feeding tests with oilseed rape have reported either a reduction in the feeding of crucifer specialists on induced tissue (Bodnaryk, 1992) or no effect (Palaniswamy & Lamb, 1993; Siemens & Mitchell-Olds, 1996). This might be due to the influence of other physico-chemical factors that change after oilseed rape is wounded. Bodnaryk & Rymerson (1994) found that, in addition to altering the glucosinolate content, wounding rape seedlings or treating them with methyl jasmonate altered cotyledon toughness, protein content, cysteine proteinase inhibitor levels and viscosity of extracts.

This study is the first to investigate how induced changes to the glucosinolate profile of oilseed rape affect the feeding of a crucifer specialist (*P. chrysocephala*). To avoid effects due to herbivore cues such as frass, the damage response of oilseed rape is induced by the application of jasmonic acid, rather than by herbivory. Jasmonates are involved in the wound-induced defences of oilseed rape and other plants (Bodnaryk, 1994; Doughty et al., 1995b). Treatment of rape seedlings with methyl jasmonate or jasmonic acid induces physiological changes, including altered glucosinolate profiles, similar to those induced by herbivore damage (Bodnaryk & Rymerson, 1994). Firstly, the effects of adult flea beetle feeding and jasmonic acid (JA) on the glucosinolate profile of oilseed rape is determined. Then the effects of the JA-altered glucosinolate profile on subsequent herbivory is tested. To determine the effects of the changes to the glucosinolate profile in the absence of other elements of the damage response, the responses of the insects to a partially purified glucosinolate fraction is also tested.

## Materials and methods

*Insects.* Cabbage stem flea beetles were collected from crops of winter rape and kept on oilseed rape plants (cv Falcon) within nylon net bags in an outside insectary.

*Effect of adult feeding on foliar glucosinolate levels.* Oilseed rape plants (cv Bienvenue) were planted in a compost mixture (Petersfield Products) in plastic pots (4″ diam), one plant to a pot. They were grown in a heated glasshouse (15–20 °C), to the sixth true leaf stage. Ten cabbage stem flea beetles, starved for 24 h, were enclosed in a muslin bag on the third true leaf of each of eight plants. Eight control plants each had the third true leaf bagged without flea beetles. After seven days all the leaves were harvested and the glucosinolates analysed.

*Feeding on whole plants after jasmonic acid induction.* Oilseed rape plants (cv Bienvenue) were glasshouse-grown as above in plastic pots (4″ diam), two plants per pot. At the second true leaf stage, one plant in each pot was treated with jasmonic acid, by applying 5 nmoles of JA in 5 $\mu$l of a solution of wetting agent (Triton 0.01%) to each cotyledon, using a pipette. The other plant in each pot was treated with 5 $\mu$l of Triton (0.01%) on each cotyledon. Four days after treatment the plants in 20 pots were harvested for glucosinolate analysis and production of glucosinolate fractions. The remaining pots of rape were transferred to a controlled environment room at L12:D12, 15/5 °C. The plants in each pot were enclosed by a hurricane lamp glass, whose lower rim fitted the pot exactly. The top of each lamp glass was covered with metal gauze. Flea beetles were introduced to 20 of the lamp glass enclosures (25 beetles per pot). The remaining 20 pots were not presented for herbivory but provided an additional control for glucosinolate analysis at the end of the experimental period. After seven hours the area ($mm^2$) of consumption was estimated by eye. All plants were then harvested for glucosinolate analysis.

*Feeding on glucosinolate fractions after jasmonic acid induction.* To produce glucosinolate fractions, a protocol similar to that of Hanley et al. (1983) was used. Lyophilised, ground plant material from the first and second true leaves was extracted in 70% methanol (3.5 ml) at 70 °C for 10 min. The extract was centrifuged at 3000 g for 10 min, after which the supernatant was stored on ice and the pellet was re-extracted in 70% methanol (3 ml) and centrifuged as before. The pellet was then discarded and the supernatants combined. As the freeze dried material absorbed liquid, the combined supernatants were made up to 5 ml in 70% methanol. DEAE Sephadex A25 (2 ml) was suspended in acetic acid (2 M) in a 1:1 ratio and loaded onto a mini column. The column was washed twice with 2 ml distilled water, the methanol extract loaded and the column washed twice with a further 2 ml distilled water. Glucosinolates were eluted from the column by washing ten times with 1 ml potassium sulphate (0.5 M). The elutes were frozen at −50 °C and freeze dried. Freeze dried material was resuspended in methanol, filtered to remove the salt and the filtrate rotary evaporated to dryness. Residue was dissolved in 1 ml ultrapure water and frozen until required. For the feeding tests, the glucosinolate fraction was stirred into agar (2%) plus sucrose (0.5%), cooled to below 50 °C. The glucosinolate fraction was made up in the same weight of agar as the fresh weight of the leaf material from which it was taken.This was poured into two circular wells (8 mm diam, 3 mm deep) cut into a perspex sheet inside a Petri dish (90 mm diameter). One well was filled with agar plus extract from JA-treated plants and the other with agar plus extract from untreated plants. Four beetles (starved for three days) were added to each dish and the dishes kept at 10 °C L12:D12 for 24 h. The number of bite marks in the agar was then counted as a measure of feeding.

*Glucosinolate analysis.* The cotyledons, first true leaves and second true leaves were bulked together within a treatment, weighed and harvested into separate bags. The bags were immediately immersed in liquid nitrogen, then stored in a freezer at −50 °C. After freeze drying, the tissue was reweighed and 200 mg of ground leaf material was extracted in 70% methanol (3.5 ml) which contained 600 $\mu$moles of

*Table 1.* Feeding by *Psylliodes chrysocephala* on untreated and jasmonic acid-treated rape in choice tests with either plant material or glucosinolate fractions

| Feeding parameter | Material tested | *n* | Untreated plants (± s.e.) | Treated plants (± s.e.) | P* |
|---|---|---|---|---|---|
| Estimated area consumed (mm$^2$) | Cotyledons | 20 | 2.20 ± 0.57 | 0.30 ± 0.16 | 0.01 |
| | True leaves | 20 | 5.90 ± 1.26 | 4.45 ± 0.87 | 0.60 |
| Number of bitemarks | Agar treated with glucosinolate fraction | 22 | 59.5 ± 10.6 | 92.5 ± 10.6 | 0.02 |

*Two-tailed Wilcoxon's test.

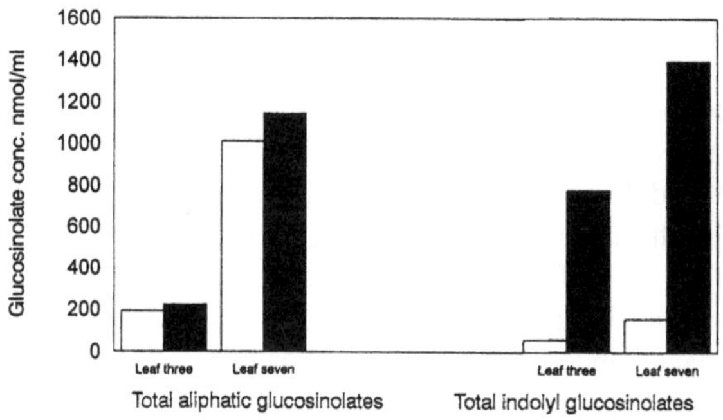


*Figure 1.* Glucosinolate levels (nmol/ml tissue water) in true leaves three and seven of oilseed rape seven days after bagging of the third true leaf with (■) and without (□) cabbage stem flea beetles.

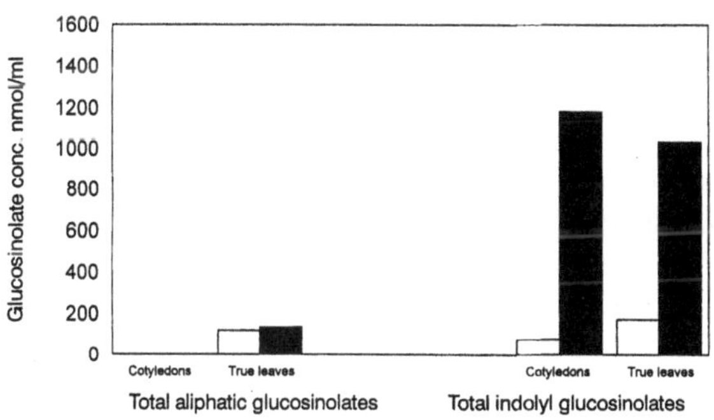


*Figure 2.* Glucosinolate levels (nmol/ml tissue water) in cotyledons and true leaves of oilseed rape at the beginning of the experimental period, four days after treatment of the cotyledons with wetter (□) and with wetter plus jasmonic acid (■).

2-propenylglucosinolate as an internal standard. The extract was heated at 70 °C for 10 min and then centrifuged at 3000 rpm for 10 min to remove particulates. The supernatant was decanted and stored on ice. The pellet was then re-extracted and centrifuged as before. The supernatants were combined and made up to 5 ml. One ml of Sephadex A25/2 M acetic acid (1:1, v/v) suspension was pipetted into a mini column (0.5 ml bed volume). The column was then washed with 6 M imidazole formate (2 ml), followed by two washes in

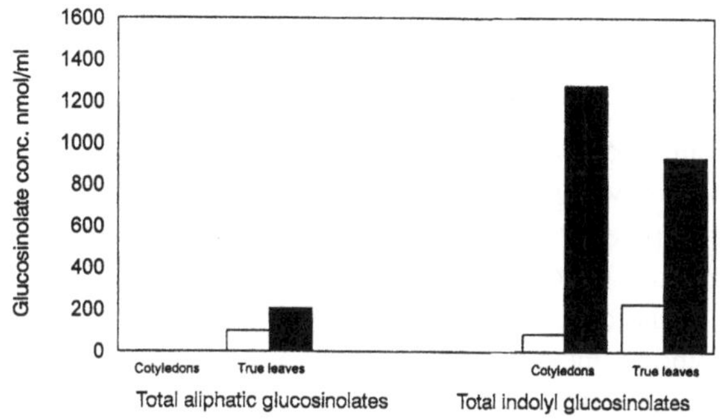


*Figure 3.* Glucosinolate levels (nmol/ml tissue water) in oilseed rape at the end of the 7-h experimental period in plants *unexposed* to herbivory by the cabbage stem flea beetle. Plants either had cotyledons treated with wetter (□) or with wetter plus jasmonic acid (■).

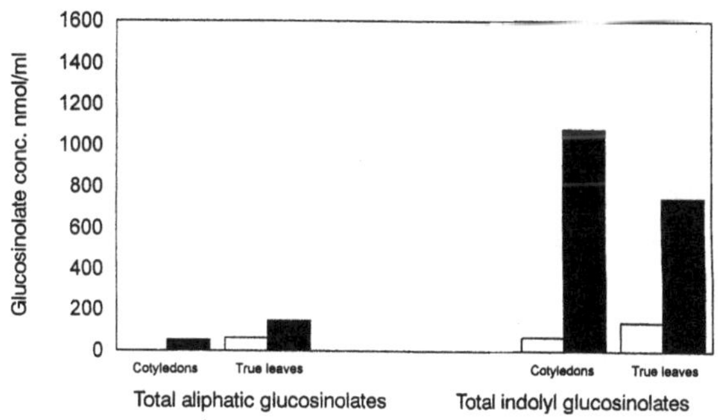


*Figure 4.* Glucosinolate levels (nmol/ml tissue water) in oilseed rape plants at the end of the 7-h experimental period in plants *exposed* to herbivory by the cabbage stem flea beetle. Plants either had cotyledons treated with wetter (□) or with wetter plus jasmonic acid (■).

2 ml of ultrapure water. One ml of the glucosinolate extract was then loaded onto the pre-equilibrated column and unbound material removed using two, 1 ml washes of 0.1 M sodium acetate (pH 4.0). This also served to adjust the pH of the resin in preparation for desulphation of the glucosinolates. Purified sulphatase (100 $\mu$l) was loaded onto the column and digestion carried out overnight. The desulphoglucosi-

nolates were eluted with three washes with 500 $\mu$l ultra pure water and stored at −20 °C until they were analysed by HPLC. For separation by HPLC, a Waters 600LC system equipped with a 996 photodiode array detector was used. Data was collected between 200 and 300 nm to establish the class of desulphoglucosinolate. The column used was a 4.6 × 250 mm $C_{18}$ column (Waters, Symmetry) with a 5 $\mu$m particle size. The mobile phase was made up of water-acetonitrile as reported in Porter et al. (1991).

## Results

The glucosinolate profile of oilseed rape was found to consist of aliphatics {mainly 2-hydroxy-3-butenyl- (progoitrin), 2-hydroxy-4-pentenyl- (gluconapoleiferin) and 4-pentenyl- (glucobrassicanapin) glucosinolates} and indolyls {3-indolylmethyl- (glucobrassicin), 1-methoxy-3-indolylmethyl-, 4-methoxy-3-indolylmethyl- and 4-hydroxy-3-indolylmethyl- glucosinolates}, with 2-phenylethyl glucosinolate (gluconasturtiin) present only in trace amounts. This profile was radically altered after herbivory by *P. chrysocephala*. Aliphatic glucosinolates did not change dramatically after induction. Some increased slightly, whilst others decreased, resulting in a slight increase in total aliphatic glucosinolate levels (Figure 1). All indolyl glucosinolates increased after the plant was damaged, leading to a dramatic increase (8.8 fold) in total indolyl glucosinolate levels. Overall, there was a 2.2 fold increase in total glucosinolates. These changes were systemic, as the profile changed, not just at the site of treatment (leaf three), but also in other true leaves (data from leaf seven are presented).

The glucosinolate profile of the plants in the JA induction experiment differed from that in the flea beetle induction experiment, probably because different leaves were sampled from plants at a different growth stage. Nevertheless, application of a low concentration of jasmonic acid to the cotyledons had a similar, systemic effect on glucosinolate levels (Figure 2). Over the plant as a whole, the concentration of indolyl glucosinolate increased 8.9 fold, whilst that of aliphatic glucosinolate did not change markedly. Overall glucosinolate concentration increased 6.4 fold in JA-treated plants. The glucosinolate profile of the plants altered little over the 7-h period of the bioassay (Figures 2 and 3). There was little difference between glucosinolate levels at the end of the experiment between plants unexposed to herbivory and those damaged by flea beetles (Figures 3 and 4), indicating that further induction of glucosinolates, due to flea beetle damage did not occur during the course of the experiment.

*P. chrysocephala* fed less on JA-treated plants than on untreated plants, the difference being significant for the cotyledons but not the true leaves (Table 1). However, the beetles fed significantly more on agar containing extract from JA-treated plants than on agar with extract from untreated plants.

## Discussion

Systemic changes to the glucosinolate profile of oilseed rape after damage from cabbage stem flea beetle adults are similar to those that occur after damage by the larvae (Koritsas et al., 1991), although they feed in a different way (chewing the leaves, rather than mining the stems). They are also similar to the changes that are induced by the application of JA.

The glucosinolate content of JA-treated plants was still substantially greater than that of the untreated plants at the end of the 7-h bioassay period, even though feeding damage by *P. chrysocephala* would eventually be expected to induce indolyl glucosinolate production in the untreated plants. Bodnaryk (1992) demonstrated that induction of indolyl glucosinolates in oilseed rape can occur within four hours of damage (24 needle punctures to the cotyledon). Feeding on untreated plants in the first few hours of this experiment must have been insufficient to produce such rapid induction.

When glucosinolate fractions were tested, the effect of the altered profile was as predicted from our knowlege of *P. chrysocephala*'s response to glucosinolates, i.e., the combined effect of increased proportions of indolyl glucosinolates and a greater overall glucosinolate concentration stimulated more feeding. However, in whole plant tests, *P. chrysocephala* did not feed more on JA-treated plants. As mentioned above (Bodnaryk & Rymerson, 1994), other elements of rape physiology are altered by plant damage or jasmonate treatment. These must have had a deterrent effect on feeding, counteracting the stimulatory effect of the altered glucosinolate profile. This is being investigated further.

Although JA induces systemic changes, the effects may have been greater at the site of treatment than in other plant parts. If so, this would explain why feeding was significantly less on treated plants than on untreated ones in the case of the cotyledons, but not in the case of the true leaves.

Damage-induced changes to the glucosinolate profile may benefit the oilseed rape plant by protecting it from fungal diseases, such as downy mildew (Doughty et al., 1995a), and generalist herbivores, such as slugs (Bartlet et al., unpubl.). They do not increase the susceptibility of the plant to *P. chrysocephala* (and possibly other crucifer-specialists) because their stimulatory effect is offset by other elements of induced defence. Furthermore, although rape plants with the altered profile have a higher glucosinolate content, they are unlikely to be more attractive to crucifer specialists. This is because only indole glucosinolates (which do not produce volatile metabolites) are induced. Thus, the emission rates of compounds such as isothiocyanates and nitriles, which are important attractants for crucifer specialists (Bartlet et al., 1997), would not be expected to increase after induction.

## Acknowledgements

Our thanks to Nigel Watts for collecting and maintaining the cabbage stem flea beetles. IACR Rothamsted receives grant aided support from the Biotechnology and Biological Sciences Research Council.

## References

Bartlet, E., D. Parsons, I. H. Williams & S. J. Clark, 1994. The influence of glucosinolates and sugars on feeding by the cabbage stem flea beetle, (*Psylliodes chrysocephala*). Entomologia Experimentalis et Applicata 73: 77–83.

Bartlet, E., M. M. Blight, P. Lane & I. H. Williams, 1997. The responses of the cabbage seed weevil *Ceutorhynchus assimilis* to volatile compounds from oilseed rape in a linear track olfactometer. Entomologia Experimentalis et Applicata 85: 257–262.

Birch, A. N. E., D. W. Griffiths, R. J. Hopkins, W. H. M. Smith & R. G. McKinlay, 1992. Glucosinolate responses of swede, kale, forage and oilseed rape to root damage by turnip root fly (*Delia floralis*) larvae. Journal of the Science of Food and Agriculture 60: 1–9.

Bodnaryk, R. P., 1992. Effects of wounding on glucosinolates in the cotyledons of oilseed rape and mustard. Phytochemistry 31: 2671–2677.

Bodnaryk, R. P., 1994. Potent effect of jasmonates on indole glucosinolates in oilseed rape and mustard. Phytochemistry 35: 301–305.

Bodnaryk, R. P. & R. T. Rymerson, 1994. Effect of wounding and jasmonates on the physico-chemical properties and flea beetle defence responses of canola seedlings *Brassica napus*. L. Canadian Journal of Plant Science 74: 899–907.

Doughty, K., R. N. Bennett, N. I. Nashaat, S. Scrijvers, G. Kiddle, B. J. Pye, S. E. Mitchell & R.M. Wallsgrove, 1995a. The response of oilseed rape (*Brassica napus* L.) seedlings to *Peronospora parasitica* and *Alternaria brassicae* following treatment with salicylic acid or methyl jasmonate. Proceedings of the Ninth International Rapeseed Congress, Cambridge 3: 971–973.

Doughty, K., G. Kiddle, B. Pye, R. Wallsgrove & J. Pickett, 1995b. Selective induction of glucosinolates in oilseed rape leaves by methyl jasmonate. Phytochemistry 38: 347–350.

Glen, D. M., H. Jones & J. K. Fieldsend, 1990. Damage to oilseed rape seedlings by the field slug *Deroceras reticulatum* in relation to glucosinolates. Annals of Applied Biology 117: 197–207.

Hanley, A. B., R. K. Heaney & G. R. Fenwick, 1983. Improved isolation of glucobrassicin and other glucosinolates. Jornal of the Science of Food and Agriculture 34: 869–873.

Isidoro, N., E. Bartlet, J. Ziesmann & I. H. Williams, 1998. Antennal contact chemosensilla in *Psylliodes chrysocephala* responding to cruciferous allelochemicals. Physiological Entomology 23: 131–138.

Koritsas, V. M., J. A. Lewis & G. R.Fenwick, 1991. Glucosinolate responses of oilseed rape, mustard and kale to mechanical wounding and infestation by cabbage stem flea beetle. Annals of Applied Biology 118: 209–221.

Larsen, M. L., J. K. Nielsen, A. Plöger & H. Sørensen, 1985. Responses of some beetle species to varieties of oilseed rape and to pure glucosinolate. In: H. Sørensen (ed.), Advances in the Production and Utilization of Cruciferous Crops. W. Junk, Dordrecht, pp. 230–244.

Palaniswamy, P. & R. J. Lamb, 1993. Wound-induced antixenotic resistance to flea beetles *Phyllotreta cruciferae* Goeze (Coleoptera: Chrysomelidae) in crucifers. Canadian Entomologist 125: 903–912.

Porter, A. J. R., A. M. Morton, G. Kiddle, K. J. Doughty & R. M. Wallsgrove, 1991. Variation in glucosinolate content of oilseed rape (*Brassica napus* L.) leaves. I. Effect of leaf age and position. Annals of Applied Biology 118: 461–467.

Roessingh, P., E. Städler, R. Baur, J. Hurter & T. Ramp, 1997. Tarsal chemoreceptors and oviposition behaviour of the cabbage root fly (*Delia radicum*) sensitive to fractions and new compounds of host leaf surface extracts. Physiological Entomology 22: 140–148.

Siemens, D. H. & T. Mitchell-Olds, 1996. Glucosinolates and herbivory by specialists (Coleoptera: Chrysomelidae, Lepidoptera: Plutellidae): consequences of concentration and induced resistance. Environmental Entomology 25: 1344–1353.

Städler, E., J. A. A. Rewick, C. D. Radke & K. Sachdev-Gupta, 1995. Tarsal contact chemoreceptors response to glucosinolates and cardenolides mediating oviposition in *Pieris rapae*. Physiological Entomology 20: 175–187.

*Entomologia Experimentalis et Applicata* **91:** 169–173, 1999.



# Pyrrolizidine alkaloid distribution in *Senecio jacobaea* rosettes minimises losses to generalist feeding

Nico J. de Boer
*Institute of Evolutionary and Ecological Sciences, Leiden University, Kaiserstraat 63, 2311 GP Leiden, The Netherlands (Phone: +31 71 527 5130; Fax: +31 71 527 4900)*

Accepted: November 30, 1998

*Key words:* pyrrolizidine alkaloids; *Senecio jacobaea*; within-plant distribution; defence; generalist herbivores; *Spodoptera exigua*; *Mamestra brassicae*

## Abstract

The within-plant distribution of pyrrolizidine alkaloids (PAs) was determined in the shoots of vegetative *Senecio jacobaea* L. Asteraceae plants. Also the distributions of feeding stimulants, soluble sugars and total nitrogen, were determined. PAs were found to decrease with leaf age, sugars were constant, and nitrogen decreased with leaf age. These results were compared with the within-plant distribution of feeding damage by the generalist herbivores *Spodoptera exigua* Hübner and *Mamestra brassicae* L. Both generalists showed a strong preference for older leaves. It was concluded, that the herbivores were forced to feed on sub-optimal parts of the plant.

## Introduction

When studying the evolution of defence against insect herbivores, the size of the participants is usually overlooked. Comparing the size of an insect meal to the size of a host plant leaf, the latter is very seldom the smaller of the two. Considering within and between leaf differences in both nutritional and detrimental compounds, most plants are clearly not a homogeneous food source (McKey, 1979; Raupp & Denno, 1983). Therefore, an insect herbivore should be selective with regard to the actual site of feeding, to optimize its rate of food intake and limit the intake of detrimental compounds (Slansky, 1993; Simpson & Raubenheimer, 1996). Often the youngest leaves of a plant are most nutritious (Slansky, 1993), but here also toxins are found to occur at higher concentrations (McKey, 1979; Hoy et al., 1998). Plants have to minimize their losses to herbivores, in order to maximize their fitness. The expected future contributions to the plant's fitness are highest from the youngest leaves (Krischik & Denno, 1983; van Dam et al., 1994), so if herbivores cannot be deterred completely, at least it would be worthwhile to deflect the damage to the least valuable leaves. Optimal defence theory (Zangerl & Bazzaz, 1992) predicts the younger leaves to be better defended, based on the value of the leaf and the risk of being found. The leaves with the highest expected contribution to plant fitness, i.e., the youngest leaves, are supposed to be the most valuable, and thus the best defended. Since specialist herbivores are often able to circumvent their hosts' defences (van der Meijden, 1996), this prediction should be tested with generalist herbivores.

Despite the long study period and numerous studies of anti-herbivore defence in *Senecio jacobaea*, the distribution of its chemical defences over the leaves has not yet been studied. Often whole plants or whole shoots are extracted and analyzed (e.g., Vrieling et al., 1993), and occasionally, the size of a single leaf allows the use of it as estimator of the whole plant's defence level (Vrieling & van Wijk, 1994). In addition, most work has been done on specialist herbivores, while generalists have received little attention. This study investigates the distribution of PAs over the leaves of vegetative *S. jacobaea* plants, and consequently tests the impact of this distribution on generalist herbivores. The nutritional quality of the leaves, expressed as nonstructural carbohydrate concentration and nitrogen concentration, is measured as a predictor for the herbivore preference in the absence of deterrent substances.

## Materials and methods

*Plant material.* To determine the distribution of PAs, *Senecio jacobaea* L. plants were grown from freshly collected seeds from the dune area 'Meijendel', near Den Haag, The Netherlands. They were grown in 13cm diameter pots in a mixture of 50% compost and 50% dune sand, and raised under artificial lighting (L15:D9, 15–25 °C, 40–80% r.h.). The plants were watered regularly and provided with 'Osmocote' slow release fertilizer (N : P : K = 15 : 11 : 13 + 2MgO, 15g/pot). Two months after planting the seedlings, the rosettes were harvested leaf by leaf, weighed and frozen to −25 °C until analysis. The four plants had 8, 8, 15, and 8 leaves respectively. The surface area of the fresh leaves was measured, to obtain an estimate of the relation between apparency of the leaves and leaf age.

Vegetative plants for total nitrogen and nonstructural carbohydrate determination were collected as rosettes in the same dune area 'Meijendel', and grown for 40 days in a climate room in the same medium (but without fertilizer) at L16:D8, 20:15 °C, >70% r.h., before harvesting them leaf by leaf. The four plants had 8, 11, 9, and 10 leaves respectively.

For measuring herbivore preference, plants from ten *S. jacobaea* genotypes were reproduced from *in vitro* stock material, originating from the same area 'Meijendel', and grown in 11 cm diameter pots in the same medium, but without fertilizer, in a climate-room at L16:D8, 20:10 °C, >70%r.h. They were offered to the herbivores at the age of eight weeks after potting.

*Insect herbivores.* *Spodoptera exigua* Hübner (Lepidoptera: Noctuidae) was raised on artificial diet (after Aerts et al., 1992) in a laboratory culture at L12:D12, 25 °C, >70% r.h. *Mamestra brassicae* L. (Lepidoptera: Noctuidae), also from a laboratory culture, was kept on Brussels Sprouts (*Brassica oleracea* L., var. *gemmifera*, cv. Titurel) in a climate room (L16:D8, 20–22 °C, 50–70% r.h.).

*Feeding damage.* The distribution of damage was determined in two separate experiments. Three third instar *S. exigua* larvae were placed per cage on one *S. jacobaea* plant, and were allowed to feed until pupation. Forty of such cages were placed in a climate room at L12:D12, 25 °C, >70% r.h. For *M. brassicae*, ten *S. jacobaea* plants were placed in a cage, and 50 third to fourth instar larvae were added. Ten cages were placed in a climate room at L12:D12, 20:15 °C, >70% rh. After eight days of feeding they were removed from the plants. Damage by both of the herbivores was estimated by eye, in classes on a 'logarithmic scale': class 0 =<1% of the leaf area damaged, 1 = 1–5%, 2 = 5–10%, 3 = 10–25%, 4 = 25–50%, and 5 = 50–100% of the leaf area removed. To compare between equally aged plants with an unequal number of leaves, all plants were divided in eight sections; whole leaves were assigned to a section based on their leaf number.

*Chemical analysis.* Total PA concentration of single leaves was determined spectrophotometrically, by acid-base extraction (Hartmann & Zimmer, 1986, adapted to small quantities), followed by a colour reaction modified after Mattocks (1967), using monocrotaline (Sigma, 98% pure) as an external standard. Total nitrogen of single leaves was determined spectrophotometrically, after a Kjelldahl destruction, followed by a colour reaction forming indophenol blue (Golterman et al., 1978). The concentration of nonstructural carbohydrates in single leaves was determined with the anthrone colour reaction after Allen (1974), adapted to small quantities. To compare between plants with unequal numbers of leaves, relative leaf numbers were calculated for each plant, by dividing the rank numbers of the leaves by its total number of leaves.

## Results

Figure 1 shows the results of both herbivore feeding experiments. Both herbivores exhibit a strong preference for the older leaves of the plants, while the younger leaves remain largely uneaten.

In Figure 2, the distributions of deterrents (PAs) and feeding stimulants (soluble sugars and nitrogen) are presented. PAs appear to decrease with leaf age. The two plants with very low PA concentrations in the youngest leaf had extremely small youngest leaves, probably exceeding the sensitivity of the determination. Apart from those, the general trend is downward: even the low PA plant, although it seems to have a flat curve, still shows an almost threefold decrease in PA concentration with leaf age. For sugars, there seems to be a slight increase of the concentration with leaf age. But the regression line $y = 0.038x + 0.10$ shows this to be a non-significant relationship ($R^2_{adj.} = 0.0$,

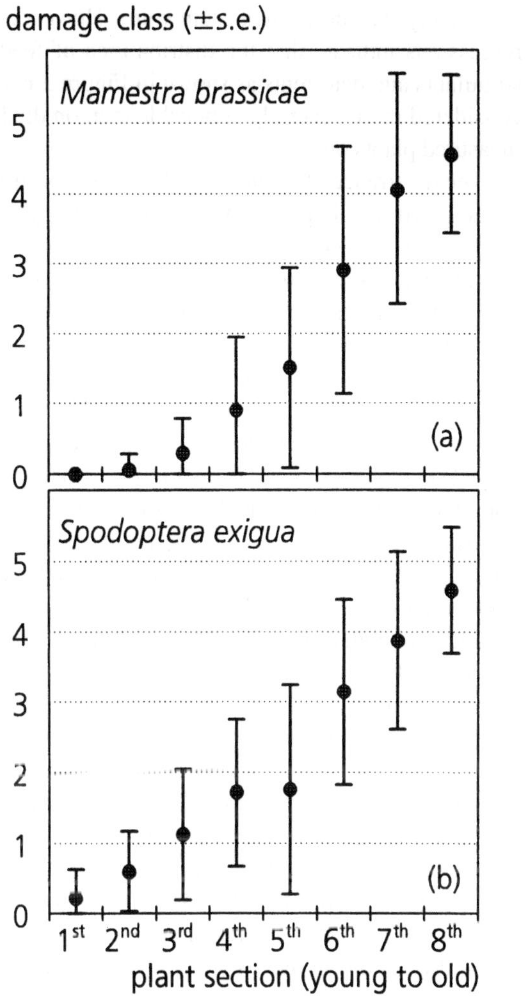


*Figure 1.* Distribution of feeding damage by the generalist herbivores *M. brassicae* (a) and *S. exigua* (b) over the leaves of vegetative *S. jacobaea* plants. The average damage class (±1 standard error) is given per plant section (40 plants for *S. exigua*, 100 plants for *M. brassicae*). Damage classes: 0 =< 1%, 1 = 1– 5%, 2 = 5–10%, 3 = 10–25%, 4 = 25–50%, 5 = 50–100% of the leaf area removed. Plant sections made by assigning whole leaves to the appropriate section.

P = 0.36). Nitrogen displays a downward trend with increasing leaf age. This is not a strong relationship, but it is significant ($y = -0.013x + 0.032$, $R^2_{adj} = 0.11$, P = 0.030). Correlations between the chemical measurements and herbivore preference cannot be calculated because of the different sets of plants for each measurement.

Figure 3 shows the relation between leaf surface area and leaf age. The largest leaves are in the middle of the plant. Both the younger and older leaves are significantly smaller (regression of leaf surface area on

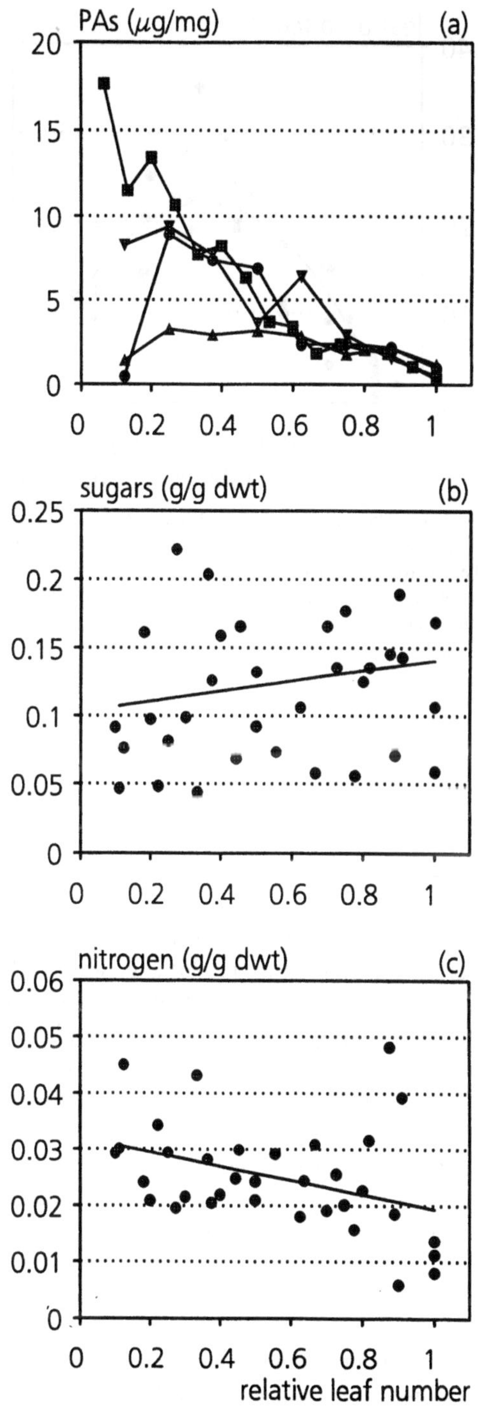


*Figure 2.* Chemical composition of the leaves of vegetative *S. jacobaea* plants. X-axis: relative leaf numbers (leaves numbered per plant, from young to old, and divided by the total number of leaves of that plant) to enable the comparison of equally aged leaves. (a) Distribution of pyrrolizidine alkaloids in µg/mg dry weight; the four plants, having 8, 8, 15, and 8 leaves respectively, are marked with different symbols. (b) Distribution of non-structural carbohydrates in g/g dry weight. Regression line: $y = 0.038x + 0.10$, $R^2_{adj.} = 0.0$. (c) Distribution of nitrogen over the leaves of 4 plants, in g/g dry weight. Regression line: $y = -0.013x + 0.032$, $R^2_{adj.} = 0.11$. The four plants of (b) and (c) had 8, 11, 9, and 10 leaves respectively.

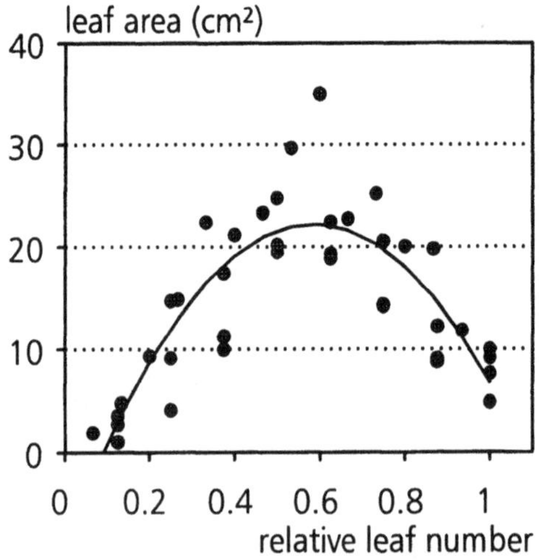


*Figure 3.* Distribution of leaf sizes of four vegetative *S. jacobaea* plants, with 8, 8, 15, and 8 leaves respectively. X-axis: relative leaf numbers (leaves numbered per plant, from young to old, and divided by the total number of leaves of that plant) to enable the comparison of equally aged leaves. Regression line: $y = -90x^2 + 105x - 8.7$, $R^2_{adj} = 0.70$.

relative leaf number: $y = -90x^2 + 105x - 8.7$, $R^2_{adj} = 0.70$, $P < 0.001$).

## Discussion

The results presented here are in agreement with a model by van Dam et al. (1996) and Iwasa & Kubo (1996). This model predicts, that leaves should be defended according to their value for the plant. This value is mainly determined by the expected returns in products of photosynthesis, which is highest for the youngest leaves. The more a leaf approaches the end of its lifetime, the less it should be defended. The distribution of PAs we found in *S. jacobaea* is clearly in agreement with this prediction. Although exact measurements of leaf lifetime and photosynthetic activity have not been made, we have observed the first leaves to die after ca. 14 weeks under lab conditions, and we assume photosynthetic activity to behave like the general trend, i.e., to decrease with leaf age (Mooney & Gulmon, 1982). The distribution of the damage by generalist herbivores then indicates that actual losses of expected profits are minimized.

It would be preferable to analyse PAs, soluble sugars and nitrogen, and feeding preference all on the very same plant material, but due to the amounts needed for analysis this was not feasible. Especially since the older leaves on many plants had been eaten completely, except for their main veins. However, the regressions indicate that the distributions of feeding stimulants and deterrents as shown in Figure 2 can be considered more generally applicable than on the four measured plants only.

The deterrence of *S. jacobaea* PAs was not further investigated in this paper. But already Bentley et al. (1984) and Dreyer et al. (1985) found *S. jacobaea* PAs to be deterrent to insect herbivores. And van Dam et al. (1995) found the structurally similar *Cynoglossum officinale* PAs to be deterrent against generalists. Therefore, we assume that *S. jacobaea* PAs are deterrent to generalist herbivores as well. Though differences in apparency between the leaves could have played a role, it is clear from Figure 3 that this is not a major factor, since the herbivores prefer smaller leaves to bigger ones. Another alternative explanation for the preference of both herbivores for old leaves over young leaves could be that lepidopteran larvae switch from high protein to high carbohydrate content in their diet when they reach the last instar (Reavey, 1993). The larvae used in the experiments were not final instars, so probably they still were limited by protein. Though sugar concentration is constant, nitrogen concentration decreases significantly with leaf age, and therefore the supposedly optimal C/N ratio for the herbivores is found in the younger leaves. We then have to conclude that they were forced to feed on sub-optimal parts of the plant, either by the deterrence of the PAs, or by another as yet unidentified cause.

## Acknowledgements

I would like to thank: Ine Derksen (Wageningen Agricultural University, Virology dept.) for *Spodoptera exigua* eggs, Leo Koopman and André Gidding (WAU, Entomology dept.) for *Mamestra brassicae* larvae, Grażyna Korbecka for developing the nitrogen and sugar determination methods, Cristina Caballero Garcia for the *Spodoptera* work, Ed van der Meijden and two anonymous referees for critically reading the manuscript.

## References

Aerts, R. J., A. Stoker, M. Beishuizen, I. Jaarsma, M. van de Heuvel, E. van der Meijden & R. Verpoorte, 1992. Detrimental effects of *Cinchona* leaf alkaloids on larvae of the polyphagous insect *Spodoptera exigua*. Journal of Chemical Ecology 18: 1955–1964.

Allen, S. E., 1974. Chemical Analysis of Ecological Materials. Blackwell Scientific Pubications, Oxford.

Bentley, M. D., D. E. Leonard, W. F. Stoddard & L. H. Zalkow, 1984. Pyrrolizidine alkaloids as larval feeding deterrents for spruce budworm, *Choristoneura fumiferana* (Lepidoptera: Tortricidae). Annals of the Entomological Society of America 77: 393–397.

Dam, N. M. van, R. Verpoorte & E. van der Meijden, 1994. Extreme differences in pyrrolizidine alkaloid levels between leaves of *Cynoglossum officinale*. Phytochemistry 37: 1013–1016.

Dam, N. M. van, L. W. M. Vuister, C. Bergshoeff, H. de Vos & E. van der Meijden, 1995. The 'Raison d'être' of pyrrolizidine alkaloids in *Cynoglossum officinale*: Deterrent effects against generalist herbivores. Journal of Chemical Ecology 21: 507–523.

Dam, N. M. van, T. J. de Jong, Y. Iwasa & T. Kubo, 1996. Optimal distribution of defences: are plants smart investors? Functional Ecology 10: 128–136.

Dreyer, D. L., K. C. Jones & R. J. Molyneux, 1985. Feeding deterrency of some pyrrolizidine, indolizidine, and quinolizidine alkaloids towards pea aphid (*Acyrthosiphon pisum*) and evidence for phloem transport of indolizidine alkaloid swainsonine. Journal of Chemical Ecology 11: 1045–1051.

Golterman, H. L., R. S. Clymo & M. A. M. Ohnstad, 1978. Methods for Physical and Chemical Analysis of Fresh Waters. 2nd ed., Blackwell Scientific, Oxford, pp. 99–101.

Hartmann, T. & M. Zimmer, 1986. Organ specific distribution and accumulation of pyrrolizidine alkaloids during the life history of two annual *Senecio* species. Plant Physiology 112: 67–80.

Hoy, C. W., G. P. Head & F. R. Hall, 1998. Spatial heterogeneity and insect adaptation to toxins. Annual Review of Entomology 43: 571–594.

Iwasa, Y. & T. Kubo, 1996. Optimal level of chemical defense decreasing with leaf age. Theoretical Population Biology 50: 124–148.

Krischik, V. A. & R. F. Denno, 1983. Individual, population, and geographical patterns in plant defense. In: R. F. Denno & M. S. McClure (eds.), Variable Plants and Herbivores in Natural and Managed Systems. Academic Press, New York, pp. 463–512.

Mattocks, A. R., 1967. Spectrophotometric determination of unsaturated pyrrolizidine alkaloids. Analytical Biochemistry 39: 443–448.

McKey, D., 1979. The distribution of secondary compounds within plants. In: G.A. Rosenthal & D. H. Janzen (eds.), Herbivores, Their Interaction with Secondary Plant Metabolites. Academic Press, New York, pp. 55–133.

Meijden, E. van der, 1996. Plant defence, an evolutionary dilemma: contrasting effects of (specialist and generalist) herbivores and natural enemies. Entomologia Experimentalis et Applicata 80: 307–310.

Mooney, H. A. & S. L. Gulmon, 1982. Constraints on leaf structure and function in reference to herbivory. BioScience 32: 198–206.

Raupp, M. J. & R. F. Denno, 1983. Leaf age as a predictor of herbivore distribution and abundance. In: R. F. Denno & M. S. McClure (eds.), Variable Plants and Herbivores in Natural and Managed Systems. Academic Press, New York, pp. 91–124.

Reavey, D., 1993. Why body size matters to caterpillars. In: N. E. Stamp & T. M. Casey (eds.), Caterpillars. Chapman & Hall, New York, pp. 29–91.

Simpson, S. J. & D. Raubenheimer, 1996. Feeding behaviour, sensory physiology and nutrient feedback: a unifying model. Entomologia Experimentalis et Applicata 80: 55–64.

Slansky, F., 1993. Nutritional ecology: the fundamental quest for nutrients. In: N. E. Stamp & T. M. Casey (eds.), Caterpillars. Chapman & Hall, New York, pp. 29–91.

Vrieling, K., H. de Vos, C. A. M. Wijk, 1993. Genetic analysis of the concentrations of pyrrolizidine alkaloids in *Senecio jacobaea*. Phytochemistry 32: 1141–1144.

Vrieling, K. & C. A. M. van Wijk, 1994. Estimating costs and benefits of the pyrrolizidine alkaloids of *Senecio jacobaea* under natural conditions. Oikos 70: 449–454.

Zangerl, A. R. & F. A. Bazzaz, 1992. Theory and pattern in plant defense allocation. In: R. S. Fritz & E. L. Simms (eds.), Plant Resistance to Herbivores and Pathogens. The University of Chicago Press, Chicago, USA, pp. 363–391.

*Entomologia Experimentalis et Applicata* **91:** 175–182, 1999.



# The population ecology of *Amorbus* Dallas (Hemiptera: Coreidae) species in Australia

Martin J. Steinbauer
*Co-operative Research Centre for Sustainable Production Forestry & CSIRO Div. of Entomology, GPO Box 1700, Canberra, ACT 2601, Australia*

Accepted: November 30, 1998

*Key words:* preference-performance linkage, life history theory, population dynamics, *Amorbus*, *Eucalyptus*

## Abstract

Species of *Amorbus* Dallas (Hemiptera: Coreidae) appear to exhibit a tight link between oviposition preference and nymphal performance: females apparently select vigorous plant modules to facilitate the development of their offspring. Such behaviour suggests that these insects should exhibit population dynamics consistent with the latent classification proposed by Price et al. (1990). While this empirical hypothesis is intuitively appealing it does not appear to incorporate current population dynamics theory. Berryman's ecodynamics (1992) offers the potential to describe a species' population dynamics which is grounded in a quantitative theoretical framework. The population ecology of *Amorbus* species from Australia is considered in relation to these hypotheses.

## Introduction

Insects which appear to impact upon and/or are seen in large numbers on trees being grown for commercial purposes are often termed 'pests' and thus are considered as warranting some form of population management. In Australia such decisions have, in some instances, been reached without having a basic understanding of a species' biology (see Clarke et al., 1998) and/or the benefit of long-term survey records (see Nothnagle & Schultz, 1987) which could allow a more judicious appraisal of their potential threat. An example of one such group of forest insects are the coreid bugs of the genus *Amorbus*.

*Amorbus* species feed exclusively on growing shoots causing a characteristic necrotic wilt. Species from temperate Australia are typically univoltine, overwintering occurs as adults. The timing of the seasonal activity of *A. obscuricornis* adults generally coincides with the onset of new season's shoot extension and leaf development in eucalypts (Steinbauer, 1997a, 1998). Thus, oviposition appears to occur at a time when newly emerged nymphs are most likely to find shoots of optimal quality (for details concerning eucalypt quality, see Steinbauer et al., 1998a). Adults and nymphs are often seen together on hosts, especially coppicing hosts. Thus, parent and offspring generations are coincident within a season. Competition for oviposition sites by females is considered unlikely on the basis of field observations (Steinbauer & Clarke, 1998; M. J. Steinbauer, pers. obs.). The specificity of these native Australian insects for species of *Eucalyptus* (Myrtaceae) has brought them to the attention of foresters following the increase in the area of eucalypt plantations. In this synthesis, recent research into their population ecology is considered in comparison to the 'latent' *versus* 'eruptive' dichotomy developed by Price et al. (1990) and 'ecodynamics' theory of Berryman (1992). In order to demonstrate how similar *Amorbus* species are to other rare herbivores the existing literature is summarised in relation to key aspects of Price's hypothesis, and new data concerning the type of plant module they prefer and the impact of these insects upon their hosts are also presented. The relevance of both works to studies of Australian forest insects is considered.

## Price: 'preference-performance linkage hypothesis'

Price et al. (1990) proposed that evolved biological traits determine a species present day ecology (hypothesis further elaborated in Price, 1994a, 1996). Specifically, phylogenetic constraints determine the adaptive syndromes exhibited by a species. The emergent properties arising from specific phylogenetic constraints and adaptive syndromes are the ecological interactions and population dynamics exhibited by a species. Price illustrates his hypothesis by comparing the ecology of species which exhibit so-called latent (rare or uncommon species) or eruptive (outbreak species) population dynamics. Basically, rare species are those for which there is a strong preference on the part of females to provide their offspring with hosts of high plant quality/vigour. Such behaviour arises because the characteristics of the species necessitates selection of such hosts. Correspondingly, the performance of the offspring is dictated by the quality of the host chosen for them. As a result of this choice the species tends to exhibit certain intraspecific and interspecific interactions. In contrast, species with eruptive population dynamics are not thought to possess such a tight linkage between female preference and offspring performance. This approach is considered 'evolutionary and explanatory and predictions on population dynamics can be made before any population data are collected based on life history traits' (P. Price, pers. comm.).

## Berryman: 'ecodynamics theory'

Ecodynamics (Berryman, 1992) considers fluctuations in the numbers of forest insects in the context of population dynamics theory. Changes in the sizes of populations are explained using a suite of elemental laws (Berryman, 1991, 1997). The models described can incorporate the influences of biotic (e.g., host plants, predators) and abiotic (e.g. climate) environmental factors. Berryman (1987, 1997) classifies insect outbreaks as either exhibiting 'gradient' or 'eruptive' dynamics. Therefore, Berryman's eruptive classification is more restrictive than that of Price as it only includes metastable species. Significantly, most of the insects Berryman has considered are not rare species.

## *Amorbus* in relation to the preference-performance linkage hypothesis

*Plant vigour and host utilisation.* Steinbauer et al. (1998b) have demonstrated that coppicing eucalypts are more attractive to *Amorbus* species than non-coppicing eucalypts. Typically, coppice is not only of higher nutritional quality (i.e., higher water content and total N and lower C:N ratio) but also grows more vigorously (Steinbauer, 1995; M. J. Steinbauer, pers. obs.). Depending upon the species the apparent differences in growth rate can be huge (Figure 1) (for a justification of the use of internode length to quantify growth rate see Wareing & Phillips, 1981). While there are significant differences in the terpene composition of juvenile and adult foliage (i.e., coppice and non-coppice foliage, respectively) (Li et al., 1994, 1995, 1996), the differential oligophagy exhibited by species of *Amorbus* (Steinbauer, 1995, 1997b) is considered to indicate that such compounds may not significantly influence utilisation of coppice but that other constituents of the coreid diet determine the response of these insects to such vegetation, for example, changes in essential amino acids (Brodbeck & Strong, 1987). This possibility has also been raised by Schaefer & Mitchell (1983) for the superfamily as a whole.

Assays of oviposition host preference under controlled conditions with species of *Amorbus* are considered to be inherently difficult because of apparent changes in behaviour in the laboratory environment (Steinbauer, 1995). Thus, evidence concerning the possible role of host quality on oviposition must be obtained by other means. For example, during 3 years of field surveying the only first instar nymph of *A. obscuricornis* to be found was located on a coppicing host (Steinbauer, 1997a) while the only eggs of an *Amorbus* species ever found *in vivo* were located on the leaves of another coppicing host (M. J. Steinbauer, pers. obs.). Both these records suggest that females oviposit on hosts with vigorous shoots. Such a suggestion is further supported by the apparently sedentary nature of most nymphs (Steinbauer & Clarke, 1998).

That weather can significantly influence the development of eucalypt shoots has been shown by Specht & Brouwer (1975). In addition, shoot development can be influenced by other environmental factors, e.g., fire (Noble, 1982; Christensen & Abbott, 1989) (see also Stein et al., 1992). The influence of weather and fire on the quality of eucalypt shoots depends upon spatial and temporal occurrences of events and plants

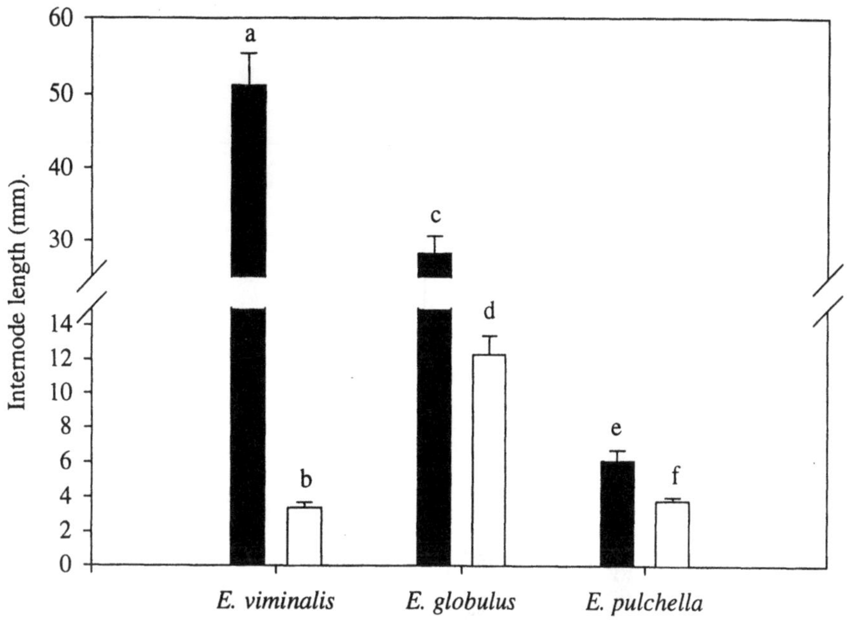


*Figure 1.* Internode lengths of coppicing ■ and non-coppicing □ *Eucalyptus* species. (Letters indicate differences between means are significantly different at $P < 0.05$, one-way ANOVA).

and results in high quality shoots being typically rare throughout the entire distribution of any particular *Amorbus* species. For example, the frequency of fire in mallee (i.e., *Eucalyptus* spp.) communities can range from 5 to 20 years between events (Hodgkinson & Griffin, 1982).

The tip wilting habit of *Amorbus* species can impact upon plant architecture (Steinbauer et al., 1997) and vegetative production (Figure 2). Despite being capable of inflicting such damage host plant mortality due to the feeding activities of these insects has never been recorded (Green, 1972; Bashford, 1993; Neumann, 1993; Stone, 1993; Steinbauer, 1995).

*Natural enemies.* There are a variety of parasitoids recorded from the eggs of *Amorbus* species (I.D. Naumann & M. J. Steinbauer, unpubl.). Rates of egg parasitisation vary considerably. For example, there was no parasitisation of the eggs of *A. obscuricornis* when placed in the field whereas approximately 53% of field collected eggs of *A. biguttatus* were parasitised (Steinbauer & Clarke, 1995; M. J. Steinbauer, unpubl.). Predation and parasitisation of nymphal and adult stages has not been commonly observed (Steinbauer & Clarke, 1995; Steinbauer, 1996), possibly because of species characteristic defensive secretions (Steinbauer & Davies, 1995). Predation by birds is not considered likely given that most nymphal stages are also aposematically coloured (Steinbauer, 1995). Thus, functional and numerical responses by predators and parasites to *Amorbus* species nymphs and adults appear weak; however, this may not be the case for the eggs. The importance of pathogen and/or viral epizootics to the population dynamics of *Amorbus* species is not known but is suspected as being of little or no significance.

*Population dynamics.* Fluctuations in populations of large amplitude (Price et al., 1990) have not been recorded for species of *Amorbus* (Green, 1972; Bashford, 1992; Steinbauer, 1997a). Typically, *Amorbus* species appear to maintain steady population densities; however, the existing data may only reflect average density and not reveal the possible scale of differences between high and low abundance occasions.

## A synthesis of Price & Berryman in relation to *Amorbus*

On the basis of the above evidence, a comparison of *Amorbus* species with the dichotomy presented in Price et al. (1990) would suggest that these forest herbivores possess traits more closely aligned with those referred to as rare herbivores than with so-called outbreak species. In particular, for species of *Amorbus* there appears to be a tight linkage between oviposition preference and nymphal performance. Inclusion

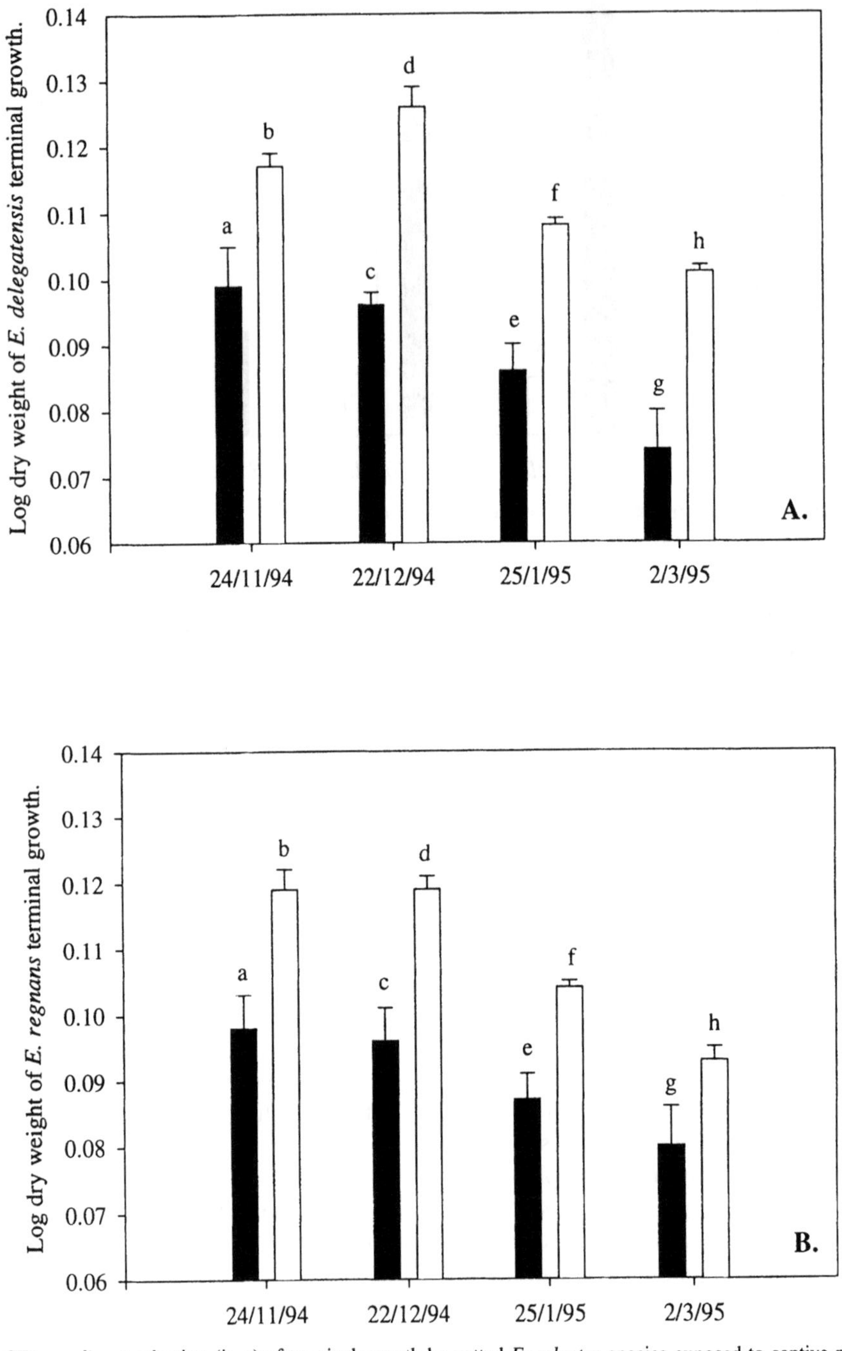


*Figure 2.* Mean (±SE) *per diem* production (in g) of terminal growth by potted *Eucalyptus* species exposed to captive populations of *A. obscuricornis* ■ ($n = 8$ plants each species) *versus* production by insect-free plants □. A. *E. delegatensis* ($n = 4$ plants). B. *E. regnans* ($n = 10$ plants). (Letters indicate differences between means are significantly different at $P < 0.05$, one-way ANOVA).

of *Amorbus* in the suite of insects exhibiting latent population dynamics would provide a non-galling life history example (Price et al., 1990). That not all galling insects attack vigorous shoots has, however, been demonstrated by de Bruyn (1995). That this hemipteran should fit with Price's latent classification is perhaps not surprising given the number of other Hemiptera exhibiting similar traits (Price et al., 1990; Price, 1994b). It would appear that the hemipteran mode of feeding (a phylogenetic trait) may result in them exhibiting characteristics which increase the likelihood of their being included in Price's latent classification. All Hemiptera possess elongated sucking mouth parts but there are quite different mechanisms for removing plant material in each group, e.g., osmotic pump in coreids, stylet sheath in aphids, lacerate and flush in lygaeids and pentatomids and macerate and flush in mirids (Miles, 1972; Miles & Taylor, 1994). How such diverse mechanisms might interact with other behaviours to result in some hemipterans exhibiting latent population dynamics remains to be addressed.

While the empirical model proposed for rare forest insects by Price describes well the population ecology of *Amorbus* species, Berryman suggests that fluctuations in numbers of such insects can be described within the current framework of population dynamics theory and does not represent a 'new paradigm'. Berryman suggests that the population dynamics exhibited by latent species such as these can be explained using a single law, i.e., competition (in the case of Price's stem-galling insects it is competition between ovipositing females and/or developing larvae) (Berryman, 1997). In proffering this explanation Berryman deviates from previous approaches to analyses of insect population dynamics, i.e., sample, analyse for apparent trends and infer a causal mechanism (N. B. Berryman indicated that essential data does not exist to derive necessary parameters to test this model). If, by knowing a little of the life history of a given species (e.g., body size, fecundity, etc. (see Stearns, 1992)), we could begin to model insect population dynamics using the laws outlined by Berryman then it appears possible to quantify the outcome of the different mechanisms which Price proposes influence 'latent' and 'eruptive' species.

Mechanistic explanations combined with population dynamics theory might facilitate prediction of the response of populations of insects to changes in their environment. For example, an unresolved question pertaining to *Amorbus* species is that concerning the response of these insects to the appearance of coppicing hosts. Ecodynamics would suggest that such a change should increase the carrying capacity of the habitat for these insects (Figure 3). Whether the attainment of a higher carrying capacity is achieved with no change to the maximum rate of increase or could be achieved because the moisture content of the shoots changes water availability in the coreid's environment requires further investigation (Figure 3) (see discussion in Leather, 1994). According to ecodynamics the linear $R$-function (or elemental logistic equation, Figure 3) is the most simplistic model for population change with negative feedback and does not take into account certain behaviours which might influence the form of the relationship. The relevance of such a model to species of *Amorbus* would require investigation of the significance of time delays (due to the response of density-dependent factors), territorial behaviour and under-population effects (Berryman, 1987, 1992). Time delays are likely given that certain *Amorbus* species have discrete generations and that ontogenetic development of shoots appears to have the potential to make coppice less favourable ($-ve$ feedback) while resource regulation (Steinbauer et al., 1998b) could make hosts more favourable ($+ve$ feedback). In this system the role of natural enemies and/or microorganisms is considered minimal. Territorial behaviour between males is considered likely in *Amorbus* species while the possibility of significant under-population effects (i.e., Allee effects) does not seem likely if attraction of migratory females by male pheromones is demonstrated (Steinbauer & Clarke, 1998). Hunter (1997) has proposed a similar model to describe the relationship between plant quality and herbivore density that incorporates typical population dynamics concepts. Similar models were first proposed by Southwood & Comins (1976) (see Price, 1997) using the $r$ and $K$-selection hypothesis rejected by Stearns (1992). While for much of the time *Amorbus* species might be considered rare, their occasional population increases might be likened to those of 'gradient' outbreak species using the classification scheme given in Berryman (1987). Whether *Amorbus* species exhibit 'sustained', 'pulse' or 'cyclical' gradient population fluctuations would appear to depend on what factor in their environment they are influenced by at the time, e.g., coppice, plantations of same age/species hosts or changes in shoot production due to weather conditions.

Wallner (1987) noted that insects have a variety of mechanisms facilitating their survival and that there

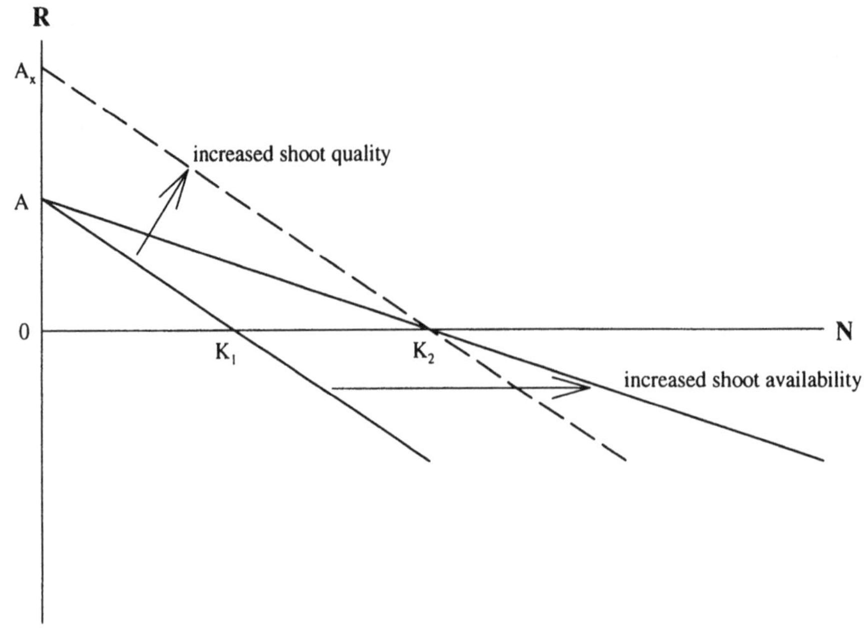


*Figure 3.* Linear Reproduction-function ($R$) for a hypothetical population of an *Amorbus* species illustrating how changes in resource quantity (solid line) and/or quality (dashed line) might influence the carrying capacity ($K$) or the maximum per-capita rate of change ($A$), respectively. ($R = \log_e(1 + B - D)$, where $B$ = the per-capita birth rate and $D$ = the per-capita death rate; $N$ is density; after Berryman, 1992.)

was no one trait of more importance in this context than any other – each species has evolved a unique suite of responses to the environment in which it lives. Based on my studies to-date I am inclined to favour this conclusion. In the case of Australian forestry, the theories of both Price and Berryman appear to have relevance to understanding the insects affecting eucalypt plantations. In the short-term the relevance of Price's ideas to a particular insect could be addressed by controlled, laboratory assays that attempt to determine the importance of foliage quality to oviposition and immature performance (see Patterson et al., 1996; Steinbauer et al., 1998a) and/or dispersal ability of offspring. Given that some Australian *Eucalyptus* herbivores are not affected by secondary chemistry (Patterson et al., 1996), host resistance through selection for leaf maturation rate and/or degree of sclerophylly could prove to be an effective practice to reduce insect outbreaks (see Berryman, 1987). In the medium to long-term, Berryman's ideas should be implemented by means of routine population monitoring protocols for each forest insect considered of relevance to the industry (Steinbauer, 1997c). Such a protocol should be simple to perform and conducted systematically both spatially and temporally to begin obtaining information that can in due course be utilised to model the population dynamics of the species (Berryman, 1994). Ultimately, some combination of the two author's ideas might allow prediction of the likelihood that a particular forest insect will exhibit latent or outbreak dynamics thereby facilitating a considered assessment of the inherent pestiferous tendency of that species. If it is not possible to combine the ideas of Price and Berryman then perhaps pest risk analyses should make more use of life history theory to achieve this aim (Stearns, 1992; Nylin & Gotthard, 1998). Pest risk analysis using life history theory and/or a preference-performance linkage/ecodynamics approach would at least provide a more objective means of determining which particular insect species should be the focus of detailed investigation in a climate where funding is typically limited.

## Acknowledgements

I thank Stephen Paterson (CRC for Sustainable Production Forestry) and John Dowse and Rex Sutherland (CSIRO Entomology) for their assistance; Prof. Alan Berryman (Washington State University) and Dr Anne Kitchener (Tasmanian Parks & Wildlife Service) for relevant discussions; Drs Mamoru Matsuki (Australian National University & CSIRO Entomology), Penny Gullan (Australian National University), Sören Nylin (Stockholm University), Prof. Peter Price

(Northern Arizona University) for comments on the manuscript and Tony Clarke (Griffith University) for his input throughout my studies.

## References

Bashford, R., 1992. Observations on the life history of two sap-feeding coreid bugs and their impact on growth of plantation eucalypts. Tasforests 4: 87–93.

Bashford, R., 1993. Insect pest problems of eucalypt plantations in Australia. 4. Tasmania. Australian Forestry 56: 375–377.

Berryman, A. A., 1987. The theory and classification of outbreaks. In: P. Barbosa & J. C. Schultz, (eds), Insect Outbreaks. Academic Press, San Diego; pp. 3–30.

Berryman, A. A., 1991. Population theory: an essential ingredient in pest prediction, management, and policy-making. American Entomologist 37: 138–142.

Berryman, A. A., 1992. Ecodynamics: Understanding, diagnosing and forecasting ecological change. Unpublished Course Notes, Washington State University, Pullman.

Berryman, A. A., 1994. Population dynamics: forecasting and diagnosis from time-series. In: S. R. Leather, A. D. Watt, N. J. Mills & K. F. A. Walters (eds), Individuals, Populations and Patterns in Ecology. Intercept, Andover; pp. 119–128.

Berryman, A. A., 1997. On the principles of population dynamics and theoretical models. American Entomologist 43: 147–151.

Brodbeck, B. & D. Strong, 1987. Amino acid nutrition of herbivorous insects and stress to host plants. In: P. Barbosa & J. C. Schultz (eds), Insect Outbreaks. Academic Press, San Diego; pp. 347–364.

Bruyn, L. de, 1995. Plant stress and larval performance of a dipterous gall former. Oecologia 101: 461–466.

Christensen, P. & I. Abbott, 1989. Impact of fire on the eucalypt forest ecosystem of southern Western Australia: a critical review. Australian Forestry 52: 103–121.

Clarke, A. R., S. C. Paterson & M. J. Steinbauer, 1998. Larval trapping as a means of sampling Paropsina (Coleoptera: Chrysomelidae: Chrysomelinae) from *Eucalyptus* forests. The Coleopterists Bulletin 52: 253–258.

Green, L. R., 1972. The biology of coreid bugs feeding on eucalypts and their interaction with the host plant. B.Agr.Sc.(Honours) Thesis, University of Tasmania, Hobart.

Hodgkinson, K. C. & G. F. Griffin, 1982. Adaptation of shrub species to fires in the arid zone. In: W. R. Barker & P. J. M. Greenslade, (eds), Evolution of the Flora and Fauna of Arid Australia. Peacock Publications, Adelaide; pp. 145-152.

Hunter, M. D., 1997. Incorporating variation in plant chemistry into a spatially explicit ecology of phytophagous insects. In: A. D. Watt, Stork N. E. & M. D. Hunter (eds), Forests and insects. Chapman & Hall, London: 81–96.

Leather, S. R., 1994. Life history traits of insect herbivores in relation to host quality. In: E. A. Bernays (ed.), Insect-Plant Interactions, Vol. V. CRC Press, Boca Raton; pp. 175–207.

Li, H., J. L. Madden & N. W. Davies, 1994. Variation in leaf oils of *Eucalyptus nitens* and *E. denticulata*. Biochemical Systematics and Ecology 22: 631–640.

Li, H., J. L. Madden & B. M. Potts, 1995. Variation in volatile leaf oils of the Tasmanian *Eucalyptus* species-1. Subgenus *Monocalyptus*. Biochemical Systematics and Ecology 23: 299–318.

Li, H., J. L. Madden & B. M. Potts, 1996. Variation in volatile leaf oils of the Tasmanian *Eucalyptus* species-2. Subgenus *Symphyomyrtus*. Biochemical Systematics and Ecology 24: 547–569.

Miles, P. W., 1972. The saliva of Hemiptera. Advances in Insect Physiology 9: 183–255.

Miles, P. W. & G. S. Taylor, 1994. 'Osmotic pump' feeding by coreids. Entomologia Experimentalis et Applicata 73: 163–173.

Neumann, F. G., 1993. Insect pest problems of eucalypt plantations in Australia. 3. Victoria. Australian Forestry 56: 370–374.

Noble, J. C., 1982. The significance of fire in the biology and evolutionary ecology of mallee *Eucalyptus* populations. In: W. R. Barker & P. J. M. Greenslade (eds), Evolution of the Flora and Fauna of Arid Australia. Peacock Publications, Adelaide; pp. 153–159.

Nothnagle, P. J. & J. C. Schultz, 1987. What is a forest pest? In: P. Barbosa & J. C. Schultz, (eds), Insect Outbreaks. Academic Press, San Diego; pp. 59–80.

Nylin, S. & K. Gotthard, 1998. Plasticity in life-history traits. Annual Review of Entomology 43: 63–83.

Patterson, K. C., A. R. Clarke, C. A. Raymond & M. P. Zalucki, 1996. Performance of first instar *Chrysophtharta bimaculata* larvae (Coleoptera: Chrysomelidae) on nine families of *Eucalyptus regnans* (Myrtaceae). Chemoecology 7: 94–106.

Price, P. W., 1994a. Patterns in the population dynamics of insect herbivores. In: S. R. Leather, A. D. Watt, N. J. Mills & K. F. A. Walters (eds), Individuals, Populations and Patterns in Ecology. Intercept, Andover; pp. 109–117.

Price, P. W., 1994b. Phylogenetic constraints, adaptive syndromes, and emergent properties: from individuals to population dynamics. Researches on Population Ecology 36: 3–14.

Price, P. W., 1996. Empirical research and factually based theory: what are their roles in entomology? American Entomologist 42: 209–214.

Price, P. W., 1997. Insect Ecology, 3rd ed. John Wiley & Sons, Inc., New York.

Price, P. W., N. Cobb, T. P. Craig, G. W. Fernandes, J. K. Itami, S. Mopper & R. W. Preszler, 1990. Insect herbivore population dynamics on trees and shrubs: New approaches relevant to latent and eruptive species and life table development. In: E. A. Bernays (ed.), Insect-Plant Interactions, Vol. II, CRC Press, Boca Raton; pp. 1–38.

Schaefer, C. W. & P. L. Mitchell, 1983. Food plants of the Coreoidea (Hemiptera: Heteroptera). Annals of the Entomological Society of America 76: 591–615.

Southwood, T. R. E. & H. N. Comins, 1976. A synoptic population model. Journal of Animal Ecology 45: 949–965.

Specht, R. L. & Y. M. Brouwer, 1975. Seasonal shoot growth of *Eucalyptus* spp. in the Brisbane area of Queensland (with notes on shoot growth and litter fall in other areas of Australia). Australian Journal of Botany 23: 459–474.

Stearns, S. C., 1992. The Evolution of Life Histories. Oxford University Press, Oxford.

Stein, S. J., P. W. Price, W. G. Abrahamson & C. F. Sacchi, 1992. The effect of fire on stimulating willow regrowth and subsequent attack by grasshoppers and elk. Oikos 65: 190–196.

Steinbauer, M. J., 1995. The biogeography and host plant utilisation of eucalypt feeding Coreidae (Hemiptera: Heteroptera). PhD Thesis, University of Tasmania, Hobart.

Steinbauer, M. J., 1996. A note on manna feeding by ants (Hymenoptera: Formicidae). Journal of Natural History 30: 1185–1192.

Steinbauer, M. J., 1997a. Seasonal phenology and developmental biology of *Amorbus obscuricornis* (Westwood) and *Gelonus tasmanicus* (Le Guillou) (Hemiptera: Coreidae). Australian Journal of Zoology 45: 49–63.

Steinbauer, M. J., 1997b. The incidence and relative abundance of *Amorbus obscuricornis* (Westwood) and *Gelonus tasmanicus* (Le

Guillou) (Hemiptera: Coreidae) in southern Tasmania and their performance on selected *Eucalyptus* species. Australian Journal of Zoology 45: 631–649.

Steinbauer, M. J., 1997c. Notes on *Psaltoda moerens* (Germar) (Hemiptera: Cicadidae) in Tasmania. Australian Entomologist 24: 169–171.

Steinbauer, M. J., 1998. Seasonal fluctuations in body weight, lipid content and the starvation-longevity of *Amorbus obscuricornis* (Westwood) and *Gelonus tasmanicus* (Le Guillou) (Hemiptera: Coreidae). Australian Journal of Entomology 37: 90–96.

Steinbauer, M. J. & A. R. Clarke, 1995. *Xenoencyrtus hemipterus* (Girault) (Hymenoptera: Encyrtidae), an egg parasitoid of Coreidae (Hemiptera) in Tasmania. Journal of the Australian Entomological Society 34: 63–64.

Steinbauer, M. J. & N. W. Davies, 1995. The defensive secretions of *Amorbus obscuricornis* (Westwood), *A. rubiginosus* (Guérin-Méneville) and *Gelonus tasmanicus* (Le Guillou) (Hemiptera: Coreidae). Journal of the Australian Entomological Society 34: 75–78.

Steinbauer, M. J. & A. R. Clarke, 1998. Field observations of dispersion, mating and development of *Amorbus obscuricornis* (Westwood) (Hemiptera: Coreidae). Australian Journal of Entomology 37: 155–157.

Steinbauer, M. J., G. S. Taylor & J. L. Madden, 1997. Comparison of damage to *Eucalyptus* caused by *Amorbus obscuricornis* and *Gelonus tasmanicus*. Entomologia Experimentalis et Applicata 82: 175–180.

Steinbauer, M. J., A. R. Clarke & J. L. Madden, 1998a. Oviposition preference of a *Eucalyptus* herbivore and the importance of leaf age on interspecific host choice. Ecological Entomology 23: 201–206.

Steinbauer, M. J., A. R. Clarke & S. C. Paterson, 1998b. Changes in eucalypt architecture and the foraging behaviour and development of *Amorbus obscuricornis* (Hemiptera: Coreidae). Bulletin of Entomological Research 88: 641–651.

Stone, C., 1993. Insect pest problems of eucalypt plantations in Australia. 2. New South Wales. Australian Forestry 56: 363–369.

Wallner, W. E., 1987. Factors affecting insect population dynamics: differences between outbreak and non-outbreak species. Annual Review of Entomology 32: 317–340.

Wareing, P. F. & I. D. J. Phillips, 1981. Growth and Differentiation in Plants, 3rd ed. Pergamon Press, Oxford.

*Entomologia Experimentalis et Applicata* **91:** 183–186, 1999.



# A galling aphid furnishes its home with a built-in pipeline to the host food supply

David Wool[1], Roni Aloni[2], Ofra Ben-Zvi[1] & Miriam Wollberg[1]
[1]*Department of Zoology and* [2]*Department of Plant Sciences, Tel Aviv University, Tel Aviv 69978, Israel*

Accepted: November 26, 1998

*Key words:* aphids, galls, tissue organisation, *Pistacia*, *Geoica wertheimae*

## Abstract

Galling insects modify the developmental pathways of their host plants and create a protected and favourable microhabitat for their offspring. Galling aphids reproduce in their galls and the resulting clones often comprise hundreds of individuals. We followed the histological changes in the host *Pistacia palaestina* (Anacardiaceae) induced by the galling aphid *Geoica wertheimae* (Aphidoidea, Pemphigidae: Fordinae). We showed that the leaf tissues are altered in the gall in a way that gives the aphids easy access to the vascular system from which they obtain their nourishment. Specifically, the cuticle that lines the normal leaf epidermal cells is absent on the surface of the inner gall cavity, and the normal palisade cells are replaced by parenchymatous tissue with numerous wide latex ducts. Associated with these ducts, many new sieve tubes (phloem elements) are formed just a few cell layers below the inner gall surface. This arrangement enables the hundreds of aphids to feed simultaneously and continuously in the gall throughout the summer.

## Introduction

Jones et al. (1997) describe organisms as 'physical ecosystem engineers' if they are 'organisms that, directly or indirectly, control the availability of resources to other organisms by causing physical state changes in biotic or abiotic materials'. There is no doubt that gall-forming insects can be described as physical ecosystem engineers. They induce the formation of a completely novel structure which houses not only their own offpsring, but often also parasites, predators and inquilines belonging to other species, genera or families.

There are several possible advantages of living in galls (Price et al., 1987), and perhaps there is some benefit to the host plant as well (e.g., Williams & Whitham, 1986). Whatever the other benefits, certainly the primary function of a gall is to guarantee the supply of food to the offspring of the causative insect. In cynipid (Hymenoptera) and cecidomyiid (Diptera) galls, a special layer of nutritive cells is formed, and the insect larvae feed on these cells (see recent examples in Shorthouse & Rohfritsch (1992)). Aphid galls differ from hymenopteran and dipteran galls in that aphids do not consume the gall tissue. Aphids are phloem-feeders and obtain their nourishment from the plant vascular system by reaching the phloem with their stylets. Histological studies on aphid galls are rare, and we are unaware of any except for Rohfritsch (1981) and Wool & Bar-El (1995).

In Israel approximately 15 species of aphids (Homoptera: Pemphigidae, Fordinae) induce galls on *Pistacia* trees (Anacardiaceae) (Koach & Wool, 1977; Wool, 1995). For this study we selected the galls induced by *Geoica wertheimae* Brown and Blackman on its primary host, *Pistacia palaestina* (Figure 1). The ecology of this species was recently described by Wool & Ben Zvi (1998).

A single fundatrix induces each gall in the spring; the fundatrix and her daughters reproduce parthenogenetically in the gall during the summer and the mature, winged aphids disperse in the fall when the galls crack open. The fundatrix nymph settles on the upper surface of a leaflet near the midvein (Figure 1). Its probing and/or feeding causes a small depression which soon deepens and encloses the aphid. On the

*Figure 1.* A young gall of *Geoica wertheimae* (arrow) on a leaf of its host tree, *Pistacia palaestina*. (The other, unmarked gall is induced by another aphid species, *Forda formicaria* von Heyden). Bar = 10 mm. Photo: A. Shoob.

underside of the leaflet the gall appears as a small globular protuberance, which increases in size during the season to 1–3 cm diameter, while the aphid clone size increases to several hundred individuals.

Here we address the questions, how do the aphids reach the phloem from within the gall and how can the many inhabitants feed together in the small space of the gall?

## Materials and methods

Galls were collected at Canada Park (about 30 km east of Tel Aviv) beginning in early April, 1997 when the first galls appeared. The galls were fixed for 48 h in formalin-aceto-alcohol (FAA) and then transferred to 70% ethanol. In preparation for the histological work, tissue samples were embedded in butanol, paraffin blocks were prepared and serial sections, 10 $\mu$m thick, were made with a rotary microtome. The sections were stained with Safranin and Fast Green (Conn, 1943-4). A special method was used to stain callose, which occurs in the phloem elements (Aloni & Sachs,1973; Aloni et al., 1995).

## Results

Examination of the tissues at the junction of the gall to the leaf revealed that the gall was tightly closed, although the epidermal cells on both sides were not fused. It is easy to convince oneself by following the tissues in the sections that the inner layer of the gall cavity is continuous with the *upper* leaflet surface. However, the gall tissues are radically different from the leaf tissue. The gall wall is 10 times thicker than the leaf (mean leaf thickness 0.19 mm; gall wall 1.6–1.9 mm; data from Wool & Ben Zvi, 1998). The cuticle which normally covers the ungalled leaf monolayer epidermis cannot be seen in the inner gall surface, and the normal, structurally distinct palisade tissue is likewise absent. The gall walls are composed of a mass of parenchymatous cells. (Figure 2). A large number of wide latex ducts is characteristic of the gall tissue. Perhaps the most important phenomenon is the appearance of a second, seemingly organized layer of latex ducts close to the gall cavity.

A detailed examination at higher magnification and a special staining technique which stains callose – a polysaccharide characteristic of sieve elements of the phloem – showed that the tissue surrounding each latex duct is phloem, with regular xylem elements developing away from the duct. Mentally transforming the situation in three dimensions, we may visualize the aphids as in a barred cage, the bars of which – thinly covered by five to six cell layers – are really feeding tubes from which they obtain nutrition (Figure 3). Moreover, sections of the thickened petiole below the galled leaflet reveal that many new vascular elements lead into the galled leaflet (the structure of the xylem and bark of petiole near a galled leaflet is asymmetrical, biased towards the leaflet carrying the gall!).

## Discussion

The induction of the gall causes a change in the design of the plant tissue: the clone of aphids in the central cavity finds easy access to the phloem in the surrounding internal gall walls, which are covered with an envelope of phloem (and xylem) elements very near the inner gall surface.

The nutritional materials arrive from the leaf in those sieve tubes that pass through the petiole and are then subdivided into a multitude of phloem elements near the inner surface of the gall. This is what a plumb-

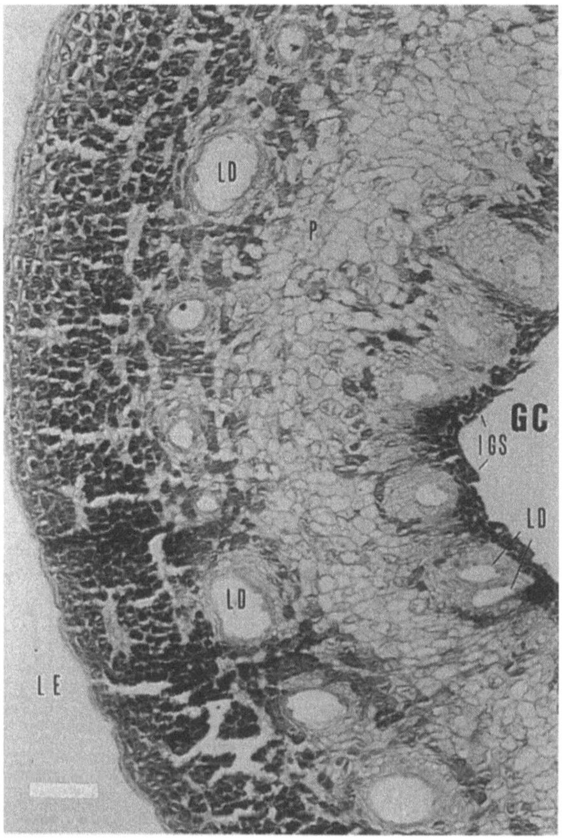


*Figure 2.* Cross section of the wall of a mature gall of *G. wertheimae*. The inner gall surface is equivalent to the upper surface of the leaf, but there is no trace of the characteristic monolayer epidermis or the palisade tissue. A small number of loose cell layers separate the aphid inhabitants from the closely-adjacent latex ducts, which are surrounded by phloem elements from which the aphids may feed. (The lower epidermis and the associated tissues are structurally similar to the lower leaf surface). LE = lower cuticle and epidermis. P = parenchymatous cells. GC = gall cavity. IGS = inner gall surface. LD = latex ducts with associated phloem. Bar = 0.2 mm.

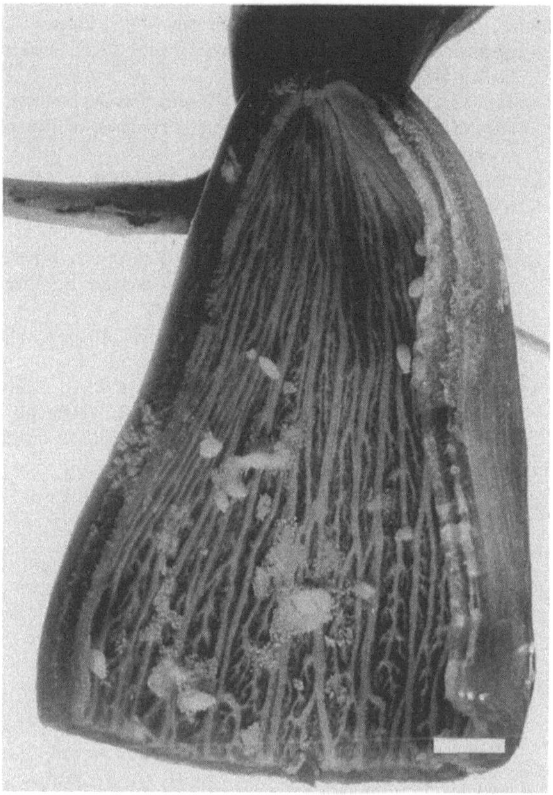

*Figure 3.* A mature *Geoica wertheimae* gall cut open to show the arrangement of the latex ducts (after fixation in FAA the thin cover of cells shrinks and the ducts are clearly visible). Some aphids can be seen in the fixed material with their mouthparts still inserted into the feeding tubes. Bar = 5 mm.

ing engineer would have designed if he had to plan a feeding system for a multitude of customers from a single pipeline. Although no trace of stylet tracks were found in the sectioned material to confirm the feeding site, when galls were fixed (Figure 3) many of their occupants were left hanging with their mouthparts inserted into the latex duct regions. Moreover, the gall is a sink (Burstein et al., 1994; Inbar et al., 1995; Larson & Whitham, 1991, 1997) which increases the flow of nutrients into the gall, improving the efficiency of the distribution system.

We show that although the aphids do not consume the gall tissue, they induce changes in tissue organization of the plant vascular system that facilitate their access to the food supply of the host tree. Together with the strong sink effect of the gall, these changes enable hundreds of individuals to feed for several months in a small, confined space.

## References

Aloni, R. & T. Sachs, 1973. The three-dimensional structure of primary phloem systems. Planta 113: 345–353.

Aloni, R., K. S. Pradel & C. I. Ullrich, 1995. The three-dimensional structure of vascular tissues in *Agrobacterium tumefaciens*-induced crown galls and in the host stems of *Ricinus communis* L. Planta 196: 597–605.

Burstein, M., D. Wool & A. Eshel, 1994. Sink strength and clone size of sympatric, gall-forming aphids. European Journal of Entomology 91: 57–61.

Conn, H. J., 1943–44. Staining procedures used by the Biological Stain Commission. Geneva, New York.

Inbar, M., A. Eshel & D. Wool, 1995. Interspecific competition among phloem-feeding insects mediated by induced host plant sinks. Ecology 76: 1506–1515.

Jones, C. G., J. H. Lawton & M. Shachak, 1997. Positive and negative effects of organisms as physical ecosystem engineers. Ecology 78: 1946–1957.

Koach, J. & D. Wool, 1977. Geographic distribution and host specificity of gall-forming aphids (Homoptera: Fordinae) on *Pistacia* trees in Israel. Marcellia 40: 207–216.

Larson, K. C. & T. G. Whitham, 1991. Manipulation of food resources by a gall aphid, the physiology of sink-source interactions. Oecologia 88: 12–21.

Larson, K. C. & T. G. Whitham, 1997. Competition between gall aphids and natural plant sinks: plant architecture affects resistance to galling. Oecologia 109: 575–582.

Price, P. W., G. W. Fernandes & G. L. Waring, 1987. Adaptive value of insect galls. Environmental Entomology 16: 15–24.

Rohfritsch, O., 1981. A 'defense' mechanism of *Picea excelsa* L. against the gall former *Chermes abietis* L. (Homoptera, Adelgidae). Zeitschrift für Angewandte Entomologie 92: 18–26.

Shorthouse, J. D. & O. Rohfritsch, 1992. Biology of Induced Galls. Oxford University Press, Oxford.

Williams, A. G. & T. G. Whitham, 1986. Premature leaf abscission: an induced plant defense against gall aphids. Ecology 67: 1619–1627.

Wool, D., 1995. Aphid induced galls on *Pistacia* in the natural forest of Israel: which, where, and how many? Israel Journal of Zoology 41: 591–600.

Wool, D. & N. Bar-El, 1995. Population ecology of the galling aphid *Forda formicaria* von Heyden in Israel: abundance, demography and gall structure. Israel Journal of Zoology 41: 175–192.

Wool, D. & O. Ben Zvi, 1998. Population ecology and clone dynamics of the galling aphid *Geoica wertheimae* (Sternorrhyncha: Pemphigidae: Fordinae). European Journal of Entomology 95: 509–518.

Entomologia Experimentalis et Applicata **91:** 187–194, 1999.



# Vive la variance: a functional oviposition theory for insect herbivores

Bernard D. Roitberg, Ian C. Robertson & Jabus G. A. Tyerman
*Behavioural Ecology Research Group and Centre for Pest Management, Department of Biosciences, Simon Fraser University, Burnaby, BC Canada V5A 1S6*

Accepted: October 21, 1998

*Key words:* oviposition, strategy, catastrophe, theory, clutch, Lepidoptera, optimization, dynamic, bet hedging

## Abstract

We developed state-dependent life-history theory to explain the variance in clutch size decisions made by insect herbivores under a variety of ecological scenarios. An important aspect of our theory is explicit representation of the distribution of host quality and frequency of occurrence. Examination of the theory suggests that clutch size decisions can be highly non-linear with respect to host quality and variability. We then use our theory to explore the potential for bet-hedging strategies to evolve as a function of unpredictable catastrophic events that decimate entire clutches. Our analysis suggests that the benefits to employing such a strategy will frequently be outweighed by costs brought on by delayed oviposition.

## Introduction

Our understanding of insect-plant interactions is greater now then it ever has been. Significant advances have been made on several fronts, including the identification of host-recognition mechanisms (Bernays & Chapman, 1994; Renwick & Chew, 1994), phylogenetic analyses of plant-insect relationships (e.g., Brown et al., 1996; Farrell et al., 1992), and learning by herbivores (Papaj & Prokopy, 1989).

Despite these advances, significantly less progress has been made toward development of functional theories of herbivore host selection and exploitation. The likely reason underlying this obstacle has been the range and diversity of potential driving variables in such systems. Thompson & Pellmyr (1991) identified 5 such parameters, including: allelochemicals (e.g., Zalucki et al., 1990), nutritional chemistry (Scriber & Slansky, 1981), plant morphology (e.g., Lawton, 1983) and natural enemies (e.g., Camara, 1997; Stamp, 1992). It is difficult to develop general theory of host-herbivore relationships when the currency associated with each of the above parameters differ both qualitatively and quantitatively. In this paper, we argue that dynamic life history theory provides a methodology for linking seemingly disparate factors into a functional interpretation of some important aspects of herbivore behaviour.

We begin by noting that the well-developed functional theory of host selection by insect parasitoids (e.g., van Alphen & Visser, 1990; Godfray, 1994) has not made significant contributions to herbivory theory, though many would agree with Price's claim that many insect herbivores could easily be classified as parasites (Price, 1980). One possible explanation for the lack of contribution by parasitoid theory is a perception that such a theory is too simple to elucidate the very complicated world of the herbivore. Here, variation in host quality can deviate by more than a few magnitudes in a nearly continuous fashion (Bernays & Chapman, 1994). We argue that degree of variation is insufficient grounds for abandoning attempts to develop heuristic theory, but when such variation is present that it should be characterized and made a focal point of any such advances. This thesis is in the same spirit as those presented by Jaenike (1978), Denno & McClure (1983), Hunter et al. (1992) and Karban et al. (1997). However we do so with some interesting twists.

Our approach to developing herbivore life history theory focuses on clutch size decisions of herbivores as a function of exogenous factors (i.e., host quality and availability) and endogenous state (i.e., eggload and expectation of life). In so doing, we ignore several important aspects of insect-plant interactions such as host range, phylogenetic constraint and direct effects of natural enemies. Further, since many herbivores including the vast majority of butterflies are single egg layers (Stamp, 1980) our clutch-size theory, at first glance, may not appear applicable. We suggest, however, that the general approach of characterizing variability in resource quality and distribution is essential to development of any theory meant to explain how a wide-range of herbivore behaviour evolves including single-egg, acceptance/rejection decisions (Jaenike, 1978). Further, since a broad range of herbivores, including many species of moths (e.g., Powell & Common, 1985) do lay clutches that vary considerably in size, our theory has broad implications.

## Our approach

We begin with a rapidly growing class of life-history models, which take various names including, stochastic dynamic, state variable or dynamic life-history models (Mangel & Clark, 1988). The models take as their starting point the supposition that natural selection favors strategies that maximize lifetime reproductive success (LRS) and then identifies those strategies under different combinations of endogenous and exogenous states. The key to development of such models is to (1) define a common currency that can be applied to a variety of decisions and situations (e.g. avoidance of predators, residence time in depleting patches, etc.) and (2) identify states that may change as a function of some behaviour (e.g., eggload and oviposition decision).

To develop a dynamic life-history theory for herbivore clutch size, we assume the following:

(1) Plants are identical in size, but may vary in quality. We set ($C_{\max}$) to ten, this being the maximum number of larvae that a plant can sustain. Per capita performance of a larva on a plant is defined by:

$$P_{PC} = (1 - (c/c_{\max})^{\delta}), \quad (1)$$

where $c$ = the number of larvae in the clutch and $\delta$ is a parameter that defines the sensitivity of larval survival to crowding. Thus, $f_c$, the return that a female herbivore receives from laying different-sized clutches is:

$$f_c = c(1 - (c/c_{\max})^{\delta}) \text{ (Roitberg \& Mangel, 1993)} \quad (2)$$

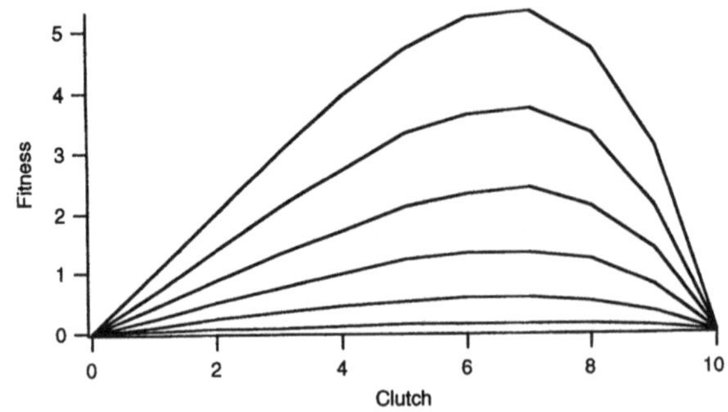


*Figure 1.* A family of curves that describes the fitness that a female herbivore accrues from laying clutches of different sizes on hosts of varying quality. For each plant class, the Lack clutch size is 7.

Plant quality is assumed to vary across plant class in a consistent manner. As such, quality differences may be a function of a limiting nutrient (Scriber & Slansky, 1981), an allelochemical (Bernays & Chapman, 1994) or water (Bernays & Chapman, 1994). For the purpose of this paper we will assume that there are seven distinct quality classes though we acknowledge that quality will frequently be a continuous quantitative trait (Rausher, 1992). We define quality ($q$) as:

$$q = (\nu/\nu_{\max})^{\gamma}, \quad (3)$$

where $\nu$ is the quality class and $\gamma$ is a shape parameter that defines the sensitivity of quality to plant class.

In this paper, we consider three different values for $\gamma$, 0.5, 1.0 and 1.5. The first value creates a concave (decelerating) curve such that the majority of plants provide good quality hosts. Karban et al. (1997) suggest that many host species fit into this category (e.g., *Helianthus annus* – Ellinger et al., 1976). When $\gamma = 1.0$ there is a linear relationship between plant class and quality. Finally, when $\gamma = 1.5$ there is a convex (accelerating) relationship between plant class and quality. Under this scenario, the majority of plant classes are of poor quality with only the highest quality classes providing good quality.

To calculate the fitness $f_{c,\nu}$ that a female accrues from laying different sized clutches ($c$) on plants of different quality classes ($\nu$), we simply take the product of Equations (2) and (3). This gives the family of curves shown in Figure 1. Several features should be noted: Regardless of plant class, the highest return that a female could accrue would be from laying a clutch of seven eggs; this is true for all plant-quality classes. This clutch is often called the Lack clutch and is frequently used in theories of avian and parasitoid clutch decisions (Charnov & Skinner, 1984). In addition to the above, it should be noted that while the

Lack clutch is identical across classes, payoffs vary dramatically across plant classes as a function of the previously described differences in quality. Further, since each of the payoff functions is parabolic then it follows that the Lack clutch is the maximum that a female should deposit on a host whereas the minimum should be situation dependent as we show below.

(2) Plants are encountered in proportion to their relative abundance or density. The seven plant classes are normally distributed with $\bar{x} = 4$ and $\sigma = 1.0$. Thus, the probability that a host from a given class $\nu$ will be encountered within a single time period is $\lambda_\nu$ and is directly proportional to its representation. In addition, we developed an algorithm that produced skewed distribution, one skewed to the left such that low-quality plants are over-represented in the plant population and another, skewed to the right where high-quality plants are over-represented (skew $= G = -0.047$ and $0.047$, respectively; Wilkinson, 1992). Such distributions are possible for a variety of reasons, including micro- and macro-geographic differences in terms of limiting nutrients (Bryant et al., 1983).

(3) Herbivores can accurately assess plant size and quality.

(4) Lifetime reproductive success (LRS) is a good index of fitness. Herbivores are capable of laying a variety of clutch sizes. We assume that they choose a clutch size (0, Lack) that maximizes LRS.

(5) Herbivores are pro-ovigenic, i.e., they harbor a full set of eggs at eclosion and do not mature new eggs thereafter. We define $\epsilon$ as the number of eggs an individual harbors and following convention (Rosenheim, 1996) refer to this quantity as eggload.

(6) Herbivores have a maximum lifespan (T) with a daily probability of dying ($\mu$).

With the preceding assumptions in place, we can now develop dynamic life history theory (Mangel & Clark, 1988) to facilitate calculation of optimal clutch size under the various ecological scenarios. Our theory assumes that several mutually exclusive events can occur with particular repercussions:

(1) The herbivore searches unsuccessfully for a host plant with probability $\lambda_0$. As a result, time advances one unit but eggload ($\epsilon$) remains the same.

(2) The herbivore locates a plant of quality $\nu$ with probability $\lambda_\nu$ and makes a clutch-size decision $c^*$ (0, Lack) that maximizes LRS. Time advances one unit and eggload is reduced by $c^*$.

The preceding events can be assembled into a dynamic life history equation as follows:

$$F(\epsilon, t, T) = [\lambda_0(1-\mu)F(\epsilon, t+1, T)] + (1-\lambda_0)\left[\sum_{\nu=1}^{\nu\max} \lambda_\nu \max(0, \text{Lack})\, c(f_{c,\nu} + (1-\mu)F(\epsilon - c, t+1, T)\right]. \tag{4}$$

Our theory is very similar in structure to those used for entomophagous insect parasitoids (e.g., Roitberg et al., 1993). The important difference however is that the parasitoid models generally consider only two types of hosts, healthy and parasitized. Our theory assumes that healthy hosts can be further subdivided into a number of quality classes and then employs explicit representation of quality and frequency of each plant class.

## Results and discussion

Examination of Equation (4) emphasizes several important points in life history theory: discounting, patterns of resource distribution and state dependence. Upon encountering a mid or low quality host, our theoretical herbivore must choose between accruing fitness now or withholding some of its eggs for future encounters with other plants. Those future events must be discounted by the likelihood of occurrence, which will depend upon adult survivorship ($\mu$) and host availability (recall our host frequency distributions). Finally, though it is obvious that eggs laid now cannot be laid later, the cost of such a decision depends upon the current eggload relative to future opportunities to exploit hosts. Thus, our theory predicts that herbivores should always 'dump' eggs (up to the Lack clutch) as they approach the end of their life. This prediction has been upheld for parasitoids under experimental conditions (Roitberg et al., 1992, 1993). A key then is to relate eggload to future opportunities.

To obtain optimal clutch $c^*$, the dynamic equation can be solved through the process of backward induction (Mangel & Clark, 1988). The solution yields optimal clutch size as a function of eggload and time. Figure 2 shows optimal clutches for a herbivore that harbors 20 eggs on the first day of her adult life when host quality varies amongst classes in a linear fashion and frequency distributions vary from skewed left (poor hosts are over-represented), to non-skewed and skewed right (good hosts are over-represented). The

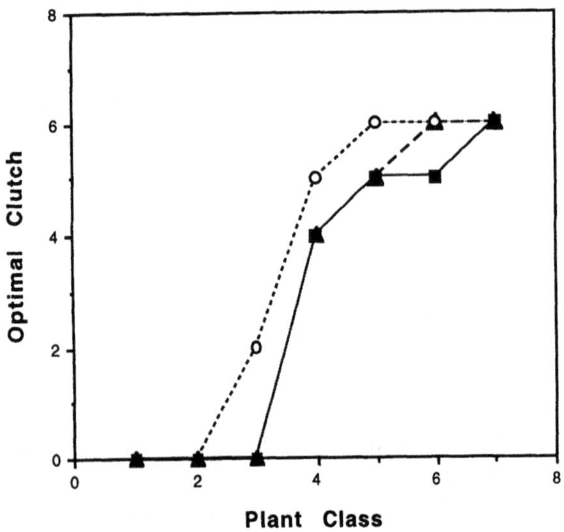


*Figure 2.* Optimal clutch size as a function of plant quality when the frequency distributions of hosts are skewed toward low quality plants (○), skewed toward high quality plants (■) or normal (Δ). Other parameter values are: eggload ($\epsilon$) = 20, time ($t$) = 1, plant quality curve slope ($\gamma$) = 1.0.

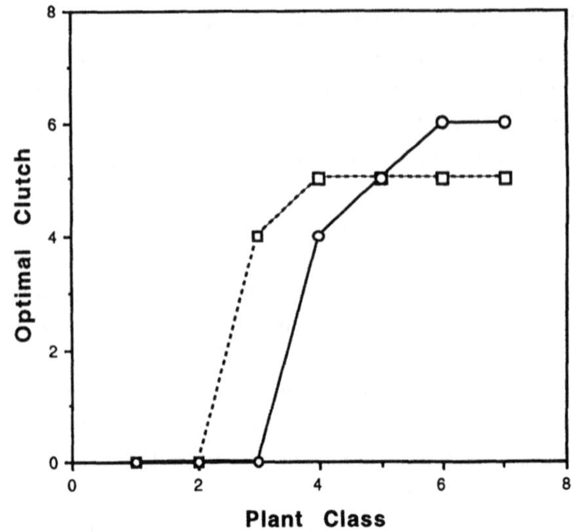


*Figure 3.* Optimal clutch size as a function of plant when plant class quality curve varies from convex ($\gamma$ = 1.5) (○) to concave ($\gamma$ = 0.5) (□). Other parameter values are: eggload ($\epsilon$) = 20, time ($t$) = 1.

results suggest that the degree of choosiness expressed can vary dramatically under different scenarios. As expected, when good hosts are more common (skewed right) the optimal decision is to be very discriminating against poor quality hosts. Likewise, degree of choosiness declines when such hosts are less common (skewed left). When the host quality curve is convex one would expect the degree of discrimination to increase even further when the difference between good and bad plants is more exaggerated. That is the case but as before, the response to plant quality is highly non-linear (Figure 3).

Our results strongly suggest that the degree of host discrimination can be influenced by ecological conditions, i.e., degree of curvature in the plant quality function as well as the distribution of plant quality classes. Few empirical studies have ever documented these distributions (see Hunter et al., 1992).

Thus far, we have used dynamic life history theory to solve for optimal clutch under various scenarios. What is not clear however is how well individuals perform when they employ sub-optimal strategies. Such information, so called fitness functions, can give insights into direction and rate of evolution for particular traits (e.g., Mangel & Ludwig, 1992; Roitberg & Mangel, 1993, 1997; Cheverud & Moore, 1994). We can easily derive these fitness functions for our model using a modified version of Equation (4) wherein clutch size decisions are allowed to vary in some consistent manner from the optimal. Here, we followed conventional wisdom (see Carrière & Roitberg, 1996) that suggests that behaviour varies among genotypes in degree of discrimination but not in terms of preference i.e., we assume that all genotypes rank hosts according to the same scheme but differ in their threshold of acceptance. Accordingly, we derived six 'mutant' types: $c^*_{\nu+1}$, $c^*_{\nu+2}$, $c^*_{\nu+3}$, $c^*_{\nu-1}$, $c^*_{\nu-2}$, $c^*_{\nu-3}$. The first three mutants overestimate plant value by one, two or three classes respectively whereas the latter three underestimate plant quality in an analogous fashion.

Figures 4 and 5 show fitness functions from two different ecological scenarios. In the first scenario, plant quality varies across plant class in a linear fashion. The fitness curves are asymmetric. Apparently, it is much more costly to underestimate plant quality (i.e., make the mistake of not exploiting good vs over-exploiting poor hosts). The rapid drop in fitness for the extreme underestimating genotypes comes from two sources: (1) not laying big clutches on very good plants and (2) not laying eggs at all. The latter occurs because the majority of hosts exist at or near the mean thus few 'suitable' hosts are encountered; given the discounting associated with future events, erroneous rejections can be very costly. Just how costly depends upon availability of future opportunities and current eggload. Finally, the curve suggests that we can expect to see considerable variation in host-selection all else being equal especially for genotypes that overestimate host quality but do not exceed Lack clutches (Roit-

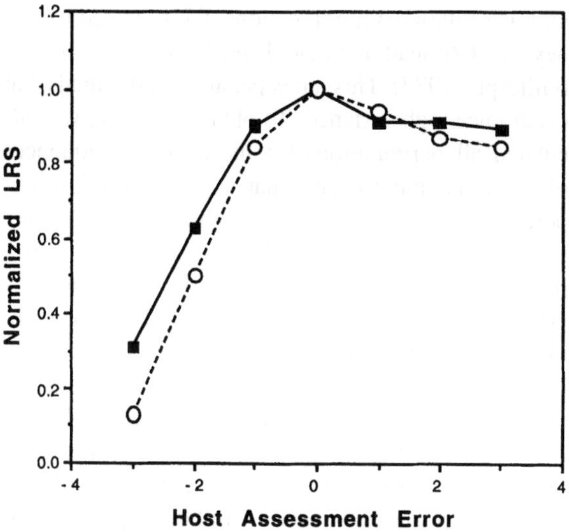


*Figure 4.* Normalized lifetime reproductive success (LRS) for herbivores that make optimal and sub-optimal plant-quality assessments when plant quality varies in a linear manner across quality classes ($\gamma = 1.0$) but the frequency distributions of hosts are skewed toward low quality (○), or high quality plants (■).

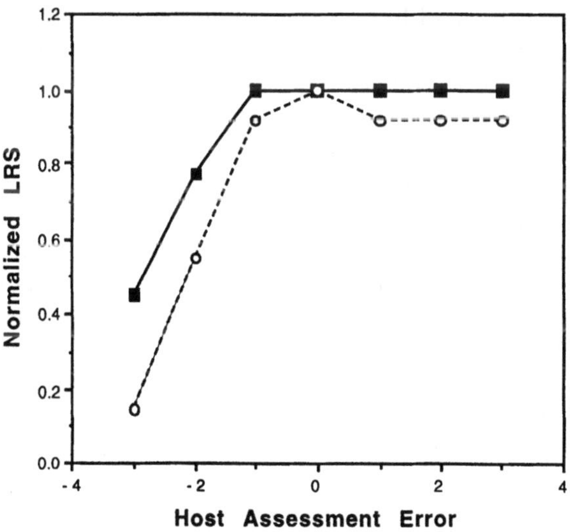


*Figure 5.* Normalized lifetime reproductive success (LRS) for herbivores that make optimal and sub-optimal plant-quality assessments when plant quality varies in a convex manner across quality classes ($\gamma = 1.5$) but the frequency distributions of hosts are skewed toward low quality (○), or high quality plants (■).

berg & Mangel, 1997). If our insight regarding plant quality distributions is correct it follows that when quality decelerates across plant classes that the fitness function should be even more asymmetric. Such is the case (Figure 5). This is particularly interesting given the evidence that concave host quality distribution curves may be common (Karban et al., 1997). Figure 5 also reinforces our thesis that it is critical to

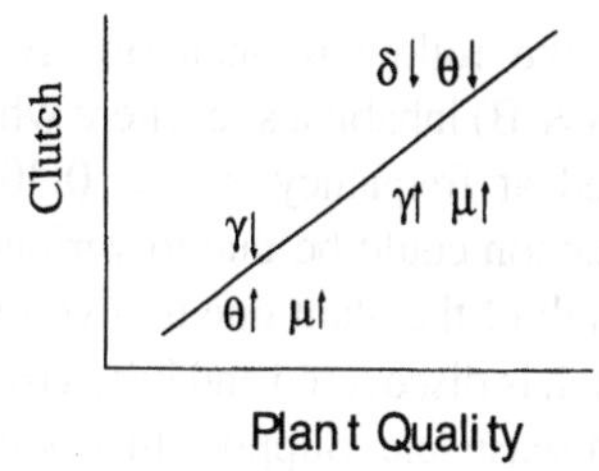


*Figure 6.* Conceptual model of selection by different agents on herbivore choosiness which is indicated by the clutch-plant quality relationship. Increases in $\delta$ (density dependence) and $\theta$ (catastrophe frequency) select for reduction in choosiness slope whereas increase in $\gamma$ (steepness of the plant quality curve) selects for increased choosiness. In addition, increase in $\mu$ (herbivore death rate) can select for increase in the intercept of the line, i.e., higher clutch size across all hosts.

characterize variance in order to understand herbivore host-selection behavior.

The theory that we developed incorporates several potentially interacting parameters each of which can have disparate effects on herbivore choosiness. Figure 6 provides a means for visualizing some of these parameters. In the figure, a diagonal line that relates clutch to plant quality represents choosiness. A very choosy herbivore would be represented by a very steep line, i.e., many more eggs are laid on good vs poor-quality plants. Several parameters are shown along with arrows that indicate the directional selection they would impose on herbivore choosiness. For example, increasing herbivore death rates ($\mu$) should select for a higher intercept for the choosiness curve (i.e., lay more eggs on all plants) whereas increasing density dependence ($\delta$) should select for decreased slope by reducing the proclivity to lay large clutches on good plants. By contrast, as the plant quality curve becomes more convex ($\gamma$) we expect herbivores should become more discriminating (i.e., increased slope). Finally, we introduce a new parameter catastrophe ($\theta$) which we suggest could act in an unusual manner by selecting against discrimination, i.e., more eggs on poor plants and fewer on good quality plants. We conclude by considering this enigmatic parameter in light of our views on life history theory.

The notion of catastrophe has been the subject of much discussion in ecology under the general heading of risk spreading or bet hedging (Seger & Brockman, 1987). The idea is that there may be circumstances under which it would pay an individual to only partially exploit one resource to reduce variance in fitness should such resources be subject to unpredictable, density-independent disastrous events.

Consider the following scenario: two herbivore genotypes (A & B) inhabit a site where whole clutches are decimated at frequency $p_c = 0.30$/generation. This phenomenon could be due to various processes, including death of the plant due to a vertebrate herbivore, the clutch is discovered and killed by a generalist predator or disease, etc. Suppose that both herbivores have eggloads of 10 and that genotype A lays all of its load in a single clutch whereas genotype B deposits two clutches of 5 each on separate hosts. Under such conditions, 70% $(1 - p_c)$ of herbivore A individuals will realize a full return of 10 eggs whereas 30% $(p_c)$ will realize zero for an average of 7 offspring per individual. By contrast, 49% $(1 - p_c)^2$ of herbivore B individuals will realize a full return from both clutches, 21% $((1 - p_c)\, p_c)$ will suffer loss of their second clutch, a further 21% $(p_c(1 - p_c))$ will suffer loss of their first clutch and 9% $(p_c)^2$ will lose both clutches. Thus, while both genotypes achieve the same arithmetic mean fitness genotype B has lower variance, i.e., it spreads the risk among two clutches.

The idea of risk spreading has had considerable appeal to ecologists for some time. In general, when risk spreading has been used to explain egg deposition patterns by insects it has been employed as a default explanation. For example, Root & Kareiva (1984) concluded that *Pieris brassicae* might be risk spreading because none of their other hypotheses could adequately explain patterns of premature patch abandonment by the butterfly. Similarly, Cronin & Strong (1993) argued that a mymarid parasitoid underexploited patches of leafhopper eggs within plants because of the danger that whole plants could fail. More recently, Freese & Zwolfer (1996) suggested that their observation of highly variable clutches by a Tephritid gallfly could be explained as a bet-hedging strategy.

Not everyone is enamoured with risk spreading as an important force in the evolution of egg dispersion. For example, Courtney (1986) showed that risk spreading could only be advantageous under a limited set of circumstances. Thus, it is not clear whether there exists significant selection against choosiness based on random, catastrophic, density-independent events. Further, recall that we cautioned that it is dangerous to analyze individual selection agents in isolation. Fortunately, life history theory provides a tool for considering risk spreading within the context of exogenous and endogenous parameters.

To evaluate the potential impact of catastrophe on choosiness traits we return to our earlier model with the following modifications. We followed Courtney's (1986) lead and calculated LRS as: $\bar{x} - 1/2\sigma^2$ (Gillespie, 1977). This index penalizes individuals that experience high variance. To obtain variance, we calculated all permutations for the different sequences of plant-encounter events that a herbivore might experience weighted by their probability. For example, a choosy moth with an eggload of seven will visit a plant of class $\nu = 7$ with probability $\lambda_7$ and deposit its entire eggload. Similarly, that same individual might visit a plant of class $\nu = 3$ with probability $\lambda_3$ and deposit one egg. On the following day, it will visit a plant of class $\nu = 5$ with probability $\lambda_5$ and deposit two eggs, etc. Each complete sequence is then subjected to catastrophe as described earlier whereupon fitness returns are evaluated. A full description of our procedures and analysis will be provided elsewhere. The key point to the modifications is that individuals who are very choosy will tend to lay fewer but larger clutches and will benefit from enhanced survival of larvae on good plants. They will, however, pay a potential cost of high variance in fitness due to greater proportions of their entire eggload being lost due to catastrophe. We can now ask if reduced choosiness will be favored as risk of catastrophe increases.

Figure 7 plots both arithmetic mean and variance as a function of catastrophe for a herbivore that is moderately choosy, i.e., it lays successively more eggs on good quality plants. Mean LRS declines as frequency of catastrophe increases. This is expected because whole clutches are lost thus LRS must decline. Variance also decreases but at a lower rate than the mean. As such, the coefficient of variation increases from 0.833 when catastrophe frequency is 0.05 versus 1.05 when catastrophe frequency is 0.25. Interestingly, however, we found that reducing the choosiness tendency rarely increased LRS. Why is that? Below, we return to our earlier themes of the need to employ a dynamic life-history perspective in addition to the importance of patterns of resource distribution.

The situation that we have considered here fits within the genre of models that Seger & Brockman (1987) called 'A bird in the hand' type. Here the animal gives up current opportunities (i.e., individuals lay fewer eggs on good hosts than they would otherwise to reduce variance in lifetime fitness). Boyce & Perrins (1987) have described similar theory for great tit laying strategies. The present situation is more complicated than the aforementioned studies, however, in that ours is state-dependent. By that, we mean that a decision to lay fewer or more eggs affects the

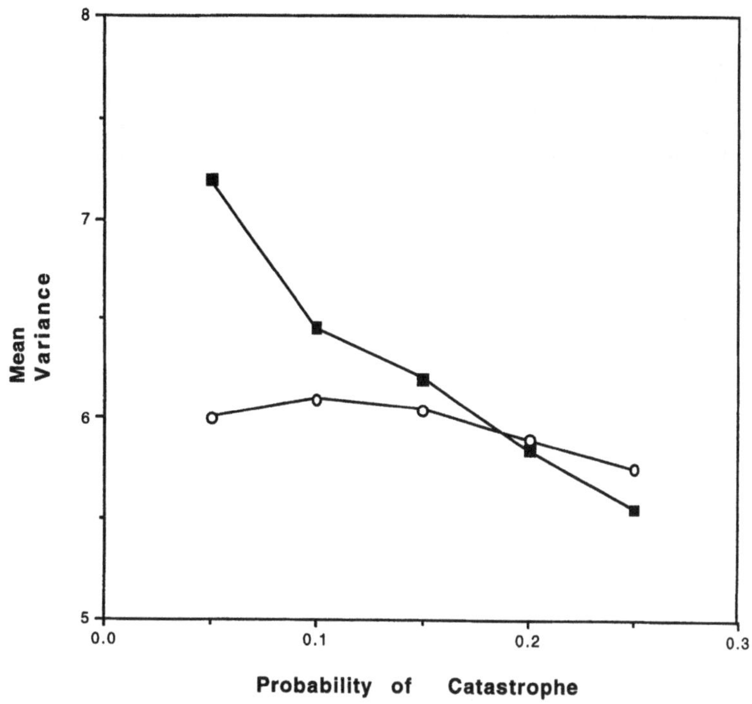


*Figure 7.* Changes in mean (■) lifetime reproductive success and associated variance (○) as a function of the probability of catastrophe (death of an entire clutch) for a choosy herbivore.

number of eggs available for use in future encounters with hosts. Our earlier discussion on dynamic life history considered the cost of withholding eggs for future opportunities. We argued that those costs might be considerable when time or hosts are in short supply. The theory of bet hedging, by contrast, is static in that it considers clutch distributions as a state-independent process. In our dynamic theory, a risk spreader is seen as an individual that withholds from depositing large clutches on good plants in order to exploit more plants. That individual will have more eggs available for future encounters with hosts but mortality and availability of such hosts must discount the future; once again, future payoffs are dependent upon distribution of plant quality and frequency. Interestingly, because plants vary in quality, bet hedging can actually increase variance in individual fitness if there is considerable variance among hosts. Thus, when resource heterogeneity is considered, choosiness in itself can be viewed as a bet hedging strategy. Freese & Zwolfer (1996) recently suggested that individual tephritid gall flies bet hedge when the costs to do so are small but 'gamble' when life-history constraints make such decisions costly. This suggestion is very much in keeping with our dynamic approach.

We conclude by referring once more to the Karban et al.'s (1997) important contribution to toxin variability in host plants. Recall that their argument rests on the shape of the plant-quality curves. Our analysis adds to a growing appreciation that characterization of variance in host quality will facilitate development of a sound theory of insect-plant interactions.

## Dedication

This paper is dedicated to Ronald J. Prokopy.

## Acknowledgements

We thank Rob Anderson, Ed Mondor, Bill Friend, Tracy Hüppelsheuser, Renee Prasad and Risa Sargent for their useful comments and suggestions. NSERC, Canada supported this work.

## References

Alphen, J. J. van & M. Visser, 1990. Superparasitism as an adaptive strategy for insect parasitoids. Annual Review of Entomology 35: 59–79.

Bernays, E. A. & R. F. Chapman, 1994. Host-Plant Selection by Phytophagous Insects. Chapman and Hall, New York.

Boyce, M. S. & C. M. Perrins, 1987. Optimizing great tit clutch size in a fluctuating environment. Ecology 68: 142–153.

Brown, J. M., W. G. Abrahamson & P. A. Way, 1996. Mitochondrial DNA phylogeography of host races of the goldenrod ball gall maker *Eurosta solidaginis* (Diptera: Tephritidae). Evolution 50: 777–786.

Bryant, J. P., F. S. Chapin & D. R. Klein, 1983. Carbon/nutrient balance of boreal plants in relation to vertebrate herbivory. Oikos 40: 357–368.

Camara, M., 1997. A recent host range expansion in *Junonia coenia* Hübner (Nymphalidae): oviposition preference, survival, growth and chemical defense. Evolution 51: 873–884.

Charnov, E. L. & S. W. Skinner, 1984. Evolution of host selection and clutch size in parasitoid wasps. Florida Entomologist 67: 5–21.

Carrière, Y. & B. D. Roitberg, 1996 Behavioural ecology and quantitative genetics as alternatives for studying evolution in insect herbivores. Evolutionary Ecology 10: 289–305.

Cheverud, J. M. & A. J. Moore, 1994. Quantitative genetics and the role of environment in behavioural evolution. In: C. R. B. Boake (ed), Quantitative Genetic Studies of Behavioural Evolution. University of Chicago Press, pp. 67–100.

Courtney, S. P., 1986. Why insects move between host patches: some comments on 'risk-spreading'. Oikos 47: 112–114.

Cronin, J. T. & D. R. Strong, 1993. Substantially sub-maximal oviposition rates by a mymarid egg parasitoid in the laboratory and field. Ecology 74: 1813–1825.

Denno, R. F. & M. S. McClure (eds). 1983. Variable Plants and Herbivores in Natural and Managed Systems. Academic Press, San Diego.

Ellinger, C. A. D., D. F. Zinkel, B. G. Chan & A. C. Waiss Jr., 1976. Diterpene acids as larval growth inhibitors. Experientia 32: 1364–1366.

Farrell, B., C. Mitter & D. J. Futuyma, 1992. Diversification at the insect-plant interface: insights from phylogenetics. Bioscience 42: 34–42.

Freese, G. & H. Zwolfer, 1996. The problem of optimal clutch size in a tritrophic system: The oviposition strategy of the thistle gallfly *Urophora cardui* (Diptera,Tephritidae) Oecologia 108: 293–302.

Gillespie, J. H., 1977. Natural selection for within-generation variance in offspring numbers: a new evolutionary principle. American Naturalist 111: 1010–1014.

Godfray, H. C. J., 1994. Parasitoids: Behavioural and Evolutionary Ecology. Princeton University Press, Princeton.

Hunter, M. D., T. Ohgushi & P. W. Price (eds), 1992. Effects of Resource Distribution in Animal-Plant Interactions. Academic Press, San Diego.

Jaenike, J., 1978. On optimal oviposition behaviour in phytophagous insects. Theoretical Population Biology 14: 350–356.

Karban, R., A. A. Agrawal & M. Mangel, 1997. The benefits of induced defenses against herbivores. Ecology 78: 1351–1355.

Lawton, J. H., 1983. Plant architecture and the diversity of phytophagous insects. Annual Review of Entomology 28: 23–39.

Mangel, M. & C. W. Clark, 1988. Dynamic Modeling in Behavioural Ecology. Princeton University Press, Princeton, NJ.

Mangel, M. & D. Ludwig, 1992. Definition and evaluation of behavioural and developmental programs. Annual Review of Ecology and Systematics 23: 507–536.

Papaj, D. R. & R. J. Prokopy, 1989. Ecological and evolutionary aspects of learning. Annual Review of Entomology 34: 315–350.

Powell, J. A. & I. F. B. Common, 1985. Oviposition patterns and egg characteristics of Australian tortricine moths (Lepidoptera: Tortricidae) Australian Journal of Zoology 33: 179–216.

Price, P. W., 1980. Evolutionary Biology of Parasites. Princeton University Press, Princeton.

Rausher, M. D., 1992. Natural selection and the evolution of insect-plant interactions. In: B. D. Roitberg & M. S. Isman (eds), Insect Chemical Ecology: an Evolutionary Approach. Chapman and Hall, New York, pp. 20–88.

Renwick, J. A. A. & F. S. Chew, 1994. Oviposition behaviour in Lepidoptera. Annual Review of Entomology 39: 377–400.

Roitberg, B. & M. Mangel, 1993. Parent-offspring conflict and life-history consequences in herbivorous insects. American Naturalist 142: 443–456.

Roitberg, B., M. Mangel, R. Lalonde, C. Roitberg, J. van Alphen &, L. Vet, 1992. Seasonal dynamic shifts in patch exploitation by parasitic wasps. Behavioural Ecology 3: 156–165.

Roitberg, B., J. Sircom, C. Roitberg, J. van Alphen & M. Mangel, 1993. Life expectancy and reproduction. Nature 364: 108.

Roitberg, B. D. & M. Mangel, 1997. Individuals on the landscape: Behaviour can mitigate landscape differences among habitats. Oikos: 80: 234–240.

Root, R. B. & P. M. Kareiva, 1984. The search for resources by cabbage butterflies (*Pieris rapae*): ecological consequences and adaptive significance of markovian movements in patchy environments. Ecology 65: 147–165.

Rosenheim, J. A., 1996. An evolutionary argument for egg limitation. Evolution 50: 2089–2095.

Scriber, J. M. & F. Slansky Jr., 1981. The nutritional ecology of immature insects. Annual Review of Entomology 26: 183–211.

Seger, J. & H. J. Brockman, 1987. What is bet hedging? Oxford Surveys In Evolutionary Biology 4: 182–211.

Stamp, N. E., 1980. Egg deposition patterns in butterflies: why do some species cluster their eggs rather than deposit them singly? American Naturalist 115: 367–380.

Stamp, N. E., 1992. Susceptibility of specialist versus generalist caterpillars to invertebrate predators. Oecologia 92: 124–129.

Thompson, J. N. & O. Pellmyr, 1991. Evolution of oviposition behaviour and host preference in Lepidoptera. Annual Review of Entomology 36: 65–89.

Wilkinson, L., 1992. Systat for Macintosh. Systat Inc. Evanston, IL.

Zalucki, M. P. , L. P. Brower, & S. B. Malcolm, 1990. Oviposition by *Danaus plexippus* in relation to cardenolide content of three *Asclepias* species in Southeastern USA. Ecological Entomology 15: 231–240.

Entomologia Experimentalis et Applicata **91:** 195–201, 1999.



# The assimilation and allocation of nutrients by symbiotic and aposymbiotic pea aphids, *Acyrthosiphon pisum*

T. L. Wilkinson* & H. Ishikawa
*Department of Biological Sciences, Graduate School of Science, University of Tokyo, Hongo, Bunkyo-ku, Tokyo 113, Japan; *Affiliated to the Bio-oriented Technology Research Advancement Institution*

Accepted: October 9, 1998

*Key words:* aphid, *Acyrthosiphon pisum*, symbionts, amino acids, sucrose, assimilation, embryos


## Abstract

The failure of a nutritionally balanced diet to ameliorate the impact of symbiont disruption in the pea aphid *Acyrthosiphon pisum* (Harris) was investigated using two approaches. The assimilation of dietary nutrients by aphids was investigated using chemically-defined diets containing $^{3}$H-labelled inulin and $^{14}$C-labelled sucrose or amino acids. Symbiotic aphids (i.e., aphids containing their bacteria) had a high sucrose demand and assimilated 72% of sucrose ingested in the diet, whereas the assimilation of sucrose by aposymbiotic aphids (in which the bacteria had been disrupted), was significantly reduced to 47%. The assimilation of individual dietary amino acids by symbiotic aphids varied between 61 and 92%, and there was no impact on the feeding or assimilation rate when the aphids were fed a 'phloem sap-like' diet containing a reduced amount of essential amino acids. Consequently, the absolute amount of each essential amino acid assimilated by symbiotic aphids feeding on a 'phloem sap-like' diet was reduced by 36–59%. Aposymbiotic aphids consistently assimilated a lower proportion of ingested amino acids, and lysine in particular was poorly assimilated from the diet. In a second experiment, the allocation of free amino acids in the haemocoel to aphid embryos was investigated following microinjection of $^{14}$C-labelled amino acids. After 2 h, radiolabel could be detected at varying levels from the embryo complement of both symbiotic and aposymbiotic aphids, indicating rapid but selective uptake by the embryos. The essential amino acids phenylalanine and lysine were incorporated into the protein fraction of embryo tissues, but the rate of incorporation per unit biomass was approximately 4-fold higher in the embryos of aposymbiotic aphids, possibly reflecting increased demand due to the lack of amino acid provisioning from the symbiotic bacteria.


## Introduction

The ability of aphids to thrive on phloem sap is attributable to the metabolic capabilities of *Buchnera*, the intracellular bacterial symbionts located in the mycetocyte cells (Buchner, 1965; Douglas, 1989; Baumann et al., 1995). Specifically, *Buchnera* supplement the low essential amino acid content of phloem sap through the synthesis of essential amino acids, which are made available to the aphid tissues (see Douglas, 1998, for a full review).

*Buchnera* can be eliminated from aphids using a mild antibiotic therapy (Houk & Griffiths, 1980; Douglas, 1989; Rahbé et al., 1994), with little or no deleterious effect on aphid tissues (Wilkinson, 1998), and the resulting 'aposymbiotic' aphids have proved a vital tool in understanding the nutritional and physiological interactions which underly the symbiosis. However, aposymbiotic aphids exhibit a generally poor 'performance': they have significantly reduced growth rates, low adult size, and a dramatically curtailed reproductive output (Prosser & Douglas, 1991; Sasaki et al., 1991; Douglas, 1992a). Moreover, the effect is similar if aposymbiotic aphids are reared on plants or chemically-defined diets (Sasaki et al., 1991; Douglas 1992a), despite the latter containing a nutritionally adequate supply of essential amino acids (50 mol%) which would be expected to overcome the shortfall in bacterial-derived essential amino acids experienced by aposymbiotic aphids. That a dietary supply of essential amino acids fails to sustain aposymbiotic aphids, and particularly fails to rescue embryogenesis in asexual parthenogenetic morphs, raises two possibilities. First, aposymbiotic aphids may not assimilate dietary essential amino acids at sufficient rates to meet the nutritional demands of the growing animal. Aposymbiotic pea aphids, *Acyrthosiphon pisum* (Harris), incorporate dietary glutamate into protein at the same rate as symbiotic aphids (at approximately 30 nmol $mg^{-1}$ aphid protein $day^{-1}$; Wilkinson & Douglas, 1996), but there are no data available concerning the assimilation of essential amino acids from the diet. Second, free amino acids in the haemocoel may not be available to the developing embryos. This possibility arises because of the spatial complexity of the symbiosis: the bacteria are harboured in mycetocyte cells located in both the maternal haemocoel and in the haemocoel of developing embryos. The relative contribution of the maternal and embryo bacteria to the nutritional demands of the developing embryos is unknown, but it is feasible that the embryos derive some or all of their requirements for essential amino acids from their own complement of bacteria. The maternal and embryo 'compartments' of pea aphids certainly do not appear to be in equilibrium: antibiotic treatment results in a far greater reduction in the growth of embryo tissue than maternal tissue during larval development (Douglas, 1996), suggesting that embryos are more dependent than the maternal tissues on the symbiotic bacteria; and the free amino acid composition of the embryo and maternal tissues is not the same (Douglas, 1996), suggesting that amino acids do not pass freely between the two pools.

To address these issues, the study reported here investigates the assimilation of dietary sucrose and amino acids by symbiotic and aposymbiotic pea aphids *A. pisum*, and the subsequent allocation of free amino acids in the haemocoel to the embryo tissues. Chemically-defined diets containing tritiated inulin and $^{14}C$-labelled nutrients were used to quantify (i) the amount of diet ingested by individual aphids, and (ii) the relative amount of $^{14}C$-labelled nutrient assimilated by each aphid. Ingested inulin passes through the gut and is excreted in the honeydew with little or no digestion or assimilation (Wright et al., 1985; T. L. Wilkinson, unpubl.), providing a convenient method to quantify the total amount of diet ingested by feeding aphids. By dual-labelling the diet with $^{14}C$-labelled nutrients the relative and absolute amount of each nutrient assimilated by individual aphids could be calculated. The nutrients chosen in this study were sucrose (the sole or predominant carbohydrate component of phloem sap) and the amino acids glutamate, tyrosine (both non-essential amino acids), phenylalanine, lysine and threonine (essential amino acids). To assess whether amino acids assimilated from the diet are made available to embryo tissues, the amount of radioactivity in the embryo fraction of symbiotic and aposymbiotic aphids was determined following microinjection of radiolabelled amino acids into the maternal haemocoel.

## Materials and methods

This study was conducted on pea aphids (*A. pisum*) clone TU9, derived from a long-term culture maintained on *Vicia faba.* at 20 °C with L18:D6 regime. Aphids of the apterous wingless morph were used exclusively in all experiments. Aposymbiotic aphids were generated by treatment with the antibiotic rifampicin over the first 2 days of larval development, as previously described (Rahbé et al., 1994; Wilkinson & Douglas, 1995). Following this pre-treatment, symbiotic and aposymbiotic aphids were transferred to chemically-defined diets or plants to continue development. The standard diet composition was formulation A of Prosser & Douglas (1992), with 150 mM amino acids and 500 mM sucrose.

The assimilation of dietary nutrients was investigated in aphids reared from birth on chemically-defined diets. On day 7 of larval development symbiotic and aposymbiotic aphids were transferred individually to diet cages (prepared as in Wilkinson & Douglas, 1995), and allowed to feed from a chemically-defined diet containing 9 $\mu$Ci $ml^{-1}$ [G-$^{3}H$]-inulin and 7 $\mu$Ci $ml^{-1}$ [U-$^{14}C$]-labelled nutrient (either an amino acid or sucrose; see results). Honeydew was collected on GF/C filter papers (Whatman) placed underneath each aphid. Preliminary experiments demonstrated that all honeydew produced by the aphids was deposited onto the filter paper. After 48 h, the filter papers were removed and the aphids were weighed individually to the nearest $\mu$g on a Shimadzu AEM 5200 microbalance. The filter papers were air-dried in scintillation vials for at least 1 h before scintillation counting using a Beckmann LS 6500 counter.

For the assimilation of amino acids, two diet compositions were used. Diet A (see above) contained essential amino acids at 50 mol% and is suitable for the long-term maintenance of pea aphids (including clone TU9; T. L. Wilkinson, unpubl.). Diet B (from Prosser & Douglas, 1992) was based on the amino acid composition of phloem sap from *Vicia faba* (Sasaki et al., 1990) and contains 20 mol% essential amino acids (see Prosser & Douglas, 1992, for full details of the diet composition). Following the addition of the radioactive components, the final concentration of amino acids and sucrose was 131 mM and 437 mM, respectively in both diet A and B.

The availability of free amino acids to aphid embryos was determined after microinjection of $^{14}$C-labelled amino acids into the haemocoel of teneral adult apterae. Symbiotic and aposymbiotic aphids were held by suction and 0.05 $\mu$l [U-$^{14}$C]-amino acid solution containing 40 pmol amino acid was injected into the ventral mesothorax. After 2 h, the aphid was immersed in 50 mm Tris-HCl pH 7.5 and the embryo complement was removed by dissection, washed three times in fresh buffer, and homogenized in 400 $\mu$l buffer. The homogenate was mixed with 4 ml aqueous scintillant and the radioactivity was counted. In a subsequent experiment, the incorporation of $^{14}$C-labelled phenylalanine and lysine by the embryo complement was determined. Aphids were injected as previously, and after 2 h the embryos from 3 symbiotic and 5 aposymbiotic aphids were pooled separately and homogenized in ice-cold 50 mm Tris-HCl pH 7.5. The embryos were fractionated by trichloroacetic acid (TCA) precipitation, as described in Douglas (1988), and incorporation into the TCA-soluble (low molecular weight) and TCA-insoluble (protein) fractions was determined by scintillation counting. The protein content of the embryos was adopted as the index of biomass in these experiments since the fresh weight of dissected embryos is unreliable (see Douglas, 1996), and was determined using the protein assay kit of the Biorad Chemical Co., following manufacturer's instructions for the microassay, with bovine serum albumen as standard.

## Results

Both symbiotic and aposymbiotic aphids settled and fed from the radioactive diets over the 48 h experimental period and continued development, as determined by the presence of exuviae in the feeding cages. The volume of diet ingested by symbiotic and aposymbiotic aphids over 48 h, as determined from the amount of $^3$H-inulin in the honeydew, was 3.53 $\pm$ 0.092 $\mu$l ($n$ = 174) and 1.16 $\pm$ 0.047 $\mu$l ($n$ = 178), respectively (mean $\pm$ s.e.). There was no difference in the volume of diet ingested by aphids feeding on diet A or diet B, and the higher ingestion rate of symbiotic aphids was entirely attributable to the difference in body size between symbiotic and aposymbiotic aphids (when fresh weight was included as a covariate in the analysis; data not shown). This is consistent with the lack of any deleterious effects of antibiotic-treatment on the feeding behaviour of pea aphids (e.g., Prosser et al., 1992; Wilkinson & Douglas, 1995). The assimilation of dietary sucrose was investigated in aphids feeding on diet A. Symbiotic and aposymbiotic pea aphids assimilated 72 $\pm$ 1.6% ($n$ = 15) and 47 $\pm$ 6.1% ($n$ = 9) of ingested sucrose, respectively (mean $\pm$ s.e.; after arcsine squareroot transformation $t_{22} = 4.9$, $P < 0.001$).

The data on the assimilation of individual amino acids from the diet are shown in Figure 1. Previous estimates have suggested that over 94% of total amino acids ingested in the diet are retained by pea aphids (e.g., Prosser & Douglas, 1991; Rhodes et al., 1996), but the results from this study clearly indicate selective uptake of amino acids by symbiotic aphids, varying between 61% and 92% of the ingested amino acid. One striking aspect of the data is the similarity in amino acid assimilation by symbiotic aphids on diet A and diet B (see Figures 1a and b). The implication is that the pea aphids in this study do not respond to the reduction in essential amino acid composition of the diet by increasing the overall assimilation of essential amino acids. Taken together with the lack of any increase in the volume of diet ingested when feeding on diet B (see above), the absolute values in Figures 1c and d indicate that the amount of individual essential amino acid assimilated by symbiotic aphids feeding on diet B is reduced by 36–59%. Aposymbiotic aphids generally assimilated a reduced proportion of individual amino acids than symbiotic aphids from both diet A and diet B. The assimilation of the amino acids lysine and glutamate was markedly reduced in aposymbiotic aphids. In particular, lysine assimilation was reduced by over 74%, and 9 out of 15 and 5 out of 18 aposymbiotic aphids feeding from diet A and B, respectively, did not assimilate any lysine from the diet.

The recovery of radioactivity from the embryos of adult aphids, following injection of equimolar amounts of each amino acid into the haemocoel, is shown in Figure 2. Radiolabel was detected in the embryos of both symbiotic and aposymbiotic aphids. The essential amino acid phenylalanine was rapidly

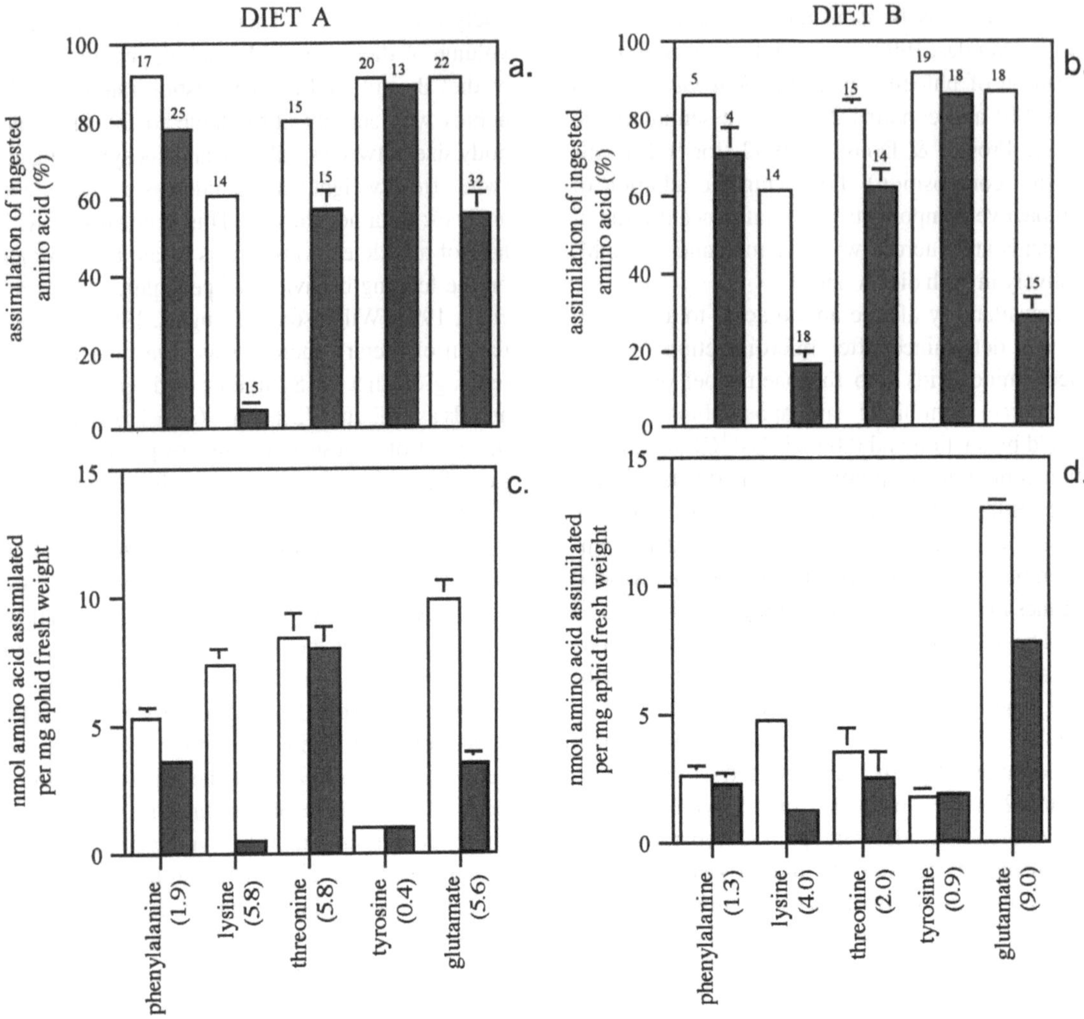


*Figure 1.* Assimilation of dietary amino acids by pea aphids. (a, b) assimilation expressed as percentage of ingested diet; (c, d) the absolute amount of each amino acid assimilated from the diet. Symbiotic aphids (open bars) and aposymbiotic aphids (closed bars) were transferred on day 7 of larval development to diets containing $^{14}$C-labelled amino acids and $^{3}$H-inulin, and honeydew was collected over 48 h. The aphids fed from diets containing either (a, c) 50 mol % essential amino acids (diet A), or (b, d) 20 mol % essential amino acids (diet B). Figures in parentheses are the concentration (in mol %) of each amino acid in the diet. Mean ± SE, sample sizes indicated in (a, b).

translocated to the embryos of symbiotic aphids such that all the injected radiolabel was recovered from the embryos at the end of the experimental period, providing further indication of the importance of this amino acid to the metabolism of aphids (see Douglas, 1996; Wilkinson & Douglas, 1996). The amount of radioactivity recovered from the embryos following injection of lysine, threonine, tyrosine and glutamate varied significantly (see amino acid term in ANOVA), indicating that there is some selectivity in the uptake of nutrients from the haemocoel by the embryos. The amount of radioactivity recovered from the embryos of aposymbiotic aphids was significantly lower than that from symbiotic aphids (see symbiosis term in ANOVA), reflecting the reduced biomass of embryos in these aphids (the mean protein content of embryos from symbiotic and aposymbiotic aphids was 0.117 $\mu$g and 0.018 $\mu$g, respectively). To account for the discrepancy in embryo biomass between symbiotic and aposymbiotic aphids, the incorporation of two essential amino acids, phenylalanine and lysine, into the embryo complement was quantified (see Table 1). Identification of the radiolabelled amino acids in the TCA-soluble fraction by thin-layer chromatography (data not shown) indicated that the majority of the radiolabel (over 80% in the case of phenylalanine) remained as the amino acid originally injected. Radiolabelled amino acids were incorporated into the

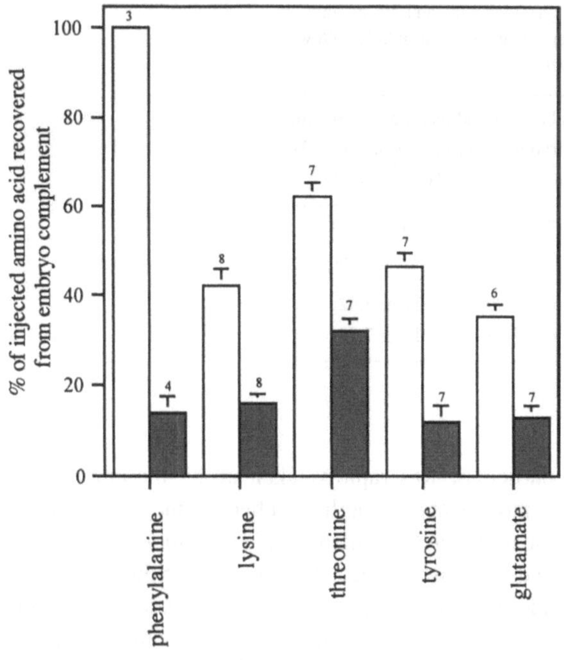


*Figure 2.* Recovery of radioactivity from the embryo fraction of symbiotic aphids (open bars) and aposymbiotic aphids (closed bars) following microinjection of $^{14}C$-labelled amino acids. Teneral apterae were injected with 40 pmol of amino acid and the amount of radioactivity in the embryo fraction after 2 h was determined by scintillation counting following removal of the embryos by dissection. Mean ± SE, sample sizes indicated on figure. ANOVA (after arcsine-squareroot transformation, omitting phenylalanine): amino acid $F_{(3,49)} = 19.6$, $P < 0.001$; symbiosis $F_{(1,49)} = 161.1$, $P < 0.001$; interaction $F_{(3,49)} = 1.4$, $P > 0.05$.

protein fractions of embryos from both symbiotic and aposymbiotic aphids, but there was a 3.5- and 4.2-fold increase in the amount of phenylalanine and lysine, respectively, incorporated by the embryos of aposymbiotic aphids.

## Discussion

The experiments described in this paper were designed to explore the inability of a nutritionally adequate diet to overcome the loss of the endogenous supply of essential amino acids in aposymbiotic aphids. The key conclusions are twofold. First, aposymbiotic aphids have a reduced capacity to assimilate sucrose, lysine and glutamate from the diet (see text and Figure 1), although the equivalent assimilation by symbiotic and aposymbiotic aphids of 3 out of 5 of the amino acids tested indicates that this is not a general metabolic dysfunction resulting from antibiotic treatment (see Wilkinson, 1998). Second, amino acids in the maternal haemocoel are taken up by the embryos of both symbiotic and aposymbiotic pea aphids, indicating that the embryos do not rely solely on their own complement of symbiotic bacteria for essential amino acid provisioning. The implications of the results from the complementary approaches adopted in this study are addressed, in turn, below.

### *Nutrient assimilation by symbiotic and aposymbiotic aphids*

Previous estimates of the amount of sucrose assimilated by *A. pisum* from chemically-defined diets are in close agreement with the data presented here: in the study of Rhodes et al. (1996), 69% of ingested sucrose was assimilated from a diet containing 0.7 m sucrose, and in the study of Mittler & Meikle (1991), 36–97% of ingested sucrose was assimilated from diets containing 0.3–1.2 M sucrose. *A. pisum* appears to have a high sucrose demand, possibly to fulfill the energy requirements for sustained embryogenesis, and the reduction in sucrose assimilation by aposymbiotic aphids may reflect the reduction in embryos which is characteristic of these aphids (Douglas, 1996). In addition, the reduced assimilation of sucrose by aposymbiotic aphids concurs with a recent hypothesis concerning osmoregulation in aphids (Wilkinson et al., 1997). Specifically, gut enzymes which mediate osmotic responses through the synthesis and excretion of oligosaccharides may be sensitive to the concentration of monosaccharides in the gut lumen. The finding that aposymbiotic pea aphids excrete oligosaccharides at lower dietary sucrose concentrations than symbiotic aphids (Wilkinson et al., 1997) may be linked to an accumulation of monosaccharides resulting from the reduced assimilation of dietary sucrose by aposymbiotic aphids.

The data on amino acid assimilation are in general agreement with our current understanding of amino acid metabolism in the pea aphid symbiosis. The reduction in the absolute amount of each essential amino acid assimilated from diet B concurs with previous evidence that the symbiotic bacteria act as a 'buffer' against dietary compositions of low amino acid quality (Prosser & Douglas, 1992; Douglas, 1992b). The implication is that the shortfall in essential amino acids experienced by symbiotic aphids on diet B may be countered by amino acid provision from the symbiotic bacteria. There is evidence to support this view: pea aphids reared on a diet lacking tryptophan produce honeydew containing tryptophan (Douglas & Prosser,

*Table 1.* Incorporation of $^{14}$C-labelled phenylalanine and lysine into the embryo complement of symbiotic and aposymbiotic pea aphids following microinjection into the maternal *haemocoel*

| Amino acid | Aphid | pmol amino acid incorporated per mg embryo protein (mean ± s.e., $n$ = 3) | |
|---|---|---|---|
| | | TCA-soluble | TCA-insoluble |
| Phenylalanine | Symbiotic | 44.7 ± 13.25 | 77.1 ± 21.01 |
| | Aposymbiotic | 120.9 ± 45.10 | 269.7 ± 133.64 |
| Lysine | Symbiotic | 32.4 ± 3.90 | 30.8 ± 5.92 |
| | Aposymbiotic | 273.4 ± 21.85 | 130.8 ± 7.75 |

1992), and a plausible explanation for the excretion of tryptophan is that the bacteria provide this, and possibly other essential amino acids, at rates in excess of the nutritional requirements of the aphid (see Douglas, 1998).

The reduced ability of aposymbiotic aphids to assimilate lysine from the diet can be expected to contribute to the poor performance of aposymbiotic aphids, although the underlying physiological mechanism inhibiting lysine assimilation is unclear. Lysine has previously been identified as an essential amino acid which is synthesised by the symbiotic bacteria (Wilkinson & Douglas, 1996), and from an analysis of the free amino acid pools of symbiotic and aposymbiotic aphids, Douglas (1996) has identified this amino acid as limiting to the growth of aposymbiotic aphids. Conversely, the reduction in glutamate assimilation by aposymbiotic aphids may simply reflect the loss of the bacterial sink for this amino acid, rather than a metabolic dysfunction. There is now substantial evidence that the bacteria synthesise essential amino acids by upgrading dietary non-essential amino acids, which predominate in the phloem sap (Sasaki et al., 1990; Douglas, 1993), and glutamate in particular may be the principal nitrogenous compound utilized by *Buchnera* (Whitehead & Douglas, 1993; Sasaki & Ishikawa, 1995).

*Allocation of amino acids to aphid embryos*

The detection of radioactivity in the embryo complement of both symbiotic and aposymbiotic aphids indicates rapid uptake of nutrients from the haemocoel, and the variability in the amount of radioactivity recovered from the embryos (see Figure 2) is a clear indication of selectivity in the allocation of nutrients derived from each amino acid. Phenylalanine, in particular, was rapidly taken up by the embryos of symbiotic aphids, perhaps reflecting the importance of the aromatic side chain for the formation of diphenols and biogenic amines (Kramer & Hopkins, 1987). In addition, the incorporation of phenylalanine and lysine into protein by the embryos of aposymbiotic aphids indicates that embryo growth in these aphids may be limited by the availability of essential amino acids. Bacterial provision of phenylalanine appears to be of particular importance to pea aphids (see Prosser & Douglas, 1991; Wilkinson & Douglas, 1996) and the most plausible explanation for the approximately 4-fold increase in incorporation of these essential amino acids by the embryos of aposymbiotic aphids is the requirement to overcome the shortfall incurred by the loss of the bacterial supply.

Surprisingly, there is little published information on the supply of nutrients to aphid embryos. Blackman (1974) demonstrated rapid incorporation of radiolabelled thymidine into the chromosomes of embryos of the aphid *Myzus persicae* (Sulzer) following injection into the haemocoel, and concluded that uptake must occur across the ovariole sheath surrounding each embryo, rather than via the trophocyte cells of the germarium. Our results are in agreement with this proposal, and taken together, the data suggest that aphid embryos are not independent of a supply of nutrients, including essential amino acids, from the maternal haemocoel. In addition to the loss of *Buchnera*-derived essential amino acids, the embryos of aposymbiotic aphids may be limited by a reduced ability of the maternal tissues to assimilate some essential amino acids, particularly lysine, from the diet. Further studies are required to determine the relative contribution of *Buchnera* from the maternal and embryo mycetocytes to the essential amino acid requirements of aphid embryos.

## Acknowledgements

Financial support was provided by the Bio-oriented Technology Research Advancement Institution, Ministry of Agriculture, Forestry and Fisheries, Japan.

## References

Baumann, P., C.-Y. Lai, D. Roubakhsh, N. A. Moran & M. A. Clark, 1995. Genetics, physiology, and evolutionary relationships of the genus *Buchnera* – intracellular symbionts of aphids. Annual Review of Microbiology 49: 55–94.

Blackman, R. L., 1974. Incorporation of thymidine into the chromosomes of aphid (*Myzus persicae*) embryos. Experientia 30: 1136–1137.

Buchner, P., 1965. Endosymbiosis of Animals with Plant Microorganisms. Wiley & Sons, New York.

Douglas, A. E., 1989. Mycetocyte symbiosis in insects. Biological Reviews 69: 409–434.

Douglas, A. E., 1992a. Requirement of pea aphids (*Acyrthosiphon pisum*) for their symbiotic bacteria. Entomologia Experimentalis et Applicata 65: 195–198.

Douglas, A. E., 1992b. Microbial brokers of insect-plant relationships. In: S. B. J. Menken, J. H. Visser & P. Harrewijn (eds), Proceedings of the 8th International Symposium on Insect-Plant Relationships. Kluwer Academic Publishers, Dordrecht, pp. 329–326..

Douglas, A. E., 1993. The nutritional quality of phloem sap utilized by natural aphid populations. Ecological Entomology 18: 31–38.

Douglas, A. E., 1996. Reproductive failure and the amino acid pools in pea aphids (*Acyrthosiphon pisum*) lacking symbiotic bacteria. Journal of Insect Physiology 42: 247–255.

Douglas, A. E., 1998. Nutritional interactions in insect-microbial symbioses: aphids and their symbiotic bacteria *Buchnera*. Annual Review of Entomology 43: 17–37.

Douglas, A. E. & W. A. Prosser, 1992. Synthesis of the essential amino acid tryptophan in the pea aphid (*Acyrthosiphon pisum*) symbiosis. Journal of Insect Physiology 38: 565–568.

Houk, E. J. & G. W. Griffiths, 1980. Intracellular symbiotes of the Homoptera. Annual Review of Entomology 25: 161–187.

Kramer K. J. & T. L. Hopkins, 1987. Tyrosine metabolism for insect cuticle tanning. Archives of Insect Biochemistry and Physiology 6: 279–301.

Mittler, T. E. & T. Meikle, 1991. Effects of dietary sucrose concentration on aphid honeydew carbohydrate levels and rates of excretion. Entomologia Experimentalis et Applicata 59: 1–7.

Prosser W. A. & A. E. Douglas, 1991. The aposymbiotic aphid: an analysis of chlortetracycline-treated pea aphid *Acyrthosiphon pisum*. Journal of Insect Physiology 37: 713–719.

Prosser W. A. & A. E. Douglas, 1992. A test of the hypothesis that nitrogen is upgraded and recycled in an aphid (*Acyrthosiphon pisum*) symbiosis. Journal of Insect Physiology 38: 93–99.

Prosser W. A., S. J. Simpson & A. E. Douglas, 1992. How an aphid (*Acyrthosiphon pisum*) symbiosis responds to variation in dietary nitrogen. Journal of Insect Physiology 38: 301–307.

Rahbé, Y., B. Delobel, G. Febvay & B. Chanterel, 1994. Aphid-specific triglycerides in symbiotic and aposymbiotic *Acyrthosiphon pisum*. Insect Biochemistry and Molecular Biology 24: 95–101.

Rhodes, J. D., P. C. Croghan & A. F. G. Dixon, 1996. Uptake, excretion and respiration of sucrose and amino acids by the pea aphid *Acyrthosiphon pisum*. Journal of Experimental Biology 199: 1269–1276.

Sasaki, T. & H. Ishikawa, 1995. Production of essential amino acids from glutamate by mycetocyte symbionts of the pea aphid, *Acyrthosiphon pisum*. Journal of Insect Physiology 41: 41–46.

Sasaki, T., T. Aoki, H. Hayashi & H. Ishikawa, 1990. Amino acid composition of the honeydew of symbiotic and aposymbiotic pea aphids *Acyrthosiphon pisum*. Journal of Insect Physiology 36: 35–40.

Sasaki, T., H. Hayashi & H. Ishikawa, 1991. Growth and reproduction of the symbiotic and aposymbiotic pea aphids, *Acyrthosiphon pisum*. Journal of Insect Physiology 41: 41–46.

Whitehead L. F. & A. E. Douglas, 1993. A metabolic study of *Buchnera*, the intracellular bacterial symbionts of the pea aphid *Acyrthosiphon pisum*. Journal of General Microbiology 139: 821–826.

Wilkinson, T. L., 1998. Elimination of intracellular symbionts from insects: a study of antibiotic-treatment in the pea aphid *Acyrthosiphon pisum*. Comparative Biochemistry and Physiology (in press).

Wilkinson, T. L. & A. E. Douglas, 1995. Aphid feeding, as influenced by the disruption of the symbiotic bacteria. Journal of Insect Physiology 41: 635–640.

Wilkinson, T. L. & A. E. Douglas, 1996. The impact of aposymbiosis on amino acid metabolism of pea aphids. Entomologia Experimentalis et Applicata 80: 279–282.

Wilkinson, T. L., D. A. Ashford, J. Pritchard & A. E. Douglas, 1997. Honeydew sugars and osmoregulation in the pea aphid *Acyrthosiphon pisum*. Journal of Experimental Biology 200: 2137–2143.

Wright, J. P., D. B. Fisher & T. E. Mittler, 1985. Measurement of aphid feeding rates on artificial diets using $^3$H-inulin. Entomologia Experimentalis et Applicata 37: 9–11.

*Entomologia Experimentalis et Applicata* **91:** 203–210, 1999.



# How nutritionally imbalanced is phloem sap for aphids?

Jonas Sandström* & Nancy Moran

*Department of Ecology & Evolutionary Biology, Center for Insect Science, University of Arizona, Tucson, AZ85721, USA; *Present address: Department of Entomology, Swedish University of Agricultural Sciences, Box 7044, S-750 07 Uppsala, Sweden*

Accepted: November 2, 1998

*Key words:* phloem sap, proteins, amino acids, aphids, symbionts, *Buchnera*, nutrition

## Abstract

Aphids harbour intracellular symbionts (*Buchnera*) that provide their host with amino acids present in low amounts in their diet, phloem sap. To find out the extent to which aphids depend on their symbionts for synthesis of individual essential amino acids, we have evaluated how well phloem sap amino acid composition matches the aphids' needs. Amino acid compositions of the ingested phloem sap were compared to amino acids in aphid body proteins and also to available information about optimal diet composition for other plant feeding insects. Phloem sap data from severed stylets of two aphid species, *Rhopalosiphum padi* (L.) (Homoptera: Aphididae) feeding on wheat, and *Uroleucon sonchi* (L.) (Homoptera: Aphididae) feeding on *Sonchus oleraceus* (L.), together with published information on phloem sap compositions from other plant species were used.

Phloem sap has in general only around 20% essential amino acids, with a range from 15–48%. Aphid body proteins and optimal diets for two other plant feeding insects have around 50%. The phloem sap of early flowering *S. oleraceus* ingested by *U. sonchi* contained 48%, which seems to be exceptional. Aphids feeding on different plants appear to be very differently dependent on their symbionts for their overall essential amino acid synthesis, due to the large variation in proportion of essential amino acids in phloem sap from different plants.

The profile of the essential amino acids in phloem sap from different plant species corresponds rather well to profiles of both aphid body proteins and optimal diets determined for plant feeding insects. However, methionine and leucine in phloem sap are in general low in these comparisons, suggesting a higher dependence on the symbiont for synthesis of these amino acids. Concentrations of several essential amino acids in phloem from different plant species seem to vary together, suggesting that levels of symbiont provisioning of different amino acids are adjusted in parallel.

## Introduction

Phloem sap has always been regarded as a highly troublesome diet for aphids due to its low nitrogen concentration and unbalanced composition. Nitrogen is mainly present in phloem sap as free amino acids with a concentration usually ranging between 50–300 mm (0.8–4.5% w/v), and a proportion of essential amino acids, of approximately 20 mol% (e.g., Sandström & Pettersson, 1994). Besides this general conclusion, very little information is available about the nitrogen balance of phloem sap for aphids.

Aphids harbour intracellular symbionts, and both physiological and genetic studies have shown that the symbionts are providing their host with certain amino acids (Douglas, 1998; Lai et al., 1994; Thao et al., 1998) which are present in low amounts in phloem sap. The symbiosis between aphids and bacteria has been considered a requisite to utilize phloem sap as a food source.

To assess the extent to which aphids depend on their symbionts for synthesis of individual amino acids, we here try to determine how well phloem sap amino acid composition matches the aphids' needs. We have made two different comparisons. First, we

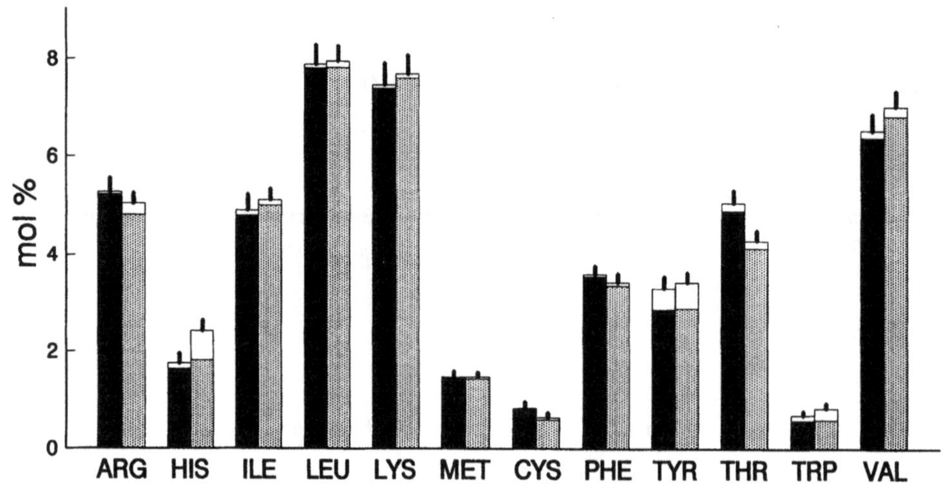


*Figure 1.* Amino acid composition in proteins and peptides (filled bottom) and in free amino acids (clear top) of two aphid species, *Rhopalosiphum padi* (left black bar, $n = 3$) *Uroleucon sonchi* (right shaded bar, $n = 3$). Relative concentration (mol%) of the 12 'essential' amino acids compared to concentration of all amino acids. Bars are means with standard errors for total amino acid content as positive lines.

compared how well the amino acid compositions in the ingested phloem sap matched with the amino acids comprising the proteins and free amino acids of the aphid. Second, we compared the above data concerning aphids with available information about optimal diet composition of amino acids in other plant-feeding insects that do not have bacterial symbionts. We used data from two aphid species, *Rhopalosiphum padi* L. feeding on wheat, and *Uroleucon sonchi* (L.) feeding on *Sonchus oleraceus* (L.). We only considered the use of amino acids for synthesis of proteins, which probably are their major function. However, amino acids can be needed for and transformed into other products, e.g., tyrosine is needed for sclerotization of the insect cuticle after moulting (Urich, 1994). Amino acids can also be utilized as an energy resource, but in aphids the latter seems to be of minor importance since carbohydrates are present in abundance and can be used for energy needs (Rhodes et al., 1996). However, it has been shown, for aphids feeding on synthetic diets, that some amino acids are used as major respiratory substrates (Febvay et al., 1995).

In insects and animals in general, ten amino acids are considered essential: arginine, histidine, isoleucine, leucine, lysine, methionine, phenylalanine, threonine, tryptophan and valine (Dadd, 1985). Additionally, tyrosine and cysteine, although they are not essential, must be considered, since they are synthesized only from essential amino acids: tyrosine from phenylalanine and cysteine from methionine. Some aphids (with symbionts) seem to be incapable of synthesizing cysteine from methionine (Dadd, 1985). For the aphid-bacteria symbiosis [*Myzus persicae* (Sulzer)] only a few amino acids are clearly essential, but for the aphid deprived of its symbionts the whole set of the ten amino acids are essential for aphid growth (Mittler, 1971). We concentrated our comparisons on the ten essential amino acids plus cysteine and tyrosine, because if they are present in insufficient amounts in the phloem sap the base for these 12 amino acids must be provided by the symbiont.

Neither the aphid nor its symbiont can fix nitrogen, and bacterial recycling of aphid nitrogenous waste products is not quantitatively important (Douglas, 1998). Therefore, the aphid has to ingest the necessary amount of nitrogen for protein synthesis from the phloem sap. Thus, the symbiont can improve aphid nutrition only by 'correcting' the composition of ingested amino acids in the phloem sap, using its broader biosynthetic capabilities.

## Materials and methods

*Aphids.* Aphid clones were established from single parthenogenetic females of *Rhopalosiphum padi*, the bird cherry-oat aphid, Arizona, USA, 1997, and *Uroleucon sonchi* from California, USA, 1996. Aphids were cultured on wheat (*Triticum aestivum* L., cv. Arapahoe) and *Sonchus oleraceus* L. (seeds collected from wild plants), respectively. Stock cultures were kept in a growth chamber at 21 ± 1 °C with a photoperiod of 16 h of artificial light.

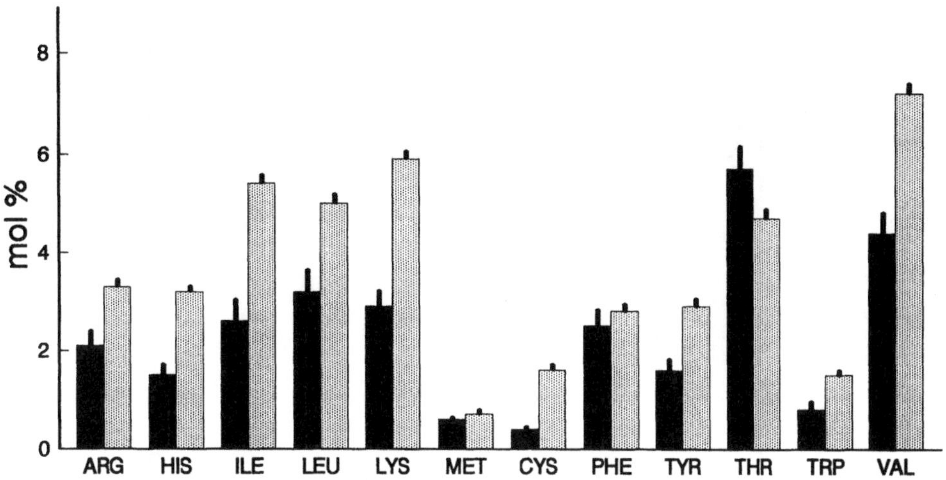


*Figure 2.* Amino acid composition in phloem sap of wheat ingested by *Rhopalosiphum padi* (left black bar, $n = 11$) and of *Sonchus oleraceus* ingested by *Uroleucon sonchi* (right shaded bar, $n = 12$). Relative concentration (mol%) of the 12 'essential' amino acids compared to concentration of all amino acids. Bars are means with standard errors as positive lines.

*Extraction and analysis of amino acids in aphids.* Aphid nymphs of mixed stages were weighed and transferred to a microfuge tube, and frozen at – 80 °C until use. The tube was frozen in liquid nitrogen and the aphids crushed with a pestle, and further homogenized after addition of ice cold buffer (50mM phosphate, pH = 7.0). After centrifugation the supernatant was hydrolyzed, and analyzed for total amino acids (proteins, peptides and free amino acids) using an ABI 420A/II amino acid analyzer. Samples were hydrolyzed by vapor-phase using 6N HCl at 155 °C for 1 h 15 min, then derivatized with phenylisothiocyanate (PITC), extracted, and transferred to an on-line HPLC. For tryptophan, prior to running, a solution of dodecanethiol was applied to sample. The addition of dodecanethiol protected tryptophan during acid hydrolysis. Samples for free amino acid content of aphids were crushed as above, but homogenized in 80% methanol. After centrifugation the supernatant was used for analyses without hydrolyzation. Free amino acid analyses were performed on a 7300 Beckman amino acid analyzer (ion-exchange column, ninhydrin method). All analyses were made on three replicate samples. Concentration in proteins and peptides was determined by subtracting free amino acids from total amino acids (proteins, peptides and free amino acids).

*Phloem sap samples.* Phloem sap was obtained from 28-day-old wheat, in the five-leaf stage, infested with *R. padi* for seven days and 80-90-day-old, early flowering, *S. oleraceus* infested with *U. sonchi* for seven days. Phloem sap was collected from severed aphid stylets according to the method described by Unwin (1978) and analyzed for free amino acids as above. The phloem sap compositions of amino acids were compared to previous studies of phloem sap from other plant species collected from severed aphid stylets. Only studies that had at least five phloem replicates, and for which analyses included at least 10 of the 12 essential amino acids, were used. If multiple treatments on the same plant species were reported, the treatment which was considered normal, and with most replicates, was used. Nine plant-aphid combinations that fulfilled these criteria are shown in Table 1.

*Statistical analyses.* The results were analyzed using SAS software (SAS institute, Cary, N.C., USA) version 6.12. (SAS Institute Inc., 1988). Data were analyzed by analysis of variance using the procedure GLM.

## Results

The total amino acid concentration (proteins, peptides, free) in both aphid species did not differ significantly ($F_{1,5} = 0.02$, $P = 0.88$); means (± s.e.) based on wet weight were 648 ± 15 nmol/mg aphid for *R. padi* and 645 ± 12 nmol/mg aphid for *U. sonchi.* These values are similar to those observed from other aphids (e.g., Febvay et al., 1988). The amino acid profiles of the proteins plus peptides of the two aphid species were

*Table 1.* Amino acid composition in phloem sap of nine different plant species

| Plant | *Medicago sativa*[1] | *Pisum sativum*[1] | *Trifolium pratense*[1] | *Vicia faba*[1] | *Avenae sativa*[2] | *Hordeum vulgare*[2] | *Oryza sativa*[3] | *Brassica juncea*[4] | *Prunus padus*[5] |
|---|---|---|---|---|---|---|---|---|---|
| Insect | *A. pisum* | *A. pisum* | *A. pisum* | *A. pisum* | *R. padi* | *R. padi* | *N. lugens* | *L. erysimi* | *R. padi* |
| Arginine | 1.9 | 5.5 | 2.9 | 4.6 | 3.1 | 3.0 | 3.5 | 4.2 | 6.9 |
| Histidine | 0.4 | 0.6 | 0.7 | 1.0 | 0.9 | 0.6 | 1.3 | 3.6 | n.m. |
| Isoleucine | 1.6 | 1.7 | 2.0 | 1.9 | 3.4 | 2.0 | 2.4 | 3.9 | 3.0 |
| Leucine | 1.4 | 1.7 | 1.9 | 2.2 | 3.9 | 2.1 | 2.5 | 2.4 | 1.9 |
| Lysine | 1.3 | 1.6 | 1.8 | 2.2 | 3.2 | 2.3 | 3.5 | 4.7 | 2.2 |
| Methionine | 0.7 | 0.7 | 0.5 | 0.7 | 0.8 | 0.01 | 0.3 | 0.2 | 2.2 |
| Cysteine | n.m. | n.m. | n.m. | n.m. | n.m. | n.m. | 1.3 | n.m. | n.m. |
| Phenylalanine | 0.7 | 1.1 | 1.3 | 1.2 | 2.5 | 1.6 | 1.1 | 2.3 | 1.7 |
| Tyrosine | 0.4 | 0.4 | 0.8 | 1.3 | 1.3 | 1.2 | 1.4 | 1.7 | 1.4 |
| Threonine | 3.6 | 2.4 | 2.8 | 3.0 | 9.0 | 5.0 | 4.9 | 6.6 | 3.3 |
| Tryptophan | 0.2 | 0.3 | 0.4 | 0.4 | 0.9 | 0.5 | 0.7 | 0.6 | 1.0 |
| Valine | 2.8 | 3.3 | 3.3 | 3.1 | 5.5 | 5.3 | 4.1 | 6.4 | 7.7 |
| Total | 14.9 | 19.4 | 18.1 | 21.6 | 34.4 | 23.6 | 27.2 | 36.6 | 31.5 |

[1]From Sandström & Pettersson (1994).
[2]Weibull et al. (1988).
[3]Kajimura, et al. (1995).
[4]Weibull & Melin (1990).
[5]Sandström (1998)[5].
All samples obtained from severed aphid stylets. Relative concentration (mol%) of the 12 'essential' amino acids compared to concentration of all amino acids analyzed. n.m. = not measured.

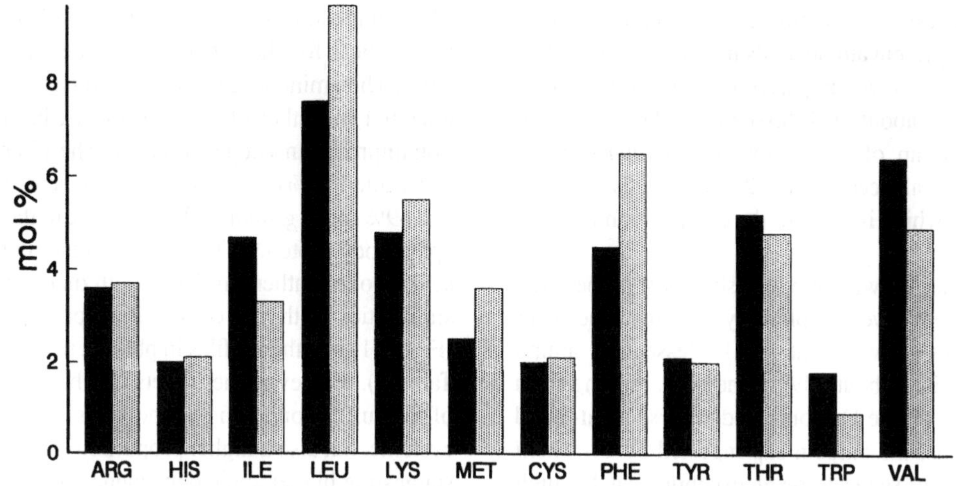


*Figure 3.* Amino acid composition in optimal synthetic diets for silk moth larvae (*Bombyx mori*) (left black bar) and the cotton weevil larvae (*Anthonomus grandis*) (right shaded bar). From Arai & Ito, (1967) and Vanderzant (1965, 1973). Relative concentration (mol%) of the 12 'essential' amino acids compared to concentration of all amino acids in diets.

very similar (lower part of bars, Figure 1) and corresponded well to compositions previously reported for other aphids (e.g., Febvay et al., 1988), and proteins from other insects (e.g., Brodbeck, et al., 1996; Barrett & Schmidt, 1991). These 12 'essential' amino acids composed 48.7% of the total 20 protein amino acids for *R. padi*, and 49.3% for *U. sonchi*.

The concentration of free amino acids in the aphids differed significantly ($F_{1,5} = 42.0$, $P = 0.003$); concentrations were 30 ± 2 nmol/mg aphid for *R. padi* and 48 ± 3 nmol/mg aphid for *U. sonchi*. The free amino acids in their body made up only a small share of the total (proteins, peptides, free), less than 8%. The composition of the free amino acids differed between the two aphid species (top of bars, Figure 1). Tyrosine was one of the major free 'essential' amino acids; abundant tyrosine is common among insects, due to its function in the sclerotization of the insect cuticle after moulting (Urich, 1994). Cysteine was present in both its monomer form, cysteine and its dimer form, cystine. A few non-protein amino acids were present in very low proportions.

The proportion of all 20 amino acids that consisted of the 12 'essential' amino acids differed significantly between the phloem sap of the two plants ($F_{1,21} = 60.7$, $P < 0.0001$). The mean percentages were 28.7 ± 3.7% for *R. padi* on wheat and 47.4 ± 1.8% for *U. sonchi* on *S. oleraceus*. Among essential amino acids, the profiles of relative abundances were generally similar for the two plants, with the main differences between the plants involving cystine and threonine and, to some degree, methionine and phenylalanine (Figure 2).

**Discussion**

The amino acid composition of body proteins in aphids is probably a conservative character that is similar among most aphids and even other insects. Likewise, the concentration of proteins in a healthy aphid is probably rather similar among aphids.

In contrast, the free amino acid pool is very variable. Large variation in free amino acid pool is a common feature in insects (Urich, 1994). In aphids this could reflect differences in the composition of the ingested phloem sap and/or differences in physiological capabilities of aphids or symbionts.

When aphid body amino acid composition (Figure 1) and phloem sap composition (Figure 2) are compared, the profiles are well matched. Major differences are that phloem sap has a somewhat lower proportion of arginine, leucine, lysine and methionine, but higher proportion of histidine, cystine (*S. oleraceus*), tryptophan, and threonine (wheat) compared to aphid protein composition. This suggests that phloem saps of these two plants are well balanced for synthesis of aphid proteins based on the profiles of these 12 amino acids. It should be emphasized that this type of comparison only allows speculative inferences (Dadd, 1985), and that total aphid amino acid requirement might differ from body protein composition.

The 12 'essential' amino acids make up 28% of the total 20 protein amino acids in phloem ingested by *R. padi*, this is low compared to aphid protein composition with about 50% 'essentials'. For *U. sonchi* the phloem sap of early flowering *S. oleraceus* is much better balanced; these 12 amino acids make up about 48%, which is close to the composition in aphid proteins.

A comparison with previous studies of phloem sap collected from severed aphid stylets from nine other plant species showed that the 12 'essential' amino acids comprised about 20% of the total, ranging from 15% to 37% (Table 1). Compared to these plant-aphid combinations, the phloem sap ingested by *U. sonchi* on *S. oleraceus* seems to be an exception: such a high proportion of 'essential' amino acid is rarely recorded in plants. In all nine plants, isoleucine, leucine, lysine, phenylalanine and valine are present in profiles similar to those in Figure 2, while arginine, histidine, methionine and threonine are more variable. Arginine, histidine and threonine are relatively high in certain plants (e.g., *Pisum sativum*, *Avena sativa*), and methionine ranges from very low to rather high. The proportions of tryptophan and tyrosine seemed to be intermediate in variation. Information about cystine/cysteine was scarce. This compilation suggests that the profiles of these 12 amino acids in phloem sap encountered by aphids feeding on different host plants seldom show any dramatic differences. Amino acid composition may differ in plants with unusual metabolism; for example, the CAM plant *Opuntia* sp. (Cactaceae) contain 70% essential amino acids in phloem (Wang & Nobel, 1995).

There have been efforts to define optimal synthetic diets for aphids with symbionts or deprived of symbionts (aposymbiotic aphids), but aphids reared on synthetic diets have a very low performance (Douglas, 1992). Performance of aposymbiotic aphids is worse than that of symbiotic aphids, regardless of synthetic diets or host plant (Douglas, 1998). Thus, it is difficult to evaluate the requirements for aposymbiotic aphids. However, optimal synthetic diets have been designed for other naturally non-symbiotic plant feeding insects, i.e., the silk moth larvae (*Bombyx mori*) (Arai & Ito, 1967) and cotton weevil larvae (*Anthonomus grandis*) (Vanderzant, 1965, 1973). These insects normally chew on plant tissue and obtain their amino acids from digested proteins. In these synthetic diets they were given free amino acids, allowing digestibility of proteins to be disregarded, as in the case of aphids which feed on free amino acids in phloem sap. The compositions of the 12 'essential' amino acids in the diets for these two insects are shown in Figure 3. The amino acids cysteine and tyrosine were not essential for either of the two insects, but their addition improved insect performance. The 12 amino acids constituted 47% of the total 20 amino acids for *B. mori* and 49% for *A. grandis*, which is close to the values for aphid body proteins. The composition of the 12 amino acids looked rather similar in both diets, and showed similarities with the composition of aphid proteins (Figure 1) and the profiles in phloem sap (Figure 2 and Table 1). However, these diets had higher proportion of the sulfur-containing amino acids, methionine and cysteine, but also a higher proportion of leucine. Thus, suggesting higher aphid dependency on symbionts for synthesis of these amino acids. Whether aposymbiotic aphids have the same nutritional requirements as the above plant-feeding larvae is unknown, but the values in Figure 3 could be a reasonable approximation. This is supported by the fact that; aposymbiotic aphids (*Acyrthosiphon pisum*) grow twice as fast on a synthetic diet with 50% essential amino acids than on a diet with only 20% essential amino acids, whereas growth of symbiotic aphids did not vary significantly with amino acid composition (Prosser & Douglas, 1992).

Although phloem sap in general has a low proportion of essential amino acids, some aphids feed on phloem with much higher proportions. Levels of essential amino acids in ingested phloem can approach those in aphid proteins and insect optimal diets, as seen for *U. sonchi* and also for some aphids that are able to manipulate their plant's phloem composition (Sandström et al., in press). Thus, aphids feeding on different plants appear to be very differently dependent on symbionts for their overall essential amino acid provisioning. The phloem profile of essential amino acids, however, similar among plants and is similar to profiles found in aphid proteins and balanced synthetic diets for other plant-feeding insects.

Some genetic evidence suggests adaptation on the part of aphid symbionts to improve provisioning of essential amino acids to hosts. In the aphid family Aphididae, genes for leucine and tryptophan biosynthesis are amplified on plasmids (Lai et al., 1994; Bracho et al., 1995; Rouhbakhsh et al., 1997). Interestingly, *U. sonchi* shows an unusual reduction in the number of functional tryptophan biosynthetic genes (*trpEG*) present on symbiont plasmids; most of the copies of the genes are nonfunctional pseudogenes (Baumann et al., 1997). A similar situation exists in *Diuraphis* noxia, which causes feeding damage inducing elevated levels of essential amino acids (Telang et al., in press), and which also has lost most functional tryptophan biosynthetic genes as pseudogenes (Lai et al., 1996). Thus, aphids that ingest phloem with unusually high proportions of essential amino acids may experience reduced selection for amino acid overprovisioning by symbionts, resulting in the loss of the symbionts extra copies of amino acid biosynthetic genes. The observation that concentrations of essential amino acids vary together suggests that symbiont provisioning of different amino acids also would be adjusted in parallel. A correlated shift in ability to provision both leucine and tryptophan is suggested by observations on *D. noxia* in which numbers of gene copies for both leucine and tryptophan production have been adjusted downwards (Thao et al., 1998).

## Acknowledgements

We thank W. Clark for help with the amino acid analyses. J. Sandström's visit at the University of Arizona was financed by a grant from the Swedish Council for Forestry and Agricultural Research. This study was also financially supported by a grant from USDA to N. Moran and P. Baumann, and a grant from the University of Arizona to J. Sandström.

## References

Arai, N. & T. Ito, 1967. Nutrition of the silkworm, *Bombyx mori*. XVI. Quantitative requirements for essential amino acids. Bulletin of Sericultural Experimental Station Tokyo 21: 373–384.

Barrett, M. & J. M. Schmidt, 1991. A comparison between the amino acid composition of an egg parasitoid wasp and some of its hosts. Entomologia Experimentalis et Applicata 59: 29–41.

Baumann, L., M. A. Clark, D. Rouhbakhsh, P. Baumann, N. A. Moran & D. J. Voegtlin, 1997. Endosymbionts (*Buchnera*) of the aphid *Uroleucon sonchi* contain plasmids with *trpEG* and remnants of *trpE* pseudogenes. Current Microbiology 35: 18–21.

Bracho, A. M., D. Martínez-Torres, A. Moya & A. Latorre, 1995. Discovery and molecular characterization of a plasmid localized in *Buchnera* sp., a bacterial endosymbiont of the aphid *Rhopalosiphum padi*. Journal of Molecular Evolution 41: 67–73.

Brodbeck, B. V., P. C. Andersen & R. F. Mizell III, 1996. Utilization of primary nutrients by the polyphagous xylophage, *Homalodisca coagulata*, reared on single host species. Archives of Insect Biochemistry and Physiology 32: 65–83.

Dadd, R. H., 1985. Nutrition: Organisms. In: G. A. Kerkut & L. I. Gilbert (eds), Comprehensive Insect Physiology Biochemistry and Pharmacology. Vol. 4, Regulation: Digestion, Nutrition, Excretion. Pergamon Press, Oxford, pp. 313–390.

Douglas, A. E. 1992. Requirement of pea aphids (*Acyrthosiphon pisum*) for their symbiotic bacteria. Entomologia Experimentalis et Applicata 65: 195–198.

Douglas, A., 1998. Nutritional interactions in insect-microbal symbioses: Aphids and their symbiotic bacteria *Buchnera*. Annual Review of Entomology 43: 17–37.

Febvay, G., B. Delobel & Y. Rahbe, 1988. Influence of the amino acid balance on the improvement of an artificial diet for a biotype of *Acyrthosiphon pisum* (Homoptera: Aphididae). Canadian Journal of Zoology 66: 2449–2453.

Febvay, G., I. Liadouze, J. Guillaud & G. Bonnot, 1995. Analysis of energetic amino acid metabolism in *Acyrthosiphon pisum*: a multidimensional approach to amino acid metabolism in aphids. Archives of Insect Biochemistry and Physiology 29: 45–69.

Kajimura, T., K. Fujisaki & F. Nakasuji, 1995. Effect of organic rice farming on leafhoppers and planthoppers. 2. Amino acid content in the rice phloem sap and survival rate of planthoppers. Applied Entomology and Zoology 30: 17–22.

Lai, C.-Y., L. Baumann & P. Baumann, 1994. Amplification of *trpEG*; adaptation of *Buchnera aphidicola* to an endosymbiotic association with aphids. Proceedings of the National Academy of Sciences in the United States of America 91: 3819–3823.

Lai C.-Y., P. Baumann & N. Moran, 1996. The endosymbiont (*Buchnera* sp.) of the aphid *Diuraphis noxia* contains plasmids consisting of *trpEG* and tandem repeats of *trpEG* pseudogenes. Applied and Environmental Microbiology 62: 332–339.

Mittler, T. E., 1971. Dietary amino acid requirements of the aphid *Myzus persicae* affected by antibiotic uptake. Journal of Nutrition 101: 1023–1028.

Prosser, W. A. & A. E. Douglas, 1992. A test of the hypotheses that nitrogen is upgraded and recycled in an aphid (*Acyrthosiphon pisum*) symbiosis. Journal of Insect Physiology 38: 93–99.

Rhodes, J. D., P. C. Croghan & A. F. G. Dixon, 1996. Uptake, excretion and respiration of sucrose and amino acids by the pea aphid *Acyrthosiphon pisum*. Journal of Experimental Biology 199: 1269–1276.

Rouhbakhsh, D., M. A. Clark, L. Baumann, N. A. Moran & P. Baumann, 1997. Evolution of the tryptophan biosynthetic pathway in (*Buchnera*) (aphid endosymbiont): studies of plasmid-associated *trpEG* within the genus *Uroleucon*. Molecular and Phylogenetic Evolution 62: 332–339.

Sandström, J., 1998. Nutritional quality of phloem sap in relation to aphid host plant use. In: J. M. Nieto Nafria & A. F. G. Dixon (eds), Aphids in Natural and Managed Ecosystems. Universidad de Leon, Leon (Spain), pp. 261–265.

Sandström, J. & J. Pettersson, 1994. Amino acid composition of phloem sap and the relation to intraspecific variation in pea aphid (*Acyrthosiphon pisum*) performance. Journal of Insect Physiology 40: 947–955.

SAS Institute Inc, 1988. SAS/STAT® User's Guide, Release 6.03 Editions. Cary, NC: SAS Institute Inc.

Thao, M. L., L. Baumann, P. Baumann & N. Moran, 1998. Endosymbionts (*Buchnera*) from the aphid *Schizaphis graminum* and *Diuraphis noxia* have different copy numbers of the plasmid containing leucine biosynthetic genes. Current Microbiology 36: 238–240.

Unwin D. M., 1978. A versatile high frequency radio microcautery. Physiological Entomology 3: 71–73.

Urich, K., 1994. Comparative Animal Biochemistry. Springer-Verlag, Berlin.

Wang, N. & P. S. Nobel, 1995. Phloem exudate collected via scale insect stylets for the CAM species *Opuntia ficus-indica* under current and doubled $CO_2$ concentrations. Annals of Botany 75: 525–532.

Weibull, J., 1988. Free amino acids in the phloem sap from oats and barley resistant to *Rhopalosiphum padi*. Phytochemistry 27: 2069–2072.

Weibull, J. & G. Melin, 1990. Free amino acid content of phloem sap from *Brassica* plants in relation to performance of *Lipaphis erysimi* (Hemiptera: Aphididae). Annals of Applied Biology 116: 417–423.

Vanderzant, E. S., 1965. Axenic rearing of the boll weevil on defined diets: amino acid, carbohydrate, and mineral requirements. Journal of Insect Physiology 11: 659–670.

Vanderzant E. S., 1973. Axenic rearing of larvae and adults of the boll weevil on defined diets: additional tests with amino acids, and vitamins. Annals of the Entomological Society of America 66: 1184–1186.

*Entomologia Experimentalis et Applicata* **91:** 211–225, 1999.



# It's about time: the evidence for host plant-mediated selection in the apple maggot fly, *Rhagoletis pomonella*, and its implications for fitness trade-offs in phytophagous insects

Jeffrey L. Feder & Kenneth E. Filchak
*Department of Biological Sciences, University of Notre Dame, Notre Dame, IN 46556, USA*

Accepted: October 17, 1998

*Key words:* apple maggot fly, sympatric speciation, negative genetic fitness trade-offs, host plant-associated selection, host races, allozymes, diapause

## Abstract

The apple maggot fly, *Rhagoletis pomonella*, Walsh (Diptera: Tephritidae), provides a unique opportunity to address the issue of host-related fitness trade-offs for phytophagous insects. *Rhagoletis pomonella* has been controversial since the 1860's when Benjamin Walsh cited the fly's shift from hawthorn (*Crataegus* spp.) to apple (*Malus pumila*) as an example of an incipient sympatric speciation event. Allozyme and mark-release-recapture studies have subsequently confirmed the status of apple and hawthorn flies as partially reproductively isolated and genetically differentiated 'host races', the hypothesized initial stage in sympatric divergence. Here, we review the ecological and genetic evidence for host-plant mediated selection in *R. pomonella*. We reach the following three major conclusions: First, although developmental timing is not everything, it is a good deal of the story. Differences in the fruiting phenologies of apple and hawthorn trees exert different selection pressures on the diapause and eclosion time characteristics of the host races. In particular, the ~3-week earlier mean fruiting phenology of apples in eastern North America appears to select for a slower rate of metabolism or deeper pupal diapause in apple than hawthorn flies. Second, host-related fitness trade-offs for *R. pomonella* may not be due to disruptive selection affecting any one specific life-history stage. Rather, it is the sum total of directional selection pressures acting across different life-stages that generates divergent selection on apple and hawthorn flies. For example, selection favors the alleles *Me 100*, *Acon-2 95* and *Mpi 37* (or linked genes) in the larval stage in both host races. However, these same alleles are disfavored in the pupal stage to follow, where they correlate with early adult eclosion, and by inference premature diapause termination. Because apple trees fruit an average of ~3 weeks earlier than hawthorn trees, this counter-balancing selection is stronger on apple-fly pupae. The net result is that the balance of selective forces is different between apple and hawthorn flies, helping to maintain the genetic integrity of the host races in sympatry in the face of gene flow. Finally, natural *R. pomonella* populations harbor a good deal of genetic variation for development-related traits. This variation allows fly populations to rapidly respond to temporal vagaries in local environmental conditions across years, as well as to broad-scale geographic differences that exist across the range of the species. Perhaps most importantly, this variation gives *R. pomonella* the flexibility to explore and adapt to novel plants. Taken together, our results underscore how difficult it can be to document host plant-related fitness trade-offs for phytophagous insects due to the need to consider details of the entire life-cycle of a phytophagous insect. Our findings also show how reproductive isolation can arise as a by-product of host-associated adaptation in insects, a central theme for models of sympatric speciation via host shifts.

## The evolutionary and ecological significance of the apple maggot

The apple maggot fly, *Rhagoletis pomonella* (Diptera: Tephritidae), highlights four fundamental issues in evolution and ecology. The first issue concerns the relative importance of selection vs drift in shaping patterns of genetic polymorphism within natural populations (Lewontin, 1974). The second is whether speciation is frequently triggered by the same forces causing phenotypic differentiation, namely divergent selection stemming from the use of alternative en-

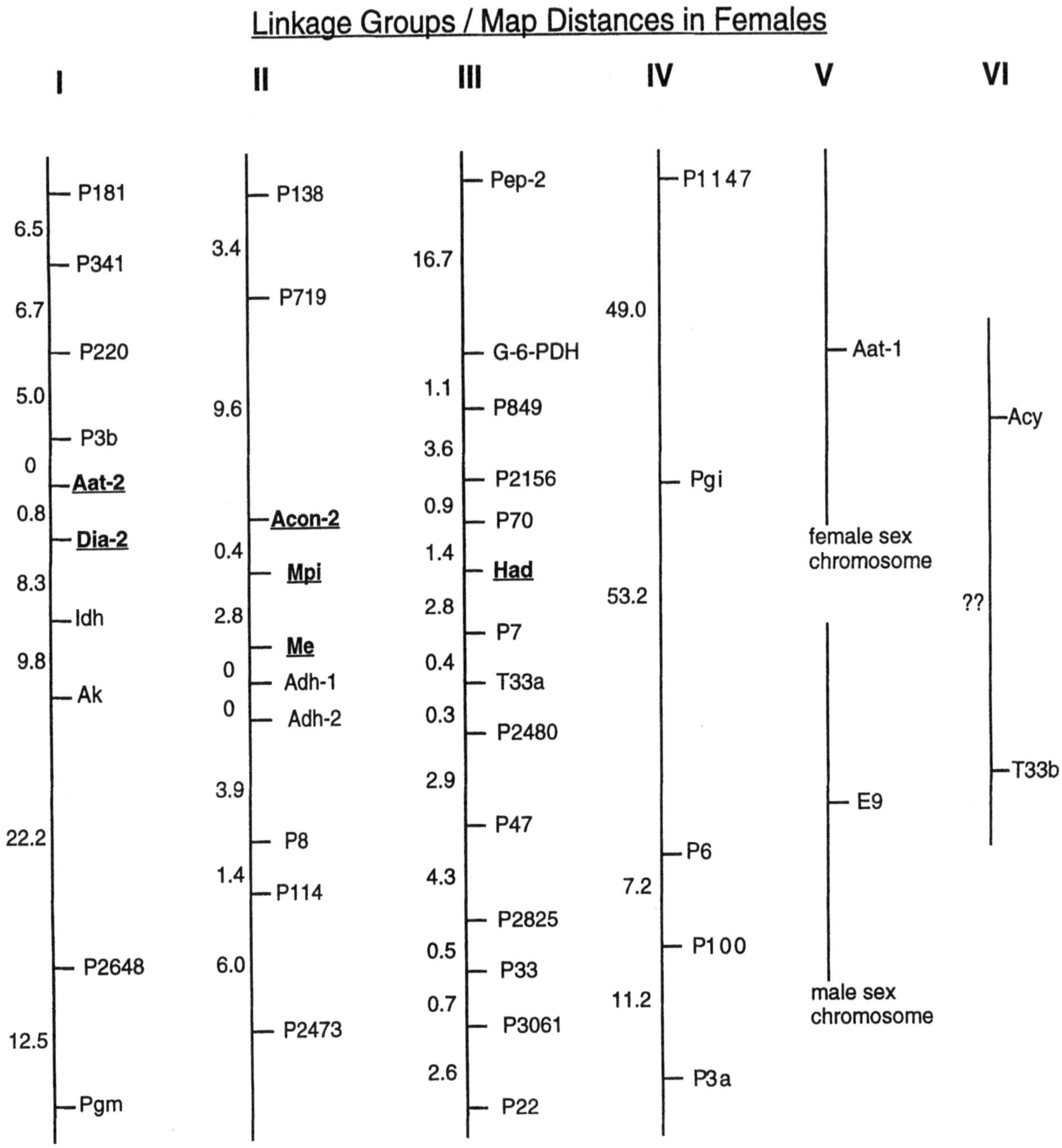


*Figure 1.* Allozyme and molecular (cDNA) linkage map for *R. pomonella*. The six allozyme loci displaying host-associated genetic differentiation appear in bold and are underlined. Map distances pertain to females only and have been corrected for the possibility of multiple exchange according to the formula of Haldane (1919), where x (recombination distance in centi-Morgans) = $-50 \ln(1-2y)$, and $y$ = the observed recombination fraction. Since recombination is limited in males, evolutionary map distances for *R. pomonella* can be approximated by halving the values given for females.

vironments (Schluter, 1999). The third question is whether allopatry (the geographic separation of populations) is a necessary precondition for speciation (Bush, 1969, 1975). In essence, can disruptive selection ever be strong enough to overcome the homogenizing effects of gene flow and result in the sympatric divergence (speciation without geographic isolation) of populations utilizing different niches/habitats? Finally, *Rhagoletis* flies also strike at the heart of a debate in the ecology community concerning insect/plant interactions. Specifically, do host plant-related fitness trade-offs exist for insects and, if so, what aspect of the host environment (plant secondary chemistry, physical characteristics such as trichomes, phenology, natural enemies, etc.) is responsible for generating these trade-offs? By trade-offs, we refer to traits that enhance the performance of an insect on one plant to its detriment on others. Such trade-offs have long provided the theoretical basis for discussions concerning insect diet-breadth and host specialization (see special features section, Ecology [1988] vol. 69: 885–915).

## The current evidence for host-related tradeoffs

Most tests of insect performance across hosts have failed to reveal the type of negative genetic correlations indicative of fitness trade-offs (Futuyma & Moreno, 1988; Rausher, 1988, 1992; Jaenike, 1990; for exceptions see Gould, 1979; Mitter et al., 1979; Fry, 1990; Karowe, 1990; Via, 1991; Mackenzie, 1996). Taken on face value, these results suggest that different plants do not usually exert strong disruptive selection pressures on insects. We may therefore have to re-evaluate the theoretical foundation upon which much of our thinking about insect-plant relationships is based (Fry, 1996; Kawecki, 1997). However, many studies testing for trade-offs in phytophagous insects have at least one of several serious shortcomings (Rausher, 1988, 1992; Fry, 1993), including problems due to limited statistical sensitivity, general vigor effects, testing for only a few components of fitness in a single life-history stage (most notably larval growth rate and pupal size), and measuring fitness under artificial laboratory conditions and not in the field. As we shall see, all of these criticisms could potentially apply to *Rhagoletis*.

## The link between trade-offs, reproductive isolation and speciation

The question of fitness trade-offs in *Rhagoletis* unites the ecological and evolutionary issues we raised above. In 1864, Benjamin Walsh proposed that many phytophagous insects speciate in the process of shifting and adapting to new host plants. A modern phrasing of Walsh's hypothesis is that reproductive isolation can evolve as a by-product (a pleiotropic side effect) of the very traits adapting insect populations to alternative host plants (Bush, 1969, 1975; Rice, 1987; Rice & Hostert, 1993). Hybrids have reduced fitness because of irreconcilable selection pressures (trade-offs) imposed by the external environment (i.e., hybrids possess phenotypes ill-suited for survival and reproduction on the host plants of parental populations). Speciation is therefore largely an ecological (adaptive) process, at least during its initial, formative stages (Dobzhansky, 1951). This ecological perspective of speciation stands in contrast to the commonly held view that reproductive isolation in animals (i.e., hybrid inviability and sterility) largely stems from internal genetic incompatibilities manifested when the genomes of diverged parental taxa are brought together in hybrids (Muller, 1940; Mayr, 1963; reviewed by Schluter, 1999).

## Sympatric speciation via host shifts

Benjamin Walsh (1864) also sparked debate on the geographic context of speciation by hypothesizing that divergence via host shifts in phytophagous insects frequently occurs in sympatry. *Rhagoletis pomonella* assumed a central position in this discussion because Walsh (1867) cited the shift of the fly, from its native host hawthorn (*Crataegus* L. spp.) to introduced domestic apple (*Malus pumila* L.), an event that occurred ~150 years ago, as a prime example of an incipient sympatric speciation event. Subsequently, the term 'host race' has been used to describe this initial stage in the sympatric speciation process, host races being partially isolated, conspecific populations that owe their isolation to host-associated adaptations (Diehl & Bush, 1984).

Recent genetic studies have supported Walsh's insights about the apple maggot fly. Six allozyme loci show significant allele frequency differences between sympatric apple and hawthorn-fly populations collected from across eastern North America (Feder et al., 1988, 1990a; McPheron et al., 1988; Feder & Bush, 1989). These six allozymes (Aspartate amino transferase-2 [*Aat-2*], NADH-Diaphorase-2 [*Dia-2*], Malic enzyme [*Me*], Aconitase-2 [*Acon-2*], Mannose phosphate isomerase [*Mpi*] and Hydroxyacid dehydrogenase [*Had*]) map to just three different regions of the *R. pomonella* genome (Figure 1; Berlocher and Smith, 1983; Feder et al., 1989; Roethele et al., 1997). In addition, significant linkage disequilibrium exists between loci within, but not among, these three chromosomal linkage groups (Feder et al., 1988, 1990a; Berlocher & McPheron, 1996). Test cross data suggest the presence of inversion polymorphism in the three regions, which could help account for the disequilibrium among linked allozymes (J. L. Feder, unpubl.).

Mark-release-recapture studies have also demonstrated the existence of host fidelity in *R. pomonella*. That is, adult flies tend to mate on and oviposit into the same species of host plant (apple vs hawthorn) that they fed within as larvae. The finding of host fidelity is important because adults mate exclusively on or near host fruits on trees (Prokopy et al., 1971, 1972). Hence, host fidelity produces assortative mating that translates directly into premating reproductive isolation, a good example of how the ecology of an organism influences population divergence. Differences in host acceptance (preference) behavior (Prokopy et al., 1988; Luna & Prokopy, 1995; Feder et al., 1994, 1998) and the timing of adult eclosion (Smith, 1988; Feder et al., 1994, 1998) both contribute to host fidelity.

However, while host fidelity is fairly strong, it is not absolute. We have estimated that on average ~6% of the adult fly population moves between apple and hawthorn trees each generation (Feder et al., 1994, 1998). Our field studies have also shown that these migrants mate with resident flies - presumably of different host origin - and that migrant females oviposit into non-natal fruits (Feder et al. 1994, 1998). These findings provide strong evidence that apple and hawthorn populations of *R. pomonella* are partially isolated races and not completely separated sibling species, as sympatric populations are still connected by gene flow (Note: mating studies suggest that ethological, sterility and viability barriers do not isolate apple from haw flies; Reissig & Smith, 1978; Smith, 1988; Bush, 1992).

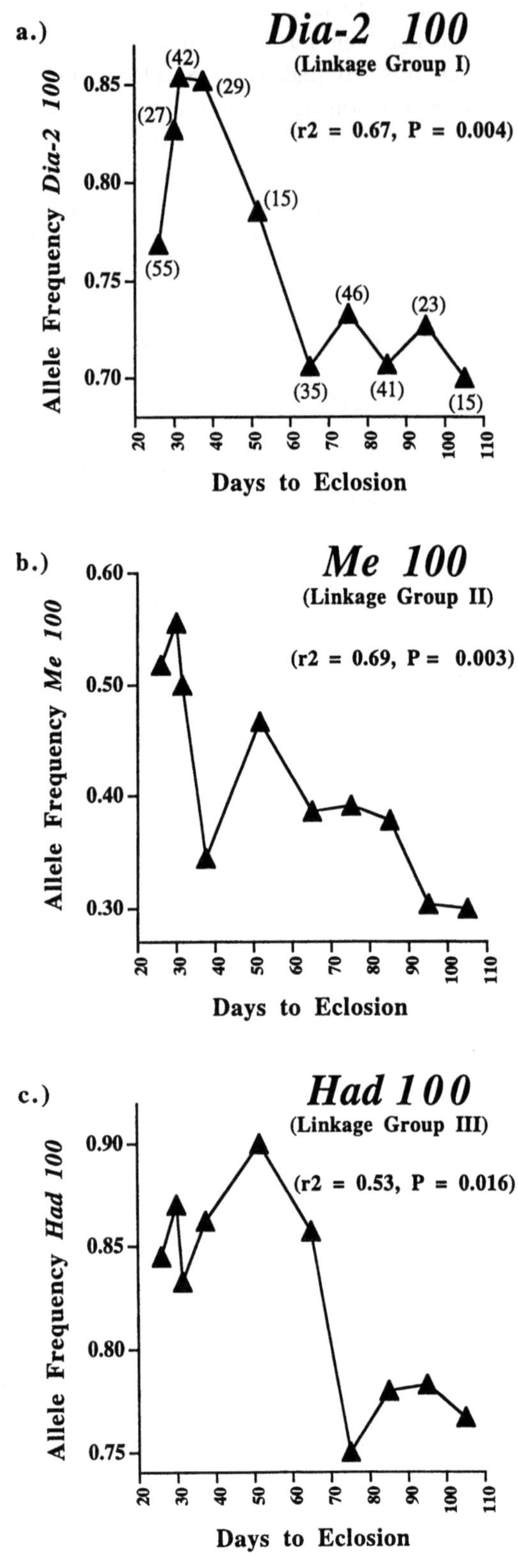


## Testing for fitness trade-offs using 'reverse genetics'

The sympatry of *R. pomonella* host races is important to the issue of fitness trade-offs because it shows that genetic differences between apple- and hawthorn-fly populations persist in the face of moderate gene flow. Some form of host plant-dependent selection (a trade-off) must therefore exist to counteract this gene flow or the *R. pomonella* host races would quickly genetically fuse in less than 20 generations (Feder, unpubl. calculations). But this is not happening, as indicated by a long-term (11 year) genetic survey of the races at a field site near Grant, Michigan, U.S.A.

*Figure 2.* Allele frequencies for (a) *Dia-2 100* (linkage group I) (b) *Me 100* (linkage group II) and (c) *Had 100* (linkage group III) in hawthorn-fly adults eclosing during 10 successive time intervals (<30 days, 30 days, 31–32 days, 33–42 days, 43–60 days, 61–70 days, 71–80 days, 81–90 days, 91–100 days and ≥101 days) following the date when pupae were removed from the cold and incubated at 21 °C. For experimental details see Feder et al. (1997b). Coefficients of determination ($r^2$) and probability levels for linear regressions of arcsin transformed allele frequencies versus time to eclosion are also presented. Sample sizes (# of flies genetically scored) are given in parentheses in (a).

(Feder et al., 1993, 1997a), the same site where our mark-release-recapture experiments were performed.

The sympatry of the host races therefore allows us to work backwards in testing for fitness trade-offs, as we already have genetic markers in three different genomic regions that are likely targets of disruptive selection. Consequently, manipulating rearing conditions for flies and testing for a genetic response at these loci provides a convenient means to pinpoint the source(s) for host-dependent selection. Our working assumption is that the allozymes themselves are not actually under selection, but are closely linked to such genes.

In the following sections, we present the results from our 'reverse genetics' approach connecting allozyme differences to phenotypes in testing for fitness trade-offs in *Rhagoletis*. We reach three main conclusions. First, host phenology (when trees fruit during the season) is a major environmental factor which *Rhagoletis* flies must adapt to or perish. In particular, the ~3 week earlier mean fruiting phenology of apples in eastern North America appears to select for a slower rate of metabolism or deeper pupal diapause in apple than hawthorn flies, at least at the Grant, MI. site. Second, host-related fitness trade-offs are not likely due to disruptive selection affecting any one life-stage. Rather, divergent selection on apple and hawthorn flies results from the sum total of varying directional selection pressures acting across the entire life-cycle. Third, natural *R. pomonella* populations harbor a good deal of genetic variation for development-related traits. It is this variation that allows fly populations to respond to vagaries in local environmental conditions and gives *R. pomonella* the potential to utilize novel plants – a process that can lead to sympatric race formation and, under certain circumstance, may trigger speciation.

## Host plant phenology and adult eclosion

One important trait for a *Rhagoletis* fly is to be developmentally synchronous with the seasonality of its host plant. *Rhagoletis pomonella* is univoltine and adult longevity is only 4–6 weeks in nature (Dean & Chapman, 1973; Boller & Prokopy, 1976; Bush, 1992). It is therefore critical that *R. pomonella* eclose and reach sexual maturity at times when host fruits are plentiful to maximize fitness. But apples and hawthorns represent different temporal resources. Fruits on apple varieties favored by *R. pomonella* generally ripen ~3 weeks earlier in the season than hawthorns (Feder et al., 1993; Feder, 1995). One result of this difference is that apple-fly adults eclose an average of ~10 days earlier than hawthorn flies at the Grant, MI. site (Feder et al., 1993; Feder, 1995). This difference is genetically based, as determined by both test crosses between the races (Smith, 1988) and eclosion time/allozyme correlations observed in the field and the lab (Feder et al., 1993, 1997a, b).

The genetic relationships between the allozymes and the timing of adult eclosion are not, however, what one would predict based on the relative order of emergence of the host races, at least not at the Grant site. Earlier eclosing adults in both host races possess higher frequencies of the alleles *Me 100, Acon-2 95, Mpi 37, Aat-2 +75, Dia-2 100* and *Had 100* (Figure 2; Feder et al., 1993, 1997a,b). These six alleles are more common in the hawthorn than the apple race in and around Grant, MI. (see site #3 in Figure 3; Feder et al., 1990a, b, 1993). This suggests that, all things being equal, hawthorn flies are genetically programmed to develop faster than apple flies. But hawthorn flies eclose later, not earlier, than apple flies in the field. How could this be?

The answer to this paradox lies in the fact that larvae finish feeding, leave abscised fruits, and pupate in the soil an average of 16 days earlier from apples than hawthorns at the Grant site (Figure 4; Feder, 1995). Therefore, while apple fly adults eclose an average of 10 days earlier than hawthorn flies, these individuals had a 16 day head start when they pupated. The net result is that apple-fly pupae actually over-winter for ~1 week longer than hawthorn-origin pupae, a difference consistent with the slower overall rate of development (deeper diapause; see next section) predicted for the apple race by the allozymes.

The finding that apple pupae at the Grant site develop more slowly than hawthorn flies shows how gene x environmental interactions could play a cen-

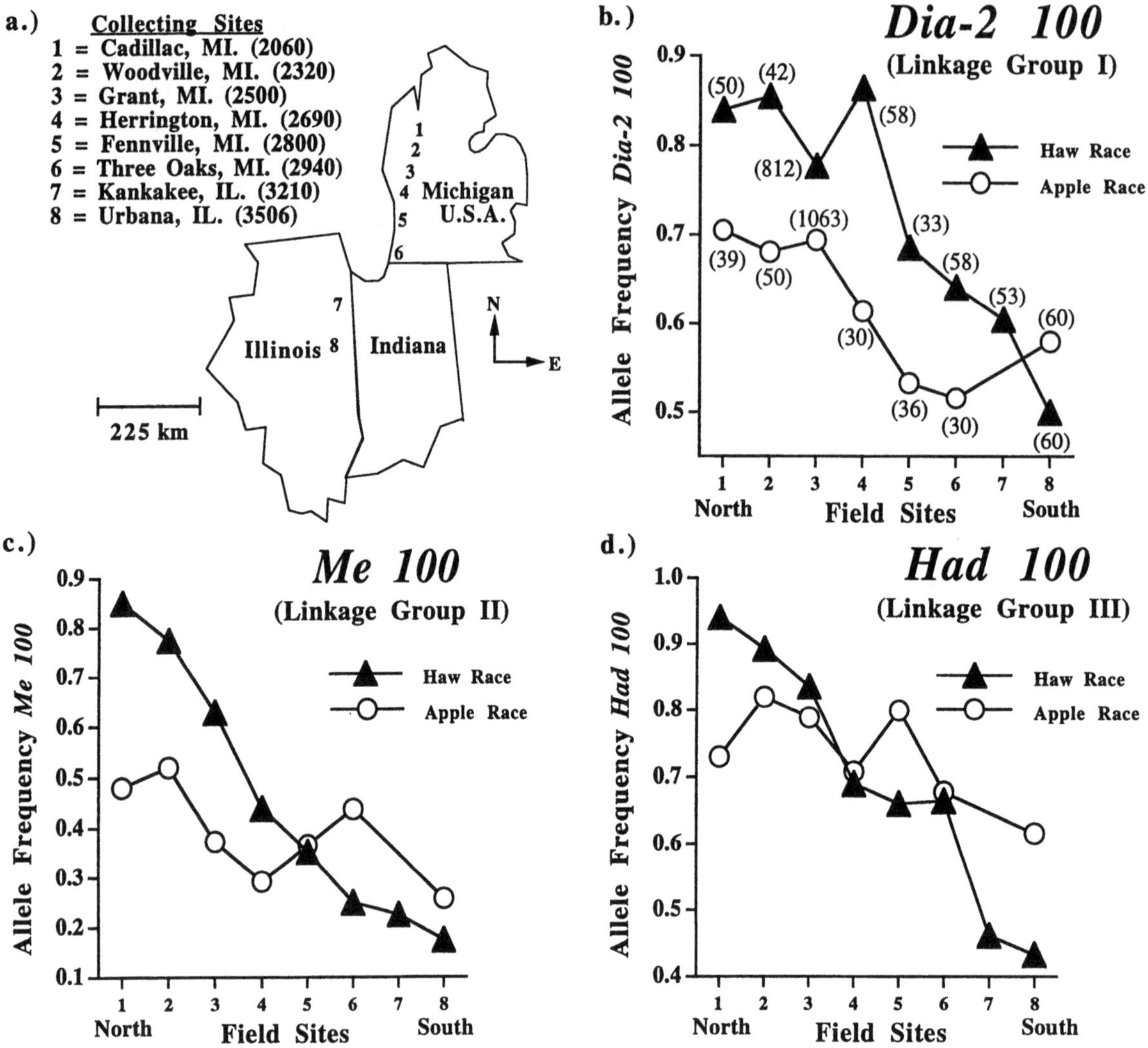


*Figure 3.* (a) Map of *R. pomonella* populations collected along a latitudinal transect in the mid-western United States. Numbers in parentheses following location names are the 50 year mean growing degree days (base 50 °F) for the entire field season (March–November) at sites. Allele frequencies for (b) *Dia-2 100* (linkage group I) (c) *Me 100* (linkage group II) and (c) *Had 100* (linkage group III) for apple and hawthorn-fly populations along the latitudinal transect. Sample sizes (*n* =number of flies genetically scored) are given in parentheses in (b).

tral role in partially reproductively isolating the host races. Under controlled laboratory conditions, F1 'hybrids' between apple and hawthorn flies should have intermediate eclosion time phenotypes between the parental races. In nature, however, these same hybrids should eclose at the extremes of the apple- and hawthorn-fly distributions. This is because hybrids that infest apples will have genotypes that cause them to eclose earlier than most apple flies, while hybrids that infest hawthorns will have genotypes that cause them to eclose later than hawthorn flies. The result would be detrimental for hybrids, as they would eclose too early in the season to optimally utilize apples and too late to attack hawthorns. Because of the difficulties in quantifying fitness effects related to adult eclosion in the field, we cannot add much more to this chapter of the *R. pomonella* story at the present time except to say that the match between eclosion time and fruit availability is an important component of host-dependent selection, a component that requires much further study.

## Host phenology and the pupal diapause in *R. pomonella*

Synchrony between adult sexual activity and fruit availability is just one life stage affected by host

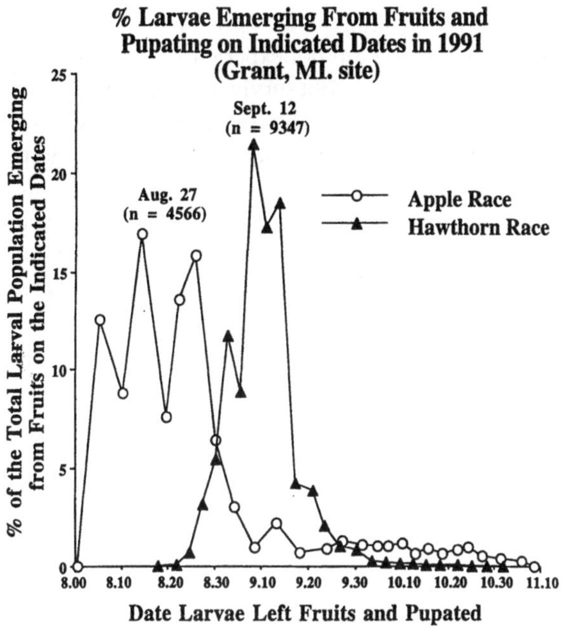


*Figure 4.* Pupation times (i.e., percentages of larvae exiting host fruits, burrowing into the soil, and pupating on indicated dates) during the 1991 field season for apple and hawthorn-fly races at the Grant, MI. site. Pupation dates ranged from 8.01 (= August 1) to 11.10 (November 10). Also given are the average pupation dates for each host race and the total number of pupae (*n*) collected in the experiments to generate the curves. See Feder (1995) for experimental details.

phenology. Equally important to a fly is its developmental status during winter. Before delving further into this topic, it is important to realize that diapause in insects is a dynamic state, such that as the season progresses, diapause depth or intensity decreases (Tauber & Tauber, 1976; Tauber et al., 1986). Diapause in most insects ends by mid-winter, not in the spring. This means that post-diapause development is prevented, often for a considerable time, until temperatures rise above some developmental threshold. Consequently, when an insect ecloses is more often a function of when it breaks diapause, rather than how fast it completes development once diapause ends (Morris & Fulton, 1970). Adaptation to temperature conditions associated with latitude is therefore typically accomplished via genetic differences affecting diapause termination (Danilevskii, 1965; Masaki, 1965; Tauber & Tauber, 1976). This led us to hypothesize that host phenology affects diapause for *R. pomonella* in a similar manner that latitude does for other insects; The early fruiting time of apple trees selects the apple race to be more recalcitrant to diapause termination than the hawthorn race. Indeed, later we will argue that both latitude and host phenology shape the genetics of *R. pomonella* through their effects on diapause.

A key factor in the life-history of *R. pomonella* is that the fly overwinters in a facultative pupal diapause (Dean & Chapmen, 1973; Boller & Prokopy, 1976). Most *R. pomonella* pupae exposed to elevated temperatures for extended periods of time before winter fail to enter or maintain a protracted diapause (Prokopy, 1968). In nature, such 'non-diapausing' flies are doomed because they eclose in late fall when suitable host fruits are no longer present. The 16 day difference in the average pupation means that apple-origin pupae are exposed to longer periods of warm weather before winter than hawthorn pupae (Figure 4). This, coupled with the facultative nature of diapause in *R. pomonella*, leads to the prediction that the earlier phenology of apples selects apple flies to have deeper pupal diapauses (or slower development rates) than hawthorn flies.

## The pre-winter experiment – A test of the 'diapause hypothesis'

To test our 'diapause hypothesis', we manipulated environmental conditions faced by pupae to see whether we could induce a genetic response at the 6 allozymes displaying host-related differences. In particular, we examined whether hawthorn flies experiencing longer pre-wintering periods simulating the earlier apple phenology evolve toward the naturally occurring allele frequencies of the apple race. The answer was yes (Feder et al., 1997a). Allele frequencies for *Me 100, Acon-2 95, Mpi 37, Aat-2 +75, Dia-2 100* and *Had 100*, the allozymes associated with earlier adult eclosion and by inference earlier diapause termination, all declined in hawthorn flies exposed to progressively longer pre-wintering periods. Selection coefficients estimated from the Pre-Winter Experiment were greater than those needed to counteract the estimated 6% gene flow from the hawthorn into the apple-fly race each generation (Feder et al., 1997a). Seasonal differences between apples and hawthorns can therefore maintain allozyme frequency differences between the fly races.

## The relevance of the pre-winter experiment to nature

Two lines of evidence connect the results from the Pre-Winter Experiment to the ecology and genetics of natural *R. pomonella* populations. First, *Aat-2 +75, Dia-2 100, Me 100, Acon-2 95, Mpi 37* and *Had 100* display latitudinal frequency clines in both apple and hawthorn races across the United States, with northern populations having higher frequencies of these alleles than southern populations (Figure 3; Feder & Bush, 1989; Feder et al., 1990a; Berlocher & McPheron, 1996). Second, allozyme frequencies in the hawthorn race at the Grant site were significantly related to ambient temperature in the spring months of the preceding year over an 11 year period beginning in 1984 (Feder et al., 1993, 1997a). Because a majority of *R. pomonella* pupae break diapause after their first winter (92–98%; Phipps & Dirks, 1933; Dirks, 1935), flies sampled in year N are representative of pupae that survived the previous season (year N-1). Hot spring temperatures in year N-1 meant an early season and correlated with increased frequencies in year N of alleles common to the apple race (or southern populations). Conversely, cold springs favored hawthorn race (or northern population) alleles.

## The effects of winter length on *R. pomonella* host races

In addition to manipulating pre-winter conditions for pupae, we have also investigated the effects of winter length on the genetics of apple and hawthorn flies (Feder et al., 1997b). Apple- and haw-origin pupae from the Grant site were chilled in a refrigerator at 4 °C for periods ranging from 1 week to 2 years. Our rationale for the 'Over-Winter Experiment' was that long winters should have the same genetic consequences as extended pre-winter periods; allele frequencies of *Aat-2 +75, Dia-2 100, Me 100, Acon-2 95, Mpi 37* and *Had 100* should decline in eclosing (surviving) adults. This is because pupae possessing these alleles should break diapause earlier than other genotypes and disproportionately die, as they exhaust critical energy reserves during extended chilling periods.

The results from the 'Over-Winter' experiments supported the diapause hypothesis. The anticipated genetic response was observed in both host races, as allele frequencies became significantly more 'apple-fly like' in eclosing adults surviving ever longer chilling periods (Feder et al., 1997b). We also found that there was an optimal over-wintering period of 15–19 weeks that maximized survivorship (Feder et al., 1997b). We inferred from this that survivorship was low for overwintering periods less than 15 weeks because many pupae did not experience adequate chilling to break diapause. But survivorship decreased after 19 weeks, because by this time a majority of pupae had broken diapause and were depleting vital energy reserves needed for adult development. Furthermore, the decline in survivorship was much more pronounced in the hawthorn than the apple race (Feder et al., 1997b). This result is consistent with our prediction that apple flies should be less sensitive to prolonged winters than hawthorn flies due to their inherently deeper pupal diapauses. We are currently testing this hypothesis by quantifying weight loss for pupae during winter and relating this to survivorship, eclosion times and the genetics of flies.

The results from the Over-Winter Experiment also suggested that flies entering winter as non-diapause pupae capable of developing given permissive temperatures were not immediately killed by cold weather. Rather, these pupae survived for up to 19 weeks in the cold (Feder et al., 1997b). This is important because it shows how pre- and over-wintering conditions interact to affect pupal survivorship. In effect, the length of the pre-winter period dictates the developmental state and energy reserves that pupae carry into winter, but does not kill them. It is the nature of the ensuing winter together with the inherent diapause characteristics of flies that determines which will survive and which will die.

## The interaction of environmental conditions preceding and during winter

We are currently engaged in an experiment investigating the interplay between pre- and over-wintering conditions. Although we have not yet genetically scored flies from this study, several observations were apparent from the eclosion curves. First, cold over-wintering temperatures hastened diapause termination. Adult eclosion curves for pupae over-wintered at 0 °C were shifted earlier than those for pupae stored at 5 °C (Figure 5a). Second, the 0 °C winter treatment inflicted higher mortality on pupae that 5 °C, and this difference became more pronounced the longer that pupae remained in the cold (Figure 5b). This is relevant because we have found that *R. pomonella* pupae regularly experience 0 °C soil temperatures during winter at the Grant site (Figure 6). Winter temperatures in nature are therefore likely to accentuate the genetic responses we induced in the our rearing experiments. Third, we observed the anticipated antagonistic effect on fly survivorship between warm pre-winter temperatures and long and cold winters. Eclosion curves were shifted earlier and fly survivorship was reduced the warmer pre-wintering temperatures were for pupae and the longer and colder the ensuing over-wintering period.

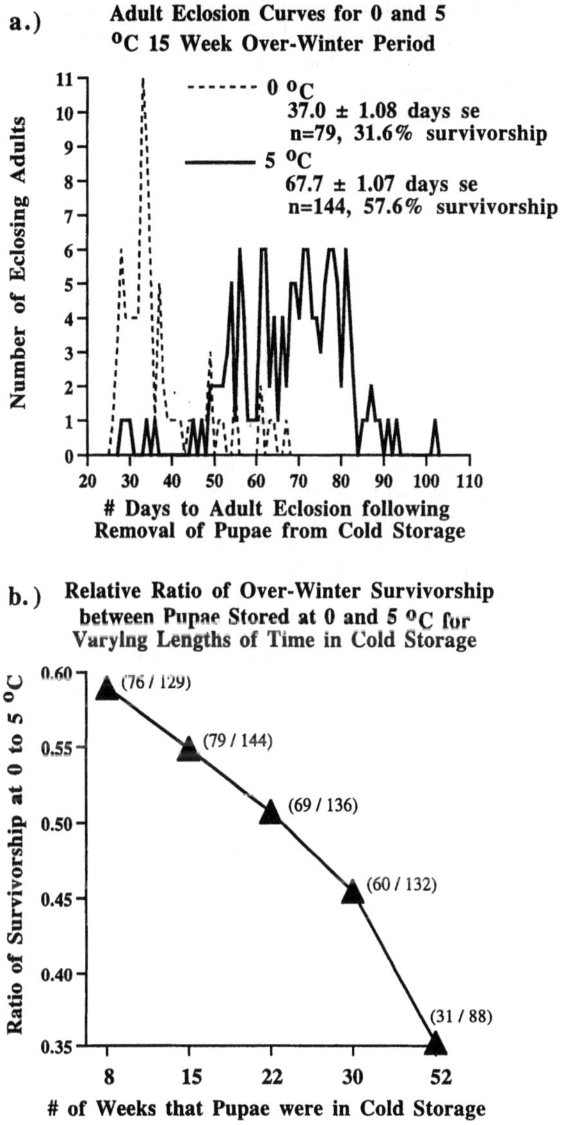


*Figure 5.* (a) Adult eclosion curves for hawthorn-fly pupae stored at 0 °C and at 5 °C for 15 weeks in a refrigerator. Average times to eclosion (± s.e.) and the total numbers and percentages of eclosing adults from an original sample size of 250 pupae are also given for the 0 °C and at 5 °C treatments. Flies were collected as larvae in infested fruits from a hawthorn tree on 112th St., near Grant, MI. on Sept. 11, 1996. (b) Ratios of hawthorn-fly pupae surviving 0 °C versus 5 °C over-wintering temperatures for varying periods of chilling (8–52 weeks). Actual numbers of eclosing adults (0 °C/5 °C) are given in parentheses from a starting sample of 250 pupae for each combination of rearing conditions.

## Selection on the larval life-history stage of *R. pomonella*

There is no reason to believe that selection in *R. pomonella* should be limited to just pupae. It seems logical that chemical and nutritional differences that surely exist between apples and haws should also differentially affect larvae feeding within fruits. Indeed, larval survivorship is central to most models of sympatric speciation, serving as a key postmating barrier to gene flow (Bush, 1992). But surprisingly, reciprocal egg transplant studies produced no evidence for feeding specialization in *R. pomonella* (Prokopy et al., 1988). Apple and hawthorn-race larvae survived equally-well in haw fruits (~52% survivorship – see Table 5, 1985 experiment of Prokopy et al., 1988) and fared equally poorly in apples (~27% survivorship). While these results are in accord with haws being the ancestral host of *R. pomonella*, they are discouraging with respect to fitness trade-offs in larvae.

However, the fruits used in Prokopy et al.'s (1988) transplant experiments were harvested immediately after they abscised from trees and stored under protective conditions. Abscised fruits quickly rot in nature, which is detrimental to larvae. We have commonly observed dead larvae drowned in water-logged fruits, and this mortality appeared to be worse in haws. This suggests that hawthorn larvae experience stronger selection pressure to quickly complete development and exit fruits or to withstand more noxious chemical conditions than apple larvae. Such an effect would have been missed in Prokopy et al.'s (1988) transplant experiments due to the removal of fruit from the field.

We therefore conducted a study in which we reared split samples of larval infested apple and hawthorn fruits in the field and in a garage (Filchak et al., 1999). Our rationale was that the more protective garage treatment should slow rates of fruit rot relative to those in the field, thereby relaxing selection pressures on larvae to quickly exit fruit. Comparisons of larval survivorship and allozyme frequencies between flies surviving the outdoor versus indoor treatments could then be made to test for host-fruit dependent selection. If the fruit necrosis hypothesis is correct, then we would expect larvae to suffer higher mortality in abscised fruits in the field than in the garage and that the difference between these two treatments should be greater for hawthorn than apple-fly larvae. We were especially interested in determining whether larvae that left host fruits the earliest possessed higher frequencies of the alleles *Aat-2 +75*, *Dia-2 100*, *Me 100*, *Acon-2 95*, *Mpi 37* and *Had 100*, the allozymes correlated with early adult eclosion in our previous studies (Feder et al., 1993, 1997a,b). If so, then this would suggest that these allozymes (or linked genes) have similar effects for both larvae and pupae, speeding rates of fly development. Two predictions for our 'Fruit Rot Experiment' would follow from this: (1) Larvae surviving field treatments should have elevated frequencies of the six aforementioned alleles, as selection for faster developing larvae that quickly leave necrosing fruits should favor these allozymes. (2) Allele frequency differences between outdoor and indoor treatments should be greater for the hawthorn than the apple race, in accord with mortality associate with necrosing fruit being higher for hawthorn than apple larvae in the field.

We found that larvae suffered higher mortality in fruits in the field than the garage, and the increase in mortality outdoors was greater in haws than in apples (Filchak et al., 1999). In addition, apple and hawthorn-fly larvae possessing the alleles *Me 100*, *Acon-2 95*, and *Mpi 37* left host fruits significantly earlier than other genotypes. Finally, allele frequencies for *Me 100*, *Acon-2 95*, and *Mpi 37* were significantly higher in both apple and hawthorn-larvae surviving field versus garage treatments (Filchak et al., 1999).

The simplest explanation for our results is that fruits kept outdoors rotted faster than those in the garage, favoring larvae that quickly left (rapidly developed in) host fruits. Because early emerging larvae tended to possess the allozymes *Me 100*, *Acon-2 95*, and *Mpi 37*, the frequencies of these alleles were elevated in the field vs the garage. The magnitudes of the genetic responses were similar, however, between the host races, despite mortality being relatively greater for hawthorn larvae in the field. This suggests that other factors, in addition to larval emergence time, contributed to the allele frequency shifts in the apple race (e.g., some unique aspect of the chemistry of apples, like a noxious secondary compound, could have affected *Me 100*, *Acon-2 95*, and *Mpi 37*). Regardless of the ultimate cause(s) for the observed allele frequency shifts, the results from the Fruit Rot Experiment imply that selection on larvae is directional and strong for *Me 100*, *Acon-2 95*, and *Mpi 37* (or linked genes) in both host races. For example, estimates of relative selection coefficients (s values) against larvae homozygous for alleles other than *Me 100*, *Acon-2 95*, and *Mpi 37* at these 3 loci ranged from 0.33 to 0.38 (Filchak et al., 1999).

*Table 1.* Common Spearman rank correlation coefficients corrected for tied ranks ($r_Z$) between indicated allozyme alleles and adult eclosion time derived from a meta-analysis (Hedges & Olkin 1985) across subsamples of apple and hawthorn flies that pupated on the same day. Correlation coefficients were z-transformed to test for significance. Flies were collected as larvae feeding within infested apple fruits at the Grant, MI. site on August 30, 1989 and in hawthorns on September 15, 1989. Total sample size ($n$ = # flies genetically scored) was 269 for the apple race and 706 for the hawthorn race

| Linkage Group | Locus/Allele | Apple race | Hawthorn race |
|---|---|---|---|
| I | *Aat-2 100* | −0.12* | −0.07* |
| | *Dia-2 100* | −0.11* | −0.08* |
| II | *Me 100* | −0.17** | −0.31**** |
| | *Acon-2 95* | −0.11* | −0.25**** |
| | *Mpi 37* | −0.01 | −0.13*** |
| III | *Had 100* | −0.13* | −0.11** |

*P ≤ 0.05; **P ≤ 0.01; ***P ≤ 0.001; ****P ≤ 0.0001. Significance level for one-tailed tests that $r_Z < 0$ according to the alternate hypothesis that flies with genotypes containing the alleles *Me 100*, *Acon-2 95*, *Mpi 37*, *Aat +75*, *Dia-2 100* and *Had 100* should eclose the earliest.

## The genetic relationship between larval and pupal life-history stages

The allozyme correlations suggest that *Me 100*, *Acon-2 95*, and *Mpi 37* affect larval and pupal development in a similar manner. However, correlations do not prove a cause and effect relationship. Test crosses are still needed to directly link the allozymes and development related traits. Nevertheless, evidence suggests that the allozyme correlations in larvae and pupae are independent and not due to a carry-over effect from a different life-stage or generation of flies. For example, significant correlations were found between *Aat-2 +75*, *Dia-2 100*, *Me 100*, *Acon-2 95* and *Had 100* and early eclosion even among apple and hawthorn flies that pupated on the same day (Table 1). Therefore, the correlations between *Me 100*, *Acon-2 95*, and *Mpi 37* and pupation date were not solely responsi-

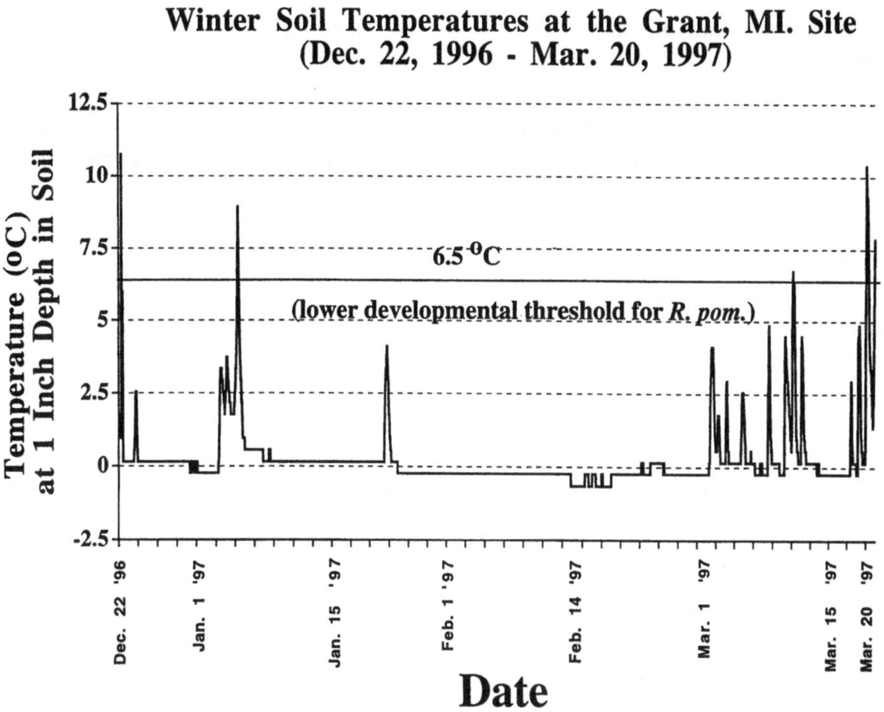


*Figure 6.* Soil temperature readings at a depth of 1 inch at the Grant, MI. site from December 22, 1996–March 20, 1997. Note how soil temperatures remain fairly constant at the site at ~ 0 °C during most of the winter. One inch is close to the mean depth that *Rhagoletis* flies pupate in the soil (Lathrop & Nickels, 1932).

ble for the relationships with eclosion. Likewise, the correlations between these allozymes and larval emergence time were not due to a seasonal trend for the frequencies of these alleles to decline with time in parental populations. Although *Aat-2 +75*, *Dia-2 100*, *Me 100*, *Acon-2 95*, *Mpi 37* and *Had 100* frequencies were higher in earlier eclosing flies at the Grant site (Feder et al., 1993), the frequencies of these alleles within sexually active adult populations did not significantly vary throughout the 1997 and 1998 field seasons (Feder et al., 1993).

## Synthesizing the experimental results – What does it all mean?

We are now at a cross-road with respect to our understanding of host-mediated selection in *R. pomonella*. We have demonstrated strong selection on allozyme loci (or linked genes) in both larval and pupal life-history stages of the fly. We have also shown that this selection is biologically relevant. In particular, that environmental conditions, as mediated by host phenology and fruit necrosis, have a major impact on the developmental profiles of apple and hawthorn flies. However, we have yet to produce definitive evidence for disruptive selection affecting any one specific life-history stage of *R. pomonella*. Rather, we have evidence for directional selection affecting many different stages.

Does this mean that host plant-related fitness trade-offs are unlikely for *R. pomonella*? The answer is no, as fitness trade-offs between the host races may stem as much, or more, from differences in the balance of directional selection pressures acting across the entire life-cycle, as from disruptive selection affecting any one stage. Allozymes *Me*, *Acon-2* and *Mpi* in linkage group II demonstrate this point most clearly. Here, we have evidence that strong directional selection on larvae favors *Me 100*, *Acon-2 95* and *Mpi 37* (or linked genes) in both host races. However, these same alleles can be disfavored in the pupal stage to follow, as they also correlate with early adult eclosion, and by inference early diapause termination (Figure 2; Feder et al., 1993, 1997a,b). Because apple trees fruit an average of ~3 weeks earlier than hawthorn trees, this counterbalancing selection is stronger on apple-fly pupae. The net result is that the balance of selective forces is different between apple and hawthorn flies, resulting in the host races residing on alternative adaptive peaks.

## Improving tests for fitness trade-offs in phytophagous insects

Our results underscore the concerns raised by Rausher (1988, 1992) about the inadequacies of many tests for fitness trade-offs. First, we had to score large numbers of flies to verify significant allele frequency shifts, even though the selection coefficients associated with the allozymes were quite large. If selection pressures were less intense, then we would have failed to detect these differences. Second, if we had performed reciprocal egg transplant experiments between apple and hawthorn fruits using a split family design (a reasonable first approach), then we may have concluded that no fitness trade-offs exist for *Rhagoletis*. Instead, we may have proclaimed a 'general vigor' effect in larval performance among families. If these transplant studies were accompanied by biochemical genetic analysis, then we might have amended our interpretation to say that directional selection favored families possessing the alleles *Me 100*, *Acon-2 95* and *Mpi 37*. But in both cases we would have failed to grasp how contrasting selection pressures between larval and pupal stages shape the pattern of genetic differentiation between the host races. Tests for host plant dependent trade-offs must therefore consider details of the entire life-cycle of phytophagous insects, and not just concentrate on feeding stages. Third, all of our experiments were performed either in the field or under simulated field conditions. The results for the Fruit Rot Experiment show how selection pressures can be relaxed or altered in the laboratory. Also, since diapause traits are important to *R. pomonella* and are sensitive to both pre- and over-wintering conditions, it can be easily seen how artificial lab environments could produce spurious results. Consequently, since we are interested in trade-offs in nature, we should test for selection in nature or under as realistic a set of field conditions as possible. The general paucity of documented fitness trade-offs for phytophagous insects may therefore be due, in part, to problems in experimental design.

## An unresolved issue and future research

Although we have learned a great deal about host-dependent selection in *R. pomonella*, we are far from a complete understanding of the system. Perhaps the most important unresolved question involves discerning why allozyme frequency clines differ between apple and hawthorn races (Figures 3a–d). Earlier, we mentioned that *Aat-2 +75*, *Dia-2 100*, *Me 100*, *Acon-2 95*, *Mpi 37* and *Had 100* display latitudinal frequency clines in both apple and hawthorn races. What we did not discuss is that the slopes of the clines are steeper among hawthorn than apple-fly populations (Figures 3a–d). As a result, the allozyme clines for the races cross, with alleles common to the hawthorn race at the Grant, MI. site present in lower frequencies in hawthorn than apple populations further south (e.g., Urbana, Illinois, site # 8 in Figures 3a–d). Nevertheless, apple trees at the Urbana site still fruit an average of $\sim$3 weeks earlier than hawthorns (S. Berlocher, pers. comm.). Our diapause hypothesis does not adequately account for the clinal differences between the host races. One possible explanation is that we have not yet characterized all of the relevant environmental factors affecting the host races. For instance, it is known that photoperiod during the larval – but not pupal – stage also influences diapause depth in *R. pomonella* (Prokopy, 1968). Variation in light intensities within apple and hawthorn fruits might therefore also affect the pattern of clinal variation. Also, we may not have identified all of the genomic regions affecting diapause in *R. pomonella*. Genetic variation in a yet to be characterized locus might therefore also help explain the crossing of the allozyme clines. We have recently developed a complimentary DNA (cDNA) recombination linkage map for *R. pomonella* (Figure 1; Roethele et al., 1997), so it will eventually be possible to address this issue and perhaps characterize the actual genes under host-dependent selection in *R. pomonella*.

## A final word

In conclusion, while timing may not be everything, it is a good deal of the story in *R. pomonella*, as seasonal differences in when apple and hawthorn trees fruit select for pupae with different diapause and eclosion time characteristics in the host races. It is worth noting that host phenology also plays a central role in several other potential examples of sympatric race formation and speciation for insects (Tauber & Tauber, 1989; Wood & Keese, 1990; Wood et al., 1990; Craig et al., 1993, 1994; Abrahamson et al., 1994; Abrahamson & Weis, 1997). In addition, we have evidence that fruit necrosis exerts strong selection for rapid larval development on both apple- and hawthorn-fly host races of *R. pomonella*. It is the balance (the net sum) of these directional selection pressures acting across the entire life-history that generates fitness trade-offs in *R. pomonella* and helps maintain the genetic integrity of the host races in the face of moderate gene flow, at least at the Grant, MI. site. The necessity to consider details of the entire life-cycle highlights how difficult it can be to document host plant-related fitness trade-offs for phytophagous insects. Nevertheless, our results show how reproductive isolation can arise as a consequence of host-associated adaptation, as predicted by models of sympatric speciation. The next question is whether the same traits that partially isolate sympatric host races ever evolve to completely separate species. Detailed systematic, genetic and behavioral analyses currently underway for closely related sibling species in the *R. pomonella* complex suggest that the answer to this question is yes (see Berlocher, 1999 for review). Regardless of the geographic issue, however, studies on *Rhagoletis* flies indicate that the basis for speciation may not always be gleamed from a genetic analysis of reproductive isolation divorced from the external environment. What is required is a 'natural history of the evolution of reproduction isolation' to complement (amend) the current heavy reliance on genetic measurements to infer mechanisms of speciation (Schluter, 1999).

## Acknowledgements

We would like to thank in alphabetical order: Hiroshi Akashi, Kristin Ardlie, Dylan Atkins, Stewart Berlocher, Andrew Berry, Guy Bush, Clarice Chen, Charley Chilcote, Scott Freeman, the Hansens, Tim Hunt, Marty Kreitman, Kristin Lewis, John MacDonald, the Nelsons, Janice O'Connell, William Perry, Naomi Pierce, the Riders, Joseph Roethele, Alan Rogers, Jim Smith, Uwe Stolz, Martin Taylor, Glenn Williams, Frank, Angela & Emma Wang, Brian Wlazlo, Edmund Wong and two anonymous reviewers for their help, comments and criticisms on this project and manuscript. This work was supported, in part, by grants from the Alfred P. Sloan Foundation, the USDA (Agri-94-37302-0540) and NSF (DEB 95-08559) to J. L. F.

## References

Abrahamson, W. G. & A. E. Weis, 1997. The Evolutionary ecology of a tritrophic-level interaction: Goldenrod, the stem gallmaker and its natural enemies. Monographs in Population Biology 29. Princeton Univ. Press, Princeton, NewJersey.

Abrahamson, W. G., W. M. Brown, S. K. Roth, D. V. Sumerford, J. D. Horner, M. D. Hess, S. T. How, T. P. Craig, R. A. Packer, & J. K. Itami, 1994. Gallmaker speciation: an assessment of the roles of host-plant characters, phenology, gallmaker competition and natural enemies. In: P. Price, W. Mattson & Y. Baranchilov (eds.), Gall-Forming Insects. General Technical Report NC-174, USDA Forest Service, North Central Experimental Station, pp. 208–122.

Berlocher, S. H., 1999. Can sympatric speciation via host or habitat shifts be proven from phylogenetic and biogeographic evidence? In: D. Howard & S. H. Berlocher (eds), Endless Forms: Species and Speciation. Oxford University Press, Oxford, U.K.

Berlocher, S. H. & D. C. Smith, 1983. Segregation and mapping of allozymes of the apple maggot fly. Journal of Heredity 74: 337–340.

Berlocher, S. H. & B. A. McPheron, 1996. Population structure of *Rhagoletis pomonella* the apple maggot fly. Heredity 77: 83–99.

Boller, E. F. & R. J. Prokopy, 1976. Bionomics and management of *Rhagoletis*. Annual Review of Entomology 21: 223–246.

Bush, G. L., 1969. Sympatric host race formation and speciation in frugivorous flies of the genus *Rhagoletis* (Diptera: Tephritidae). Evolution 23: 237–251.

Bush, G. L., 1975. Sympatric speciation in phytophagous parasitic insects. In: P. W. Price (ed), Evolutionary Strategies of Parasitic Insects and Mites. Plenum, New York. pp. 187–206.

Bush, G. L., 1992. Host race formation and sympatric speciation in *Rhagoletis* fruit flies (Diptera: Tephritidae). Psyche 99(4): 335–358.

Craig, T. P., J. K. Itami, W. G. Abrahamson & J. D. Horner, 1993. Behavioral evidence for host-race formation in *Eurosta solidaginis*. Evolution 47: 1696–1710.

Craig, T. P., J. K. Itami, J. D. Horner & W. G. Abrahamson, 1994. Host shifts and speciation in gall-forming insects. In: P. Price, W. Mattson & Y. Baranchilov (eds), Gall-Forming Insects. General Technical Report NC-174, USDA Forest Service, North Central Experimental Station. pp. 194–207.

Danilevsky, A. S., 1965. Photoperiodism and Seasonal Development of Insects (English transl.). Oliver and Boyd Publications, London, 282 pp.

Dean, R. W. & P. J. Chapman, 1973. Bionomics of the apple maggot in eastern New York Search Agricultural Entomology Geneva No. 3., Geneva, N.Y.

Diehl, S. R. & G. L. Bush, 1984. An evolutionary and applied perspective of insect biotypes. Annual Review of Entomology 29: 471–504.

Dirks, C. O., 1935. Larval production and adult emergence of the apple fruit fly in relation to apple varieties. Journal of Economic Entomology 28: 198–203.

Dobzhansky, T., 1951. Genetics and the Origin of Species, 3rd ed. Columbia University Press, New York, New York.

Feder, J. L., 1995. The effects of parasitoids on sympatric host races of the apple maggot fly, *Rhagoletis pomonella* (Diptera: Tephritidae). Ecology 76: 801–813.

Feder, J. L., C. A. Chilcote & G. L. Bush, 1988. Genetic differentiation between sympatric host races of *Rhagoletis pomonella*. Nature 336: 61–64.

Feder, J. L. & G. L. Bush, 1989. Gene frequency clines for host races of *Rhagoletis pomonella* (Diptera: Tephritidae) in the midwestern United States. Heredity 63: 245–266.

Feder, J. L., C. A. Chilcote & G. L. Bush, 1989. Inheritance and linkage relationships of allozymes in the apple maggot fly. Journal of Heredity 80: 277–283.

Feder, J. L., C. A. Chilcote & G. L. Bush, 1990a. The geographic pattern of genetic differentiation between host associated populations of *Rhagoletis pomonella* (Diptera: Tephritidae) in the eastern United States and Canada. Evolution 44: 570–594.

Feder, J. L., C. A. Chilcote & G. L. Bush, 1990b. Regional, local and micro-geographic allele frequency variation between apple and hawthorn populations of *Rhagoletis pomonella* in western Michigan. Evolution 44: 595–608.

Feder, J. L., T. A. Hunt & G. L. Bush, 1993. The effects of climate, host plant phenology and host fidelity on the genetics of apple and hawthorn infesting races of *Rhagoletis pomonella*. Entomologia Experimentalis et Applicata 69: 117–135.

Feder, J. L., S. Opp, B. Wlazlo, K. Reynolds, W. Go & S. Spisak, 1994. Host fidelity is an effective pre-mating barrier between sympatric races of the apple maggot fly. Proceedings of the National Academy of Sciences, USA 91: 7990–7994.

Feder, J. L., J. B. Roethele, B. Wlazlo & S. H. Berlocher, 1997a. Selective maintenance of allozyme differences between sympatric host races of the apple maggot fly. Proceedings of the National Academy of Sciences, USA 94: 11417–11421.

Feder, J. L., S. H. Berlocher & S. B. Opp, 1998. Sympatric host race formation and speciation in *Rhagoletis* (Diptera: Tephritidae): A tale of two species for Charles D. In: S. Mopper & S. Strauss (eds), Genetic Structure and Local Adaptation in NaturalInsect Populations. Chapman & Hall, New York, pp. 408–441.

Feder, J. L., Stolz, U., Lewis, K. M., Perry, W., Roethele, J. B. & A. Rogers, 1997b. The effects of winter length on the genetics of apple and hawthorn races of *Rhagoletis pomonella* (Diptera: Tephritidae). Evolution 51: 1862–1876.

Filchak, K. E., J. L. Feder, J. B. Roethele & U. Stolz, 1999. A field test for host-plant dependent selection on larvae of the apple maggot fly, *Rhagoletis pomonella*. Evolution, 53: 187–220.

Fry, J. D., 1990. Trade-offs in fitness on different hosts: Evidence from a selection experiment with a phytophagous mite. American Naturalist 136: 569–580.

Fry, J. D., 1993. The 'general vigor' problem: Can antagonistic pleiotropy be detected when genetic covariances are positive? Evolution 47: 327–333.

Fry, J. D., 1996. The evolution of host specialization: are trade-offs overrated? American Naturalist 148: S84–S107.

Futuyma, D. J. & G. Moreno, 1988. The evolution of ecological specialization. Annual Review of Ecology and Systematics 19: 207–233.

Gould, F., 1979. Rapid host range evolution in a population of the phytophagous mite *Tetranycchus urticae* Koch. Evolution 33: 791–802.

Haldane, J. B. S., 1919. The combination of linkage values, and the calculation of distance between loci of linked factors. Journal of Genetics 8: 299–309.

Hedges, L. V. & I. Olkin, 1985. Statistical Methods for Meta-analysis. Academic Press Orlando, FL.

Jaenike, J., 1990. Host specialization in phytophagous insects. Annual Review of Ecology and Systematics 21: 243–273.

Karowe, D. N., 1990. Predicting host range evolution: Colonization of *Coronnilla varia* by *Colias philodice* (Lepidoptera: Pieridae). Evolution 44: 1637–1647.

Kawecki, T. J., 1997. Sympatric speciation via habitat specialization driven by deleterious mutations. Evolution 51: 1751–1763.

Lathrop, F. H. & C. B. Nickels, 1932. The biology and control of the blueberry maggot in Washington County Maine. Technical Bulletin 275 United States Dept. of Agriculture. Washington, D.C., 77 pp.

Lewontin, R. C., 1974. The Genetic Basis of Evolutionary Change. Columbia Univ. Press, New York, New York.

Luna, I. & R. J. Prokopy, 1995. Behavioral differences between hawthorn-origin and apple-origin *Rhagoletis pomonella* flies in patches of host trees. Entomologia Experimentalis et Applicata 74: 277–282.

Mackenzie, A., 1996. A trade-off for host plant utilization in the Black Bean Aphid, *Aphis fabae*. Evolution 50: 155–162.

Masaki, S., 1965. Geographic variation in the intrinsic incubation period: A physiological cline in the Emma field cricket (*Teleogryllus*) Bulletin of the Faculty of Agriculture Hirosaki University 11: 59–90.

Mayr, E., 1963. Animal Species and Evolution. Harvard University Press, Cambridge, Mass. 797 pp.

McPheron, B. A., D. C. Smith & S. H. Berlocher, 1988. Genetic differences between *Rhagoletis pomonella* host races. Nature 336: 64–66.

Mitter, C., D. J. Futuyma, J. C. Schneider & J. D. Hare, 1979. Genetic variation and hostplant relations in a parthenogenic moth. Evolution 33: 770–790.

Morris, R. F. & W. C. Fulton, 1970. Models for the development and survival of *Hypantria cunea* in relation to temperature and humidity. Memoirs of the Entomological Society of Canada 70: 1–60.

Muller, H. J., 1940. Bearings of the *Drosophila* work on systematics. In: J. S. Huxley (ed), The New Systematics. Clarendon Press, Oxford. pp. 185–268.

Phipps, C. R. & C. O. Dirks, 1933. Notes on the biology of the apple maggot fly. Journal of Economic Entomology 26: 349–358.

Prokopy, R. J., 1968. The influence of photoperiod, temperature and food on the initiation of diapause in the apple maggot. Canadian Entomologist 100: 318–329.

Prokopy, R. J., E. W. Bennett & G. L. Bush, 1971. Mating behavior in *Rhagoletis pomonella* (Diptera: Tephritidae). I. Site of assembly. Canadian Entomologist 103: 1405–1409.

Prokopy, R. J., E. W. Bennett & G. L. Bush, 1972. Mating behavior in *Rhagoletis pomonella* (Diptera: Tephritidae). II. Temporal organization. Canadian Entomologist 104: 97–104.

Prokopy, R. J., S. R. Diehl, & S. S. Cooley, 1988. Behavioral evidence for host races in *Rhagoletis pomonella* flies. Oecologia 76: 138–147.

Rausher, M. D., 1988. Is coevolution dead? Ecology 69: 898–901.

Rausher, M. D., 1992. Natural selection and the evolution of plant-insect interactions. In: B. D. Roitberg & M. B. Isman (eds). Insect Chemical Ecology: An Evolutionary Approach. Chapman and Hall Publ., New York.

Reissig, W. H. & D. C. Smith, 1978. Bionomics of *Rhagoletis pomonella* in *Crataegus*. Annals of the Entomological Society of America 71: 155–159.

Rice, W. R., 1987. Speciation via habitat specialization. Evolutionary Ecology 1: 301–314.

Rice, W. R. & E. E. Hostert, 1993. Laboratory experiments on speciation: what have we learned in 40 years? Evolution 47: 1637–1653.

Roethele, J. B., J. L. Feder, S. H. Berlocher, M. E. Kreitman & D. Lashkari, 1997. Towards a molecular genetic linkage map for the apple maggot fly, *Rhagoletis pomonella* (Diptera:Tephritidae): a comparison of alternative strategies. Annals of the Entomological Society of America 90: 470–479.

Schlüter, D., 1999. Ecological causes of speciation. In: D. Howard & S. H. Berlocher (eds), Endless Forms: Species and Speciation. Oxford University Press, Oxford, U.K.

Smith, D. C., 1988. Heritable divergence of *Rhagoletis pomonella* host races by seasonal asynchrony. Nature 336: 66–67.

Tauber, M. J. & C. A. Tauber, 1976. Insect seasonality: Diapause maintenance, termination, and postdiapause development. Annual Review of Entomology 21: 81–107.

Tauber, C. A. & M. J. Tauber, 1989. Sympatric speciation in insects: perception and perspective. In: D. Otte & J. A. Endler (eds), Speciation and Its Consequences. Sinauer, Sunderland, Mass., pp. 307–344.

Tauber, M. J., C. A. Tauber & S. Masaki, 1986. Seasonal Adaptations of Insects. Oxford University Press, Oxford, UK, pp. 411.

Via, S., 1991. The genetic structure of host plant adaptation in a spatial patchwork: Demographic variability among reciprocally transplanted pea aphid clones. Evolution 45: 827–852.

Walsh, B. J., 1864. On phytophagous varieties and phytophagous species. Proceedings of the Entomology Society of Philadelphia 3: 403–430.

Walsh, B. D., 1867. The apple-worm and the apple maggot. Journal of Horticulture 2: 338–343.

Wood, T. K. & M. C. Keese, 1990. Host-plant-induced assortative mating in *Enchenopa* treehoppers. Evolution 44: 619–628.

Wood, T. K., K. C. Olmstead & S. I. Guttman, 1990. Insect phenology mediated by host-plant water relations. Evolution 44: 629–636.

*Entomologia Experimentalis et Applicata* **91:** 227–232, 1999.



# Behavioural correlates of genetic divergence due to host specialization in the pea aphid, *Acyrthosiphon pisum*

Marina C. Caillaud
*Department of Entomology, Cornell University, Ithaca NY 14853, USA*

Accepted: October 18, 1998

*Key words:* Aphid, EPG, host races, local adaptation, feeding behaviour


## Abstract

Host specialization plays a central role in the diversification of herbivorous insects and yet we know very little about the evolution of this trait. Populations of the pea aphid, *Acyrthosiphon pisum* (Harris), are specialized and locally adapted to either alfalfa or clover. Preference for either plant produces assortative mating, and gene flow between alfalfa and clover populations is consistently restricted (Via, 1991a,b, 1994). Behavioural studies of freely-moving aphids on the plant surface for 30 min suggest that host preference is chemically mediated, involves chemoreception, and is most likely due to secondary compounds located in the epidermal or mesophyll cells. Pea aphids do not distinguish between hosts and non-hosts at a distance but determine whether the plant is suitable or not after only a short probe. Thirty-minute recordings of the activities of aphid stylets using an electrical monitoring system (EPG), where aphids are attached to a gold wire with silver paint, provide a different picture, suggesting that EPG experiments do not accurately reflect natural behaviour during the first 30 min of the aphid-plant interaction.


## Introduction

Plant-feeding insect species make up over one quarter of all multicellular organisms (Strong et al., 1984) and although the ability to feed on many plants would provide obvious advantages, most of them have a restricted diet breadth (Bernays & Chapman, 1994; Futuyma, 1991; Jaenike, 1990). What determines diet breadth and why are so many herbivorous insects feeding specialists? Because host specificity can impose assortative mating and reproductive isolation between locally adapted populations, these questions are pivotal in understanding insect speciation (Bush, 1994; Mopper & Strauss, 1997; Rice, 1987). Also, some phytophagous insect species became major pests because of their high degree of specialization on crops (e.g., brown planthopper), or because of a shift or expansion in their host range (e.g., Colorado potato beetle, apple maggot) (Diehl & Bush, 1984). A clear understanding of the determinants of the evolution of host-plant specialization is therefore also crucial in agricultural systems.

An important unresolved question that consistently limits our ability to predict changes in host specificity over time, including host shifts, is the mode of inheritance of host specificity (Via, 1990; Futuyma, 1991). Knowledge of the genetic basis of the complex trait ‘host use’ would both permit us to understand how host use has evolved in the past and help us predict future evolutionary events related to host specificity. Debates over host shifts and sympatric speciation in herbivorous insects arise directly from differing assumptions concerning the genetic basis of host plant preference (Diehl & Bush, 1989; Kondrashov & Mina, 1985). Currently, these debates rage on in the absence of data on the genetic architecture of host specialization.

Populations of the pea aphid, *Acyrthosiphon pisum* (Harris), that are found in closely adjacent fields of clover and alfalfa are genetically differentiated, and locally adapted, to their host plant: they exhibit a high fitness on their home plant (for instance: alfalfa for genotypes found in alfalfa fields), and a low fitness on the away plant (clover) (Via, 1991a,b, 1994). Gene flow is significantly restricted across crops relative to among fields of the same crop and this is mainly due to host-plant fidelity (S. E. Via, 1999). If released in large cages and given the choice between alfalfa and clover, 92 to 100% of winged aphids are found on the plant where the performance of their offspring is the highest after 48 h. Since pea aphids mate on the host plant on which they feed, host preference produces assortative mating and is the first and major cause of gene flow restriction between alfalfa and clover populations. However, the mechanisms and genetic basis of host plant preference in the pea aphid are not known, which consistently limits our understanding of the current determinants and possible origins of premating isolation in this system.

Host preference is a highly integrated and complex trait, the genetic basis of which may be hard to decipher without a means of dissecting the behaviour into simple components. For example, analyses of dance communication in the honeybee showed that small elements of this complex behaviour were regulated by subsets consisting of simple genic systems (Rinderer & Beaman, 1995). A dissection of the trait 'host preference' requires the consideration of the four successive steps involved in the acceptance of a plant as a host in aphids: landing and settlement of the winged individuals (step 1), brief probing and evaluation of the surface/subepidermic plant characteristics (step 2), deep probing and search for the nutritional tissues (step 3), and evaluation of the phloem sap as a suitable nutrient for ingestion (step 4). In a previous study, I found no evidence that pre-landing events (step 1) were important for understanding host specialization in the pea aphid since all genotypes, whatever their specialization, walk or fly to clover and alfalfa within 7 min (M. C. Caillaud, unpubl.). In this paper, I examine step 2 of host-plant acceptance using two methods. First, I study the activities of aphids moving freely onto the leaf. Although this situation is close to natural conditions, it does not provide information about the mouthpart activities inside the plant tissues. Thus, I also record aphid stylet activities using an electrical monitoring system, where aphids are attached to a gold wire and are restricted in their behaviour by the length of the wire. For both methods, the aphid behaviour was followed continuously for 30 min, in order to examine only the interactions between aphid stylets and the first layers of the plant tissue (epidermis, mesophyll cells).

## Materials and methods

I considered an alfalfa specialist (genotype L9) and a clover specialist (genotype NC1) collected in North America (New York State, Tompkins County). Experiments involved 3-6-day-old apterous adults. At this age, behavioural variance among individuals of the same genotype is small, and thus between genotype variances appear more clearly. Individuals were starved for 3 h prior to experiments in order to (1) increase their reactivity to the plant and (2) standardize the pre-experimental physiological conditions. Experiments were performed at the same time every day to remove any effect of diurnal rhythm on the results (5 p.m. local time $\pm$ 30 min), at room temperature (humidity of 60%) and under artificial illumination (one 40W Daylight tube and one 40W plant grow light were used to match the spectrum of natural daylight, 5.2 W/m$^2$). Clover and alfalfa plants were grown in a growth chamber at 20 °C and were 5 to 6 weeks old at the time of the experiments described below.

In a first experiment, 4 aphids were placed individually on different leaves of either clover or alfalfa plants. All their activities were recorded for 30 min continuously, using a computer-based event recorder ('The Observer', Noldus Information Technology, Wageningen, NL). I recorded every penetration, bout of walking or immobility and whether aphids produced offspring. Contrary to the black-bean aphid, *Aphis fabae* (Hardie et al., 1992), antennal and body movements are not reliable indicators of stylet penetration in the pea aphid. Here, penetration was scored when immobile aphids had their (black) rostrum perpendicular to their body and when their antennae made an angle of less than 90° with their abdomen. Aphids that did not probe within 15 min (rare) or that were in a position preventing the observer from seeing their rostrum (not frequent), were discarded. From these observations, several parameters were evaluated: time before the first probe, duration of the first probe, number of probes over 30 min, mean duration of a probe, total time spent probing the plant tissues, number of larvae produced over 30 min, percentage of aphids remaining at the end of the 30-min observation period, and time before aphids left the plant (eventually). Twelve to 15 replicates per genotype and per plant were made . Each replicate was the mean of the behaviour of four aphids (or three aphids when one of the four was discarded for the reasons outlined earlier). I was aware of possible behavioural interactions between aphids (Ibbotson & Kennedy, 1959 ) but I thought that (1) both host races were studied using the same protocol and comparisons were thus fully valid, and (2) the mean of three or four aphids was a more accurate measure of the behavioural phenotype than the single measure of one aphid.

In a second experiment, aphid stylet activities during plant penetration were recorded individually with an electrical monitoring system for 30 min (DC-EPG or EPG, Tjallingii, 1978, 1988) following the protocol of Caillaud et al. (1995). In the DC-EPG, the aphid and the plant are part of an electrical circuit. The penetration of the aphid stylets in the plant modifies the characteristics (voltage) of the electrical signal recorded and these changes provide information about the behaviour (ingestion, salivation) and the stylet tip position (phloem, xylem,etc.) of the aphid during plant penetration (Tjallingii, 1978, 1988, 1990). For L9 and NC1, monitored for 30 min only, six different electrical waveforms were distinguished in the EPG signal. Waveforms A, B and C are considered as pathway phase and represent the aphid searching for target tissues (xylem or phloem). Potential drops (pd's) are brief intracellular punctures performed by the aphid stylets during their progression towards the feeding sites. Many plant allelochemicals are stored in the cell vacuole and aphids might be able to perceive the presence of these secondary compounds during these short pd's. NP represents non-penetration activities (resting or walking) and F is still of unknown significance. All parameters described in experiment 1 but the last two were measured, because aphids are attached to a thin gold wire with silver paint during EPG recording and thus are not able to leave the plant. Thirteen to 18 individuals per genotype and plant were recorded.

For each experiment, the aphid behaviour was analysed by Multivariate Analysis of variance, MANOVA, with fixed and crossed factors (Proc Glm; SAS, 1989). Analyses were first performed using a complete model (genotype, plant, genotype∗plant effects) then the data set was partitioned and each plant habitat was analysed separately (genotype effect only). Also, within each plant habitat, and using canonical correlation coefficients generated in the MANOVA analysis, the sets of behaviour that best differentiate between the alfalfa and the clover specialist were identified. In the absence of significant differences in the MANOVA analysis, differences among genotypes were analysed separately for each behavioural parameter by two-way ANOVA. The residuals for two behavioural parameters, the total number of probes and the number of larvae produced, were problematic. I tried many transformations to make the data more normal, but the results of the statistical analyses were comparable with the results using the raw data. Thus, I present here the analyses performed on raw data since their interpretation is easier.

## Results

As described for aphid fitness (Via, 1991a), there is significant genotype∗plant interaction for feeding behaviour in the pea aphid (Table 1a). The overall behaviours of freely-moving individuals of L9 and NC1 differs significantly in each plant habitat (Tables 1b and 1c). These differences are best explained, in the order given by the standardized canonical coefficients extracted from the 'by plant' MANOVA, by the total time spent probing plant tissues, the duration of the first probe, and the mean duration of a probe. L9 and NC1 probed host and non-host plants within 60 s (Table 2) suggesting that cues prior to stylet penetration, such as plant trichomes, do not play an important role in plant acceptance (no significant genotype∗plant interaction: $F = 3.17$, $P = 0.08$). On the host plant (alfalfa for L9 for instance), aphids performed a long first probe (300 to 700 s), did not change location very frequently, and produced an average of 0.42 to 0.52 larvae over 30 min. On the non-host plant, the first probe was significantly shorter than on the preferred host (87 s for L9 on clover, 167 s for NC1 on alfalfa)(significant genotype∗plant interaction: $F = 6.95$, $P = 0.01$). It was followed by numerous and increasingly shorter probes. Most aphids left the non-host plant after about 20 min (54% and 57% of L9 and NC1 did so), without depositing any larvae.

In contrast to the observations performed on freely-moving aphids, L9 and NC1 individuals studied by EPG for 30 min did not exhibit significantly different behaviour on either clover or alfalfa (Tables 3b and 3c). Unlike free individuals, tethered aphids probed more rapidly on hosts (L9 on alfalfa) than on non-hosts (significant genotype∗plant interaction, $F = 9.04$, $P = 0.004$) and did not differ in the length of their first penetration ($F = 1.13$, $P = 0.29$) or in the total number of penetrations over 30 min ($F = 0.37$, $P = 0.54$) (Table 4). The first probe of L9 individuals on their preferred host lasted between 12 s and 1776 s (390 s on average) and thus was not found significantly different from the first probe on clover which lasted between 8 s and 59 s (14 s in average). Wired aphids did not produce any larvae. Interestingly, the first penetration on

*Table 1.* The behaviour of freely-moving individuals of the specialized genotypes L9 and NC1 (alfalfa and clover specialists, respectively). Manova and canonical correlation analysis

| Source of variation | Wilks' | F | Pr>F | Total canonical correlation |
|---|---|---|---|---|
| *1a. Complete model* | | | | |
| Genotype | 0.6 | 2.57 | 0.06 | 0.62 |
| Plant | 0.64 | 2.22 | 0.09 | 0.59 |
| Genotype*Plant | 0.04 | 85.15 | 0.0001 | 0.97 |
| *1b. On alfalfa* | | | | |
| Genotype | 0.05 | 31.21 | 0.0001 | 0.97 |
| *1c. On clover* | | | | |
| Genotype | 0.006 | 211.5 | 0.0001 | 0.99 |

*Table 2.* The behaviour of freely-moving individuals on alfalfa and clover. L9 and NC1 are specialized on alfalfa and clover, respectively. Means and (SD)

| Freely-moving aphids | L9 on host | L9 on non-host | NC1 on host | NC1 on non-host |
|---|---|---|---|---|
| Time to 1st penetration (s) | 39 (34) | 30 (37) | 59 (63) | 37 (35) |
| Duration of 1st penetration (s) | 389 (384) | 87 (83) | 643 (547) | 167 (90) |
| Total number of penetrations | 4.75 (3.4) | 16.5 (4.4) | 2.9 (1.7) | 12.5 (3.6) |
| Total time spent penetrating (s) | 1606 (129) | 318 (63) | 1600 (115) | 632 (213) |
| Mean duration of a penetration (s) | 704 (432) | 23 (3) | 915 (447) | 81 (56) |
| No. of larvae produced | 0.42 (0.4) | 0 | 0.52 (0.27) | 0 |
| % adults remaining on plant after 30 min | 100 | 54 (27) | 100 | 57 (38) |
| Time before plant was left (s) | – | 1262 (222) | – | 1067 (225) |

*Table 3.* Probing behaviour, as recorded by DC-EPG, of the specialized genotypes L9 and NC1 (alfalfa and clover specialists, respectively). Manova and canonical correlation analysis

| Source of variation | Wilks' | F | Pr>F | Total canonical correlation |
|---|---|---|---|---|
| *3a. Complete model* | | | | |
| Genotype | 0.99 | 0.11 | 0.98 | 0.11 |
| Plant | 0.67 | 4.89 | 0.001 | 0.57 |
| Genotype*Plant | 0.79 | 2.58 | 0.003 | 0.45 |
| *3b. On alfalfa* | | | | |
| Genotype | 0.78 | 1.15 | 0.36 | 0.46 |
| *3c. On clover* | | | | |
| Genotype | 0.75 | 1.6 | 0.19 | 0.49 |

*Table 4.* Probing behaviour of the specialized genotypes L9 and NC1 as recorded by DC-EPG (30-min recordings). Means and (SD)

| EPG observations | L9 on host | L9 on non-host | NC1 on host | NC1 on non-host |
|---|---|---|---|---|
| Time to 1st penetration (s) | 37 (35) | 189 (240) | 62 (56) | 240 (342) |
| Duration of 1st penetration (s) | 390 (640) | 14 (4.2) | 76 (151) | 127 (222) |
| Total number of penetrations | 7.81 (5.3) | 5.6 (3.2) | 6.05 (2.77) | 7.92 (4.3) |
| Total time spent penetrating (s) | 1144 (527) | 377 (567) | 729 (567) | 755 (559) |
| Mean duration of a penetration (s) | 120 (135) | 70 (80) | 284 (391) | 156 (202) |
| No. of larvae produced | 0 | 0 | 0 | 0 |

non-hosts always included at least one potential drop (pd) soon after stylet penetration started.

## Discussion

The behaviour of freely-moving aphids and tethered aphids, during the first 30 min of the insect-plant interaction, differed in many ways. Tethering of aphids increases the manipulation of the individuals and can potentially induce behavioural change. Also, tethering causes behavioural restrictions as individuals cannot either explore the whole plant or leave it (Tjallingii, 1986). In 1993, Powell and colleagues examined the behaviour of aphids connected to an EPG amplifier for 15 min on leaves treated or not with polygodial, a feeding deterrent. Polygodial in this situation caused no significant behavioural differences in terms of time taken to initiate a first penetration, or the number of penetrations, whereas clear differences were shown by observations performed on freely-moving aphids (Powell et al., 1993). Wilkinson & Douglas (1998) also drew attention to the inadequacies of the EPG technique when host affiliation is mediated by factors associated with the leaf surface, and suggested that the tethering effect masked the aphid response to these surface cues. Thus, the EPG technique should not be used to analyse the first moments of the aphid-plant interaction. Its use should be restricted to long recordings (>1h) where it provides incomparable and extremely detailed information about stylet activity as they progress towards the nutritional tissues.

The behaviour of freely-moving aphids on alfalfa and clover mirrors very well their performances on these plants (S. E. Via, unpubl.). NC1, for instance, in which nymphs die before reaching maturity on alfalfa, reject this plant within 30 min (most individuals leave the plant), while they readily accept clover, a plant where there is almost no larval mortality and a lifetime fecundity of up to 80 nymphs. There is an extremely clear correlation between host specificity and the events associated with plant surface exploration. The first penetration is consistently shorter on non-hosts, when compared to hosts, suggesting that some allelochemical(s), either deterrents that are present or stimulants that are lacking, were perceived by aphids during this first probe. EPG recordings showed that the first penetration always included a pd. During a pd, it is believed that the tip of the aphid stylets enters briefly (a few ms) inside a cell, briefly enough to induce no damage to the cell, but long enough to detect the presence of allelochemicals (Tjallingii & Hogen Esch, 1993). It is tempting to infer that the allelochemical(s) correlated with host specialization in the pea aphid are located inside the epidermal cells or in the vacuoles of the first row of mesophyll cells. The perception of these chemicals results in a strong and rapid rejection of the nonhost. Under natural conditions, winged individuals which land on the 'wrong' host, i.e., alfalfa if they are clover specialists, presumably take off again shortly afterwards to find a more favourable plant. The rapidity of this rejection can increase their chances to escape natural enemies foraging on plants (Bernays & Wcislo, 1994).

Rejection is not modified if the aphid stays on the plant for more than 30 min, as shown by another experiment in which step 3 and 4 of host-plant acceptance were studied by DC-EPG (M. C. Caillaud & S. E. Via, unpubl.). Pea aphids never reached the phloem vessels or ingested phloem sap on the non host, and performed only a few short probes on that plant. Complete rejection presumably causes starvation on non-host plants (no ingestion of nutrients from the phloem), from which the aphid derives its low fitness. The identification of the plant compounds involved is in progress.

## Acknowledgements

This work was supported by the Institut National de la Recherche Agronomique (INRA, France) and a North Atlantic Treaty Organisation (NATO) post-doctoral fellowship. I am very grateful to Sara Via (University of Maryland) for providing the aphid genotypes L9 and NC1 and for hosting my postdoctoral fellowship. I also thank Pamela Mole for her help with the analysis of EPG recordings. Bryan Danforth of Cornell University made helpful comments on the manuscript.


## References


Bernays, E. A. & R. F. Chapman, 1994. Host-Plant selection by phytophagous insects. Chapman & Hall, New York, 312 pp.

Bernays, E. A. & W. T. Wcislo, 1994. Sensory capabilities, information processing, and resource specialization. Quarterly Review of Biology 69: 187–204.

Bush, G. L., 1994. Sympatric speciation in animals: new wine in old bottles. Trends in Ecology and Evolution 9: 285–288.

Caillaud, C. M., J. S. Pierre, B. Chaubet & J. P. Di Pietro, 1995. Analysis of wheat resistance to the cereal aphid *Sitobion avenae* using electrical penetration graph and flow charts combined with correspondence analysis. Entomologia Experimentalis et Applicata 75: 9–18.

Diehl, S. R. & G. L. Bush, 1984. An evolutionary and applied perspective of insect biotypes. Annual Review of Entomology 29: 471–504.

Diehl, S. R. & G. L. Bush, 1989. The role of habitat preference in adaptation and speciation. In: D. Otte & J.A. Endler (eds), Speciation and its Consequences. Sinauer Press, NY, pp. 345–365.

Futuyma, D. J., 1991. Evolution of host specificity in herbivorous insects: genetic, ecological and phylogenetic considerations. In: P. W. Price, T. M. Lewinsohn, G. W. Fernandes & W. W. Benson (eds), Plant-Animal Interactions: Evolutionary Ecology in Tropical and Temperate Regions. Wiley, New-York, pp. 431–454.

Hardie, J., M. Holyoak, N. J. Taylor & D. C. Griffiths, 1992. The combination of electronic monitoring and viseo-assisted observations of plant penetration by aphids and behavioural effects of polygodial. Entomologia Experimentalis et Applicata 62: 233–239.

Ibbotson, A. & J. S. Kennedy, 1959. Interaction between walking and probing in *Aphis fabae* Scop. Journal of Experimental Biology 36: 377–390.

Jaenike, J., 1990. Host specialization in phytophagous insects. Annual Review of Ecology and Systematics 21: 243–273.

Kondrashov, A. S. & M. V. Mina, 1985. Sympatric speciation: when is it possible? Biological Journal of the Linnean Society London 27: 201–223.

Mopper, S. & S. Y. Strauss, 1997. Genetic structure and local adaptation in natural insect populations. Effects of Ecology, Life History and Behaviour. Chapman & Hall, New York, 449 pp.

Powell, G., J. Hardie & J. A. Pickett, 1993. Effects of the antifeedant polygodial on plant penetration by aphids, assessed by video and electrical recording. Entomologia Experimentalis et Applicata 68: 193–200.

Rice, W. R., 1987. Speciation via habitat specialization: the evolution of reproductive isolation as correlated character. Evolutionary Ecology 1: 301–314.

Rinderer, T. E. & L. D. Beaman, 1995. Genic control of honey bee dance language dialect. Theoretical and Applied Genetics 91: 727–732.

SAS Institute Inc., 1989. SAS/StAT User's guide, version 6, Fourth edition, Volume 1, SAS Institute Inc., Cary, NC , 943 pp.

Strong, D. R., J. H. Lawton & R. Southwood, 1984. Insects on Plants: Community Patterns and Mechanisms. Blackwell, Oxford, UK.

Tjallingii, W. F., 1978. Electronic recording of plant penetration behaviour by aphids. Entomologia Experimentalis et Applicata 24: 521–530.

Tjallingii, W. F., 1986. Wire effects on aphids during electrical recording of stylet penetration. Entomologia Experimentalis et Applicata 40: 89–98.

Tjallingii, W. F., 1988. Electrical recording of stylet penetration activities. In: A. K. Minks & P. Harrewijn (eds), Aphids, Their Biology, Natural Enemies and Control, vol. 2B. Elsevier, Amsterdam, pp. 95–108.

Tjallingii, W. F., 1990. Continuous recording of stylet penetration activities by aphids. In: R. K. Campbell & R. D. Eikenbary (eds), Aphid-Plant Genotype Interactions. Elsevier, Amsterdam, pp. 89–99.

Tjallingii, W. F. & Hogen Esch, T. 1993. Fine structure of aphid stylet routes in plant tissues, correlation with EPG signals. Physiological Entomology 18: 317–328.

Via, S. E., 1990. Ecological genetics and host adaptation in herbivorous insects: the experimental study of evolution in natural and agricultural systems. Annual Review of Entomology 35: 421–446.

Via, S. E., 1991a. The genetic structure of host plant adaptation in a spatial patchwork: demographic variability among reciprocally transplanted pea aphid clones. Evolution 45: 827–852.

Via, S. E., 1991b. Specialized host plant performance is not altered by experience. Ecology 72: 1420–1427.

Via, S. E., 1994. Population structure and local adaptation in a clonal herbivore. In: L. Real (ed.), Ecological Genetics. Princeton University Press, Princeton, pp. 58–85.

Via, S.E., 1999. Reproductive isolation between sympatic races of pea aphids: I. gene flow restoction and habitot choice. Evolution, in press.

Wilkinson, T. L. & A. E. Douglas, 1998. Plant penetration by pea aphids (*Acyrthosiphon pisum*) of different plant range. Entomologia Experimentalis et Applicata 87: 43–50.

Entomologia Experimentalis et Applicata **91:** 233–236, 1999.



# Using hybrid and backcross larvae of *Papilio canadensis* and *Papilio glaucus* to detect induced phytochemical resistance in hybrid poplar trees experimentally defoliated by gypsy moths

J. Mark Scriber, Kirsten Weir, Dylan Parry & Jessica Deering
*Department of Entomology, Michigan State University, East Lansing, MI 48824, USA*

Accepted: November 10, 1998

*Key words:* hybrid genotypes, herbivore defoliation, phytochemical induction, differential growth rates, indirect competition, phenolic glycoside detoxification, *Lymantria dispar*, Papilionidae, tiger swallowtail butterflies

## Abstract

Sub-plots of hybrid poplars were experimentally defoliated using 10 million gypsy moth larvae. Half of the defoliated (and undefoliated control) plots were fertilized to see if this would ameliorate the predicted induction of carbon-based phenolic defenses in the regrowth leaves. In order to bioassay the leaves of the four different treatments, we employed a continuum of genotypes (different hybrids and backcrosses of two different species of tiger swallowtail butterflies) with different abilities to detoxify these allelochemicals. Based on our previous studies with phytochemicals from the Salicaceae plant family, *Papilio canadensis* was likely to consume and process all *Populus* spp treatments, whereas *P. glaucus* predicted to either not consume or else quickly die on all *Populus* treatment leaves. Hybrid and backcross larvae of these two butterfly species are known to have intermediate levels of esterase detoxication enzymes and would therefore be likely to provide a continuum or at least varying degrees of sensitivity in bioassays for even the most subtle induction responses in the regrowth leaves. This presumption was supported in the feeding and growth studies conducted at different times post-defoliation during the 1997 growing season in Michigan.

## Introduction

Detection of induced phytochemical resistance (rapid or delayed) after insect defoliation will depend upon the species (or genotype) of plant and the abiotic environment, including nitrogen availability and the C/N balance (Bryant et al., 1983; Herms & Mattson, 1992). However, perhaps most significantly, it depends upon the particular herbivore species or genotype encountering this post-defoliation plant, and the degree of behavioral or physiological inducibility in its response capabilities (Dankert et al., 1997).

With plants in the Salicaceae family (aspens, poplars, willows, etc.), we have observed well-adapted herbivores (e.g., *Papilio canadensis* and *P. rutulus*) with esterase detoxification systems (Lindroth et al., 1988; Scriber et al., 1991). Other very closely related congeneric species have low esterase activities and very poor abilities to survive on any Salicaceae species (e.g., *P. glaucus*, *P. alexiares*, and *P. multicaudatus*; Scriber et al., 1995; Scriber, 1996).

Interspecific hybrids and backcross larvae of the Salicaceae-adapted *P. canadensis* and the unadapted *P. glaucus* had intermediate esterase levels and correspondingly intermediate growth rates when challenged with phenolic glycosides from aspen (Scriber et al., 1989). This continuum of swallowtail butterfly genotypes might provide a differentially sensitive bioassay technique with which to detect even subtle changes in phytochemical suitability (induced defense) after severe defoliation of *Populus* hybrids by gypsy moth larvae.

As part of the design, half of the defoliated plots were fertilized to assess the degree to which induced resistance might be ameliorated by N-fertilization. For

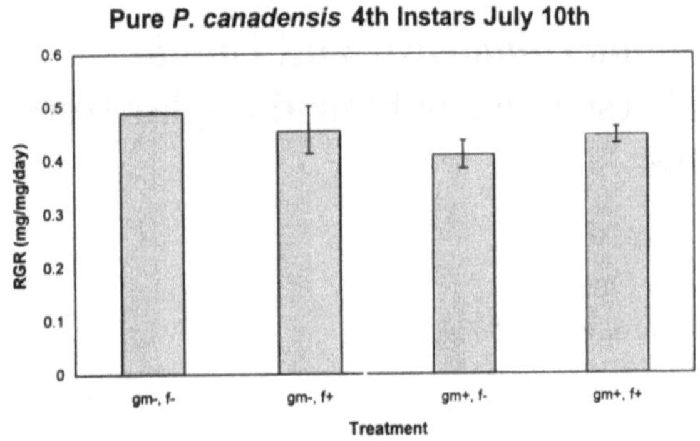


*Figure 1.* The relative growth rates (mg $\times$ $mg^{-1}$ $\times$ $d^{-1}$) of penultimate instar *P. canadensis* fed hybrid *Populus* from four experimental treatments (+ with defoliation, + with fertilization, etc.). No significant difference exists for defoliation (F = 1.39, P = 0.30) or fertilization (F = 1.23, P = 0.33).

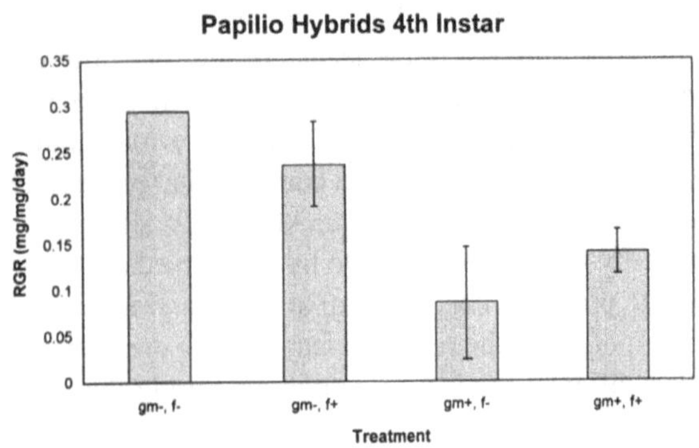


*Figure 2.* The relative growth rates of penultimate instar hybrid larvae (*P. canadensis* female $\times$ *P. glaucus* male) fed four *Populus* treatments.

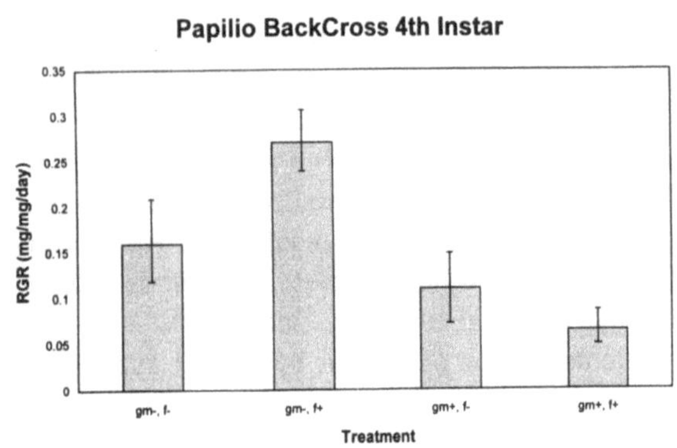


*Figure 3.* The relative growth rates of penultimate instar backcross larvae (hybrid female $\times$ *P. glaucus* male) fed four *Populus* treatments.

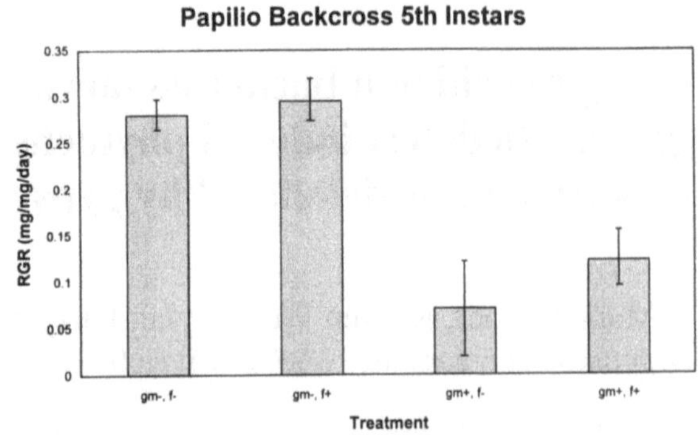


*Figure 4.* The relative growth rates of final instar backcross larvae (hybrid female X *P. glaucus* male) fed four *Populus* treatments.

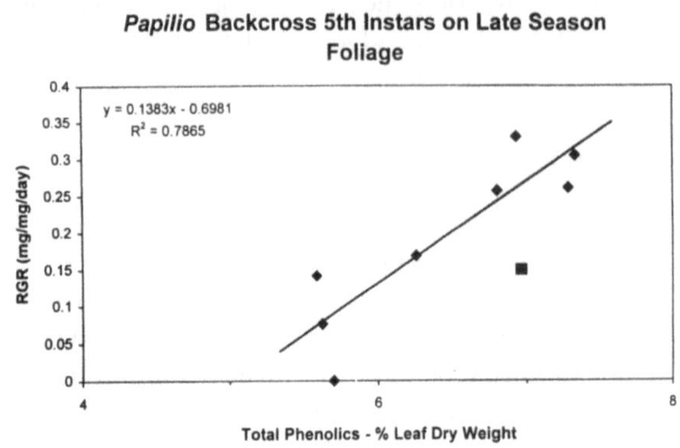


*Figure 5.* Final instar growth rates of backcross larvae as a function of total foliar phenolics.

comparison, half of the undefoliated plot was also fertilized.

*Figure 6.* A photograph of the defoliated versus non-defoliated treatments (1997).

## Materials and methods

*Creating outbreak (defoliation) and control (undefoliated) plots.* In the hybrid poplar plantation plots of Michigan State University's Kellogg Biological Station's L.T.E.R. 'Agroecology' Site, we supplemented natural gypsy moth infestations in 1996 (2 million formalin-treated eggs) and 1997 (10 million treated eggs) in order to simulate outbreak dynamics, but also to create experimentally controlled adjacent plots without defoliation. The formalin killed the transovariole virus (NPV) that would likely have had major disruptive effects on the gypsy moth damage levels we were trying to create. Two large plots (600 $m^2$) were used in a randomized split plot design with defoliation in the main plots and fertilization (100 kg $ha^{-1}$ of N each year) in the subplots.

Defoliation-free plots were created without the use of pesticides by using tanglefoot protective sticky barriers, and intensive caterpillar removal from burlap-banded control trees as well as general caterpillar 'wrangling' and transport back to defoliation plots. Frass fall in traps correlated very closely with gypsy moth defoliation levels in all 16 treatment plots ($R^2 = 0.75$, $P < 0.01$; see Parry et al., 1998). Defoliation in these two plots was estimated at 30–40% in 1996 and 95–100% in 1997.

*Leaf chemical analyses.* Leaves (1997) for chemical analyses were collected June 28 and August 8, 1997. Each plot and subplot had 14 leaves sampled from upper crown branches using pole pruners. Leaves were immediately flash frozen in liquid nitrogen and transported in ice chests from the field. A methanolic extract of phenolics was quantified using a Folin-Denis procedure and a Rapid–Flow Analyzer. Condensed tannins were quantified using a modified sulfuric acid procedure (J. N. Nitao et al., unpublished) on the same equipment and are included in the total phenolics.

*Larval rearing and bioassays.* Larval bioassays were conducted in individual Petri dishes (150 × 25 mm) with a Poplar leaf from one of the 4 treatments supported in a water-filled aqua-pic in a controlled environment chamber at 25 °C and an 18:6 photoperiod. Neonate larval survival was recorded at 6 days. Later instars (4th and 5th) were reared on black cherry (a favorite common host for both *P. canadensis* and *P. glaucus*) until the experimental treatment leaves were introduced to freshly molted caterpillars. Standard gravimetric techniques were used to assess rates and efficiencies of larval consumption and growth for 72 h. Virtually identical size (weight) larvae were used to avoid some of the potential gravimetric technique problems described by Simpson & Raubenheimer (1995) and Scriber (1996b). Hybrids and backcrosses were obtained by hand-pairing lab-reared adult females with lab or field males. *Papilio glaucus* were obtained from Georgia (Clarke Co.) and *P. canadensis* from northern Michigan (Cheboygan, Charlevoix, and Emmet Counties). Oviposition was achieved as described by Scriber (1993).

## Results

Pure *Papilio canadensis* larvae fed leaves from the 4 treatment plots in mid-July (3 weeks after peak defoliation by gypsy moths) showed no significant treatment differences in relative growth rates (RGR = $mg \times mg^{-1} \times day^{-1}$) for either the penultimate (4th; Figure 1) or final (5th) instars. In contrast 4th instar offspring larvae of *P. glaucus* × *P. canadensis* exhibited poor growth (RGR = 0.085) and the highest 3-day mortality (66%) on July 4 leaves from the Defoliated/Unfertilized ($gm^+$, $f^-$) plots (Figure 2). In neonate studies, none of the *P. glaucus* larvae survived. There was a significant defoliation effect detected in F1 hybrid larvae ($F = 7.3$, $P = 0.05$), but not pure *canadensis* which were bioassayed simultaneously ($F = 1.7$, $P = 0.26$). Unlike the hybrids, these *P. canadensis* larvae showed 100% survival on all treatments. In mid-August, 6–7 weeks after defoliation, backcross progeny (Pc × Pg) × Pg showed significant reduction in growth on the two defoliated treatments compared to the two control plots for both penultimate and final instars (Figure 3 penultimate instar $F = 8.5$, $P = 0.04$: Figure 4 final instar $F = 42.6$, $P = 0.003$). Defoliation caused a significant elevation in the levels of total phenolics ($F_{1,3} = 59.6$; $P < 0.005$) and condensed tannins ($F_{1,3} = 23.4$; $P < 0.02$). Fertilizer had neither a direct nor interactive effect on phenolics or total foliar tannins (Parry et al.,1998). At the time near peak defoliation (July 3, 1997) relative growth rates of 4th instar hybrids are negatively correlated with total foliar phenolics ($R^2 = 0.808$). However, at 6–7 weeks post-defoliation (August 11, 1997) the growth of backcross larvae in the different plots was positively correlated with total foliar phenolics with the lowest phenolic levels in the defoliated plots (penultimate instar $R^2 = 0.567$; final instar $R^2 = 0.787$; Figure 5).

## Discussion

Severe defoliation of hybrid poplars by gypsy moths negatively altered the suitability of regrowth leaves, especially for hybrid and backcross *Papilio* genotypes. This held true early in the season (July 3 just after peak plot defoliation) and at 6–7 weeks post-defoliation (mid-August). All of the pure *Papilio glaucus* larvae (Salicaceae-unadapted) died on all poplar treatment leaves. In contrast, essentially all of the Salicaceae-adapted *P. canadensis* larvae survived and showed little, if any, suppression of growth rates on leaves of any of the 4 defoliation/fertilizer treatment plots. Reduced survival and growth of hybrids in early July correlated negatively with total phenolic concentrations (suggesting that leaves from the defoliated plots may be poorer for this reason). However, this would not explain our observations that later in the season (mid-August) the hybrid backcrosses grew the best on the undefoliated (control) plot leaves which had the highest levels of total phenolics (Figure 5). Leaf water content of leaves did not vary among treatments from mid-July to mid-September. However, immediately after regrowth (July 3, 1997) the defoliated plot leaves had higher water contents (80–82%, compared to 70–72% in the control plots). It is interesting that the fastest larval growth of hybrids at this time was on the undefoliated plot leaves in spite of their lower water content. We suspect that differences in nitrogen content of leaves or specific phenolic glycosides such as tremulacin (Ayres et al., 1997) may largely explain the *Papilio* backcross larval growth rate suppression on the defoliated (with lower total leaf phenolics) versus the undefoliated controls, however we have not yet completed these analyses.

## Acknowledgements

This project was supported by the NSF (LTER, BSR-8702332), (Ecosystems, DEB-9510044) and the REU Program, as well as the MAES, College of Agriculture and Natural Resources, College of Natural Sciences of Michigan State University. We thank Kevin Kosola for extensive help in the field plots. We also thank those helping with the gypsy moth defoliation, field plots, and lab studies: Mark Deering, Joshua Edwards, Carrie Fetzner, Piera Giroux, Heather Govenor, Gabe Ording, Christina Schoen, and Aram Stump. We are grateful for help with the chemical analyses from: Bruce Birr, Bill Mattson, Muralee Nair, and James Nitao.

## References

Ayres, M. P., T. P. Clausen, S. F. MacLean, Jr., A. M. Redman & P. B. Reichardt. 1997. Diversity of structure and antiherbivore activity in condensed tannins. Ecology 78: 1696–1712.

Bryant, J. P., F. S. Chapin III, P. Reichert & T. Clausen, 1983. Plant carbon/nutrient balance of boreal plants in relation to vertebrate herbivory. Oikos 40: 357–368.

Dankert, B. A., D. A. Herms, D. Parry, J. M. Scriber & L. A. Haas, 1997. Mediation of interspecific competition between folivores through defoliation-induced changes in host quality. In: A. Raman (ed), Ecology and Evolution of Plant-Feeding Insects in Natural and Man-made Environments. Backhus Publishers, Leiden, pp. 71–88.

Herms, D. & W. J. Mattson, 1992. The dilemma of plants: to grow or defend. Quarterly Review of Biology 67: 283–335.

Lindroth, R. L., J. M. Scriber & M. T. S. Hsia, 1988. Chemical ecology of the tiger swallowtail: mediation of host use by phenolic glycosides. Ecology 69: 814–822.

Parry, D., D. A. Herms, K. Kosola & J. M. Scriber, 1998. Examining the role of delayed-induced resistance in gypsy moth outbreaks. Proceedings of 1998 Interagency Research Forum in Gypsy Moth. USDA-NE Forest Service Publication.

Scriber, J. M., 1993. Absence of behavioral induction in oviposition preferences of *Papilio glaucus* (Lepidoptera: Papilionidae). Great Lakes Entomologist 26: 81–95.

Scriber, J. M., 1996. Tiger tales: natural history of native North American swallowtails. American Entomologist 42: 19–32.

Scriber, J. M., 1996b. Generalist North American silkmoth *Antherea polyphemus*, grows more efficiently and rapidly than its specialized relative, *A. pernyi*: (Lepidoptera: Saturniidae). Holarctic Lepidoptera. 3: 23–30.

Scriber, J. M., R. L. Lindroth & J. Nitao, 1989. Differential toxicity of a phenolic glycoside from quaking aspen leaves by *Papilio glaucus* subspecies, their hybrids, and backcrosses. Oecologia 81: 186–191.

Scriber, J. M., R. L. Lindroth & J. Nitao, 1991. Detoxification of phenolic glycosides from quaking aspen in the western swallowtail butterfly, *Papilio rutulus* (Lepidoptera: Papilionidae). Great Lakes Entomologist 24: 173–180.

Scriber, J. M., R. C. Lederhouse & R. V. Dowell, 1995. Hybridization studies with North American swallowtails. In: J. M. Scriber, Y. Tsubaki, & R. C. Lederhouse (eds), Swallowtail Butterflies: Their Ecology and Evolutionary Biology. Scientific Publishers, Gainesville, FL, pp. 269–281 .

Simpson, S. J. & D. Raubenheimer, 1995. The geometric analysis of feeding and nutrition: a user's guide. Journal of Insect Physiology 41: 545–553.

Entomologia Experimentalis et Applicata **91:** 237–244, 1999.



# *Ostrinia* spp. in Japan: their host plants and sex pheromones

Yukio Ishikawa, Takuma Takanashi, Choong-gon Kim, Sugihiko Hoshizaki, Sadahiro Tatsuki & Yongping Huang
*Laboratory of Applied Entomology, Graduate School of Agricultural and Life Sciences, The University of Tokyo, Tokyo 113-8657, Japan*

Accepted: November 30, 1998

*Key words:* host plant range, sex pheromone, *Ostrinia furnacalis*, *Ostrinia latipennis*, *Ostrinia nubilalis*, *Ostrinia orientalis*, *Ostrinia palustralis*, *Ostrinia scapulalis*, *Ostrinia zaguliaevi*, *Ostrinia zealis*, Lepidoptera, Pyralidae

## Abstract

To contribute to the understanding of the genus *Ostrinia* (Lepidoptera; Pyralidae) in Japan, we collected larvae of *Ostrinia* spp. from known host plants and plants not recorded as hosts, and we examined the morphology and sex pheromones of the adults obtained. Consequently, the host plant ranges of the 7 *Ostrinia* spp. in Japan were clarified, and the sex pheromones of the 5 species *O. scapulalis*, *O. zealis*, *O. zaguliaevi*, *O. palustralis* and *O. latipennis* were identified in addition to that of the Asian corn borer *O. furnacalis*. The phylogenetic relationships of Japanese *Ostrinia* spp., with reference to the European corn borer *O. nubilalis*, are discussed based on these findings and results of molecular phylogenetic analyses.

## Introduction

Among the 20 *Ostrinia* spp. (Lepidoptera: Pyralidae) in the world (Mutuura & Munroe, 1970), the 7 species *O. furnacalis* (Asian corn borer), *O. orientalis*, *O. scapulalis*, *O. zealis*, *O. zaguliaevi*, *O. palustralis* and *O. latipennis* currently inhabit Japan. Among these species, the former 5 are morphologically very similar and form a '*furnacalis* species complex.' In this complex, the identification of the species of individual insects is impossible at the larval stage, and very difficult at the adult stage. Host plants have provided important information regarding the identification of species, but this is not entirely reliable because there are substantial overlaps of host plants among species. Therefore, studies of the ecology and host-plant relationships of *Ostrinia* species in Japan have been hindered. To counter these difficulties, we have recently started research on the biology of *Ostrinia* species in Japan employing multiple methodologies that include morphology, molecular phylogenetics and chemical ecology.

At the start of the present study, the sex pheromones of only 2 *Ostrinia* species, *O. nubilalis* and *O. furnacalis* were reported. The European corn borer, *O. nubilalis*, uses (*E*)- and (*Z*)-11-tetradecenyl acetates (*E*11- and *Z*11-14:OAc) as sex pheromone components, and is known to be polymorphic with respect to the component ratio: *E*-strain females produce an *E:Z* blend of 99:1, whereas *Z*-strain females produce a 3:97 blend (Klun et al., 1973; Kochansky et al., 1975). In contrast to this species, the Asian corn borer, *O. furnacalis*, employs the positional isomers *E*12- and *Z*12-14:OAc as its sex pheromone components, at a ratio varying from 1:1 (Cheng et al., 1981; Klun et al., 1980; Kou et al., 1992) to 2:3 (Ando et al., 1980) depending upon the geographic population. In addition to these species, the smartweed borer, *O. obumbratalis*, is known to be attracted in the field by a mixture of *E*11- and *Z*11-14:OAc at a ratio of 1:1, although the natural composition in the sex pheromone gland is unknown (Klun & Robinson, 1972).

In the present study, the host plants and sex pheromones of the 7 *Ostrinia* species in Japan were investigated. The phylogenetic relationships of these species are discussed based on the present findings and results of molecular phylogenetic studies. Although the European corn borer *O. nubilalis* is not distributed in Japan, this species was included in the analyses because of its very close relationship to the Asian *O. orientalis*.

## Materials and methods

*Insects.* The larvae of *Ostrinia* infesting known host plants and plants not recorded as hosts were collected at more than 10 locations ranging from Obihiro (43.0° N, 143.2° E) to Nishigoshi (32.9° N, 130.7° E) in Japan in 1993 – 1997. Larvae were reared on a commercial diet for insects (Insecta LF, Nihon Nosan, Yokohama, Japan) at 25 °C and 50–60% r.h. with a L15:D9 photocycle. The pupae were sexed based on the morphology of the abdominal terminal segments of the pupae, and maintained under the same conditions of larval rearing. Identification of the adults to species was conducted based on the morphology of male genitalia and mid-tibia, and on the presence of a hair tuft on the abdominal sternites (Mutuura & Munroe, 1970; Yoshiyasu, unpubl.).

*Observation of calling.* For the observation of female calling behavior, females were put individually in plastic cups (200 ml) placed upside down on a damp towel to keep the inside of the cup at a high humidity. Observations of 10 virgin females (2 to 4 days old) were made in red light at 1-hr intervals. For the observation of copulation, 10 pairs each of the same age were put into two screen cages (25 × 25 × 25 cm), and copulating pairs were checked at 30-min intervals.

*Sex pheromone extraction.* Sex pheromone was usually extracted from individual females. During the calling period, the abdominal tip including the sex pheromone gland was excised from a 2- to 4-day-old virgin female, and immersed for 30 min in 8 $\mu$l hexane containing tridecyl acetate as an internal standard (1 ng/$\mu$l). The extract was then concentrated to <2 $\mu$l under a nitrogen stream and injected into a gas chromatograph.

*Analysis and identification of sex pheromone.* The chemical identification of the sex pheromone components were conducted using gas chromatograph (GC) and GC combined with mass spectrometry (GC-MS) with fused silica capillary columns (DB-Wax and DB-1, 30 m × 0.25 mm ID, J&W Scientific, Folsom, CA). As a bioassay, GC with electroantennographic detection (GC-EAD, Struble & Arn, 1984) and wind-tunnel experiments were conducted in each species. For the final confirmation, field trapping experiments were conducted in most of the species. For experimental details regarding the respective species, refer to the references listed in Table 3.

## Results and discussion

*Phylogeny of* Ostrinia *spp. inhabiting Japan.* The phylogenetic relationship of Japanese *Ostrinia* spp. and *O. nubilalis* proposed by Mutuura & Munroe (1970) is shown in Figure 1. Based on the intricacy of the male genitalia, the genus *Ostrinia* was largely classified into three species groups, i.e., the *penitalis* group (Group I), the simple or bifid uncus species group (Group II) and the trilobed uncus species group (Group III). The trilobed uncus species group is further divided into 'large-tibia,' 'medium-tibia' and 'small-tibia' subgroups, the last subgroup considered to be most primitive. *Ostrinia furnacalis*, *O. nubilalis* and *O. orientalis* belong to the 'small-tibia' subgroup, and *O. scapulalis* is included in the 'large-tibia' subgroup together with *O. zealis* and *O. zaguliaevi*. Mutuura & Munroe (1970) recognized the close relationships between *O. nubilalis* and *O. orientalis*, and between *O. zealis* and *O. zaguliaevi*. Particularly, the former two species cannot be distinguished definitely by morphological data only. Mutuura & Munroe (1970) treated these two as different species because of their discontinuous geographic distribution and clear difference in the host plant ranges.

For comparison, the phylogenetic relationship of Japanese *Ostrinia* spp. deduced from a mitochondrial DNA (mtDNA) cytochrome oxidase subunit II gene (COII, 682 bp) sequence analysis is also shown in Figure 1. The grouping of species into Groups II and III is supported by the mtDNA analysis, although the distance between *O. latipennis* and *O. palustralis* is as large as that between Groups II and III. The most significant difference between the 2 phylogenetic trees is that the subgroups in Group III are not supported by the mtDNA analysis; i.e., species with small tibia and those with large tibia do not form separate clades. A very close relationship among *O. orientalis* (small tibia), *O. nubilalis* (small tibia) and *O. scapulalis* (large tibia) was indicated by the mtDNA analysis. In particular, the distance between *O. nubilalis* and *O. orientalis* is as small as the intraspecific variation (difference is 1 bp, 0.15%). In contrast, *O. furnacalis* is rather distantly situated from the other Group III

*Table 1.* Preferred host plants of *Ostrinia* spp. inhabiting Japan[a]

| Species | | Preferred host plants | Plant family[b] |
|---|---|---|---|
| *O. latipennis* | oligophagous | *Polygonum* spp. (knotweeds) | Polygonaceae |
| *O. palustralis* | oligophagous | *Rumex* spp. (docks) | Polygonaceae |
| *O. furnacalis* (Asian corn borer) | polyphagous | *Zea mays*, *Sorghum bicolor*, *Zingiber officinalis*, *Rumex* spp. (maize) (sorghum) (ginger) (docks) | Gramineae, Gingiberaceae Solanaceae, Polygonaceae |
| *O. orientalis* | oligophagous | *Xanthium* spp., *Senecio* sp., *Rumex* spp. (cockleburs) (docks) | Compositae, Polygonaceae |
| *O. scapulalis* (adzuki bean borer) | polyphagous | *Vigna* spp., *Humulus lupulus*, *Xanthium* spp., *Rumex* spp. (legumes) (hops) (cockleburs) (docks) | Leguminosae, Moraceae Compositae, Polygonaceae |
| *O. zealis* | oligophagous | *Arctium lappa*, *Cirsium* spp., *Dahlia* spp. (burdock) (thistles) (dahlia) | Compositae |
| *O. zaguliaevi* | monophagous | *Petacites japonicus* (butterbur) | Compositae |

[a]Produced from Hattori & Mutuura (1987) and findings by S. Ohno (Hokkaido Univ.), Y. Yoshiyasu (Kyoto Pref. Univ.) and ourselves.
[b]Plant family in which at least one species is recorded as the host plant for respective *Ostrinia* species.

*Table 2.* The calling and mating times of female *Ostrinia* spp., under a photocycle of L15:D9

| Species | D1 | D2 | D3 | D4 | D5 | D6 | D7 | D8 | D9 | L1 | L2 | L3 | L4 | L5 | L6– |
|---|---|---|---|---|---|---|---|---|---|---|---|---|---|---|---|
| *latipennis* | | | | ○ | ○ | ◯ | ◯ | ◯ | ◯ | | | | | | |
| *palustralis* | | | | ○ | ○ | ◯ | ◯ | ◯ | ◯ | | | | | | |
| *furnacalis* | | | ○ | ○ | ○ | ◯ | ◯ | ◯ | ◯ | | | | | | |
| *orientalis* | | | | | | | | | | ○ | ◯ | ◯ | ◯ | ◯ | ○ |
| *scapulalis* (Matsudo A) | | | | ○ | ○ | ○ | ◯ | ◯ | ◯ | | | | | | |
| *scapulalis* (Matsudo B) | | ◯ | ◯ | ○ | ○ | ○ | ○ | ○ | ○ | | | | | | |
| *scapulalis* (Amami) | | | | | | | | | | ◯ | ◯ | ◯ | ◯ | ○ | |
| *zealis* | | | | | | ○ | ◯ | ◯ | ◯ | ○ | ○ | ○ | | | |
| *zaguliaevi* | | | | ○ | ○ | ◯ | ◯ | ◯ | ◯ | | | | | | |

○calling. ◯calling and mating.

*Table 3.* Female sex pheromone components and their ratios in *Ostrinia* spp. in Japan

| Species | Z9-14:OAc | *E*11-14:OAc | Z11-14:OAc | *E*12-14:OAc | Z12-14:OAc | *E*11-14:OH | | References |
|---|---|---|---|---|---|---|---|---|
| *latipennis* | – | – | – | – | – | 100 | b* | Takanashi, unpublished |
| *palustalis* | – | 99 | 1 | – | – | – | b | Huang et al. (1998c) |
| *furnacalis* (Matsudo)** | – | – | – | 38 | 62 | – | b | Huang et al. (1998b) |
| *furnacalis* (Nishigoshi)*** | – | – | – | 46 | 54 | – | – | Huang et al. (1998b) |
| *orientalis* | – | + | + | – | – | – | – | Huang, unpublished |
| *scapulalis* (*E* morph) | – | 99 | 1 | – | – | – | a | Takanashi, unpublished |
| *scapulalis* (*Z* morhp) | – | 3 | 97 | – | – | – | b | Huang et al. (1997) |
| *scapulalis* (hybrid) | – | 63 | 37 | – | – | – | a | Takanashi, unpublished |
| *zealis* | 70 | 22 | 8 | – | – | – | b | Ishikawa et al. (1999) |
| *zaguliaevi* | 45 | 5 | 50 | – | – | – | a | Huang et al. (1998a) |

*a: identification based on wind-tunnel bioassay. b: based on both wind-tunnel and field trap experiments.
**: Matsudo (35.8° N, 139.9° E).
*** Nishigoshi (32.9° N, 130.7° E). A similar component ratio is reported for the populations from the Philippines (Klun et al., 1980), from the People's Republic of China (Cheng et al., 1981) and from Taiwan (Kou et al., 1992).

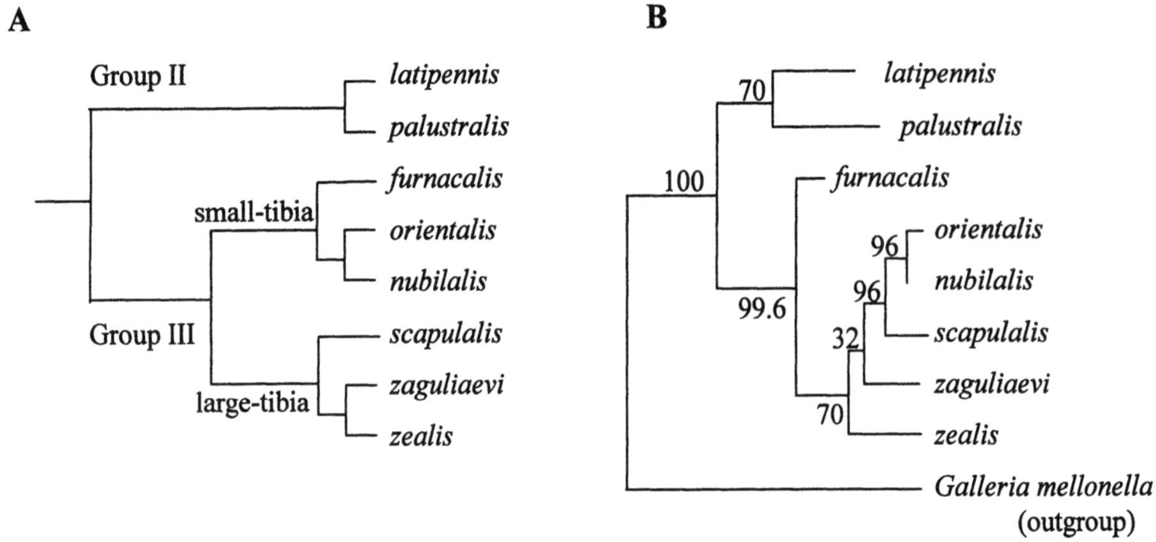


*Figure 1.* Proposed phylogenetic relationships of *Ostrinia* spp. inhabiting Japan with reference to the European corn borer *O. nubilalis*. (A) Based on morphological data (Mutuura & Munroe, 1970). (B) Based on mitochondrial COII gene (682 bp) sequence analyses (Kim, 1997). Group II: simple or bifid uncus species group. Group III: trilobed uncus species group. The species that belong to Group I (*penitalis* group) and 'medium-tibia' subgroup in Group III are not present in Japan. Branch lengths are arbitrary in tree (A). Tree (B) was produced by the neighbor-joining method using the computer software program PHYLIP (Felsenstein, 1993). Branch lengths in B indicate phylogenetic distances. Values at the nodes indicate bootstrap confidence levels.

members. Despite the very close relationship between *O. nubilalis* and *O. orientalis*, distinct differences in calling and mating time were found in our study in addition to the differences in host plant range and geographic distribution, as described below.

*Host plants.* The preferred host plants of the *Ostrinia* spp. inhabiting Japan are shown in Table 1. Although the two most important pests, *O. nubilalis* and *O. furnacalis,* feed on maize, Gramineae feeding is not common in *Ostrinia*. Host plants of the 'primitive' species *O. palustralis* and *O. latipennis* are limited to Polygonaceae. In contrast, the host plants of the species with the most complex genitalia articulations, *O. zealis* and *O. zaguliaevi*, are limited to Compositae. Between these two extremes are two polyphagous species (*O. furnacalis* and *O. scapulalis*). Although *O. furnacalis* and *O. scapulalis* are polyphagous, substantial differences in their host plant ranges exist (Table 1). The host plant range of *O. scapulalis* is very wide, but host plants belonging to Gramineae are very few. The infestation of maize by *O. scapulalis* occurs occasionally, but it seems to be accidental. In contrast, *O. furnacalis* seems to be basically a Gramineae feeder, although host plants from several plant families are reported. *Ostrinia nubilalis* has a host preference similar to that of *O. furnacalis*, i.e., maize (Gramineae) feeding. However, the analyses of mtDNA indicated that the two species are not very close phylogenetically (Figure 1B). This suggests that Gramineae feeding was acquired by the two species independently. *Ostrinia orientalis*, the species most closely related to the European corn borer, *O. nubilalis*, interestingly does not feed on maize. The host plants of *O. orientalis* are virtually limited to *Xanthium* spp. (Compositae) and *Rumex* spp. (Polygonaceae).

*Geographic distribution.* Information on the geographic distribution of *Ostrinia* spp. is scarce and not organized. The approximate distributions of *O. nubilalis*, *O. furnacalis*, *O. orientalis* and *O. scapulalis* in the world are shown in Figure 2. The distribution of *O. nubilalis* was originally restricted to Europe, but as is well known, this species was imported to the United States and is now wide spread in North America (Brindley & Dicke, 1963). The distribution of *O. orientalis* appears to be limited to Japan and the easternmost part of the Asian continent. The distribution of *O. scapulalis* in Asia is similar to that of *O. orientalis*, although this species is also found in east Europe. The distributions of these two species in the central and southern regions of the Asian continent should be investigated, however, before making conclusions about their distribution areas. In contrast

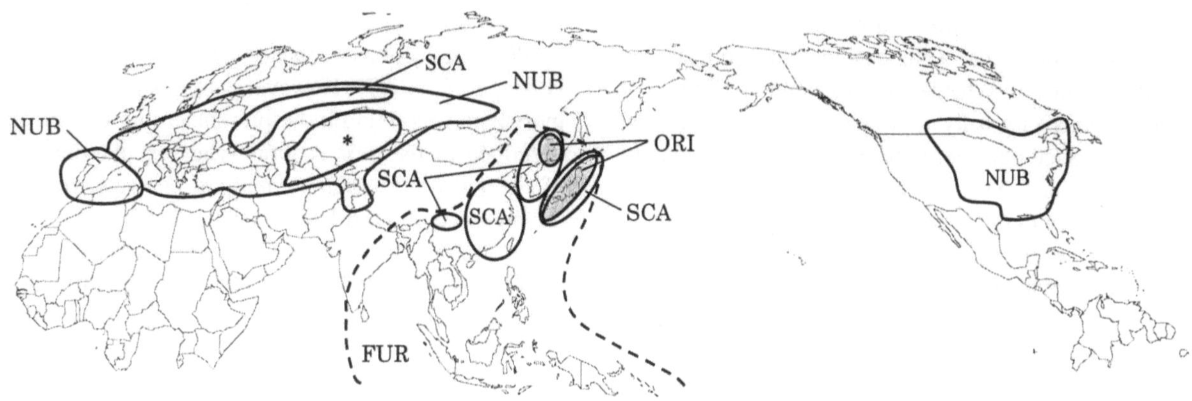


*Figure 2.* Approximate distributions of *Ostrinia furnacalis* (FUR), *O. nubilalis* (NUB), *O. orientalis* (ORI) and *O. scapulalis* (SCA). Prepared from Kozhanchikov (1938), Mutuura & Munroe (1970), Kozakevich (1978), Frolov (1994) and the 'European corn borer home page' (http://www.ent.iastate.edu/pest/cornborer/). ∗This area seems to be free from borers due to a dry climate.

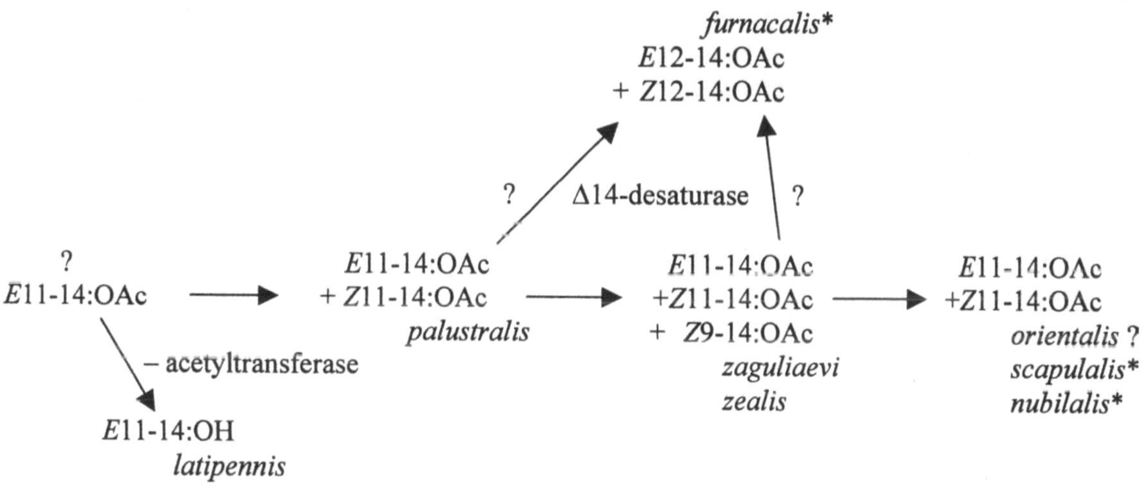


*Figure 3.* Possible transition of female sex pheromone composition in *Ostrinia* spp. ∗:polymorphic.

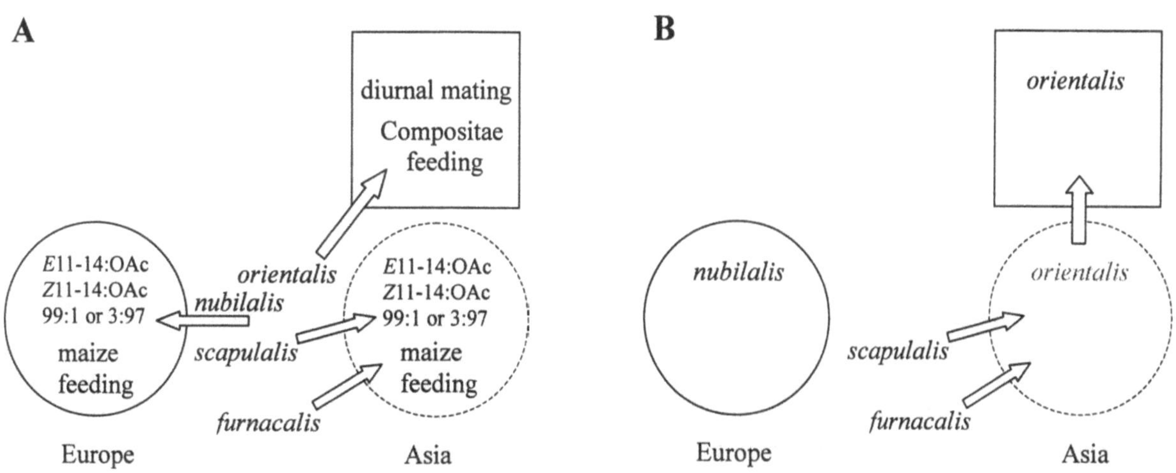


*Figure 4.* Schematic representation of the relationships of *O. nubilalis*, *O. orientalis*, *O. furnacalis* and *O. scapulalis*. In (A), *O. orientalis*, *O. furnacalis* and *O. scapulalis* have evolved completely independently from each other. The Asian niche equivalent to that occupied by *O. nubilalis* happened to be utilized by *O. scapulalis* and *O. furnacalis*. In (B), *O. orientalis* 'escaped' from the original niche equivalent to that occupied by *O. nubilalis*. *Ostrinia furnacalis* and *O. scapulalis* have occupied the vacated niche.

to these two species, the distribution of *O. furnacalis* in Asia is very wide, being extended to Southeastern Asian countries. Whether the distribution of *O. furnacalis* was originally wide or the distribution has been extended along with the extension of maize cultivation is not clear. In the former case, since the cultivation of maize in Japan, for instance, started only 4 centuries ago, *O. furnacalis* is considered to have fed on other host plants for the most part of its history.

*Variation in calling and mating time.* Under a L15:D9 photocycle, the calling of female *Ostrinia* spp. typically started from D4 and continued during the scotophase (Table 2). Copulation occurred frequently late in the scotophase (D6–D9). Among the 7 species, *O. orientalis* is characterized by diurnal mating behavior. Although the two species *O. marginalis* and *O. peregrinalis* distributed in the boreal regions of North America and Siberia, respectively, are reported to be diurnal, these species are characterized by reduced eyes and are considered to be phylogenetically very distant from the *furnacalis* species complex (Mutuura & Munroe, 1970). Thus, the diurnal mating behavior of *O. orientalis* was at first considered to be very exceptional in the *furnacalis* species complex. However, *O. scapulalis* collected from Amami, Japan (28.3° N, 129.4° E) was found to be diurnal; calling and mating occurred at the beginning of the photophase (Table 2). In addition to these diurnal species and strain, *O. scapulalis* collected at Matsudo, Japan (35.8° N, 130.7° E) was found to include a strain in which mating occurs at the beginning of the scotophase (D2–D3). Calling behavior of *O. zealis* started in the late scotophase, and interestingly, it continued after the light-on until L3 (Table 2). These findings indicate that the time for calling behavior is not rigidly fixed in *Ostrinia.* The genetic mechanism(s) that control the times for calling and mating behavior is of great interest.

*Sex pheromone composition.* Sex pheromone components and their ratios of *Ostrinia* spp. are shown in Table 3. *E*11-14:OAc is commonly found in every species except for *O. latipennis* and *O. furnacalis. Ostrinia latipennis* is characterized by the use of an alcohol, *E*11-14:OH in place of the acetate, and *O. furnacalis* is characterized by the use of positional isomers, *E*12-and *Z*12-14:OAc. The sex pheromones of *O. palustralis* and *O. scapulalis* are composed of two components, *E*11- and *Z*11-14:OAc. The sex pheromone of the adzuki bean borer, *O. scapulalis*, was first found to be similar to that of Z-strain *O. nubilalis* in both components and ratio ($E : Z = 3 : 97$, Huang et al., 1997). However, sex pheromone polymorphism was found to be present in this species as described in the next section. The sex pheromones of *O. zealis* and *O. zaguliaevi* are characterized by the presence of a third component, *Z*9-14:OAc in addition to *E*11- and *Z*11-14:OAc. The analysis of the sex pheromone of *O. orientalis* has met with difficulty because the amount of pheromone in the female sex pheromone gland is very small. However, the presence of *E*11- and *Z*11-14:OAc has been confirmed. The genus *Ostrinia* thus has evolved one-component, 2-component and 3-component sex pheromone systems.

*Sex pheromone polymorphism.* A sex pheromone polymorphism very similar to that of *O. nubilalis* was observed in the adzuki bean borer, *O. scapulalis* (Table 3). The *Z* morph *O. scapulalis* produces an *E:Z* blend of 3:97, while the *E* morph produces a 99:1 blend. The hybrid between these two morphs produces an intermediate blend of 63:37. In this species, the genetic mechanism controlling the pheromone blend also seems to be similar to that of *O. nubilalis* (Roelofs et al., 1987; Takanashi, unpublished data). The sex pheromone of *O. furnacalis* from the southernmost part of Japan was found to be a 46:54 mixture of *E*12- and *Z*12-14:OAc. Since *O. furnacalis* from the other regions of Japan has a 38:62 blend, it is evident that *O. furnacalis* also has sex pheromone polymorphism (Table 3). The presence of sex pheromone polymorphism in the 3 species (*O. nubilalis, O. scapulalis* and *O. furnacalis*) in the genus *Ostrinia* may indicate that polymorphic pheromones is not an unusual phenomenon in this genus. Rather, a basic framework for the occurrence of pheromone polymorphism was probably acquired relatively early in the evolution of the *furnacalis* species complex.

*Evolution of sex pheromone biosynthesis system.* The biosynthesis of Δ11-14:OAc is considered to follow the following steps; 16:Acyl is first chain-shortened to 14:Acyl by limited $\beta$-oxidation, desaturated to Δ11-14:Acyl by Δ11-desaturase, reduced to Δ11-14:OH by acyl reductase, and acetylated to Δ11-14:OAc by acetyltransferase (Bjostad et al., 1987; Zhu et al., 1996). It seems that in *O. latipennis*, the enzyme that works in the final step of Δ11-14:OAc synthesis (acetyltransferase) is dysfunctional, and thereby *E*11-14:OH becomes the final product. The acquisition of a new component, *Z*9-14:OAc in *O. zaguliaevi* and

*O. zealis* can be most plausibly explained by a variant Δ11-desaturase with substrate specificity for 16:Acyl (16:Acyl → Δ11-16:Acyl → Δ9-14:Acyl → Δ9-14:OH → Δ9-14:OAc). The production of anomalous Δ12-14:OAc in *O. furnacalis* has been ascribed to the advent of Δ14-desaturase (16:Acyl → Δ14-16:Acyl → Δ12-14:Acyl → Δ12-14:OH → Δ12-14:OAc; Zhao et al., 1990). A plausible explanation of the changes in sex pheromone biosynthesis in *Ostrinia* is shown in Figure 3.

*Why not* O. orientalis*?* *Ostrinia furnacalis*, *O. orientalis*, *O. nubilalis* and *O. scapulalis* are particularly interesting objects in the study of evolution and speciation in the genus *Ostrinia*. The ecological niche in Asia equivalent to that occupied by *O. nubilalis* in Europe is occupied not by the species most closely related to *O. nubilalis* (*O. orientalis*), but rather by *O. furnacalis*, a rather distantly related species (Figures 1 and 2). Regarding the sex pheromone, almost the same sex pheromone system as that of *O. nubilalis*, including the occurrence of polymorphism, is utilized by *O. scapulalis* (Table 3). Thus the Asian counterpart of the European corn borer *O. nubilalis* is the Asian corn borer *O. furnacalis* from the standpoint of feeding habit, and is the adzuki bean borer *O. scapulalis* from the standpoint of the sex pheromone. However, phylogenetically, *O. orientalis* is the latent counterpart of the European corn borer (Figure 1). Why is the ecological niche of *O. orientalis* so different from that of *O. nubilalis*? Two scenarios can be proposed. (A) *O. orientalis* happened to evolve a Compositae-feeding habit and diurnal mating behavior completely independently from the other species, or (B) *O. orientalis* vacated the original niche for some unknown reason (Figure 4). As to the second scenario, the reason could be competition between *O. orientalis* and *O. furnacalis* for food and/or competition between *O. orientalis* and *O. scapulalis* regarding the use of the sex pheromone. At present we do not have sufficient information to estimate the validity of these two scenarios.

## Acknowledgements

We are grateful to Dr H. Honda of Tsukuba University for his cooperation in the identification of the sex pheromones. We thank Dr Y. Yoshiyasu of Kyoto Prefectural University for the identification of the species, and Dr A. N. Frolov of All-Russian Institute of Plant Protection for providing information on the distributions of *Ostrinia* spp. in Europe. This research was sponsored by the Japanese Society for the Promotion of Science (JSPS), a Grant-in-Aid for Co-operative Research (A) (No. 07306003) and a Grant-in-Aid for Developmental Research (B) (No. 10556008) from the Ministry of Education, Science and Culture of Japan.

## References

Ando, T., O. Saito, K. Arai & N. Takahashi, 1980. (*Z*)- and (*E*)-12-Tetradecenyl acetates: Sex pheromone components of oriental corn borer (Lepidoptera: Pyralidae). Agricultural and Biological Chemistry 44: 2643–2649.

Bjostad, L. B., W. A., Wolf & W. L. Roelofs, 1987. Pheromone biosynthesis in Lepidopterans: desaturation and chain shortening. In: G. D. Prestwich & G. J. Blomquist (eds), Pheromone Biochemistry. Academic Press, New York, pp. 77–120.

Brindley, T. A. & F. F. Dicke, 1963. Significant developments in European corn borer research. Annual Review of Entomology 8: 155–176.

Cheng, Z. Q., J. C. Xiao, X. T. Huang, D. L. Chen, Q. J. Li, Y. S. He, S. R. Huang, Q. C. Luo, C. M. Yang & T. H. Yang, 1981. Sex pheromone components isolated from China corn borer, *Ostrinia furnacalis*, Guenée (Lepidoptera: Pyralidae), (*E*)- and (*Z*) 12-tetradecenyl acetates. Journal of Chemical Ecology 7: 841–851.

Felsenstein, J., 1993. PHYLIP (Phylogeny Inference Package), Version 3.5c. Department of Genetics, University of Washington, Seattle, Washington.

Frolov, A. N., 1994. The geographic variation of population structure in *Ostrinia* spp. living on dicotyledonous host plants and the factors determining the variation. Zoologicheskii Zhurnal 73: 47–59 (in Russian).

Hattori, I. & A. Mutuura, 1987. Identification of Japanese species belonging to the genus *Ostrinia* with the host relationships. Plant Protection 41(2): 24–31 (in Japanese).

Huang, Y., H. Honda, Y. Yoshiyasu, S. Hoshizaki, S. Tatsuki & Y. Ishikawa, 1998a. Sex pheromone of the butterbur borer, *Ostrinia zaguliaevi*. Entomologia Experimentalis et Applicata 89: 281–287.

Huang, Y., T. Takanashi, S. Hoshizaki, S. Tatsuki, H. Honda, Y. Yoshiyasu & Y. Ishikawa, 1998b. Geographic variation in the sex pheromone of Asian corn borer, *Ostrinia furnacalis*, in Japan. Journal of Chemical Ecology 24: 2079–2088.

Huang, Y., S. Tatsuki, C. Kim, S. Hoshizaki & Y. Ishikawa, 1998c. Identification of the sex pheromone of *Ostrinia palustralis*. Entomologia Experimentalis et Applicata 86: 313–318.

Huang, Y., S. Tatsuki, C. Kim, S. Hoshizaki, Y. Yoshiyasu, H. Honda & Y. Ishikawa, 1997. Identification of sex pheromone of adzuki bean borer, *Ostrinia scapulalis*. Journal of Chemical Ecology 23: 2791–2802.

Ishikawa, Y., T. Takanashi & Y. Huang, 1999. Comparative studies on the sex pheromones of *Ostrinia* spp. in Japan: the burdock borer, *Ostrinia zealis*. Chemoecology 9: 25–32.

Kim, C., 1997. Molecular phylogenetic studies on evolution and speciation in the Asian corn borer and its allied species. Ph.D. Dissertation (in Japanese). Tokyo University.

Klun, J. A., B. A. Bierl-Leonhardt, M. Schwarz, J. A. Litsinger, A. T. Barrion, H. C. Chiang & Z. Jiang, 1980. Sex pheromone of the Asian corn borer moth. Life Sciences 27: 1603–1606.

Klun, J. A., D. L. Chapman, K. C. Mattes, P. W. Wojtkowski, M. Beroza & P. E. Sonnet, 1973. Insect sex pheromones: minor amount of opposite geometrical isomer critical to attraction. Science 181: 661–663.

Klun, J. A. & J. F. Robinson, 1972. Olfactory discrimination in the European corn borer and several pheromonally analogous moths. Annals of the Entomological Society of America 65: 1337–1340.

Kochansky, J., R. T. Cardé, J. Leibherr & W. L. Roelofs, 1975. Sex pheromone of the European corn borer, *Ostrinia nubilalis* (Lepidoptera: Pyralidae), in New York. Journal of Chemical Ecology 1: 225–231.

Kou, R., H. Y. Ho, H. T. Yang, Y. S. Chow & H. J. Wu, 1992. Investigation of sex pheromone components of female Asian corn borer, *Ostrinia furnacalis* (Hübner) (Lepidoptera: Pyralidae) in Taiwan. Journal of Chemical Ecology 18: 833–840.

Kozakevich, Z. M., 1978. Brush-legged borer – *Ostrinia scapulalis* Walker (Lepidoptera, Pyraustidae) in Ukraine. Vestnik Zoologii 2: 27–30 (in Russian).

Kozhanchikov, I. V., 1938. Geographical distribution and physiological characters of *Pyrausta nubilalis* Hbn. Zoologicheskii Zhurnal 17: 246–259 (in Russian).

Mutuura, A. & E. Munroe, 1970. Taxonomy and distribution of the European corn borer and allied species: Genus *Ostrinia* (Lepidoptera: Pyralidae). Memoirs of the Entomological Society of Canada 71: 1–112.

Roelofs, W. L., T. Glover, X. H. Tang, I. Sreng, P. Robbins, C. Eckenrode, C. Löfstedt, B. S. Hansson & B. O. Bengtsson, 1987. Sex pheromone production and perception in European corn borer moths is determined by both autosomal and sex-linked genes. Proceedings of the National Academy of Science, USA 84: 7585–7589.

Struble, D. L. & H. Arn, 1984. Combined gas chromatography and electroantennogram recording of insect olfactory responses. In: H. E. Hummel & T. A. Miller (eds), Techniques in Pheromone Research. New York, Springer-Verlag, pp. 161–178.

Zhao, C. H., C. Löfstedt & X. Y. Wang, 1990. Sex pheromone biosynthesis in the Asian corn borer *Ostrinia furnacalis* (II): Biosynthesis of (*E*)- and (*Z*)-12-tetradecenyl acetate involves Δ-14 desaturation. Archives of Insect Biochemistry and Physiology 15: 57–65.

Zhu, J. W., C. H. Zhao, F. Lu, M. Bengtsson & C. Löfstedt, 1996. Reductase specificity and the ratio regulation of *E/Z* isomers in pheromone biosynthesis of the European corn borer, *Ostrinia nubilalis* (Lepidoptera: Pyralidae). Insect Biochemistry and Molecular Biology 26: 171–176.

*Entomologia Experimentalis et Applicata* **91:** 245–250, 1999.



# Genetic basis for established and novel host plant use in a herbivorous ladybird beetle, *Epilachna vigintioctomaculata*

Hideki Ueno, Naoyuki Fujiyama*, Kiyoshi Irie, Yuji Sato & Haruo Katakura*
*Laboratory of Biology, Faculty of Education, Niigata University, Niigata 950-2181, Japan; *Laboratory of Systematics and Evolution, Graduate School of Science, Hokkaido University, Sapporo 060-0810, Japan*

Accepted: December 4, 1998

*Key words:* genetic correlations, *Epilachna vigintioctomaculata*, *Solanum tuberosum*, *Schizopepon bryoniaefolius*

## Abstract

Genetic trade-offs in host plant use are thought to promote the evolution of host specificity. Experiments on a range of herbivorous insects, however, have found negative genetic correlation in host plant use in only a limited number of species. To account for the general lack of negative genetic correlation, recent hypotheses advocate that different stages in evolution of host use must be distinguished: initial performance on a novel host in comparison with the established host, and performance on both hosts after the insect population has interacted with both hosts for a long time. The hypotheses suggest that genetic correlation may not necessarily be negative at the initial stage. The present study examines growth performance on both the established and a novel host in a herbivorous ladybird beetle, *Epilachna vigintioctomaculata* Motschulsky (Coccinellidae, Epilachninae). The results show that traits of growth performance across hosts were positively or neutrally correlated, but there was no evidence of a negative genetic correlation. In addition, significant genetic variance of growth performance on each host was detected, suggesting that *E. vigintioctomaculata* can potentially respond to selection for increased performance on both plant species. These results and similar results from experiments on other herbivores suggest that host expansion may not be constrained genetically, at least at the initial stage of host range evolution.

## Introduction

Genetic correlations between insect performance across host plants, have important implications for the determination of the host range of herbivorous insects (Rausher, 1984a; Diehl & Bush, 1989). If the performance traits on different host plants are negatively correlated, selection for an increase in a performance trait on one host plant will result in a decrease in the analogous trait on the other plant species. Genetic trade-offs in host plant use are, therefore, thought to promote the evolution of host specificity (Via, 1990).

Experiments estimating the genetic correlation in different host use have been carried out on a range of herbivorous insects (Jaenike, 1990). However, negative correlations have been found in only a limited number of species (Gould, 1979; Fry, 1990; Via, 1991; Mackenzie, 1996). Instead of a negative genetic correlation, positive or no genetic correlations were commonly reported in the previous studies (Rausher, 1984b; Via, 1984; Hare & Kennedy, 1986; Futuyma & Philippi, 1987; James et al. 1988; Jaenike, 1989; Karowe, 1990; Fox, 1993; Lu et al. 1997). To account for this general lack of negative genetic correlations, it has been suggested that antagonistic pleiotropic effects reflecting underlying constraints may be masked by several factors that increase genetic variation in general performance across hosts (Service & Rose, 1985; Houle, 1991; Price & Schluter, 1991; Fry, 1993).

Other recent hypotheses highlight the sequential outcome of genetic correlation in the course of host range evolution (Thompson, 1994; Joshi & Thompson, 1995). These hypotheses suggest that the history of selection experienced by the insect population on the hosts can influence the current genetic structure and therefore that different stages in evolution of host use should be distinguished: initial performance on a novel host in comparison with the established host, and performance on both hosts after the insect population has become adapted to both hosts. Genetic correlation may not necessarily be negative at the initial stage, as alleles affecting the growth performance on the novel host are expected to be randomly associated with alleles affecting the performance on the established host (Joshi & Thompson, 1995). Therefore, in studying genetic correlation between insect performances on different hosts, it is important to consider host plant affiliations in the herbivore population. The experiments reported in this paper represent an initial attempt to examine growth performance on both the established and a novel host in a herbivorous ladybird beetle, *Epilachna vigintioctomaculata*.

*E. vigintioctomaculata* mainly feeds on *Solanum tuberosum* L. (Solanaceae). Although some additional plants in the Solanaceae family are subsidiarily used by adult beetles, larvae develop almost entirely on *S. tuberosum* (Katakura, 1981). Frequently, beetle populations in Hokkaido (the main northern island of Japan) are found on *Schizopepon bryoniaefolius* Maxim. (Cucurbitaceae) but not in Honshu (the main central island) (Katakura, 1981), exhibiting interpopulational variation in host plant use. The present study compared growth performance on *S. tuberosum* and *S. bryoniaefolius* in a Honshu population. The particular interest in this study was to establish whether genetic factors are responsible for the absence of the beetles on *S. bryoniaefolius* in Honshu populations.

## Materials and methods

*E. vigintioctomaculata* is a univoltine species. Overwintered adults were collected on *S. tuberosum* in a suburb of Niigata City (Honshu) in May 1997. Rearing conditions were 22 °C and L16:D8 throughout this experiment. In excess of 22 males and females were confined as pairs in plastic cases for mating. Virtually all females would have copulated and stored sperm before overwintering (Katakura, 1982) and additional copulations could have also been possible in the late spring before collection. Hence we started the experiment after we confirmed at least three copulations for each female. This was to ensure last male paternity; last male sperm precedence (P2) in closely related species (*E. pustulosa*) was reported as 0.651–0.827 (Nakano, 1985). Assuming the same sperm precedence pattern for the last three successive copulations, the expected fertilization success of previous males are $0.349^3 = 0.0425$ at most. Thus, although the experiment was arranged as a split brood design, full-sib family would have been virtually expected. In the following statistical calculations larvae from each female were regarded as full-sib progenies, assuming complete last male paternity.

For each of the females, 40 eggs were collected from several egg clutches. Of these 40 eggs, 20 were reared on *S. tuberosum* and 20 on *S. bryoniaefolius*. Larvae were reared individually. Fresh host plant leaves were supplied every other day. Larval period was checked daily. On the day of pupation, fresh weight was measured to within 0.1 mg with a microbalance.

A mixed model ANOVA was performed to detect the contribution of factors to the total variation in growth performance, setting family and family × host interaction as random, and sex and host as fixed effects. With this statistical procedure, the variance component due to the family main effect detects the covariance of the performance across the host plants. When divided by the interaction, the family main effects with F-values above 97.5% indicate significant positive genetic covariation, while those with F-values below 2.5% indicate significant negative genetic covariation, at the 5% level (Fry, 1992). Two further ANOVAs were also performed to evaluate the among-family variance of performance on each host plant separately. The analyses were performed to detect the potential response to the plant environment. All ANOVAs were performed using PROC GLM in SAS.

The proportion of phenotypic variation in growth performance attributable to variables was calculated using the restricted maximum likelihood option (REML in PROC VARCOMP, SAS), which estimates the proportion of variables due to each random factor. Because asymptotic variance of each estimate is appropriate for only large samples, significance of the components was determined by the probability obtained from the analogous model ANOVA. Broad sense heritabilities were calculated using the variance components as $h^2 = 2\sigma^2$ (Family)/[$\sigma^2$(Family)+$\sigma^2$ (Error)]. Standard errors for the heritabilities were estimated after Becker (1984). Between full-sib family variances contain non-additive sources of genetic variation, including dominance and maternal effects. These values should, therefore, be regarded as approximations of genetic properties but they provide an upper limit of genetic variation.

Genetic correlation of the same performance trait across the hosts was estimated using family means

*Table 1.* Percentage survival and means (±s.e.) for larval period and pupal weight in a herbivorous ladybird beetle, *Epilachna vigintioctomaculata* reared on *Solanum tuberosum* or *Schizopepon bryoniaefolius*. Sample sizes are given in parentheses

| | Host | |
|---|---|---|
| | *Solanum tuberosum* | *Schizopepon bryoniaefolius* |
| Larval period (days) | | |
| Male | 14.9 ± 0.1 (168) | 17.5 ± 0.2 (141) |
| Female | 14.9 ± 0.1 (148) | 17.8 ± 0.1 (137) |
| Pupal weight (mg) | | |
| Male | 40.2 ± 0.4 (168) | 33.8 ± 0.3 (141) |
| Female | 43.9 ± 0.4 (148) | 36.9 ± 0.3 (137) |
| Survival to adult (%) | 71.8 | 63.2 |

*Table 2.* Results of log-likelihood-ratio analysis for survivorship in a herbivorous ladybird beetle, *Epilachna vigintioctomaculata* reared on *Solanum tuberosum* or *Schizopepon bryoniaefolius*

| Source | d.f. | G | P |
|---|---|---|---|
| Host | 1 | 7.50 | 0.006 |
| Family | 21 | 62.4 | <0.001 |
| Family × Host | 21 | 22.4 | 0.375 |

as $r_m = \text{Cov}(X, Y)/[\text{Var}(X)\text{Var}(Y)]^{1/2}$, where $\text{Cov}(X, Y)$ is the covariance of family mean of the trait reared on one host plant and family mean of the same trait reared on the other host, and $\text{Var}(X)$ is the variance of the family mean on one host plant (Via, 1984). Genetic correlation was also estimated using variance components obtained from REML in PROC VARCOMP adopted to the mixed model. Here genetic correlation was estimated as $r_G = \sigma^2(\text{Family})/[\sigma^2(\text{Family} \times \text{Host}) + \sigma^2(\text{Family})]$ (Fry, 1992). Jackknife procedure was used to calculate the correlation with their standard errors (Sokal & Rohlf, 1995). Here the correlation was jackknifed with each family omitted once so that the total number of iterations was equal to the number of families. The correlation coefficients were z-transformed for jackknifing procedures and transformed back to the coefficient scale after obtaining 95% confidence intervals for z-values.

## Results

Totally, 594 beetles were obtained from 22 families. Log-likelihood-ratio analysis detected a significant effect of host plant on survivorship. Individuals reared on *S. bryoniaefolius* showed a higher mortality rate than those reared on *S. tuberosum* (Tables 1, 2). Different survivorship rates among families were detected, but there was no evidence for family × host interaction (Table 2). Larval period and pupal weight were also significantly influenced by host plants (Table 1). Those reared on *S. tuberosum* showed a shorter larval period than those reared on *S. bryoniaefolius*. Larvae which were reared on *S. tuberosum* achieved a heavier pupal weight than their siblings which were raised on *S. bryoniaefolius*. This between plant difference in overall tendency was observed in both males and females. Females were significantly heavier than male siblings, while larval periods were not significantly different between the sexes (Tables 1, 3).

The pattern of contribution of factors to the variation in performance was different for the larval period and for pupal weight. The mixed model ANOVAs detected significant family × host interaction for the larval period but not for pupal weight. By contrast, the family main effect in the mixed model ANOVAs, which indicates genetic covariance across hosts, was not significant for the larval period but was significantly positive for pupal weight with a two-tailed test (Table 3).

The genetic correlation calculated with variance components reflected the above results of the mixed model ANOVAs (Table 4). The genetic correlations for both components of growth performances were

*Table 3.* ANOVAs to compare variations in larval period and pupal weight of *Epilachna vigintioctomaculata* reared on *Solanum tuberosum* or *Schizopepon bryoniaefolius*. $r_2$ shows the fit of the overall model

| | Larval period | | | | Pupal weight | | | |
|---|---|---|---|---|---|---|---|---|
| Source | d.f. | MS | F | P | d.f. | MS | F | P |
| Sex | 1 | 0.007 | 1.202 | 0.2735 | 1 | 0.962 | 46.249 | <0.0001 |
| Host | 1 | 3.599 | 287.227 | <0.0001 | 1 | 3.901 | 133.974 | <0.0001 |
| Family | 21 | 0.021 | 1.629 | 0.1359 | 21 | 0.076 | 2.570 | 0.0179 |
| Family × Host | 21 | 0.013 | 2.160 | 0.0021 | 21 | 0.029 | 1.416 | 0.1036 |
| Error | 549 | 0.006 | | | 549 | 0.021 | | |
| $r^2$ | 0.587 | | | | 0.392 | | | |

*Table 4.* Genetic correlation estimates of *Epilachna vigintioctomaculata* between the same characters across different plant species with their 95% confident limits in parentheses. Genetic correlations were estimated using variance and covariance of the family means ($r_m$) and using variance components of the mixed model ($r_G$)

| | Larval period | Pupal weight |
|---|---|---|
| $r_m$ | 0.320 (−0.111, 0.649) | 0.560 (0.015, 0.849) |
| $r_G$ | 0.378 (−0.045, 0.686) | 0.641 (0.181, 0.871) |

positive. In particular, the genetic correlation was highly positive for pupal weight. Estimated values of family mean genetic correlations ($r_m$) were in agreement with those based on variance components ($r_G$).

From the ANOVA performed for growth performance on *S. tuberosum* and *S. bryoniaefolius* separately, significant among-family variation was detected in each case (Table 5). ANOVA detected significant among-family variance for both the larval period and pupal weight. Variance components attributable to family were larger for beetles reared on *S. bryoniaefolius* than for those reared on *S. tuberosum* (Table 6). Broad sense heritabilities were moderate for performance traits on *S. bryoniaefolius*, while rather lower heritabilities were obtained for both the larval period and pupal weight on *S. tuberosum* (Table 6).

## Discussion

The purpose of this study was to see how traits of growth performance on the established host are genetically correlated with analogous traits on a novel host. The results showed that performance traits were significantly positively or neutrally correlated, suggesting that selection for increased performance on one host will not constrain, but will rather facilitate evolution of increased performance on the other host. Despite the difference in host plants, the present study found no evidence for genetic trade-offs in host plant use.

In addition, significant among-family variance on each host was detected when statistical analyses were made for each host plant, suggesting that *E. vigintioctomaculata* can potentially respond to selection for increased performance on both plant species. Rather small among-family variance components were found on *S. tuberosum* when compared to those on *S. bryoniaefolius*. This result suggests that, to a substantial extent, selection on *S. tuberosum* has been imposed and that the response to further selection on the established host will evolve less readily due to reduced genetic variation. In contrast, larger among-family variance components on *S. bryoniaefolius*, suggest that the present population would easily respond to selection for increased performance on the novel host.

The above results indicate that the ability to exploit the novel host is not likely to be genetically constrained either in terms of genetic correlation or of genetic variation. When experiments were made on established and novel host plants, the results from previous experiments on other herbivorous insects indicate positive or no genetic correlation (Rausher, 1984b; Via, 1984; Hare & Kennedy, 1986; James et al., 1988; Karowe, 1990; Fox, 1993), and significant genetic variation in growth performance on each plant (Rausher, 1984b; Hare & Kennedy, 1986). These and the present results suggest that host expansion may not be constrained genetically at least at the initial stage of host range evolution.

It is important to note, however, that the experimental conditions may have been responsible for the absence of genetic constraints. In the studies com-

*Table 5.* ANOVAs to compare variations in larval period and pupal weight of *Epilachna vigintioctomaculata* reared on *Solanum tuberosum* or *Schizopepon bryoniaefolius*. $r_2$ shows the fit of the overall model

| *Solanum tuberosum* | Larval period | | | | Pupal weight | | | |
|---|---|---|---|---|---|---|---|---|
| Source | d.f. | MS (×100) | F | P | d.f. | MS | F | P |
| Sex | 1 | 0.002 | 0.005 | 0.9455 | 1 | 0.436 | 14.069 | 0.0002 |
| Family | 21 | 0.685 | 1.621 | 0.0438 | 21 | 0.057 | 1.841 | 0.0149 |
| Error | 293 | 0.422 | | | 293 | 0.031 | | |
| $r^2$ | 0.104 | | | | 0.156 | | | |
| *Schizopepon bryoniaefolius* | Larval period | | | | Pupal weight | | | |
| Source | d.f. | MS | F | P | d.f. | MS | F | P |
| Sex | 1 | 0.017 | 2.133 | 0.1454 | 1 | 0.535 | 58.325 | <0.0001 |
| Family | 21 | 0.025 | 3.187 | <0.0001 | 21 | 0.051 | 5.507 | <0.0001 |
| Error | 255 | 0.008 | | | 255 | 0.009 | | |
| $r^2$ | 0.217 | | | | 0.409 | | | |

*Table 6.* Variance components (×100) for analyses in Table 5, and broad sense heritabilities calculated using the variance components. Variance components were estimated using REML with host and sex as fixed effects. Significance tests from the ANOVA in Table 5 were used to mark variance components that are greater than zero. Percentages of the variance components and standard errors of the heritabilities are given in parentheses

| | Larval period | | Pupal weight | |
|---|---|---|---|---|
| | *Solanum tuberosum* | *Schizopepon bryoniaefolius* | *Solanum tuberosum* | *Schizopepon bryoniaefolius* |
| Family | 0.018 (4.0)* | 0.142 (15.2)** | 0.180 (5.5)* | 0.353 (27.7)** |
| Error | 0.424 (96.0) | 0.788 (84.8) | 3.102 (94.5) | 0.918 (72.3) |
| Heritability | 0.081 (0.066) | 0.305 (0.120) | 0.110 (0.073) | 0.555 (0.155) |

*P < 0.05; **P < 0.0001.

paring growth performance on established and novel hosts, novel hosts on which insects have been usually tested are the hosts of the conspecific or congeneric population. It is possible that an absence of genetic constraint is what permits the use of these hosts in the conspecific or congeneric population. Therefore, that the hosts of related population can potentially be incorporated into host range, may not necessarily indicate the absence of genetic constraints that prevent the incorporation of other plants.

One aim of this study was to establish whether genetic factors are responsible for the absence of the beetles on *S. bryoniaefolius* in Honshu populations. In this study, there was no evidence for genetic constraints for the Niigata population in expanding its host range to include *S. bryoniaefolius*. The results suggest that behavioural and ecological factors may be the primary influence on host plant use. Investigations of these factors in natural populations will be necessary in future studies.

## Acknowledgements

We express our sincere thanks to A. Ueno for assistance throughout this experiment and to M. Bradley for comments on the manuscript. Two anonymous referees improved the manuscript very much. This study was supported in part by a Grant-in-Aid for Scientific Research (No. 09440258) from the Ministry of Education, Science and Culture, Japan.

## References

Becker, W. A., 1984. Manual of Quantitative Genetics. 4th ed. Academic Enterprises, Pullman.

Diehl, S. R. & G. L. Bush, 1989. The role of habitat preference in adaptation and speciation. In: D. Otte & J. A. Endler (eds), Speciation and its Consequences, Sinauer, Sunderland, pp. 345–365.

Fox, C. W., 1993. A quantitative genetic analysis of oviposition preference and larval performance on two hosts in the bruchid beetle, *Callosobruchus maculatus*. Evolution 47: 166–175.

Fry, J. D., 1990. Trade-offs in fitness on different hosts: evidence from a selection experiment with a phytophagous mite. American Naturalist 136: 569–565.

Fry, J. D., 1992. The mixed model analysis of variance applied to quantitative genetics: biological meaning of the parameters. Evolution 46: 540–550.

Fry, J. D., 1993. The 'general vigor' problem: can antagonistic pleiotropy be detected when genetic covariances are positive? Evolution 47: 329–333.

Futuyma, D. J., & T. E. Philippi, 1987. Genetic variation and covariation in responses to host plants by *Alsophila pometaria* (Lepidoptera: Geometridae). Evolution 41: 269–279.

Gould, F., 1979. Rapid host range evolution in a population of the phytophagous mite *Tetranychus urticae* Koch. Evolution 33: 791–802.

Hare, J. D. & G. G. Kennedy, 1986. Genetic variation in plant-insect associations: survival of *Leptinotarsa decemlineata* populations on *Solanum carolinense*. Evolution 40: 1031–1043.

Houle, D., 1991. Genetic covariance of fitness correlates: what genetic correlations are made of and why it matters? Evolution 45: 630–648.

Jaenike, J., 1989. Genetic population structure of *Drosophila tripunctata*: patterns of variation and covariation of traits affecting resource use. Evolution 43: 1467–1482.

Jaenike, J., 1990. Host specialization in phytophagous insects. Annual Review of Ecology and Systematics 21: 243–273.

James, A. C., J. Jackubczak, M. P. Ridley & J. Jaenike, 1988. On the causes of monophagy in *Drosophila quinaria*. Evolution 42: 626–630.

Joshi, A. & J. N. Thompson, 1995. Trade-offs and the evolution of host specialization. Evolutionary Ecology 9: 82–92.

Karowe, D. N., 1990. Predicting host range evolution: colonization of *Coronilla varia* by *Colias philodice* (Lepidoptera: Pieridae). Evolution 44: 1637–1647.

Katakura, H., 1981. Classification and evolution of the phytophagous ladybirds belonging to *Henosepilachna vigintioctomaculata* complex (Coleoptera, Coccinellidae). Journal of the Faculty of Science, Hokkaido University, Series VI, Zoology 22: 301–378.

Katakura, H., 1982. Long mating season and its bearing on the reproductive isolation in a pair of sympatric phytophagous ladybirds (Coleoptera, Coccinellidae). Kontyû 50: 599–603.

Lu, W., G. G. Kennedy & F. Gould, 1997. Genetic variation in larval survival and growth and response to selection by Colorado potato beetle (Coleoptera: Chrysomelidae) on tomato. Environmental Entomology 26: 67–75.

Mackenzie, A., 1996. A trade-off for host plant utilization in the black bean aphid, *Aphis fabae*. Evolution: 50 155–162.

Nakano, S., 1985. Sperm displacement in *Henosepilachna pustulosa* (Coleoptera, Coccinellidae). Kontyû 53: 516–519.

Price, T. & D. Schluter, 1991. On the low heritability of life history traits. Evolution 45: 853–861.

Rausher, M. D., 1984a. The evolution of habitat selection in subdivided populations. Evolution 38: 596–608.

Rausher, M. D., 1984b. Trade-offs in performance on different hosts: evidence from within- and between-site variation in the beetle, *Deloyala guttata*. Evolution 38: 582–595.

Service, P. M. & M. R. Rose, 1985. Genetic covariation among life history components: the effect of novel environments. Evolution 39: 943–945.

Sokal, R. B. & F. J. Rohlf, 1995. Biometry. 3rd ed. Freeman, New York.

Thompson, J. N., 1994. The Coevolutionary Process. University of Chicago Press, Chicago.

Via, S., 1984. The quantitative genetics of polyphagy in an insect herbivore. II. Genetic correlations in larval performance within and among host plants. Evolution 38: 896–905.

Via, S., 1990. Ecological genetics and host adaptation in herbivorous insects: the experimental study of evolution in natural and agricultural systems. Annual Review of Entomology 35: 421–446.

Via, S., 1991. The genetic structure of host plant adaptation in a spatial patchwork: demographic variability among reciprocally transplanted pea aphid clones. Evolution 45: 827–852.

Entomologia Experimentalis et Applicata **91:** 251–257, 1999.



# Host-plant choice and larval growth in the cinnabar moth: do pyrrolizidine alkaloids play a role?

Klaas Vrieling & Nico J. de Boer
*Institute of Evolutionary and Ecological Sciences, Leiden University, Kaiserstraat 63, 2311 GP Leiden, The Netherlands (Fax: +31 71 5274900; Phone: +31 71 5275136; E-mail: Vrieling@RULSFB.LeidenUniv.nl)*

Accepted: December 4, 1998

*Key words: Senecio jacobaea, Senecio erucifolius, Tyria jacobaeae*, pyrrolizidine alkaloids, oviposition preference, chemotypes

## Abstract

Witte et al. (1992) described two distinct chemotypes of *Senecio jacobaea* L. Asteraceae, a chemotype with jacobine as one of the major pyrrolizidine alkaloids (PAs) and a chemotype with erucifoline as one of the major PAs. We hypothesized that the presence of erucifoline might be the factor responsible for the lack of success of the cinnabar moth on *Senecio erucifolius* L. Asteraceae and the *S. jacobaea* erucifoline chemotype. We performed a survey of the distribution of the two chemotypes in the Netherlands and compared this with the distribution map of *Tyria jacobaeae* L. Lepidoptera, Arctiidae. The distribution of the two chemotypes in the Netherlands is poorly correlated with the distribution of the cinnabar moth. The jacobine chemotype occurs along the coast and the erucifoline chemotype predominantly inward.

An oviposition experiment showed that the cinnabar moth did not discriminate between the two chemotypes of *S. jacobaea* and *S. erucifolius*. Larval performance did not differ between the two chemotypes and species. Although the distribution of *S. jacobaea* jacobine chemotype is loosely associated with the abundance of the cinnabar moth the oviposition and growth experiments indicate that other factors than the presence of erucifoline play a role in this association.

The absence of recordings of *S. erucifolius* as a foodplant for the cinnabar moth might be explained by the phenology of the foodplant. Ovipositing females of the univoltine cinnabar moth prefer flowering plants for oviposition. *S. erucifolius* starts flowering about 1–2 month later than *S. jacobaea* just after the peak density of moths.

## Introduction

The diversity in secondary metabolites is very large (Harborne, 1993). Within thoroughly investigated plant species always more than one class of compounds is reported. In general, if a certain compound is present it is accompanied by many 'sister' compounds with a more or less similar basic chemical structure, but differing in hydroxylation, methylation, (un) saturated bonds, etc. (Hartmann, 1995). Despite these chemical similarities their biological activity may be very different. Berenbaum et al. (1986, 1989) found that the toxicity of the angular furanocoumarins differed highly from the chemically closely related linear furanocoumarins in their effects on the oligophagous parsnip webworm.

More than thirty years ago, Ehrlich & Raven (1964) proposed the hypothesis that the driving force for diversification in secondary metabolites is the selective pressure of specialist herbivores. Each time a new secondary metabolite evolves in a plant genotype, this will be selected for by a decreased attack by specialist herbivores which are unable to adapt to this new metabolite. A potential example of such a new defence is the occurrence of the particular pyrrolizidine alkaloid (PA) erucifoline in a chemotype of *Senecio jacobaea*. *Tyria jacobaeae* (cinnabar moth) is recorded from a number of host plants. The most preferred host plant of the cinnabar moth is *S. jacobaea* (South, 1973; Skinner, 1984; Heath & Emmet, 1979). Other host plant recordings of the cinnabar moth are *Senecio vulgaris, Petasites officinale, Tussilago farfara* and *Senecio* species in general (South, 1973; Skinner, 1984; Heath & Emmet, 1979). All these species contain PAs (Hartmann & Witte, 1995). *Senecio erucifolius* is assumed to be the closest relative of *S. jacobaea* (Jeffrey, 1979; Knox & Palmer, 1995) and has about two third of its PAs in common with *S. jacobaea* (Witte et al., 1992), but has to our knowledge never been recorded as a host for the cinnabar moth. The abundance of the cinnabar moth in The Netherlands is low on inland populations of *S. jacobaea* while it flourishes on the coastal populations. Witte et al. (1992) described two distinct chemotypes of *S. jacobaea*. One chemotype with jacobine as one of the the major PAs and another chemotype with erucifoline as one of the major PAs. They found that the jacobine types occurred predominantly in the western part of Europe while the erucifoline type dominated in the eastern part. Mass rearing in our lab of larvae of the cinnabar moth suggested that larvae performed less well on erucifoline chemotypes of *S. jacobaea*. The erucifoline chemotype of *S. jacobaea* has its major PA erucifoline in common with the major PA of *S. erucifolius* (Witte et al., 1992). Both species and chemotypes have most of the minor PAs in common. The chemotype distribution and host-plant recordings made us hypothesize that the presence of erucifoline might be the factor responsible for the lack of success of the cinnabar moth on *S. erucifolius* and the *S. jacobaea* erucifoline chemotype. The possibility that the absence of jacobine as a necessary stimulating factor could account for the observed pattern was considered unlikely since *T. jacobaeae* is reported to feed on jacobine-free species such as *Senecio vulgaris, Petasites officinale* and *Tussilago farfara.* To test this hypothesis we made a survey of the chemotypes of *S. jacobaea* in the Netherlands and compared these with available data on the distribution of the cinnabar moth. To test whether erucifoline influences the oviposition preference and larval performance, we performed a laboratory experiment using the two chemotypes of *S. jacobaea* and *S. erucifolius*.

## Materials and methods

*Survey of* S. jacobaea *chemotypes in the Netherlands.* Seeds of plants were collected at 14 locations in the Netherlands and on 2 locations in Belgium. The seeds were germinated and plants were grown in a climate room (20/15 °C, L16:D8, r.h. >70%) in 50/50 dune-sand/peat mixture. After eight weeks the total shoot of one plant per location was harvested and dried. Total PA content was determined spectrophotometrically, modified after Mattocks (1967). Individual PAs were quantified by GC-FID, modified after Vrieling et al. (1993), on a chrompack model 437A equipped with a WCOT fused silica column, 50 m × 0.25 mm ID, 0.4 $\mu$m (Chrompack CP-Sil 8); split ratio 1:60 (1:15), oven 265 °C, injection and detection 300 °C, FID. Methyl-nonadecanoate was used as an internal standard, carrier gas $H_2$, injection volume 0.4 $\mu$l. Three distinctive samples, together containing all the different PAs, were also analyzed by GC-MS by Witte according to Witte et al. (1992) for comparison and confirmation.

*Distribution of* T. jacobaeae *in the Netherlands.* Data on the distribution of *T. jacobaeae* in the Netherlands were obtained from the Dutch Butterfly Conservation in Wageningen. These data are based on the number of sightings of adults since 1950.

*Distribution of* S. jacobaea *and* S. erucifolius *in the Netherlands.* *S. jacobaea* is predominantly distributed along the coast and in the southern parts of the Netherlands (Figure 1). *Senecio erucifolius* is mainly found along the rivers Rhine, Maas and IJssel and the delta of these rivers (van der Meijden et al., 1989).

*Origin of plants and animals for oviposition and larval growth experiment.* *Senecio jacobaea* jacobine chemotype plants were grown from seeds collected in the sand dune area Meijendel near The Hague (the Netherlands). *S. jacobaea* erucifoline chemotypes were grown from seeds from five plants in the area of Braunschweig (Germany). *S. erucifolius* plants were obtained form various botanical gardens and institutes from Belgium, France and the Netherlands which collected these species in the wild at four sites: Ouffet and Longeau (Prov. de Liège, B), Moulins-lès-Metz (Moselle, F), Wageningen (NL). The plants from different sites were used randomly in the experiments. From *S. erucifolius* 2 chemotypes are reported (Witte et al., 1992). The chemotype of the *S. erucifolius* plants used in this experiment are unknown.

*Tyria jacobaeae* adults from the dunes of Meijendel were used for the oviposition experiments. The egg

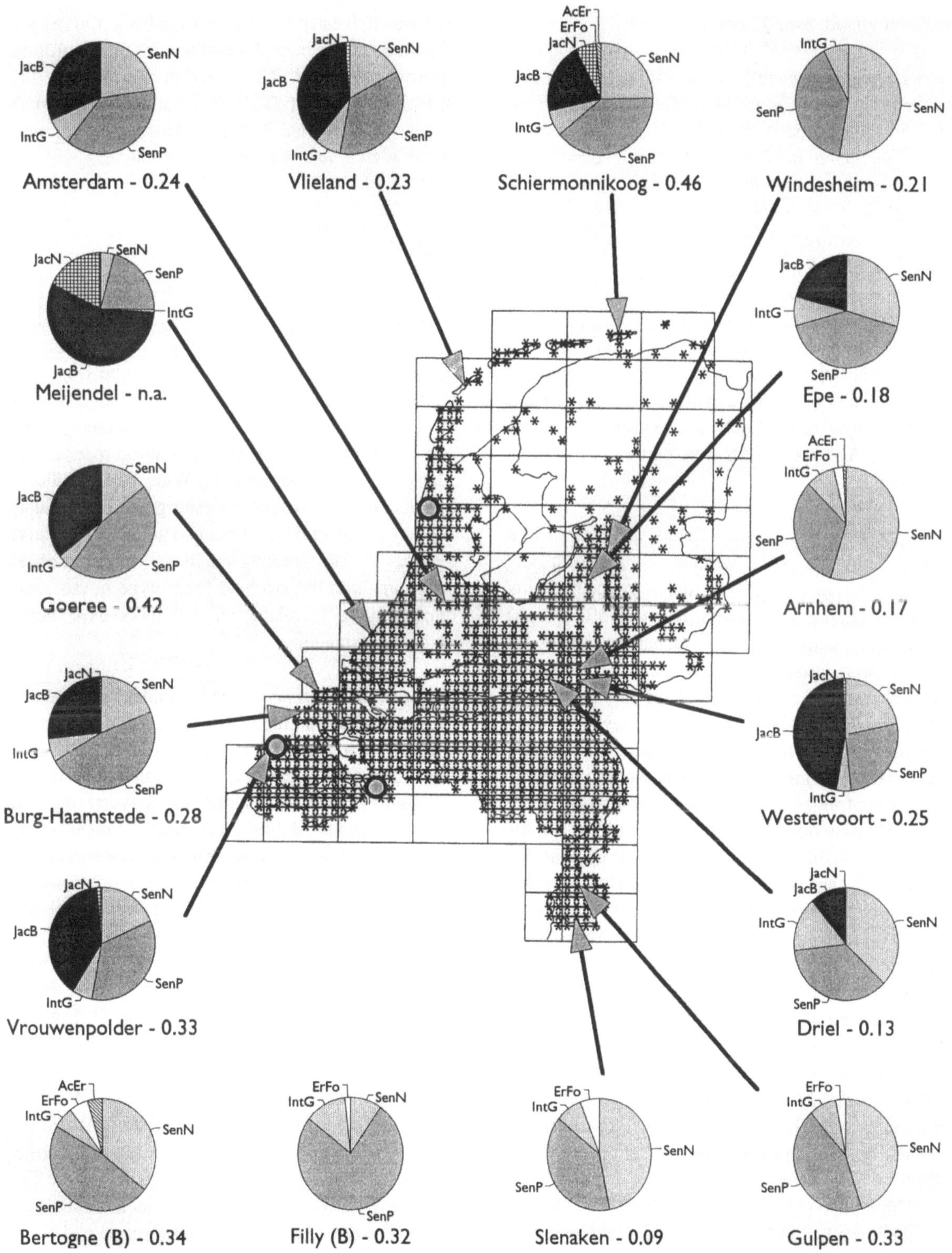


*Figure 1.* Distribution of the chemotypes of *S. jacobaea*. The presence of *S. jacobaea* since 1950 is indicated by an asterisk (reprinted with permission from van der Meijden et al. (1989)). Arrows indicate the geographic location of the sampled populations. Dots indicate the three populations analysed by Witte et al. (1992). These were all *S. jacobaea* jacobine chemotype. Pie-charts: PA spectrum; Figure below location: total PA ($\mu$g/g d.wt.) SenN: senecionine, SenP: seneciphylline, IntG: integerrimine, PAs specific to jacobine chemotypes: JacB: jacobine, JacN: jaconine, PAs specific to erucifoline chemotypes ErFo: erucifoline, AcEr: Acetylerucifoline. The two populations Bertogne and Filly are located in Belgium. At each location one randomly chosen plant was analysed for its PA spectrum.

batches from the oviposition experiment were used for the larval growth experiments.

*Oviposition experiments.* Oviposition experiments were carried out in transparant plastic cages (70 × 70 × 50 cm). In each cage six plants were placed: two *S. jacobaea* jacobine chemotypes, two *S. jacobaea* erucifoline chemotypes and two *S. erucifolius.* Plants of similar size were chosen for the experiment. All plants were reared in a climate room (20/15 °C, L16:D8, r.h. 70%) in pots containing a mixture of 90/10 dune-sand/peat. In four trials in each cage six female and two male, and in two trials nine female and one male moths caught in Meijendel were released. After four days the number of egg batches and number of eggs in each batch were counted. The experiment was repeated six times. Each time newly caught adults were used. All experiments were carried out in early June 1996.

*Larval growth experiments.* Plants were reared in a climate room (20/15 °C, L16:D8, r.h. 70%) in pots containing a mixture of 90/10 dune-sand/peat. Plants were placed in plastic cylinders (30 cm ⌀, 50 cm high) with tops and bottoms covered with gauze. Three types of plants were used as food plants: *S. jacobaea* jacobine chemotype, *S. jacobaea* erucifoline chemotype and *S. erucifolius.* On each plant one newly hatched larva was placed. The larvae originated from the egg batches laid on *S. jacobaea* jacobine chemotype, *S. jacobaea* erucifoline chemotype and *S. erucifolius* in the oviposition experiment. Immediately (within 12 h) after hatching on the leaves of the plants the egg batches were laid on, the larvae were transferred to the experimental plants. Larvae from at least two egg batches laid on *S. jacobaea* jacobine chemotype were evenly distributed over the three types of plants. In a similar way larvae from egg batches from *S. jacobaea* erucifoline chemotype and *S. erucifolius* were evenly distributed over the three plant types. The experiment started with 10 larvae per treatment. Unfortunately the tiny first instar larvae were able to escape from the cage. This resulted in a loss of about 50% of the larvae. Escape from the cages appeared to be random with respect to food plant ($\chi^2 = 0.95$, df = 2, NS) and oviposition plant ($\chi^2 = 2.32$, df = 2, NS).

Pupal weight, and number of days till pupation were recorded. Pupal weight and developmental time were recorded because pupal weight is highly correlated with fecundity and developmental time gives an estimate how long larvae in the field are exposed to predation. From each plant one fully expanded leaf was harvested for total PA analysis. Larvae were supplied with a new food plant of the appropriate type when needed. The experiment was carried out in a climate room (20/15 °C, L16:D8, r.h. 70%). All *S. erucifolius* started flowering during the experiment whereas all *S. jacobaea* stayed vegetative.

## Results

*Distribution of* S. jacobaea *chemotypes and* T. jacobaeae. All samples but one from the coastal dunes belong to the jacobine chemotype, while 5 out of 9 inland samples have the erucifoline type (Figure 1). One sample 'Schiermonnikoog' is of a mixed type. For the inland population Windesheim neither jacobine nor erucifoline could be detected in the rosette stage. The populations analyzed by Witte et al. (1992) in the Netherlands (Bergen, Doesburg and Hoogerheide) support the general view obtained in the current survey (Figure 1). The greatest abundance of *T. jacobaeae* co-occurs with the jacobine chemotype in the coastal dunes (Figure 2). Inland populations do exist, but are on average smaller.

*Oviposition experiments.* The moths laid a total of 50 egg batches ranging in size from 1 to 182 eggs with a mean batch size of 30 eggs. On average each female laid 1.2 egg batch during the experimental period.

Moths made no distinctions between the three types of plants offered. The total number of egg batches laid did not deviate from the expected uniform distribution ($\chi^2 = 1.2$, df = 2, NS). Furthermore the size of the egg batches did not differ between the plant types (Anova F = 0.089, df = 2, 49, NS).

*Larval growth experiment.* A two-way anova showed that plant type from which the egg batch originated had an effect on pupal weight and developmental time (Table 1). The type of plant upon which the larvae fed had no significant effects (Figure 3). Total PA concentration differed between the plant types (Anova F = 6.544, df = 2, 46, $P < 0.01$, data log transformed) with *S. jacobaea* jacobine chemotype having the highest levels: 1.36% dry weight. The concentration of PA in the *S. jacobaea* erucifoline chemotype and *S. erucifolius* is 0.59% and 0.41% dry weight respectively. PA concentration was not significantly correlated with any of the larval performance measures (pupal weight:

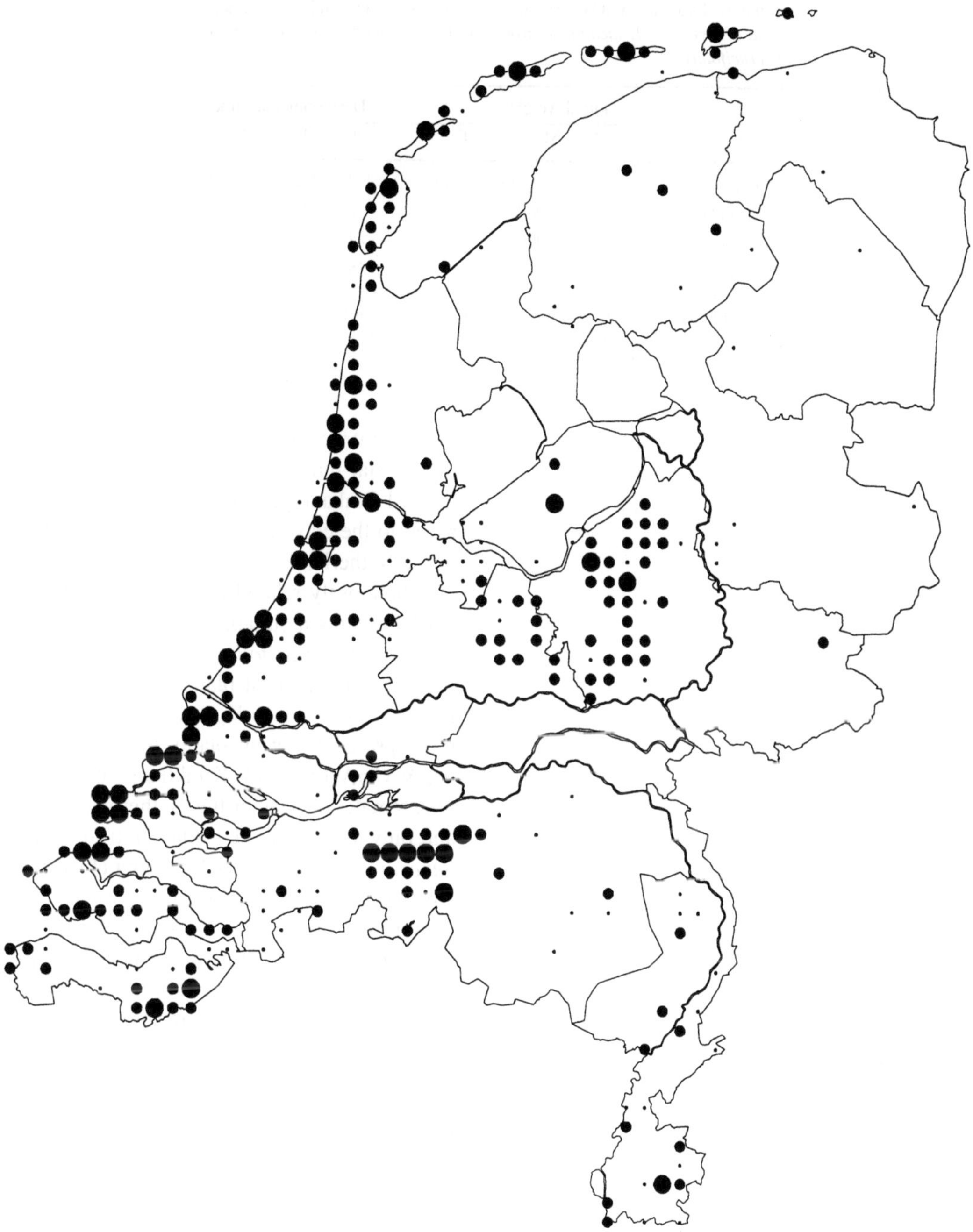

*Figure 2.* Distribution of *T. jacobaeae*. Number of sightings since 1950. Small dots = 1, medium dots = 2–10 and large dots >10 observations. Data were kindly supplied by the Dutch Butterfly Conservation Wageningen.

$r = -0.12$, $n = 41$, NS; developmental time: $r = 0.23$, $n = 41$, NS; PA concentration log transformed).

## Discussion

A clear trend is observed in the distribution of the chemotypes over the Netherlands and part of Belgium. The general trend is that the jacobine chemotype is present along the coast and that more land inward the erucifoline chemotype dominates. Although this is

*Table 1.* Two-way ANOVA (model I) on pupal weight and developmental time of larvae of *T. jacobaeae* grown on two *S. jacobaea* chemotypes and *S. erucifolius*

| | Pupal weight | | | Developmental time | | |
|---|---|---|---|---|---|---|
| | df | MS | F | df | MS | F |
| Oviposition plant | 2 | 0.00111 | 4.46* | 2 | 10.59 | 3.52* |
| Food plant | 2 | 0.00065 | 2.61 | 2 | 1.04 | 0.34 |
| Interaction | 4 | 0.00055 | 2.22 | 4 | 4.78 | 1.59 |
| Error | 35 | 0.00025 | | 35 | 3.01 | |

Oviposition plant: type of plant on which the eggs were oviposited. Food plant: type of plant used as food plants. df = degrees of freedom. MS = mean squares, * = P < 0.05.

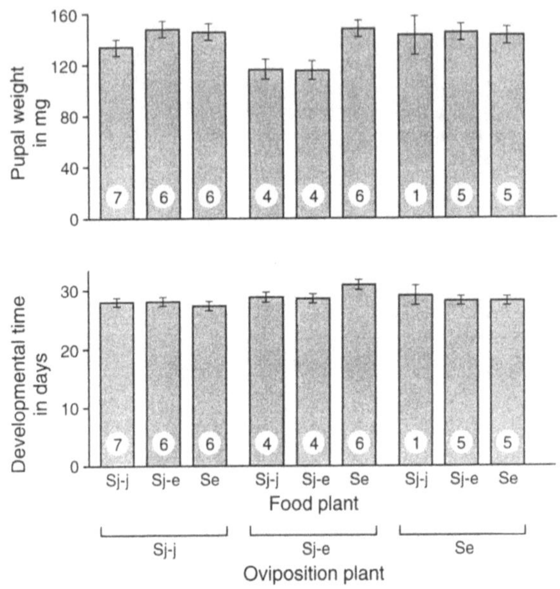


*Figure 3.* The effect of type of plants on which the eggs were oviposited and type of plant upon which *T. jacobaeae* larvae were reared. Upper graph: pupal weight, lower graph: developmental time. Sj-j: *S. jacobaea* jacobine chemotype, Sj-e: *S. jacobaea* erucifoline chemotype, Se: *S. erucifolius*. Food plant: Plant type on which larvae were reared, Oviposition plant: Plant on which the egg batch was oviposited. Numbers in the bars indicate the number of replicates. Standard errors based on the total data set are indicated on top of the bars.

loosely correlated with the abundance of the cinnabar moth, the oviposition and feeding experiments show that this is not caused by the presence of erucifoline in the food plants of the cinnabar moth. The distribution of the cinnabar moth needs to be explained by other factors and it is clear that the cinnabar moth is not a selective force favouring a distinct chemotype of *S. jacobaea*. As yet it is unclear what the selective forces are, that explain the occurrence and distribution of the two chemotypes of *S. jacobaea*. An explanation might be that pupal survival in sandy soils of the coastal dunes, mid- (Veluwe) and lower central (Noord-Brabant) part of the Netherlands is better compared to the clay and peat soils in the remaining part of the Netherlands.

The closely related erucifoline containing *S. erucifolius* seems a suitable host plant for the cinnabar moth. That *S. erucifolius* to our knowledge has never been reported as host plant might be caused by differences in its phenology. Females of the cinnabar moth prefer flowering plants for oviposition (Vrieling & van Wijk, 1994; Klinkhamer et al., 1997). The first females appear in the Netherlands in late April and only a few individuals last until the end of June. *Senecio erucifolius* starts flowering at the end of July whereas *S. jacobaea* starts flowering approximately 2 months earlier (van der Meijden, 1996). Between the *S. erucifolius* plants from different sites no differences in larval performance were observed, which suggests that chemotype did not matter or that all populations were of the same chemotype.

As reported by Witte et al. (1992) the chemotypes of *S. jacobaea* could be distinguished by analysing leaf material. Only in one case neither jacobine nor erucifoline could be detected (Windesheim). The distribution of the jacobine chemotype of *S. jacobaea* is mainly concentrated along the coast but also inland populations are present. Although *S. erucifolius* started flowering during the larval growth experiment, larval performance was not affected.

The finding of a significant effect of oviposition plant on the performance of larvae may be explained by the use of a limited number of egg batches. Soldaat & Vrieling (1992) found that there were large (genetic) differences in performance between egg batches from different females on *S. jacobaea* jacobine chemotypes. In this experiment at least two egg batches were used to provide larvae for each food plant. However, the restricted use of the number of egg batches might have caused this result through sampling error. In line with earlier research (Vrieling et al., 1991) no correlations between larval performance and PA-concentration were found. The cinnabar moth is perfectly adapted to PAs in its diet.

## Acknowledgements

We kindly thank 'Dutch Butterfly Conservation Wageningen' and the 'Rijksherbarium' for providing respectively the distribution map on sightings of cinnabar moths and the distribution of *S. jacobaea* in the Netherlands. We thank Prof. T. Hartmann for providing us with seeds of *S. jacobaea* erucifoline chemotype, and Dr L. Witte for the analysis of three of the chemotypes. We thank Eddy van der Meijden for commenting on earlier versions of this article and Edwin Poels for technical assistance.

## References

Berenbaum, M., A. R. Zangerl & J. K. Nitao, 1986. Constraints on chemical coevolution: wild parsnips and the parsnip webworm. Evolution 40: 1215–1228.

Berenbaum, M., A. R. Zangerl & K. Lee, 1989. Chemical barriers to adaptation by a specialist herbivore. Oecologia 80: 501–506.

Ehrlich, P. R. & P. H. Raven, 1964. Butterflies and plants: A study in coevolution. Evolution 18: 586–608.

Harborne, J. B., 1993. Introduction to Ecological Biochemistry. 4th ed., Academic Press, London.

Hartmann, T., 1995. Pyrrolizidine alkaloids between plants and insects: a new chapter of an old story. Chemoecology 5/6, 3/4: 139–146.

Hartmann, T. & L. Witte, 1995. Chemistry, biology and chemoecology of the pyrrolizidine alkaloids. In: S. W. Pelletier (ed), Alkaloids: Chemical and Biological Perspectives, Vol. 9, Pergamon, GB-Trowbridge, pp. 155–233.

Jeffrey, C., 1979. Generic and sectional limits in *Senecio* (Compositae): II Evaluation of some recent studies. Kew Bulletin 34: 49–58.

Heath. J. & A. M. Emmet, 1979. The Moths and Butterflies of Great Britain and Ireland, Vol. 9, Curwen Books, London.

Klinkhamer, P. G. L., T. Kubo & Y. Iwasa, 1997. Herbivores and the evolution of the semelparous perennial life-history of plants. Journal of Evolutionary Biology 10: 529–550.

Knox, E. B. & J. D. Palmer, 1995. The origin of *Dendrosenecio* within the Senecioneae (Asteraceae) based on chloroplast DNA evidence. American Journal of Botany 82: 1567–1573.

Mattocks, A. R., 1967. Spectrophotometric determination of unsaturated pyrrolizidine alkaloids. Analytical Chemistry 34: 443–447.

Meijden, R. van der, 1996. Heukels' Flora van Nederland. 22nd ed., Wolters-Noordhoff, Groningen.

Meijden, R. van der, C. L. Plate & E. J. Weeda, 1989. Atlas van de Nederlandse Flora. 3. Minder zeldzame en algemene Soorten. RHHB Leiden/CBS Voorburg.

Skinner, B., 1984. Colour Identification Guide to Moths of the British Isles (Macrolepidoptera). Viking, London.

Soldaat, L. L. & K. Vrieling, 1992. The influence of nutrional and genetic factors on larval performance of *Tyria jacobaeae* under laboratory conditions. Entomologia Experimentalis et Applicata 62: 29–36.

South, R., 1973. The Moths of the British Isles. I. New Edition. Frederick Wayne & Co, London.

Vrieling K., L. Soldaat & W. Smit, 1991. The influence of pyrrolizidine alkaloids of *Senecio jacobaea* on *Tyria jacobaeae*, *Brachycaudus cardii* and *Haplothrips senecionis*. Netherlands Journal of Zoology 41: 228–239.

Vrieling, K., H. de Vos & C. A. M. van Wijk, 1993. Genetic analysis of the concentration of pyrrolizidine alkaloids of *Senecio jacobaea*. Phytochemistry 32: 1141–1144.

Vrieling, K. & C. A. M. van Wijk, 1994. Estimating costs and benefits of the pyrrolizidine alkaloids of *Senecio jacobaea* under natural conditions. Oikos 70: 449–454.

Witte L., L. Ernst, H. Adam & T. Hartmann, 1992. Chemotypes of two pyrrolizidine alkaloid-containing *Senecio* species. Phytochemistry 31: 559–565.

*Entomologia Experimentalis et Applicata* **91:** 259–265, 1999.



# It's all in the neurones

R. F. Chapman
*Division of Neurobiology and Center for Insect Science, University of Arizona, Tucson, AZ 85721, USA*

Accepted: October 17, 1998

## Abstract

This paper views some of the work presented in the symposium from a neural perspective. It is argued that any change in hostplant selection depends on a change in the peripheral or central nervous systems. Any changes in hostplant use and, ultimately, the evolution and phylogeny of phytophagous insects have a neural basis.

## Introduction

The wealth of new ideas and data, and the breadth of material presented at this meeting is so great that I feel it impossible to give a general overview. Instead, I will focus on the nervous system and try to put some of the papers in a neural context, even though that was not the author's intention in many cases. My justification for doing so arises from the fact that the nervous system must ultimately be involved in all aspects of the interactions between insects and their hostplants. This is obviously true in the case of behaviour, but remember that a host switch can only occur if the insect's behaviour, and something in its nervous system, changes. Such changes lead to evolution and so we can think of evolution as stemming from changes, which become genetically fixed, in the nervous system. Ultimately, we develop phylogenies. Here, too, each branch of a phylogenetic tree will have arisen through a behavioral/neural change.

I apologize to those of you at the meeting, in fact the majority of participants, whose work I do not refer to. This is in no way a reflection on the work, or my interest in it, rather that I cannot do justice to it.

## Insect chemosensory systems

The contrast in the organization of the olfactory and gustatory systems was well made on the first morning of the symposium. Laurent and Hildebrand both described the organization of the olfactory system with its huge numbers of sensory afferents converging on to a limited number of glomeruli with a relatively small number of output neurones connecting with the mushroom bodies. By contrast, the afferents from gustatory receptors terminate in the ganglion of the segment in which they occur and there is no known discrete gustatory centre. This is what makes the work described by Rogers & Simpson so challenging. Looking for the interneurones that integrate gustatory inputs is like looking for a needle in a haystack.

The arrangement of the olfactory system and the contrasting lack of a discrete centre for the gustatory system are common to all animals (e.g., Hildebrand, 1995) and reflect their different functions. The olfactory system is concerned with the perception of a broad array of chemical stimuli which may have ecological relevance. The gustatory system, by contrast, is concerned with the recognition of suitable food or an oviposition substrate. Basically it is concerned with the questions 'is this food?' and 'is it toxic?'. All animals need sources of carbohydrate and protein and there is no basic need to differentiate between 'toxic' stimuli, all one needs to know is that they are toxic. Thus the gustatory system can be relatively hard-wired because the stimuli to which it responds and messages it needs to convey are invariant. Of course, the system becomes more refined than I have suggested, but I think these are the basic principles.

## The olfactory system

*Coding.* Laurent and Hildebrand presented two contrasting papers, the former emphasizing the possible roles of oscillations of electric fields within the brain as neuronal firing comes in and out of synchrony, the latter the probable specificity of the glomeruli and the presence of labeled lines providing the insects with direct pathways for the transmission of important information. These approaches are not mutually incompatible. The two groups are working on different insects, Laurent on the locust, Hildebrand on a moth, and there are big differences between the olfactory lobes of these insects. The locust has about 1000 rather poorly defined glomeruli, *Manduca*, by contrast, has 64 well-defined glomeruli. Perhaps more important are the time scales over which these authors are working. The patterns observed by Laurent occur over periods of seconds, while Hildebrand's group is interested in responses that occur within milliseconds. We know that odour 'plumes' really consist of wisps of odorous air separated by regions of 'clean' air. As a result, a flying insect encounters the odour of a mate, and probably also a plant at some distance from the source, as a series of very short-lived stimuli and the olfactory system of moths, as described by Hildebrand and others, is well-suited to respond to such a pattern of stimulation. Laurent's longer term neural patterns, however, seem better suited to periods of sustained stimulation, such as might occur when an insect is sitting on a leaf. Perhaps this is a way of ensuring that the insect pays attention to a stimulus as Bernays (1998) has suggested is so important.

*The importance of learning.* For many organisms, some specific odours habitually have a key role in governing behaviour. Pheromones are the obvious example, but hostplant-specific odours are also in this category. We would expect the neural circuitry for these odours to be built in to the insect's nervous system (we say they are hard-wired). Many other odours, however, may only occasionally be relevant and their diversity probably makes it impossible to have evolved the appropriate circuitry for all of them. Instead, the individual develops an association between an odour and the relevant situation in which it was encountered. This is widespread across many animal taxa and the olfactory system seems to be designed to facilitate it. For example, the chemical characteristics ('odour') of the river in which young salmon develop becomes imprinted on to their olfactory systems (Dittman et al., 1997) and foetal rabbits become imprinted to feed on juniper once they are weaned if their mother feeds on juniper during late pregnancy (Semke et al., 1995). And there are similar examples from amphibia and mammals together with a multitude of studies on associative learning that involves the olfactory system.

In insects, neurones from the olfactory lobe connect directly with the mushroom bodies that are the principal site of the brain for learning and memory. Consequently, and by analogy with the vertebrate findings, we might expect that insects have the facility to make associations between particular events, such as oviposition or feeding, and the odour which they experienced at the time. In the section on tritrophic interactions we heard several examples of parasitoids making associations between the odour of a damaged leaf and the presence of a host (and see e.g., Vet & Dicke, 1992). This should not surprise us, in fact, we ought to be surprised if this were not the case, and if it occurs with parasitoids I think we should also expect it to occur in relation to oviposition and feeding in phytophagous insects, especially perhaps in oligophagous species. Such effects, if they occurred repeatedly or consistently, might become genetically fixed in populations and might be the basis of speciation events as seems to be occurring, for example, in the evolution of *Rhagoletis* which was described by Feder (see Frey et al., 1992).

*Olfaction at close range.* Because of the obvious importance of odours in enabling insects to finds their hosts (or mates) from a distance, the likelihood that odours may have a significant role in governing the insect's behaviour once it has reached its host is often overlooked. The papers by de Jong & Städler and Feeny & Haribal (see elsewhere in this volume) both made reference to such behaviour, however. Both dealt with the importance of odour in oviposition behaviour once the insects have arrived at the host, the first with the cabbage root fly, the second with the zebra swallowtail butterfly. Feeny et al. (1989) have previously demonstrated a similar effect in the black swallowtail butterfly, and I believe there is every reason to think that this is a common phenomenon. This is probably true of feeding as well as oviposition. Why are caterpillars relatively well-endowed with olfactory sensilla, and why are their antennae in a position which normally keeps them close to the leaf surface? No one considered this aspect at the present meeting, but it is important not to forget it. See papers by de Boer (1991) and Saxena & Schoonhoven (1982).

## The gustatory system

*Neural control of feeding.* Rast & Bräunig (1997) have shown that the motor neurones innervating the mandibular adductor and abductor muscles of *Locusta* fire in a pattern that resembles their activity during feeding even in an isolated suboesophageal ganglion, when the ganglion is bathed in pilocarpine. Not only are the mandibular motor neurones activated, so are those innervating the salivary glands and some neurosecretory cells that are normally active during feeding (Schachtner & Bräunig, 1993, 1995). These results indicate that mandibular movement and some associated activities are driven by a pattern generator in the mandibular neuromere of the suboesophageal ganglion. Separate, but presumably linked, pattern generators are present in the maxillary and labial neuromeres. There is also evidence that mandibular activity in caterpillars is driven by a pattern generator (Griss et al., 1991; Rohrbacher, 1994) and it is likely to be a general phenomenon, just as the pattern of leg movements during walking and the basic wing movements in flight arise from built-in pattern generators of the thoracic ganglia (Robertson & Pearson, 1985; Ryckebusch & Laurent, 1993). Pattern generators are also known to occur in the abdominal ganglia where, in locusts, they govern the alternating movements of the ovipositor valves during oviposition (Thompson, 1986).

As with other pattern generators in the insect nervous system, the pattern of output in the mandibular nerves from the isolated ganglion is normal, but the cycle period is longer than in the intact insect. The cycle period of the wings is modulated by proprioceptive input especially, in the locust, from the forewing stretch receptors (Pearson & Ramirez, 1990). Proprioceptive feedback from the mandibles probably modulates the basic cycle period of the mandibular pattern generator (see Chapman, 1995a).

Thus, once the insect has arrived on its host plant, it is possible to think of feeding, and perhaps also oviposition, as being determined by the switching on and off of a pattern generator.

It is well known that some nutrients, especially sugars, stimulate feeding; that is, they initiate mandibular and associated swallowing movements. At this meeting, Hanson (this volume) gave a good example of the sugar alcohol, inositol, acting as a phagostimulant for larval *Manduca*. Other compounds, deterrents, inhibit feeding. This was the effect of KCl in *Manduca* (Hanson) and of azadirachtin in larval *Pieris* (van Loon & Schoonhoven) and it has long been clear that feeding (or the activation of the mandibular pattern generator) depends on a balance between phagostimulatory and deterrent inputs from the periphery (Chapman, 1995b; Schoonhoven et al., 1998).

However, understanding exactly what these inputs are proves to be more complicated than we had thought, as Städler explained. For many years, integration had been thought to occur only in the central nervous system so that a knowledge of the responses of the individual neurones in a sensillum to their particular stimulating chemicals would enable us to understand the chemosensory basis of food selection behaviour. Over the last 15 years, however, there have been increasing numbers of examples of interactions occurring at the periphery within each sensillum so that the responses of the individual neurones are not independent and the activity of any one cell is determined by the suite of chemicals reaching the dendrites of the whole ensemble (e.g., Dethier & Bowdan, 1989; Mitchell, 1988; brief reviews in Chapman, 1995b; Schoonhoven et al., 1998). Thus, it is now clear that integration occurs at the peripheral receptors as well as in the CNS and this is a common, if not a general, phenomenon when an insect encounters food.

How is this input integrated within the CNS? Obviously the axons from peripheral chemoreceptors must synapse with a series of interneurones that ultimately switch the pattern generators on or off, but because of the dispersed nature of gustatory integration (see above) it has been extremely difficult to obtain anatomical or physiological information on the details. Rogers & Simpson have given us some pieces of the jigsaw. They studied interneurones in the suboesophageal ganglion of *Locusta*. All the cells they recorded from responded to stimulation of the palp-tip sensilla with both wheat (an acceptable food) and cabbage (unacceptable) although the changes in membrane potential of the interneurones differed according to the stimulus. They did not find cells that responded only to wheat or only to cabbage suggesting that the sensory pathways from phagostimulatory and deterrent sensory cells soon become mixed. All the cells also responded to mechanical stimulation of the sensilla indicating that chemical and mechanical information converges on to these cells although several different response types were noted.

*Possible neural basis of food selection.* We can now discuss the neural bases of some of the behavioural observations in other papers on the basis of this model. Van Loon & Schoonhoven (this volume) showed that larval *Pieris* have a specialized deterrent cell in the lateral styloconic sensilla that is very sensitive to cardenolides. This cell complements the deterrent cell in the medial styloconic sensilla that responds to a wide range of deterrent compounds (see Schoonhoven et al., 1998). When *Pieris* caterpillars were given a choice between cabbage treated with the cardenolide and untreated cabbage, they preferred the latter, but when cardenolide-treated and azadirachtin-treated cabbage were paired, the cardenolide-treated cabbage was preferred. We do not yet know the neurophysiological patterns of interactions responsible for the behaviour, but assuming that phagostimulatory input is similar in all three treatments, the differences lie in the different levels of deterrence: none (or very little) on untreated cabbage, a large amount on azadirachtin-treated cabbage and an intermediate amount on cardenolide-treated cabbage. In the azadirachtin/cardenolide comparison we must suppose that the balance between phagostimulation and deterrence favours the cardenolide, so that it will be fed upon. It is important to recognize that these experiments do *not* imply that cardenolides are deterrent in one circumstance and phagostimulatory in another.

A most significant paper (in this volume) was that of Renwick & Lopez describing experiments with larvae of *Pieris rapae*. Neonate larvae would feed on cowpea, but larvae that had previously eaten cabbage would not do so unless the cowpea was treated with glucosinolates. The cowpea was shown not to contain deterrents for the larvae. In my simplified neural scheme, it appears that the phagostimulants in cowpea could drive feeding in neonate larvae but, after an experience with food containing glucosinolates, they were inadequate unless a glucosinolate-sensitive cell (which we know is present) is also active. Some change from acceptance to rejection of cowpea occurs after exposure to cabbage for only 6 h, and is complete after 30 h exposure. This is ample time for changes to occur in receptor sites in the peripheral neurones or at synapses, or for anatomical changes at synapses. A somewhat similar situation was described by Yamamoto (1974) with neonate larvae of *Manduca sexta* which he regarded as polyphagous until exposed to solanaceous plants. In this species the critical period was in the first instar larva. There are also many example of induced preference in later stage caterpillars (Bernays & Weiss, 1996; Jermy, 1987) and, interestingly, Saxena and Schoonhoven (1982) showed that induced preference in larval *Manduca sexta* may have an olfactory component.

The process of induction of preference is in some ways analogous to the examples of imprinting described in relation to the olfactory system although 'imprinting' is classically defined as occurring early in an animal's development.

Withers described experiments to test the applicability of Courtney's hierarchy-threshold model (Courtney et al., 1989) to the feeding behaviour of the beetle *Zygogramma bicolorata*. Immediately after a meal the insects would not feed even on a highly-ranked host (one that they would normally accept readily) and we can presume that this was primarily due to the inhibition of feeding as a consequence of gut distension and other feeding-related factors as has been described in some other insects (Bernays & Simpson, 1982). An hour or two later, the beetles fed on a highly-ranked host, but not on a less acceptable species. Presumably by this time volumetric inhibition has declined to a low level, but deterrents in the less palatable plant remain dominant. With increasing food deprivation, the probability of feeding on this plant increases although the total time feeding remains much less than on the highly-ranked host even after 6 days without food. Similar changes have been recorded in other insects (e.g., Chapman & Bernays, 1977) and could arise from an increase in sensitivity of the peripheral nutrient (phagostimulant) receptors, while the sensitivity of deterrent cells is unaltered, or it could arise from a change in the sensitivity of interneurones. This could be achieved by nutritional feedbacks directly or acting via neuromodulators and might eventually lead to a situation where mechanostimulation alone provides an adequate switch for the pattern generator. When, as a matter of observer convenience, locusts were habitually starved for 24 h or more before their feeding behaviour was examined, they were described as 'indiscriminate biters' (Dadd, 1963) and the anecdotes of locusts eating window curtains or the sails from a ship probably arise from similar behaviour.

There is good evidence that locusts and caterpillars can adjust their nutrient intake of artificial diets in relation to their own nutritional status (Raubenheimer, this volume). This implies the presence of receptors capable of distinguishing between different nutrients and several species are known to have separate neurones in mouthpart receptors that respond to sugars, amino acids and inorganic salts (Chapman, 1995b; Schoonhoven et al., 1998). Earlier work by Simpson and his associates (Simpson et al., 1991) has shown that the sensitivity of these cells varies with the nutrient content of the haemolymph. For example, the sensitivity of sugar-sensitive cells in *Locusta* is greatly reduced when the sucrose content of the haemolymph is high. At least with amino acids, this change is the result of a direct feedback of the haemolymph amino acids on the sensory neurones (Simpson & Simpson, 1992). Wright et al. have taken a significant step forwards by extending these studies to real plants and, in fact, doing it in the field. They showed that, while acceptance of leaves of the tree *Cedrela odorata* by the weevil *Exopthalmus jekelanus* was dependent on the absence (or low levels) of a limonoid, over 50% of the variability in meal duration could be accounted for by variation in the leaf-levels of sucrose and nitrogenous compounds. So far they have only correlated meal size with nutrient composition of the foliage. What we all now want to know is how this is affected by the nutrient status of the insect.

*Evolution of feeding patterns.* The neural changes described above, and especially those associated with increasing periods of food deprivation, may result in the insect accepting a much wider range of plants than is normally the case and, if the neural changes become genetically fixed, could result in a change in the food selection behaviour of the species. Such a change is described by Y. Asaoka (unpubl. results) in his experiments with mutant silkworms. Wild type silkworms only feed on mulberry or artificial diet which includes mulberry, the mutants, however, will accept other food plants and are loosely called polyphagous (although the extent of their polyphagy is not known). Asoaka has shown that the mutant larvae are behaviourally much less sensitive to two representative deterrents (strychnine nitrate and salicin) than normal larvae, but found no differences in the inputs from the deterrent cell in the medial styloconic sensillum. The implication is that the mutation produces a change in the central nervous system, perhaps as simple as a reduction in gain at a synapse, leading to a reduction in deterrent input relative to phagostimulatory input. Interestingly, working with other mutants, Y. Asaoka did find that other polyphagous mutants had deterrent cells with reduced sensitivity to some compounds. These experiments demonstrate that small changes in either the peripheral or central nervous systems of the insects can produce large changes in the range of hostplants accepted by the insects simply by altering the level of sensitivity to deterrents as Bernays & Chapman suggested (1994).

In another relevant paper (in this volume), Roessingh et al. investigated the genetics of the sensory response to the sugar alcohols, dulcitol and sorbitol, in *Yponomeuta* species. They have shown elsewhere (P. Roessingh et al., unpubl. results) that the receptor proteins for these two compounds, which are stereoisomers, are located on a homologous neurone in the styloconic sensilla of the caterpillars of different species within the genus. Thus here we see a relatively simple genetic change apparently leading to a change in the form of the receptor molecules that accompanied, and presumably contributed to, the switch from Celastraceae to Rosaceae as hostplants. The former are characterized by dulcitol, the latter by sorbitol.

De Yong & Nielsen described experiments (in this volume) with near-isogenic lines of flea beetles. They found that the readiness of *Phyllotreta nemorum* to feed on *Barbarea vulgaris* was regulated by different genes in different insect lines. The special interest here is in demonstrating that there are different ways in which a species may evolve to achieve a host switch.

Van Loon & Schoonhoven suggested that neurones specifically tuned to indicator chemicals in hostplants, such as the glucosinolates in Brassicaceae, might be derived initially from deterrent receptor cells with a broad specificity These might become more specialized, like the cardenolide-sensitive cells in *Pieris* larvae, presumably by the enhancement of one type of receptor protein at the expense of others. Such a specialized cell might then easily change from deterrent to phagostimulatory by a synaptic switch from inhibitory to excitatory. Thus a deterrent might become a sign stimulus. It could be imagined that some such switch occurred in the evolution of the Pieridae. There are numerous examples of phytophagous insects with deterrent responses to glucosinolates (e.g., Shields & Mitchell, 1995) even in species that commonly feed on brassicas. In most of these insects, the deterrent cells probably respond to a wide range of compounds, but a narrowing of the response range to glucosinolates followed by change in sign at a synapse could lead to an insect that normally fed only when glucosinolates were present.

The studies by Feeny and his associates (see Feeny, 1991), exemplified here in the paper by Feeny & Haribal show that the phylogeny of the swallowtail butterflies (Papilionidae), which reflects changes in hostplant use, also reflects differences in the use of specific chemical cues for oviposition. Subfamilies or tribes respond to particular classes of chemical, in the case of Papilioninae to hydroxycinnamic acid derivatives, and within this subfamily different tribes use different compounds. The zebra swallowtail, *Eurytides marcellus*, in the tribe Graphini, is stimulated to oviposit by 3-caffeoyl-muco-quinic acid, whereas 5-caffeoylquinic acid, which is an oviposition stimulant for members of the tribe Papilionini, is not. This series of studies promises to give us our most complete understanding of the association between hostplant chemistry and insect phylogeny.

## Conclusions

I have tried to show how changes in insect behaviour that may lead to hostplant switches and ultimately to the evolution of new species, may be related to simple changes in the nervous system. Perhaps the most important advance that we have seen beginning to emerge at this meeting is in our understanding of some of the genetic processes that regulate neural responses. It seems that even a single gene mutation could have a profound effect. But across the whole field, from sensory physiology to phylogeny, we are beginning to be able to fit the pieces into a coherent story.

I cannot conclude my brief survey of the meeting without drawing attention to two papers that I cannot fit into my scheme even though the findings obviously arise from neural events. The first is the paper by Potter & Held showing that the Japanese beetle, *Popillia japonica*, did not exhibit food aversion learning after feeding on the flowers of geranium which caused it to become temporarily paralyzed. We should not infer from this that the beetle might not show aversion learning of other foods, but my curiosity about this specific example makes me want to know more.

The second is the paper by Dobler and her associates on the phytophagous silphids. Here we have a genus of phytophagous beetles, apparently specialists on Chenopodiaceae, that is derived from essentially predaceous ancestors via detritivory. What has happened to the insects' sensory and digestive physiology that makes such a transfer possible?

I hope these authors and everyone else who was at the meeting will be able to continue in our quest for understanding. Thank you all for your contributions to this splendid meeting.

## References


Bernays, E. A., 1998. The value of being a resource specialist: behavioral support for a neural hypothesis. American Naturalist 151: 451–464.

Bernays, E. A. & R. F. Chapman, 1994. Host-plant Selection by Phytophagous Insects. Chapman & Hall, New York.

Bernays, E. A & S. J. Simpson, 1982. Control of food intake. Advances in Insect Physiology 16: 59–118.

Bernays, E. A & M. R. Weiss, 1996. Induced food preferences in caterpillars: the need to identify mechanisms. Entomologia Experimentalis et Applicata 78: 1–8.

Boer, G. de, 1991. Effect of diet experience on the ability of different larval chemosensory organs to mediate food discrimination by the tobacco hornworm, *Manduca sexta*. Journal of Insect Physiology, 37: 763–769.

Chapman, R. F., 1995a. Mechanics of food handling by chewing insects. In: R. F. Chapman & G. de Boer (eds), Regulatory Mechanisms in Insect Feeding. Chapman & Hall, New York, pp. 3–31.

Chapman, R. F., 1995b. Chemosensory regulation of feeding. In: R. F. Chapman & G. de Boer (eds), Regulatory Mechanisms in Insect Feeding. Chapman & Hall, New York, pp. 101–136.

Chapman, R. F. & E. A. Bernays, 1977. The chemical resistance of plants to insect attack. Scripta Varia 41: 603–633.

Courtney, S. P., G. K. Chen & A. Gardner, 1989. A general model for individual host selection. Oikos 55: 55–65.

Dadd, R. H., 1963. Feeding behaviour and nutrition in grasshoppers and locusts. Advances in Insect Physiology 1: 47–109.

Dethier, V. G. & E. Bowdan, 1989. Effects of alkaloids on sugar receptors and the feeding behaviour of the blowfly. Physiological Entomology 14: 127–136.

Dittman, A. H., T. P. Quinn, G. A. Nevitt, B. Hacker & D. R. Storm, 1997. Sensitization of olfactory guanylyl cyclase to a specific imprinted odorant in Coho salmon. Neuron 19: 381–389.

Feeny, P., 1991. Chemical constraints on the evolution of swallowtail butterflies. In: P. W. Price, T. M. Lewinsohn, G. W. Fernandes & W. W. Benson (eds), Plant-Animal Interactions. Wiley, New York, pp. 315–340.

Feeny, P., E. Städler, I. Ahman & M. Carter, 1989. Effects of plant odor on oviposition by the black swallowtail butterfly, *Papilio polyxenes* (Lepidoptera: Papilionidae). Journal of Insect Behavior 2: 803–827.

Frey, J. E., T. J. Bierbaum & G. L. Bush, 1992. Differences among sibling species *Rhagoletis mendax* and *R.pomonella* (Diptera: Tephritidae) in their antennal sensitivity to host fruit compounds. Journal of Chemical Ecology 18: 2011–2024.

Griss, C., S. J. Simpson, J. Rohrbacher & C. H. F. Rowell, 1991. Localization in the central nervous system of larval *Manduca sexta* (Lepidoptera: Sphingidae) of areas responsible for aspects of feeding behaviour. Journal of Insect Physiology 37: 477–482.

Hildebrand, J. G., 1995. Analysis of chemical signals by nervous systems. Proceedings of the National Academy of Sciences of the United States of America 92: 67–74.

Jermy, T., 1987. The role of experience in host-plant selection by phytophagous insects. In: R. F. Chapman, E. A. Bernays & J. G. Stoffolano (eds), Perspectives in Chemoreception and Behavior. Springer-Verlag, New York, pp. 143–158.

Mitchell, B. K., 1988. Adult leaf beetles as models for exploring the chemical basis of host-plant recognition. Journal of Insect Physiology 34: 213–225.

Pearson, K. G. & J. M. Ramirez, 1990. Influence of input from the forewing stretch receptors on motorneurones in flying locusts. Journal of Experimental Biology 151: 317–340.

Rast, G. F. & P. Bräunig, 1997. Pilocarpine-induced motor rhythms in the isolated locust suboesophageal ganglion. Journal of Experimental Biology 200: 2197–2207.

Robertson, R. M. & K. G. Pearson, 1985. Neural circuits in the flight system of the locust. Journal of Neurophysiology 53: 110–128.

Rohrbacher, J., 1994. Fictive chewing activity in motor neurons and interneurons of the suboesophageal ganglion of *Manduca sexta* larvae. Journal of Comparative Physiology A 175: 629–637.

Ryckebusch, S. & G. Laurent, 1993. Rhythmic patterns evoked in locust leg motor neurons by the muscarinic agonist pilocarpine. Journal of Neurophysiology 69: 1583–1595.

Saxena, K. N. & L. M. Schoonhoven, 1982. Induction of orientational and feeding preferences in *Manduca sexta* larvae for different food sources. Entomologia Experimentalis et Applicata 32: 173–180.

Schachtner, J. & P. Bräunig, 1993. The activity pattern of identified neurosecretory cells during feeding behaviour of the locust. Journal of Experimental Biology 185: 287–303.

Schachtner, J. & P. Bräunig, 1995. Activity pattern of suboesophageal ganglion cells innervating the salivary glands of the locust *Locusta migratoria*. Journal of Comparative Physiology A 176: 491–501.

Schoonhoven, L. M., T. Jermy & J. J. A. van Loon, 1998. Insect-Plant Biology. Chapman & Hall, London.

Semke, E., H. Distel & R. Hudson, 1995. Specific enhancement of olfactory receptor sensitivity associated with foetal learning of food odours in the rabbit. Naturwissenschaften 82: 148–149.

Shields, V. D. C. & B. K. Mitchell, 1995. The effect of phagostimulant mixtures on deterrent receptor(s) in two crucifer-feeding lepidopterous species. Philosophical Transactions of the Royal Society of London B 347: 459–464.

Simpson, S. J. & C. L. Simpson, 1992. Mechanisms controlling modulation by haemolymph amino acids of gustatory responsiveness in the locust. Journal of Experimental Biology 168: 269–287.

Simpson, S. J., S. James, M. S. J. Simmonds & W. M. Blaney, 1991. Variation in chemosensitivity and the control of dietary selection behaviour in the locust. Appetite 17: 141–154.

Thompson, K. J., 1986. Oviposition digging in the grasshopper. I. Functional anatomy and the motor program. Journal of Experimental Biology 122: 387–411.

Vet, L. E. M. & M. Dicke, 1992. Ecology of infochemical use by natural enemies in a tritrophic context. Annual Review of Entomology 37: 141–172.

Yamamoto, R. T., 1974. Induction of hostplant specificity in the tobacco hornworm, *Manduca sexta*. Journal of Insect Physiology 20: 641–650.

# Index of authors

# General index

Terms are indexed to the first page of the relevant paper.

# Listing of oral and poster presentations from SIP 10

The following has been based on the symposium programme, with certain presentations reordered into the most appropriate session. Underlined authors were those making the presentation.

**Opening address**

Louis M. Schoonhoven

## Session 1: Central neural bases of host plant recognition

**Oral presentations**

Gilles Laurent, Synchronization and temporal codes for odour perception

J.G. Hildebrand, J.R. King, W. Mechaber, W. Roessler, K. Selchow, V.D.C. Shields & T.A. Christensen, Plant-odour processing in the olfactory system of *Manduca sexta*

Stephen M. Rogers & Stephen J. Simpson, Chemo-discriminatory neurones in the sub-oesophageal ganglion of *Locusta migratoria*

Georg F. Rast & Peter Braunig, Locust feeding motor patterns elicited by pilocarpine

## Session 2: Chemosensory bases of host plant recognition

**Oral presentations**

Erich Städler, Chemosensory bases of host-plant recognition

Joop J.A. van Loon & Louis M. Schoonhoven, Specialist deterrent chemoreceptors enable *Pieris* caterpillars to discriminate between chemically different deterrents

J. Alan A. Renwick and Kimberly Lopez, Experience-based food consumption by larvae of *Pieris rapae*: addiction to glucosinolates?

Ruurd de Jong & Erich Städler, The role of olfaction in host-plant acceptance by the cabbage root fly, *Delia radicum* (Diptera: Anthomyiidae)

L.J. Wadhams, J.A. Bennison, M.A. Birkett, A. Hooper, M. Manjunatha, I.S. Segers, J.A. Pickett, E.M. Pow & C.M. Woodcock, Responses of western flower thrips to flower volatiles

**Poster presentations**

Stefan Mikus & Manfred Kaib, The importance of contact chemoreceptors during food choice by *Schedorhinotermes lamanianus* (Isoptera: Rhinotermitidae): an ethological and neurophysiological approach

C. P. Keathley, D.A. Potter & R.L.Houtz, Freezing-altered palatability of Bradford pear to Japanese beetle: evidence for decompartmentalization and enzymatic degradation of feeding deterrents

Nicole Kalberer, Ted Turlings & Martine Rahier, The effect of plant volatiles in host plant discrimination by an alpine leaf beetle, *Oreina cacaliae* (Coleoptera: Chrysomelidae)

Julio C. Rojas & Tristram D. Wyatt, Host-finding behaviour of the cabbage moth: effects of long and short range host odour cues

Florian P. Schiestl, Manfred Ayasse, Hannes F. Paulus, Christer Loefstedt, Bill Hansson, Stefan Francke & Wittko Francke, Male attracting odour compounds in *Andrena nigroaenea* (Hymenoptera: Apoidea) and in the sexually deceptive orchid *Ophrys sphegodes*

E. Thibout, D. Pierre, J-C. Biemont, J. Pouzat & J. Auger, Host-finding by the monophagous tephritid *Platyparea poeciloptera,* the asparagus fly

A. Sen, Neural basis of feeding behavior in *Rhapidopalpa foevicollis* (Coleoptera: Chrysomelidae)

Charles Marriott, Alison N. Hulme & K. Andrew Evans, Isolation and identification of semiochemicals from wheat attractive to larvae of the wheat bulb fly

M. Solinas & M. Rebora, Tarsal chemosensilla of the olive fruit fly, *Bactrocera oleae* Gmel. (Diptera: Tephritidae)

Kiyoshi Asaoka, Taste in the 'non-preference' mutant of *Bombyx mori*, NPS

Stefan Schütz, Bernhard Weiábecker & Hans E. Hummel, Potato plants infested by *Phytophtora infestans* show increased attractivity to *Leptinotarsa decemlineata*

Frank Hanson, Carrie McManaman, Michelle Olaes & Lisa Pierre, Binary mixtures as feeding stimuli for *Manduca*: synergism and suppression

**Session 3: Integrative studies of insect behaviour**

**Oral presentations**

David Raubenheimer & S. J. Simpson, Integrating nutrition: a geometrical approach

Elizabeth Bernays & Daniel Funk, Do specialists make faster decisions than generalists? A test with aphids

Sanford D. Eigenbrode, Nelson Kabalo & Reinhard Jetter, Adhesion by an insect predator to plant waxes: principles and ecological implications

Glen Powell, Sangita P. Maniar, John A. Pickett & Jim Hardie, Aphid responses to non-host epicuticular lipids

Sylvie Derridj, G. Touraud & A. Kollmann, Glucosinolates on the leaf surface of *Brassica napus* L., and their effects on egg-laying by *Ostrinia nubilalis* (Lepidoptera)

W. Fred Tjallingii & Beata Gabryś, Hesitant phloem access by two aphids?

G.A. Wright, S.J. Simpson, D. Raubenheimer & P. Stevenson, Application of a taste model to a field study of the feeding behaviour of the weevil, *Exopthalmus jekelanus*, on the tropical tree, *Cedrela odorata*

Lindsay Barton-Browne & David Raubenheimer, Ontogenetic changes in growth and feeding by larvae of *Helicoverpa armigera* (Lepidoptera: Noctuidae)

Toni M. Withers, Examining the hierarchy threshold model in a no-choice feeding assay

C.B. Hoffmann-Campo, J. A. Ramos Neto, M. C. N. Oliveira & L. J. Oliveira, Behavioural and physiological effects of rutin on *Anticarsia gemmatalis*

**Poster presentations**

Flavio Roces, The control of crop filling in the nectar-feeding ant *Camponotus rufipes*

Flavio Roces, Energy expenditure during harvesting and foraging decisions in leaf-cutting ants

M. Kielkiewicz, Structures of tomato leaf surface in response to the carmine spider mite (*Tetranychus cinnabarinus* Boisd.)

M. Kielkiewicz & D. Goszczyñska, The ethylene production by tomato plants infested by the carmine spider mite, (*Tetranychus cinnabarinus* Boisd.)

Beata Gabryś & Magdalena Pawluk, The acceptability of different species of Brassicaceae as hosts for the cabbage aphid *Brevicoryne brassicae* (L.) (Homoptera, Aphididae)

Jan Scheirs & Luc De Bruyn, The influence of monocotyl leaf anatomy on the feeding pattern of a grass-mining agromyzid (Diptera)

Claudio C. Ramírez & H.M. Niemeyer, The effect of the experience on host selection and feeding by a cereal aphid

Claudio C. Ramírez & H.M. Niemeyer, Salivation into the sieve elements in relation to plant chemistry: the case of the aphid *Sitobion fragariae* (Walker) and the winter wheat *Triticum aestivum* L.

Stuart Warwick, David Raubenheimer & Stephen J. Simpson, The regulation of nuptial gift consumption in*Gryllodes sigillatus*: an exceptionally long meal?

Peter East & Gillian Colebatch, Organisation of digestion in the mirid bug *Creontiades dilutus*

Stephen A. Jones & David Raubenheimer, Mechanisms of nutritional compensation in the German cockroach, *Blattella germanica*

Scott M. Gouveia, David Raubenheimer & Stephen J. Simpson, Metabolic costs associated with processing plant allelochemicals by generalist and specialist locust species

## Session 4: Insect behaviour in tritrophic systems

### Oral presentations

Marcel Dicke, Arthropod behaviour in tritrophic interactions

Ted C. J. Turlings, Sandrine Gouinguené & Thomas Degen, Variability in herbivore-induced odour emissions in plants: does it leave room for specific signals?

Felix Wäckers, Herbivory induces extrafloral nectar production: optimization of indirect defence?

G. M. Poppy & T.H. Schuler, Tritrophic interactions on insect-resistant transgenic plants

Roel P.J. Potting, Nicolene E. Vermeulen & Des E. Conlong, Active defence of herbivorous hosts against parasitism: adult parasitoid mortality risk involved in attacking a concealed stemboring host

Letizia Mattiacci, Ellen Hütter & Silvia Dorn, The influence of plant odours on host handling and reproduction of a parasitoid of *Cydia pomonella*

Torsten Meiners & Monika Hilker, The influence of plant odours on host handling and reproduction of a parasitoid of *Cydia pomonella*

G. M. Gurr & L. R. Baggen, Flowers in tri-trophic systems: mechanisms allowing selective exploitation by insect natural enemies for conservation biological control

### Poster presentations

Sam Elliot, Maurice Sabelis, Arne Janssen & Leo van der Geest, Can plants use arthropod pathogens as bodyguards?

Torsten Meiners & Monika Hilker, Specifity of kairomones for two egg parasitoids of Chrysomelidae

Akiko Fukui & Takayuki Ohgushi, Gall-making aphids increase impacts of stink bugs on the mountain ash

P. Garcia & L. Oliveira, Host plant selection by egg parasitoids: avoidance of interspecific competition?

H.A. Bell, J.P. Edwards, J.A.Gatehouse& A.M.R.Gatehouse, The effect of snowdrop lectin (GNA) delivered via artificial diet and transgenic plants on *Eulophus pennicornis* (Hymenoptera: Eulophidae), a parasitoid of the tomato moth *Lacanobia oleracea*

T. H. Schuler, G. M. Poppy, I. Denholm, L. C. Birnie, K. Vahed, P. T. Lynch, D. J. James & B. R. Kerry, Interactions between transgenic plants and beneficial insects

Eduardo Fuentes-Contreras & Hermann M. Niemeyer, Indirect effects of wheat resistance on population dynamics of parasitoids of cereal aphids

G. M. Gurr, P. C. E. Bailey, L.R. Baggen, C. Fernando & J. Gooderham, Behaviour of *Copidosoma koehleri* (Hymenoptera: Encyrtidae) in a tritrophic system with *Phthorimaea operculella* (Lepidoptera: Gelechiidae) and solanaceous plants

**Session 5: Plant defences**

**Oral presentations**

M. Rahier, P. Ballabeni & D. Conconi, Host use and chemical defence against natural enemies in related *Oreina* leaf beetle species (Coleoptera, Chrysomelinae)

K. Konno, C. Hirayama, H. Yasui, S. Okada, M. Nakamura & H. Shinbo, Glycine in digestive juice: a defence molecule of herbivorous insects against plant phenolics activated by plant enzyme

Maarten A. Jongsma, Paul Bijkerk, Willem Stiekema & Dirk Bosch, Regulation of gut serine protease genes of *Spodoptera exigua* in response to plant protease inhibitors

M.H. Pham-Delègue, A.L. Picard-Nizou, M. Le Métayer, C. Girard, M. Bonnadé-Bottino, O. Pons, C. Hennequet & L. Jouanin, Effects of transgenic plants expressing proteinase inhibitors on honeybees

G. Guillet, C. Podeszfinski, C. Regnault-Roger, J.T. Arnason & B.J.R. Philogene, Behavioural and biochemical adaptations of generalist *vs* specialist herbivorous insects on *Hypericum perforatum*

Elspeth Bartlet, Guy Kiddle, Andrew Martin, Ingrid Williams & Roger Wallsgrove, Wound-induced increases in the indolyl glucosinolate content of oilseed rape and their effect on subsequent herbivory

Christian Kjær & Jens Kvist Nielsen, Impacts of herbicide drift and deposition on plant defence

Robert Bradburne & Richard Mithen, A single gene in *Brassica oleracea* that controls the attraction of the aphid parasitoid *Diaretiella rapae* to Brassicas and oilseeds

V. Ossipov, E. Haukioja, S. Ossipova& K. Pihlaja, Defensive role of birch leaf tannins against insects

Nico J. de Boer, Pyrrolizidine alkaloid distribution in *Senecio jacobaea* rosettes minimizes losses to generalist feeding

P. Scutareanu, J. Boon, G. v. d. Doelen, A. Johanssona & M. W. Sabelis, Systemic and nonsystemic induced responses to psyllids damage in three pear tree varieties

**Poster presentations**

Sueli S. Martinez & H. F. van Emden, Effect of sublethal concentrations of azadirachtin on the development of *Spodoptera littoralis*

Lise Jouanin, Francesca De Leo, Michel Bonadé-Bottino, Luigi Ceci & Raphaele Gallerani, Opposite effects on *Spodoptera littoralis* larvae of low and high expression levels of trypsin proteinase inhibitor in transgenic plants

Cécile Girard & Lise Jouanin, Cloning and characterization of digestive enzymes in a cruciferous-feeding pest, *Pheadon cochleariae* (Coleoptera: Chrysomelidae)

Linsey Stapley, The interaction of thorns and symbiotic ants as an effective defence mechanism of swollen-thorn acacias

Wolf Haberer & Susanne Dobler, Sequestration of pyrrolizidine alkaloids in some species of the genus *Longitarsus* (Chrysomelidae: Alticinae)

P-A. Calatayud, B. Delobel, J. Guillaud& Y. Rahbé, A new phenolic compound detected in the cassava mealybug *Phenacoccus manihoti*

A. Urbanska, B. Leszczynski & H.Matok, Effect of cereal phenolics on probing behaviour and enzymatic defence of grain aphid

S. Ossipova, V. Ossipov& E. Haukioja, Proanthocy anidins of birch leaves: chemical properties and effect on plant – insect relationships

A. Couty, G. Poppy, W. Powell & A.M.R Gathehouse, Direct and indirect effects of transgenic potatoes expressing the snowdrop lectin (GNA) on the aphid parasitoid, *Aphidius ervi*

Meena Haribal & J. Alan A. Renwick, Variation in *Alliaria petiolata* flavonoids that affect the feeding behavior of larvae of *Pieris napi oleracea*

Jean-Luc Boev & Erik Jacobs, Sequestration of iris metabolites by larvae of the sawfly *Rhadinoceraea micans*

Thomas Degen, Sandrine Gouinguené & Ted C. J. Turlings, Herbivore-induced volatile emission by maize plants: dose-response curves obtained with different induction methods

Sandrine Gouinguené, Thomas Degen & Ted Turlings, Induction of odour emissions in maize plants by different larval instars of *Spodoptera littoralis* (Lepidoptera: Noctuidae)

Caroline Müller & Monika Hilker, Do tansy-feeding tortoise beetle larvae use repellents from their host plant for defence against predators?

**Session 6: Insect life histories and plant growth responses**

**Oral presentations**

Sören Nylin, Life history perspectives on insect-plant interactions

Eddy van der Meijden & Peter G.L. Klinkhamer, Conflicting interests of plants and the natural enemies of herbivores

Martin J. Steinbauer, *Amorbus dallas* species (Hemiptera: Coreidae) in Australia: latent or eruptive forest herbivores of eucalypts?

E. Elle, N. M. van Dam & J. D. Hare, Cost of herbivore resistance in *Datura wrightii*

David Wool, Roni Aloni, Ofra Ben-Zvi & Miriam Wollberg, A galling aphid furnishes its home with a built-in pipeline to the host food supply

Luc De Bruyn, Growth responses of *Phragmites australis* (Poaceae) to gall formation by *Lipara lucens* (Diptera, Chloropidae)

Robert K. Vander Meer & Robert G. Shatters, Jr, The effect of fire ants on soybean germination and growth

**Poster presentations**

Parviz Shishehbor & P. A. Brennan, Demographic analysis of castor whitefly, *Trialeurodes ricini* Misra, on four castor cultivars

John C. Reese, Ken Kofoid, Diane Lapp & Esau Formusoh, Recurrent selection for sorghum tolerance to greenbug (Homoptera: Aphididae) feeding damage

S.S.Krishna, Impact of *Allium sativum* (L.) bulb volatiles on the biology of *Earias vittella* (F.) (Lepidoptera: Noctuidae)

Rinny E. Kooi & Paul M. Brakefield, Effect of food on the reaction norm for wing pattern induction in the tropical butterfly *Bicyclus anynana*

B. Jozwiak, B. Leszczynski & A.F.G. Dixon, Variation in content of amino acids during host-plant alternation of bird cherry-oat aphid

Moshe Inbar, Hamed Doostdar & Richard T. Mayer, Effect of host plant growing conditions on insect herbivores: a comparison between feeding guilds

Bernd F. Hägele & Stephen J. Simpson, Phase and hatching time differ vertically within desert locust *(Schistocerca gregaria)* egg pods: implications for the evolution of gregaristion

Martijn Egas, Maurice W. Sabelis & Steph B.J. Menken, Host specialisation in phytophagous arthropods: when will general flexibility pay off?

E. Despland & Stephen J. Simpson, Effects of small-scale vegetation structure on desert locust gregarisation

L. J. Oliveira, M. A. Garcia, M. C. N. de Oliveira & C. B. Hoffmann-Campo, Offspring performance of *Pyllophaga cuyabana* on different plants

**Session 7: Choosing host plants: mechanisms and evolution**

**Oral presentations**

Bernard D. Roitberg, On the evolution of herbivore oviposition behaviours: Vive la variance!

Marion O. Harris & M. Sandanayake, How good is the relationship between oviposition behaviour and offspring survival and growth in the hessian fly?

Ronald J. Prokopy, Socially facilitated oviposition behaviour in tephritid fruit flies

D.A. Potter & D. Held, Geranium-induced intoxication and absence of food-aversion learning in a polyphagous insect, the Japanese beetle (*Popillia japonica* Newman)

Paul Feeny & Meena Haribal, Oviposition stimulant for the zebra swallowtail butterfly, *Eurytides marcellus*, from *Asimina triloba*

Rufus Isaacs, Matthew Cahill & David N. Byrne, Behavioural insight into mechanisms of host choice in whitefly

Stefan Schütz, Bernhard Weiábecker & Hans E. Hummel, Impact of oxidative stress on host plant finding and host plant choice by *Leptinotarsa decemlineata*

G. G. Grant, D. Langevin & B. Zhao, Contrasting effects of phenolic acids and flavonoids versus aromatic and aliphatic acids on oviposition of the spruce budworm, *Choristoneura fumiferana* (Lepidoptera: Tortricidae)

**Poster presentations**

Hiroo Kanno & Marion Harris, Oviposition behaviour of the hessian fly, *Mayetiola destructor* (Diptera: Cecidomyiidae): the role of structural features on the leaf surface

R.W.H.M. van Tol, J.H. Visser& M.W. Sabelis, Semiochemicals of the black vine weevil (*Otiorhynchus sulcatus*)

L.D. Galanihe & M. O. Harris, Plant volatiles mediate host-finding behaviour of the apple leafcurling midge

D. Crook, J. Cross, R. Brennan, N. Birch & A. J. Mordue (Luntz), Interactions of the blackcurrant leaf curling midge, *Dasineura tetensi* (Diptera: Cecidomyiidae) with resistant and susceptible plant cultivars

Michihiro Ishihara, Host-plant selection by the willow leaf beetle, *Plagiodera versicolora*

Peter Anderson & Hans Alborn, Oviposition response of the moth, *Spodoptera littoralis*, to cotton plants subjected to herbivory by conspecific larvae

Takashi A. Inoue, Kazuaki Seta, Sawako Niki, Kanako Hashimoto & Kiyoshi Asaoka, The response of female butterflies of *Papilio memnon* to the extracts of a host and 7 non-host Rutaceae plants

**Session 8: Microbial partners in insect-plant associations**

**Oral presentations**

Nancy A. Moran, The evolution of aphid endosymbionts

Alex Mira & David C. Krakauer, Bottlenecks in endosymbiont transmission: a mechanism for reducing deleterious mutations is presented

Andrew J. Storer, David L. Wood & Thomas R. Gordon, Effects of ecological disturbance, caused by an introduced tree pathogen, on co-evolved insect-plant relationships

Dereje Dugassa-Gobena & Stefan Vidal, Biochemical aspects of the systemic interaction between a fungal endophyte and herbivorous insects

T.L. Wilkinson & H. Ishikawa, Assimilation and partitioning of nutrients by pea aphids

**Poster presentations**

J. Sandström & N. Moran, How imbalanced is phloem sap for aphids?

Chin Yu, Probing the structure of the $\alpha5$ subunit of the insecticidal $\delta$-endotoxin using two-dimensional proton NMR

Anja Hildebrands & Stefan Vidal, Factors influencing the host evaluation and host suitability of members of the *Aphis fabae*-complex for a polyphagous parasitoid, *Lysiphlebus testaceipes* (Hymenoptera: Aphidiidae)

**Session 9: Genetic bases of host-plant associations**

**Oral presentations**

Jeffrey L. Feder, It's about time: the evidence for host plant-mediated selection in the apple maggot fly

Peter W. de Jong & Jens K. Nielsen, The use of near-isogenic flea beetle lines in studies of insect host- plant relationships

P. Ballabeni & M. Rahier, Quantitative genetics of larval performance of the leaf beetle *Oreina elongata* on two normal hosts

Joakim Hjältén, Resistance of hybrid and parental willows to herbivores and pathogens: the role of plant secondary chemistry

J. Mark Scriber, Kirsten Weir, Dylan Parry & Jessica Deering, Using hybrid and backcross larvae of *Papilio canadensis* and *P. glaucus* to detect induced chemical resistance in hybrid poplar trees experimentally defoliated by gypsy moths

Steffi Knoll & Martine Rahier, Spatial population structure of the alpine leaf beetle *Oreina cacaliae* (Coleoptera, Chrysomelidae)

Marina C. Caillaud & Sara E. Via, Genetic architecture of host-plant specialization in two incipient species

**Poster presentations**

Y. Ishikawa, T. Takanashi & Y. Huang, *Ostrinia* spp. in Japan: their host plants and sex pheromones

Hideki Ueno, Naoyuki Fujiyama, Kiyoshi Irie, Yuji Sato & Haruo Katakuru, Genetic basis for normal and novel host plant use in a herbivorous ladybird beetle, *Epilachna vigintioctomaculata*

Steffi Knoll , M. Hossaert-McKey & D. McKey, Evolution of plant-insect associations: linking plant demography, insect demography and insect population structure in the *Sesbania-eudiagogus* system

**Session 10: The phylogeny of insect-plant associations**

**Oral presentations**

Brian Farrell, The phylogeny of insect-plant associations

Susanne Dobler, Josef Müller & Claudia Gack, How to become a vegetarian? - from meat to beets in silphid beetles

Peter Roessingh, Katja Hora, Steph Menken& Joop van Loon, Speciation in small ermine moths: investigating the role of the sugar-alcohols ducitol and sorbitol

George Beccaloni & Fernley Symons, Diet breadth of butterfly larvae in relation to host plant growth form: an analysis of two faunas

Klaas Vrieling & Nico de Boer, Host-plant choice and larval growth in the cinnabar moth: do pyrrolizidine alkaloids play a role?

**Poster presentation**

Susanne Dobler & Wolf Haberer, Evolution of host affiliation in *Longitarsus* leaf beetles - plant compounds as guide posts and life history as a constraint?

**Closing address**

Reginald F. Chapman

# List of registered participants

Inger Ahman, Nilsson-Ehle Laboratory, Svalof Weibull AB, Svalov SE-26881, Sweden
*E-mail: inger.ahman@swseed.se*

Peter Anderson, Dept. of Plant Protection Sciences, Swedish Univ. of Agriculture, Box 44, Alnarp S-23053, Sweden
*E-mail: peter.anderson@vsv.slu.se*

Kiyoshi Asaoka, Dept. of Insect Physiology and Behaviour, National Inst. of Sericultural & Entomological Science, 1-2 Ohwashi, Tsukuba 305-8634, Japan
*E-mail: asaoka@nises.affrc.go.jp*

Pierluigi Ballabeni, Inst. de Zoologie, Univ. de Neuchâtel, Rue Emile Argand 11, Neuchâtel CH-2007, Switzerland
*E-mail: pierluigi.ballabeni@zool.unine.ch*

Elspeth Bartlet, Dept. of Entomology & Nematology, IACR Rothamsted, Rothamsted Experimental Station, Harpenden, AL5 2JQ, UK
*E-mail: ebartlet@bbsrc.ac.uk*

Lindsay Barton Browne, Dept. of Entomology, CSIRO, Private Bag No 3, PO Indooroopilly, Brisbane QLD, 4068, Australia
*E-mail: lindsay.barton.browne@brs.ento.csiro.au*

George Beccaloni, Dept. of Entomology, The Natural History Museum, Cromwell Rd., South Kensington, London SW7 5BD, UK
*E-mail: gwb@nhm.ac.uk*

Spencer Behmer, Dept. of Zoology, Oxford Univ., South Parks Rd., Oxford OX1 3PS, UK
*E-mail: spencer.behmer@zoo.ox.ac.uk*

Howard Bell, Pest Management Studies, Central Science Laboratory, Sand Hutton, York YO41 1LZ, UK
*E-mail: h.bell@csl.gov.uk*

Steve Belmain, Food Security Dept., Natural Resources Inst., Central Av., Chatham Maritime, Kent ME4 4TB, UK
*E-mail: s.r.belmain@gre.ac.uk*

Elizabeth Bernays, Dept. of Entomology, Univ. of Arizona, Forbes 410, Tucson, AZ 85721, USA
*E-mail: schistos@ag.arizona.edu*

Jean-Luc Boeve, Dept. Entomologie, IRSNB, Rue Vautier, Brussels B-1000, Belgium
*E-mail: boeve@kbinirsnb.be*

Steve Bonnage, ADC BioScientific, 12 Spurling Works, Pindar Rd., Hoddesdon, Herts, EN11 0DB, UK

Robert Bradburne, Dept. of Brassica & Oilseeds, John Innes Centre, Cohney Lane, Norwich NR4 7UH, UK
*E-mail: bradburne@bbsrc.ac.uk*

Toby Bruce, Dept. of Pest Management, Natural Resources Inst., Central Av., Chatham ME4 4TB, UK
*E-mail: btll@gre.ac.uk*

Bill Budenberg, Tracksys Ltd., Vernon House, 18 Friar Lane, Nottingham, NG1 6DQ, UK

Guy Bush, Dept. of Zoology, Michigan State Univ., East Lansing, MI 48824-1115, USA
*E-mail: bushfly@pilot.msu.edu*

Marina Caillaud, Dept. of Entomology, Cornell Univ., Comstock Hall, Ithaca, NY 14853, USA
*E-mail: cmc27@cornell.edu*

Paul-Andre Calatayud, International Centre of Tropical Agriculture, Cali AA 6713, Columbia
*E-mail: p.calatayud@cgnet.com*

Reginald Chapman, ARL Division of Neurobiology, Univ. of Arizona, , P.O.Box 210077, Tucson, AZ-85721-0077, USA
*E-mail: chapman@neurobio.arizona.edn*

Desmond Conlong, Dept. of Entomology, South African Sugar Association Exp. Station, P. Bag X02, Mount Edgecombe 4300, South Africa
*E-mail: xentdc@sugar.org.za/conlong@iafrica.com*

Aude Couty, Dept. of Entomology & Nematology, IACR Rothamsted, Rothamsted Experimental Station, Harpenden, AL5 2JQ, UK
*E-mail: aude.couty@bbsrc.ac.uk*

Damon Crook, Dept. of Zoology, Univ. of Aberdeen, Dept. of Entomology, HRI, East Malling, Kent ME19 6BJ, UK
*E-mail: damon.crook@hri.ac.uk*

Nico de Boer, Dept. of Plant Ecology, I.E.E.S. Leiden Univ., P.O. Box 9516, Leiden NL-2300 RA, The Netherlands
*E-mail: sbu4nb@rulsfb.leidenuniv.nl*

Luc de Bruyn, Dept. of Biology, Evolutionary Biology Grp, Univ. of Antwerp (RUCA), Groenenborger Laan 171, Antwerp 2020, Belgium
*E-mail: debruyn@ruca.ua.ac.be*

Kitty de Jager, Nunhems Zaden b.v., P.O. Box 4005, Haelen 6080 AA, The Netherlands
*E-mail: k.de.jager@nunhecom*

Ruurd de Jong, Swiss Federal Research Station, Wädenswil CH - 8820, Switzerland
*E-mail: ruurd.de.jong@mbox.faw.admin.ch*

Thomas Degen, Inst. of Zoology, Univ. of Neuchâtel, Rue Emile Argand 11, Neuchâtel CH-2007, Switzerland
*E-mail: thomas.degen@zool.unine.ch*

Sylvie Derridj, Dept. of Phytopharmacy, INRA, Route de St Cyr, Versailles Cedex 78026, France
*E-mail: derridj@versailles.inra.fr*

Emma Despland, Dept. of Zoology, Oxford Univ., South Parks Rd., Oxford OX1 3PS, UK
*E-mail: edesplan@jesus.ox.ac.uk*

Marcel Dicke, Dept. of Entomology, Wageningen Agricultural Univ., P.O. Box 8031, Wageningen 6700 EH, The Netherlands
*E-mail: marcel.dicke@medew.ento.wau.nl*

Susanne Dobler, Dept. of Biology I, Universität Freiburg, Hauptstr. 1, Freiburg D-79104, Germany
*E-mail: doblers@ruf.uni-freiburg.de*

Angela Douglas, Dept. of Biology, Univ. of York, PO Box No 373, York, UK
*E-mail: aed2@york.ac.uk*

Rachel Drake, Dept. of Plant Biology, Zeneca Agrochemicals, Jealott's Hill Research Station, Bracknell, Bracknell RG42 6ET, UK
*E-mail: rachel.drake@aguk.zeneca.com*

Dereje Dugassa-Gobena, Inst. for Plant Disease and Plant Protection, Hannover Univ., Herrenhäuser Str.2, Hannover 30419, Germany
*E-mail: dugassa@mbox.lpp.uni-hannover.de*

Peter East, Dept. of Biotechnology, CSIRO Entomology, GPO Box 1700, Canberra ACT 2601, Australia
*E-mail: peter.east@ento.csiro.au*

Martijn Egas, Inst. of Systematics and Population Biology, Univ. of Amsterdam, Kruislaan320, Amsterdam SM 1098, The Netherlands
*E-mail: egas@bio.uva.nl*

Sanford Eigenbrode, Dept. of Plant, Soil & Entomological Sciences, Univ. of Idaho, Moscow, ID 83844-2339, USA
*E-mail: sandforde@uidaho.edu*

Simon Elliot, Section Population Biology, Inst. for Systematics and Population Biology, Univ. of Amsterdam, Kruislan 320, Amsterdam 1098 SM, The Netherlands
*E-mail: elliot@bio.uva.nl*

Brian Farrell, Museum of Comparative Zoology, Harvard Univ., Cambridge, Cambridge MA 02138, USA

Jeffrey Feder, Dept. of Biological Sciences, Univ. of Notre Dame, Notre Dame IN 46556, USA
*E-mail: feder.2@nd.edw*

Paul Feeny, Dept. of Ecology and Systematics, Cornell Univ., Corson Hill, Ithaca, NY 14853, USA
*E-mail: ppf1@cornell.edu*

Elena-Maria Fritzsche, Inst. of Zoology, Univ. of Neuchâtel, Rue Emile Argand 11, Neuchâtel CH-2007, Switzerland
*E-mail: maria.fritzsche@zool.unine.ch*

Eduardo Fuentes-Contreras, Dept.o de Ciencias Ecologicas, Universidad de Chile, Casilla 653, Santiego, Chile
*E-mail: leche@abulfia.ciencias.uchile.cl*

Akiko Fukui, Inst. of Low Temperature Science, Hokkaido Univ., Nishi 8, Kita 19, Kita-Ku, Sapporo 060-0819, Japan
*E-mail: aki@pop.lowtem.hokudai.ac.jp*

Beata Gabryś, Dept. of Agricultural Entomology, Agricultural Univ., Cybulskiego 32, Wroclaw 50-205, Poland
*E-mail: begab@ozi.ar.wroc.pl*

Patricia Ventura Garcia, Biologia Dept., Acores Univ., Rua da Mae de Deus 58, Ponta Delgada 9500, Portugal
*E-mail: patricia@alf.uac.pt*

Cecile Girard, Dept. Biologie Cellulaire, INRA Versailles, Route de St Cyr, Versailles F-78026, France
*E-mail: girard@versailles.inra.fr*

Sandrine Gouinguene, Laboratory of Animal Ecology & Entomology, Univ. of Neuchâtel, Rue Emile Argand 11, Neuchâtel CH-2007, Switzerland
*E-mail: sandrine.gouinguene@zool.unine.ch*

Scott Gouveia, Dept. of Zoology, Univ. of Oxford, Room E23, South Parks Rd., Oxford OX1 3PS, UK

Gary Grant, Canadian Forest Service, P.O. Box 490, 1219 Queen St. E., Sault Ste. Marie P6A 5M7, Canada
*E-mail: ggrant@nrcan.gc.ca*

Geoff Gurr, Dept. of Plant Sciences, Orange Agricultural College, The Univ. of Sydney, P.O. Box 883, Orange NSW 2800, Australia
*E-mail: geoff.gurr@oac.usyd.edu.au*

Wolf Haberer, Dept. of Biologie I, Univ. Freiburg, Hauptstr.1, Freiburg D-79104, Germany
*E-mail: haberer@ruf.uni-freiburg.de*

Bernd F. Hägele, Dept. of Zoology, Univ. of Oxford, South Parks Rd., Oxford OX1 3PS, UK
*E-mail: bernd.hagele@zoo.ox.ac.uk*

Lars Monrad Hansen, Dept. of Crop Protection, Danish Inst. of Agricultural Sciences, Research Centre, Flakhebjerg, Slagelse DN-4200, Denmark
*E-mail: lmh@aps.sd.dk*

Frank Hanson, Dept. of Biology, UMBC, 1000 Hillcrest Circle, Baltimore, MD 21250, USA
*E-mail: hanson@umbc.edu*

Jim Hardie, Dept. of Biology, Imperial College, Silwood Park, Ascot, Berks SL5 7PY, UK
*E-mail: j.hardie@ic.ac.uk*

Daniel Hare, Dept. of Entomology, Univ. of California, Riverside, CA 92521, USA
*E-mail: harejd@citrus.ucr.edu*

Meena Haribal, Boyce Thompson Inst. / Ecology & Systematics, Cornell Univ., Tower Rd., Ithaca, NY 14853, USA
*E-mail: mmh3@cornell.edu*

Marion Harris, Insect Science Group, Hort Research, Private Bag 92169, Auckland, New Zealand
*E-mail: mharris@hort.cri.nz*

John Hildebrand, ARL Division of Neurobiology, Univ. of Arizona, P.O.Box 210077, Tucson, AZ 85721-0077, USA
*E-mail: jgh@neurobio.arizona.edu*

Anja Hildebrands, Inst. of Plant Diseases and Plant Protection, Univ. of Hannover, Herrenhauser Str.2, Hannover 30419, Germany
*E-mail:*
*hildebrands@mbox.ipp.uni-hannover.de*

Monika Hilker, Inst. für Zoologie, Freie Universität Berlin, Haderslebenerstr. 9, Berlin 12163, Germany
*E-mail: Hilker@zedat.fu-berlin.de*

Joakim Hjalten, Dept. of Animal Ecology, Swedish Univ. of Agricultural Sciences, Umea S-90183, Sweden
*E-mail: joakim.hjalten@szooek.slu.se*

Clara Beatriz Hoffmann-Campo, Dept. of Entomology, Embrapa Soja, Caixa Postal 231, Londrina PR 86001-970, Brazil
*E-mail: hoffmann@cnpso.embrapa.br*

Paul Innocenzi, Dept. of Pest Management, Natural Resources Inst., Univ. of Greenwich, Central Av., Chatham ME4 4TB, UK

Takashi Inoue, Dept. of Neurophysiology, National Inst. of Sericultural & Entomological Science, Ibaraki, Tsukuba, Ohwashi, Nises 305-8634, Japan
*E-mail: inoueatp@ss.nises.affrc.go.jp*

Rufus Isaacs, Dept. of Entomology, Michigan State Univ., 203, Pesticide Research Center, East Lansing, MI 48824, USA
*E-mail: isaacsr@pilot.msu.edu*

Michihiro Ishihara, Inst. of Low Temperature Science, Hokkaido Univ., Nishi 8, Kita 19, Kita-Ku, Sapporo 060-0019, Japan
*E-mail:*
*mishiha@orange.lowtem.hokudai.ac.jp*

Yukio Ishikawa, Laboratory of Applied Entomology, Graduate School of Agriculture & Life Sciences, Univ. of Tokyo, Bunkyo-Ku, Tokyo 113-8657, Japan
*E-mail: ayucky@hongo.ecc.u-tokyo.ac.jp*

Stephen Jones, Dept. of Zoology, Univ. of Oxford, South Parks Rd., Oxford OX1 3PS, UK
*E-mail: stephen.jones@zoo.ox.ac.uk*

Maarten Jongsma, Dept. of Molecular Biology, CPRO-DLO, P.O. Box 16, Wageningen 6700 AA, The Netherlands
*E-mail: m.a.jongsma@cpro.dlo.nl*

Lise Jouanin, Dept. Biologie Cellulaire, INRA Versailles, Route de St Cyr, Versailles F-78026, France
*E-mail: jouanin@versailles.inra.fr*

Nicole Kalberer, Dept. of Animal Ecology & Entomology, Univ. of Neuchâtel, Inst. of Zoology, Rue Emile Argand 11, Neuchâtel 2007, Switzerland
*E-mail: nicole.kalberer@zool.unine.ch*

Hiroo Kanno, Dept. of Insect Science, Hort Research, Mt. Albert Research Centre, Private

Bag, Auckland 92169, New Zealand
*E-mail: hkanno@hort.cri.nz*

Cathy Kennedy, Oxford University Press, Great Clarendon Street, Oxford , OX2 6DP, UK
*E-mail: kennedyc@oup.co.uk*

Margaret Kielkiewicz, Dept. of Applied Entomology, Warsaw Agricultural Univ., Nowoursynowska 166, Warsaw 02 - 787, Poland
*E-mail: kielkiewicz@alpha.sggw.waw.pl*

Christian Kjaer, Dept. of Terrestrial Ecology, National Environmental Research Inst., Vejlsovej 25, Silkeborg DK-8600, Denmark
*E-mail: ckj@dmu.dk*

Paula Knighton, Desert Demonstration Garden, 504 Calypso Dr., Henderson, NV 89015, USA
*E-mail: knighton@vegas.infi.net*

Stephanie Knoll, Equipe Coevolution CEFE/CNRS, CEFE/CNRS, 1919 Route de Mendex, Montpellier Cedex 5, F-34293, France
*E-mail: knoll@cefe.cnrs-mop.fr*

Kotaro Konno, National Inst. of Sericultural & Entomological Science, 1-2 Ohwashi, Tsukuba, Ibaraki 305-8634, Japan
*E-mail: konno@nises.affrc.go.jp*

Rinny E. Kooi, Evolutionary Biology, Univ. of Leiden, P.O. Box 9516, Leiden 2300 RA, The Netherlands
*E-mail: kooi@rulsfb.leidenuniv.nl*

Gilles Laurent, California Inst. of Technology, Biology Division, 139-74 Pasadena, CA 91125, USA
*E-mail: laurentg@cco.caltech.edu*

Bogumil Leszczynski, Dept. of Biochemistry, Agricultural & Pedagogic Univ., UL. B Prusa 12, Siedlce PL-08110, Poland
*E-mail: leszczb@wsrp.siedlce.pl*

Richard Lilley, Dept. of Entomology, Horticultural Research Int, East Malling, Kent ME19 6BJ, UK

Jennifer Jund, Dept. of Zoology, Univ. of Oxford, South Parks Rd., Oxford OX13PS, UK
*E-mail: jennifer.lund@oriel.ox.ac.uk*

Charles Marriott, Inst. of Ecology & Research Management, Univ. of Edinburgh, Kings Buildings, West Mains Rd., Edinburgh EH9 3JG, Scotland, UK
*E-mail: c.marriott@ed.sac.ac.uk*

Sueli Martinez, Area de Progao de Plantas, Inst.o Agronomico do Parana, Cx. Postal 481, Londrina PR 86001-970, Brazil
*E-mail: suemart@sercomtel.com.br*

Letizia Mattiacci, Dept. of Applied Entomology / Plant Science, Swiss Federal Inst. of Technology (ETH), Clausiusstr.25 / NW, Zürich 8092, Switzerland
*E-mail: letizia.mattiacci@ipw.agrl.ethz.ch*

Richard Mayer, Laboratory Director, USDA ARS, USHRL, 2120 Camden Rd., Orlando, FL 32803-1419, USA
*E-mail: rmayer@ix.netcom.com*

Torsten Meiners, Angewandte Zoologie / Oekologie der Tiere, Freie Universität Berlin, Haderslebenerstr.9, Berlin 12163, Germany
*E-mail: meito@zedat.fu-berlin.de*

Steph Menken, Dept. of Systematics & Ecology, Univ. of Amsterdam, P.O. Box 94766, Amsterdam 1090 GT, The Netherlands
*E-mail: menken@bio.uva.nl*

Stefan Mikus, Dept. of Animal Physiology, Univ. of Bayreuth, Bayreuth D-95440, Germany
*E-mail: Stefan.Mikus@uni-bayreuth.de*

Alex Mira, Dept. of Zoology, Oxford Univ., South Parks Rd., Oxford OX1 3PS, UK
*E-mail: mira@ermine.oxford.ac.uk*

Nancy Moran, Dept. of Ecology & Evoloutionary Biology, Univ. of Arizona, Biological Sciences West, Tucson, AZ 85721, USA
*E-mail: nmoran@u.arizona.edu*

Jennifer Mordue (Luntz), Dept. of Zoology, Univ. of Aberdeen, Tillydrone Av., Aberdeen AB24 2TZ, UK
*E-mail: a.j.mordue@abdn.ac.uk*

Caroline Müller, Inst. für Angewandte Zoologie / Oekologie der Tiere, Freie Universität Berlin, Haderslebenerstr.9, Berlin D-12163, Germany
*E-mail: asterlin@zedat.fu-berlin.de*

Jens Kvist Nielsen, Chemistry Dept., Royal Vet. & Agricultural Univ., Thorvaldsensvej 40, Frederiksberg C DK - 1871, Denmark
*E-mail: jeni@kvl.dk*

Takaaki Nishioka, Graduate School of Agricultural Sciences, Kyoto Univ., Sakyo-ku, Kyoto 606-8502, Japan
*E-mail: nishioka@kais.kyoto-u.ac.jp*

Sören Nylin, Dept. of Zoology, Stockholm Univ., Stockholm 10691, Sweden
*E-mail: snylin@zoologi.su.se*

Takayuki Ohgushi, Inst. of Low Temperature Science, Hokkaido Univ., Nishi 8, Kita 19, Kita-Ku, Sapporo 060-0189, Japan

*E-mail: ohgushi@orange.lowtem.hokudai.ac.jp*

Maria Luisa Oliveira, Biologia Dept., Acores Univ., Rua de Mae de Deus 58, Ponta Delgada 9500, Portugal
*E-mail: ziza@alf.uac.pt*

Lenita Jacob Oliveira, Dept. of Entomology, National Soybean Research Centre / Embrapa, Caixa Postal 231, Londrina 86001-970, Brazil
*E-mail: lenita@cnpso.embrapa.br*

Vladimir Ossipov, Dept. of Biology, Section Ecology, Univ. of Turku, Turku 20014, Finland
*E-mail: ossipov@utu.fi*

Svetlana Ossipova, Dept. of Biology, Section Ecology, Univ. of Turku, Turku 20014, Finland
*E-mail: ossipov@utu.fi*

Jacques Pasteels, Biologie Animale et Cellulaire, Univ. Libre de Bruxelles, CP 160/12, 50 av.F.D.Roosevelt, Brussels 1050, Belgium
*E-mail: jmpastee@ulb.ac.be*

Minh-Ha Pham-Delegue, Laboratoire Neurobiologie Componee des Invertebres, INRA, BP 23, Bures / Yvette 91440, France
*E-mail: pham@jouy.inra.fr*

Bernard Philogene, Dept. of Biology, Univ. of Ottawa, P.O. Box 450, Ottawa K1N 6N5, Canada
*E-mail: bphilog@science.uottawa.ca*

Dainius Plepys, Dept. of Ecology, Lund Univ., Animal Ecology, Ecology Building, Lund 223 62, Sweden
*E-mail: dainius.plepys@zooekol.lu.se*

Anja Pool, Dept. of Resistance, Ryk Zwaan B.V., P.O. Box 40, De Lier 2678 ZG, The Netherlands
*E-mail: a.pool@rykzwaan.nl*

Guy Poppy, Dept. of Entomology & Nematology, IACR Rothamsted Experimental Station, West Common, Harpenden AL5 2JQ, UK
*E-mail: poppy@bbsrc.ac.uk*

Daniel Potter, Dept. of Entomology, S-225 Agricultural Science Building - North, Univ. of Kentucky, Lexington, KY 40546-0091, USA
*E-mail: dapotter@ca.uky.edu*

Roel Potting, Dept. of Entomology & Nematology, IACR Rothamsted, West Common, Harpenden AL5 2JQ, UK
*E-mail: roel.potting@bbsrc.ac.uk*

Glen Powell, Dept. of Biology, Imperial College at Silwood Park, Ascot, Berks SL5 7PY, UK
*E-mail: g.powell@ic.ac.uk*

Ronald Prokopy, Dept. of Entomology, Univ. of Massachusetts, Amherst, MA 01003, USA
*E-mail: propkopy@ent.umass.edu*

Martine Rahier, Inst. de Zoologie, Univ. of Neuchâtel, Emile Argand 11, Neuchâtel CH-2007, Switzerland
*E-mail: martine.rahier@zool.unine.ch*

Claudio Ramirez, Dept.o de Ciencias Ecologicas, Univ. de Chile, Casilla 653, Nunoa, Santiego , Chile
*E-mail: claudio@abulafia.ciencias.uchile.cl*

Georg Rast, Dept. of Zoology, Technische Univ., Lichtenbergstr.4, Garching D-85747, Germany
*E-mail: gr@cip1.zoo.chemie.tu-muenchen.de*

David Raubenheimer, Dept. of Zoology, Univ. of Oxford, South Parks Rd., Oxford OX1 3PS, UK
*E-mail: david.raubenheimer@zoo.oxford.ac.uk*

John Reese, Dept. of Entomology, Kansas State Univ., Waters Hill, Manhattan, KS 66506, USA
*E-mail: jreese@oz.oznet.ksu.edu*

Alan Renwick, Boyce Thompson Inst., Tower Rd., Ithaca, NY 14853, USA
*E-mail: jar14@cornell.edu*

Flavio Roces, Dept. of Zoology, Univ. of Würzburg, Am Hubland, Zoologie II, Theodor-Bovert-Inst., Würzburg 97074, Germany
*E-mail: roces@biozentrum.uni-wuerzburg.de*

Peter Roessingh, Inst. of Systematics and Population Biology, Univ. of Amsterdam, Kruislaan 320, Amsterdam 1098 SM, The Netherlands
*E-mail: roessingh@bio.uva.nl*

Stephen Rogers, School of Biological Sciences, Univ. of Southampton, Bassett Crescent East, Southampton SO16 7PX, UK
*E-mail: smr1@soton.ac.uk*

Bernard Roitberg, Dept. of Biosciences, Simon Fraser Univ., 8888 Univ. Drive, Burnaby V5A 156, Canada
*E-mail: roitberg@sfu.ca*

Miriam Rothschild, Ashton Wold, Peterborough PE8 5LZ, UK

Jonas Sandström, Dept. of Ecology and Evolutionary Biology, Univ. of Arizona, Tucson, 85721 AZ, USA
*E-mail: jonass@u.arizona.edu*

Florian Schiestl, Dept. of Evolutionary Biology, Inst. of Zoology, Univ. of Vienna, Althanstrasse 14, Vienna A-1090, Austria
*E-mail: A8917722@unet.enivie.ac.at*

Louis Schoonhoven, Dept. of Entomology, Agricultural Univ., P.O. Box 8031, Wageningen 6700 EH, The Netherlands
*E-mail: office@adm.ento.wau.nl*

Stefan Schütz, Biological & Biotechnical Plant Protection, Justus-Liebig Univ., Ludwigstrasse 21 B, Giessen D-35390, Germany
*E-mail: stefan.schuetz@agrar.uni-giessen.de*

Mark Scriber, Dept. of Entomology, Michigan State Univ., East Lansing, MI 48824, USA
*E-mail: scriber@pilot.msu.edu*

Petru Scutareanu, Inst. of Systematics and Population Biology, Univ. of Amsterdam, Kruislaan 320, Amsterdam 1098 SM, The Netherlands
*E-mail: scutereanu@bio.uva.nl*

Alok Sen, Dept. of Entomology Research Inst., Loyola College, Chennai 600034, India
*E-mail: asen@erilc.ren.nic.in*

Stella Shackel, School of Biological & Earth Science, John Moore's Univ., Byron Street, Liverpool, L3 3AF, UK

Parviz Shishehbor, Dept. of Plant Protection, Faculty of Agriculture, Shahio Chamran Univ., Mollasani, Ahwaz, Iran

Steve Simpson, Dept. of Zoology and Univ. Museum of Natural History, Oxford Univ., South Parks Rd., Oxford OX1 3PS, UK
*E-mail: stephen.simpson@zoo.ox.ac.uk*

Mario Solinas, Agricultural Entomology Inst., Univ. of Perugia, Borges XX Giugno, 74, Perugia 06121, Italy
*E-mail: msolinas@unipg.it*

Richard Southwood, Dept. of Zoology, Oxford Univ., South Parks Rd., Oxford OX1 3PS, UK

Erich Städler, Eidg. Forschungsanstalt, Schloss 334, Wädenswil CH-8820, Switzerland
*E-mail: Erich.Staedler, faw.wae.admin.ch*

Linsey Stapley, Dept. of Zoology, Univ. of Cambridge, Downing Street, Cambridge CB2 3EJ, UK
*E-mail: ls208@cus.cam.ac.uk*

Martin Steinbauer, Co-Operative Research Centre for Sustainable Production Forestry & CSIRO Entomology, GPO Box 1700, Canberra ACT 2601, Australia
*E-mail: martin.steinbauer@ento.csiro.au*

Andrew Storer, Division of Insect Biology, Univ. of California at Berkeley, 201 Wellman Hall, Berkeley, CA 94720, USA
*E-mail: storer@nature.berkeley.edu*

Eric Thibout, Faculty of Sciences, Univ. F. Rabelais, parc de Grandmont, Tours 37200, France
*E-mail: thibout@univ-tours.fr*

Fred Tjallingii, Dept. of Entomology, Agricultural Univ. Wageningen, P.O. Box 8031, Wageningen 6700 EH, The Netherlands
*E-mail: freddy.tjallingii@medew.ento.wau.nl*

Ted Turlings, Inst. of Zoology, Univ. of Neuchâtel, Rue Emile Argand 11, Neuchâtel CH-2007, Switzerland
*E-mail: ted.turlings@zool.unine.ch*

Hideki Ueno, Laboratory of Biology, Faculty of Education, Niigata Univ., Ikarashi-Hi-No-Cho, Niigata 950-2181, Japan
*E-mail: hueno@ed.niigata-u.ac.jp*

Eddy van der Meijden, Inst. of Evolutionary & Ecological Sciences, Univ. of Leiden, P.O. Box 9516, Leiden 2300 RA, The Netherlands

Joop van Loon, Dept.of Entomology, Wageningen Agricultural Univ., P.O. Box 8031, Wageningen 6700, The Netherlands

Rob van Tol, Dept. of Crop Protection, Research Station for Nursery Stock, P.O. Box 118, Boskoop 2770 AC, The Netherlands
*E-mail: R.W.H.M.VAN.TOL@bpo.agro.nl*

Robert Vander Meer, Agriculture Research Service, US Dept. of Agriculture, 1600 SW 23rd Drive, Gainesville FL 32608, USA
*E-mail: bobvm@nervm.nerdc.ufl.edu*

Joe Veltre, AEI Technologies, ADC BioScientific, 12 Spurling Works, Pindar Rd., Hoddesdon, Herts, EN11 0DB, UK

Klaas Vrieling, Dept. of Plant Ecology, Inst. of Evolutionary and Ecological Sciences, Univ. of Leiden, P.O.Box 9516, Leiden 2300 KA, The Netherlands

Lester Wadhams, Dept. of Biological & Ecological Chemistry, IACR - Rothamsted, Rothamsted Experimental Station, Harpenden, Harpenden AL5 2JQ, UK
*E-mail: Lester.Wadhams@bbsrc.ac.uk*

Felix Wäckers, Dept. of Applied Entomology, ETH - Z, Clausiusstr. 25, Zürich 8092, Switzerland
*E-mail: felix.waeckers@ipw.agrl.ethz.ch*

Stuart Warwick, Dept. of Zoology, Univ. of Oxford, South Parks Rd., Oxford OX1 3PS, UK
*E-mail: stuart.warwick@sjc.ox.ac.uk*

Tom Wilkinson, Dept. of Biological Sciences, Graduate School of Science, Univ. of Tokyo, Hongo, Bunkyo-Ku, Tokyo 113, Japan
*E-mail: tlw@biol.s.u-tokyo.ac.jp*

Toni Withers, Dept. of Plant Health, New Zealand Forest Research Inst., P.B.3020, Rotorua, New Zealand
*E-mail: witherst@fri.cri.nz*

David Wool, Dept. of Zoology, Tel Aviv Univ., Tel Aviv 69978, Israel
*E-mail: dwool@ccsg.tau.ac.il*

Geraldine Wright, Dept. of Zoology, Univ. of Oxford, South Parks Rd., Oxford OX1 3PS, UK
*E-mail: geraldine.wright@zoo.ox.ac.uk*

Tristram Wyatt, Dept. of Zoology, Oxford Univ., South Parks Rd., Oxford OX1 3PS, UK
*E-mail: tristram.wyatt@zoo.ox.ac.uk*

# SERIES ENTOMOLOGICA

1. M. Fischer: *Revision der indo-australischen Opiinae* (Hymenoptera, Braconidae). 1966 ISBN 90-6193-111-8
2. H. Reiss and W.G. Tremewan: *A systematic catalogue of the Genus* Zygaena *Fabricus* (Lepidoptera: Zygaenidae). 1967 ISBN 90-6193-112-6
3. P. Starý and E.I. Schlinger: *A revision of the Far East Asian Aphidiidae* (Hymenoptera). 1967 ISBN 90-6193-113-4
4. M.S. Mani: *Ecology and biogeography of high altitude insects.* 1968 ISBN 90-6193-114-2
5. K. Harz: *Die Orthopteren Europas / The Orthoptera of Europe*, Vol 1. 1969 ISBN 90-6193-115-0
6. P. Starý: *Biology of aphid parasites* (Hymenoptera: Aphidiidae) *with respect to integrated control.* 1970 ISBN 90-6193-116-9
7. E. Zielke: *Revision der Muscinae der äthiopischen Region.* 1971 ISBN 90-6193-117-7
8. G.C.D. Griffiths: *The phylogenetic classification of* Diptera cyclorrhapha, *with special reference to the structure of the male postabdomen.* 1972 ISBN 90-6193-118-5
9. K.A. Spencer: *Agromyzidae* (Diptera) *of economic importance.* 1973 ISBN 90-6193-119-3
10. V.M. Dirsh: *Genus Schistocerca* (Acridomorpha, Insecta). 1974 ISBN 90-6193-120-7
11. K. Harz: *Die Orthopteren Europas / The Orthoptera of Europe*, Vol. 2. 1975 ISBN 90-6193-121-5
12. K. Harz and A. Kaltenbach: *Die Orthopteren Europas / The Orthoptera of Europe*, Vol. 3. 1976 ISBN 90-6193-122-3
13. J.D. Holloway: *The Lepidoptera of Norfolk Island.* Their biogeography and ecology. 1977 ISBN 90-6193-124-X
14. N. Papavero: *The world Oestridae* (Diptera), *mammals and Continental Drift.* 1977 ISBN 90-6193-123-1
15. J.D. Holloway: *A survey of the Lepidoptera, biogeography and ecology of New Caledonia.* 1979 ISBN 90-6193-125-8
16. A. Berríos-Ortiz and R.B. Selander: *Skeletal musculature in larval phases of the beetle* Epicauta segmenta (Coleoptera, Meloidae). 1979 ISBN 90-6193-126-6
17. R. Rosen and P. DeBach: *Species of* Aphytis *of the world* (Hymenoptera: Aphelinidae). 1979 ISBN 90-6193-127-4
18. J. Feltwell: *Large White Butterfly.* The biology, biochemistry and physiology of *Pieris brassicae* (L.). 1982 ISBN 90-6193-128-2
19. V. Labeyrie (ed.): *The ecology of Bruchids attacking legumes (pulses).* Proceedings of an International Symposium (Tours, France, 1980). 1981 ISBN 90-6193-883-X
20. G.P. Moretti (ed.): *Proceedings of the 3rd International Symposium on Trichoptera* (Perugia, Italy, 1980). 1981 ISBN 90-6193-130-4
21. R. Rozkošný: *A biosystematic study of the European Stratiomyidae* (Diptera), Part 1: Introduction, Beridinae, Sarginae and Stratiomyinae. 1982 ISBN 90-6193-132-0 *For Part 2, see Volume 25.*
22. L. Herrera: *Catalogue of the Orthoptera of Spain / Catlogo de los Ort¢pteros de España.* 1982 ISBN 90-6193-131-2
23. V.K. Brown and I. Hodek (eds.): *Diapause and life cycle strategies in insects.* 1983 ISBN 90-6193-133-9
24. H. Malicky: *Atlas of European Trichoptera / Atlas der europäischen Kcherfliegen / Atlas des Trichoptères d'Europe.* 1983 ISBN 90-6193-134-7
25. R. Rozkošný: *A biosystematic study of the European Stratiomyidae* (Diptera), Part 2: Clitellariinae, Hermetiinae, Pachygasterinae and bibliography. 1983 ISBN 90-6193-135-5
26. A. Szujecki: *Ecology of forest insects.* 1987 ISBN 90-6193-136-3
27. M. Canard, Y. Séméria and T.R. New (eds.): *Biology of Chrysopidae.* 1984 ISBN 90-6193-137-1
28. S. Endrödi: *The Dynastinae of the world.* 1985 ISBN 90-6193-138-X
29. P. Skidmore: *The biology of the Muscidae of the world.* 1985 ISBN 90-6193-139-8
30. J. C. Morse (ed.): *Prcoeedings of the 4th International Symposium on Trichoptera* (Clemson, S.C., USA, 1983). 1984 ISBN 90-6193-003-0
31. D.J. Carter: *Pest Lepidoptera of Europe*, with special reference to the British Isles. 1984 ISBN 90-6193-504-0
32. T.R. New: *Insect conservation.* An Australian perspective. 1984 ISBN 90-6193-507-5
33. G.E. Ball (ed.): *Taxonomy, phylogeny and zoogeography of beetles and ants.* A volume dedicated to the memory of Philip Jackson Darlingotn, Jr. (1904-1983). 1985 ISBN 90-6193-511-3
34. J.F. Duffels and P.A. van der Laan: *Catalogue of the Cicadoidea* (Homoptera, Auchenorhyncha), 1956-1980. 1985 ISBN 90-6193-522-9
35. I. Hodek (ed.): *Ecology of Aphidophaga.* Proceedings of the 2nd Symposium (Zvíkovské Podhradí, Czechoslovakia, 1984). 1986 ISBN 90-6293-537-7
36. M.S. Mani: *Butterflies of the Himalaya.* 1986 ISBN 90-6193-545-8
37. A. Neboiss: *Atlass of Trichoptera of the Southwest Pacific - Australian Region.* 1986 ISBN 90-6193-575-X
38. J.E. Wright and A. Retnakaran (eds.): *Chitin and benzoylphenyl ureas.* 1987 ISBN 90-6193-612-8

# SERIES ENTOMOLOGICA

41. V. Labeyrie, G. Fabres and D. Lachaise (eds.): *Insects - Plants*. Proceedings of the 6th International Symposium on Insect-Plant Relationships (Pau, France, 1986). 1987 ISBN 90-6193-642-X
42. P.H. Jolivet, E. Petitpierre and T.H. Hsiao (eds.): *Biology of Chrysomelidae*. 1988 ISBN 90-6193-655-1
    *See also Volume 50.*
43. H. Steinmann: *World catalogue of Dermaptera*. 1989 ISBN 0-7923-0096-3
44. I.C. Campbell (ed.): *Mayflies and Stoneflies: Life histories and biology*. Proceedings of the 5th International Ephemeroptera Conference and the 9th International Plecoptera Conference (Marysville, Australia, 1987). 1990 ISBN 0-7923-0289-3
45. K.A. Spencer: *Host specialization in the world Agromyzidae* (Diptera). 1990 ISBN 0-7923-0402-0
46. K. Fujii, A.M.R. Gatehouse, C.D. Johnson, R. Mitchel and T. Yoshida (eds.): *Bruchids and Legumes: Economics, ecology and coevolution*. Proceedings of the 2nd International Symposium (Okayama, Japan, 1989). 1990 ISBN 0-7923-0701-1
47. M.R. Wagner, S.K.N. Atuahene and J.R. Cobbinah: *Forest entomology in West Tropical Africa: Forest insects of Ghana*. 1991 ISBN 0-7923-1026-8
48. J.P.N. Pathak (ed.): *Insect immunity*. 1993 ISBN 0-7923-2086-7
49. S.B.J. Menken, J.H. Visser and P. Harrewijn (eds.): *Proceedings of the 8th International Symposium on Insect-Plant Relationships* (Wageningen, The Netherlands, 1992). 1992 ISBN 0-7923-2099-9
50. P.H. Jolivet, M.L. Cox and E. Petitpierre (eds.): *Novel aspects of the biology of Chrysomelidae*. 1993 ISBN 0-7923-2185-5
51. K. Desender, M. Dufršne, M. Loreau, M.L. Luft and J.-P. Maelfait (eds.): *Carabid beetles: Ecology and evolution*. 1994 ISBN 0-7923-2464-1
52. H.V. Danks (ed.): *Insect life-cycle polymorphism*. Theory evolution and ecological consequences for seasonality and diapause control. 1994 ISBN 0-7923-2828-0
53. E. Städler, M. Rowell-Rahier and R. Baur (eds.): *Proceedings of the 9th International Symposium on Insect-Plant Relationships*. 1996 ISBN 0-7923-4127-9
54. I. Hodek and A. Honěk: *Ecology of Coccinellidae*. 1996 ISBN 0-7923-4177-5
55. J. Bruin, L.P.S. van der Geest and M.W. Sabelis (eds.): *Ecology and Evolution of the Acari*. 1999 ISBN 0-7923-5658-6
56. S.J. Simpson, A.J. Mordue and J. Hardie (eds.): *Insect-Plant Relations*. Proceedings of the 10th Symposium on Insect-Plant Relationships. 1999 ISBN 0-7923-5773-6

KLUWER ACADEMIC PUBLISHERS – DORDRECHT / BOSTON / LONDON

MIX
Papier aus verantwortungsvollen Quellen
Paper from responsible sources
FSC® C105338

If you have any concerns about our products,
you can contact us on
**ProductSafety@springernature.com**

In case Publisher is established outside the EU,
the EU authorized representative is:
**Springer Nature Customer Service Center GmbH**
**Europaplatz 3, 69115 Heidelberg, Germany**

Printed by Libri Plureos GmbH
in Hamburg, Germany